ANTIBODIES IN HUMAN DIAGNOSIS AND THERAPY

Antibodies in Human Diagnosis and Therapy

Royal Society of Medicine Foundation Series

Editors

Edgar Haber, M.D.
Cardiac Unit
Massachusetts General Hospital
Department of Medicine
Harvard Medical School
Boston, Massachusetts

Richard M. Krause, M.D.
National Institute of Allergy and Infectious Diseases
National Institutes of Health
Bethesda, Maryland

Raven Press ▪ New York

Raven Press, 1140 Avenue of the Americas, New York, New York 10036

Made in the United States of America

International Standard Book Number 0-89004-089-3
Library of Congress Catalog Card Number 75-32089

Foreword

This volume is based on a symposium on *The Future of Antibodies in Human Diagnosis and Therapy,* one of a series of Anglo-American meetings sponsored in the United States and the United Kingdom by the Royal Society of Medicine Foundation, Inc., New York City, and the Royal Society of Medicine, London. The topics are of current major interest to scientists trained in the basic biomedical disciplines and to physicians concerned with medical care on both sides of the Atlantic, and are presented by speakers from the two countries, with full participation by the registrants.

The Royal Society of Medicine is well known by reason of its long history and its international membership. The Royal Society of Medicine Foundation, Inc. was established in New York in 1967 to promote cooperation between members of the medical profession in the United States and in Britain, and also to enable the academic, scientific, and practical experience of the profession in both countries to be shared in a manner that will create better doctors, better medicine, and better relations between the English-speaking peoples. The funds it collects are being used to organize these Anglo-American conferences; to send young American doctors to the United Kingdom on a Traveling Fellowship basis with the opportunity of living at the Society's Domus Medica for part of that time to see British medicine at work; to plan Anglo-American symposia, followed by publication and wide distribution on subjects of outstanding medical importance; to help involve American physicians and scientists in medical meetings taking place in Great Britain; and to strengthen the Society's links with medical organizations in North America and to broaden its membership there.

The affairs of the Royal Society of Medicine Foundation, Inc. are in the hands of leading members of the American medical, legal, and business professions. Additionally, the president of the Royal Society of Medicine is *ex officio* a member of the Foundation's Board of Directors, and the Society's Executive Director is Vice President of the Foundation.

Sir Gordon E. W. Wolstenholme
Harry J. Robinson
July, 1976

Preface

The concept that an infectious disease may be prevented by deliberate *active* immunization dates back many centuries to Jenner's pioneering work in smallpox vaccination. The earlier decades of the 20th century witnessed the application of *passive* immunization in the treatment of infectious diseases. It was discovered that when certain diseases were already clinically manifest and it was too late for active immunization, sera of previously immunized animals proved to be effective therapeutic agents. Notable successes were registered, particularly in the treatment of diphtheria and the prevention of tetanus. During the same period, the remarkable selectivity of the antibodies contained within these sera was recognized. Landsteiner showed that almost any compound on the organic chemist's shelf could be recognized specifically by the appropriate antiserum. Much was learned about the kinetics of antibody production in relation to antigen exposure, but antibody itself remained the proverbial black box until, in the mid-1950s, Rodney Porter and Gerald Edelman showed that it was a protein molecule, not too different in its structural-functional relationships from an enzyme. It could be dissected into component parts and some of its functions divided from each other. The next 20 years witnessed an explosion of activity by immunologists, protein chemists, and X-ray crystallographers. The structural determinants of antibody specificity and variability began to be understood. The intimate features of antigen binding were uncovered. The parts of the molecule responsible for its biological functions were defined.

Now that the antibody molecule has emerged as a visible, perhaps even malleable, entity, it seems appropriate to revisit some of the early promise of the therapeutic antisera. Can antibodies be considered very specific drugs for a wide variety of applications extending far beyond their use in the treatment or prophylaxis of infectious diseases?

This book derives from a meeting held to answer this question. The conference was designed to bring together scientists looking at the antibody problem from a variety of viewpoints. First, the protein chemists examine the primary structure of antibody and its relationship to antigenic specificity. The X-ray crystallographers display the molecule in three dimensions and show how the antigen neatly fits into a combining cavity. Peptide chemists then ask what was only recently a heretical question: If we know the primary structure of an antibody, and if three-dimensional structure is really the consequence of primary structure, should it not be possible simply to make an antibody by the sequential assembly of amino acids? Perhaps this specialized organic chemistry is the most efficient way to create these exquisitely specific drugs of the future. The cell biologists counter by pointing out that it may not be necessary to bypass nature's own very efficient machinery. Although penicillin can be synthesized by the chemists, fermentation remains the most effective means of commercial production. Why not harness techniques for growing antibody-producing cells *in vitro* for large-scale manufacture? Another alternative discussed, but for the moment discarded, is the use of the cell's protein synthetic machinery without the cell itself.

Present or future applications of antibodies in human diagnosis or therapy are discussed in relation to cancer, organ transplantation, hemolytic disease of the newborn, reversal of drug toxicity, and prevention of infectious diseases. These chapters leave us with the impression that, while progress in these areas is impressive, they are representative of only a small part of all the possible clinical uses of antibodies. Clearly, what is lacking are the antibodies themselves. We await the fulfillment of the promise made by the fundamental investigators, to produce them in quantity, with the desired specificity, and in a form that does not give rise to immune reactions in the recipient.

Edgar Haber
Boston, Massachusetts

Acknowledgments

The Royal Society of Medicine Foundation, Inc. and the Royal Society of Medicine wish to acknowledge the donation made specifically in support of this conference by the Merck Sharp & Dohme Research Laboratories.

They also take this opportunity of thanking those who support their other endeavors in the promotion of Anglo-American relations in medicine, namely the members of the Royal Society of Medicine in the United States and the following corporate donors: Abbot Laboratories Fund, Adair Fund for Cancer Research Inc., Mary Reynolds Babcock Foundation Inc., Becton Dickinson & Co., Bristol Myers Products, Bristol Myers Fund, Burroughs Wellcome Co., Burroughs Wellcome Fund, The Commonwealth Fund, Walter Ames Compton Foundation, Cyanamid International, Dudley Foundation Inc., Endo Foundation Inc., Exxon Corporation, General Telephone and Electronics Foundation, Grant Foundation Inc., Hoffman-LaRoche Inc., Hoffman-LaRoche Foundation, IBM World Trade Corporation, Interim Faculty Club Committee of the University of Alabama Medical Center, Johnson & Johnson Associated Industries Fund, Eli Lilly & Co., Eli Lilly & Co. Foundation, William Muir Manger Research Foundation, Inc., Merck & Co. Inc., Merck Co. Foundation, Merck Sharp & Dohme Postgraduate Program, Merck Sharp & Dohme Research Laboratories, Merrell-National Laboratories, Miles Laboratories Inc., Miles Laboratories Foundation, Mobil Foundation Inc., Mobil Oil Corporation, Parke Davis & Co., SB Penick Foundation, Gustavus & Louise Pfeiffer Research Foundation, Pfizer International Inc., Pfizer Pharmaceuticals, Rockefeller Foundation, Rohlf Memorial Clinic Foundation, Rohrer Foundation Inc., Sandoz-Warner Inc., GD Searle & Co., Searle Foundation, Smith Kline & French Foundation, Smith Kline Corporation, ER Squibb & Sons Inc., Sterling Drug Inc., Henry M. and Lillian Stratton Foundation, Syntex Corporation, Syntex International, Syntex Pharmaceuticals Ltd., University Associates in Psychiatry, Wallace Pharmaceuticals, Warner-Lambert Co., Warner-Lambert Charitable Foundation, Warner-Lambert Pharmaceutical Co., William C. Whitney Foundation.

The editors wish to thank Ms. Jean Brumbaugh for her superior editorial assistance.

Contents

Contributors

B. Asted
The Rockefeller University
New York, New York 10021

E. E. Abola
Division of Biological and Medical Research
Argonne National Laboratory
Argonne, Illinois 60439

L. Mario Amzel
Department of Biophysics
The Johns Hopkins University School of Medicine
Baltimore, Maryland 21205

Paul H. Black
Medical Services
Massachusetts General Hospital
and
Department of Medicine
Harvard Medical School
Boston, Massachusetts 02114

Thomas Braciale
Department of Pathology
University of Pennsylvania Medical School
Philadelphia, Pennsylvania 19174

Dietmar G. Braun
Basel Institute for Immunology
Grenzacherstrasse 487
Basel 4058, Switzerland

James Burton
Cardiac Unit
Massachusetts General Hospital
and
Department of Pathology
Harvard Medical School
Boston, Massachusetts 02114

Vincent P. Butler, Jr.
Department of Medicine
Columbia University College of Physicians and Surgeons
New York, New York 10032

L. Edward Cannon
Cardiac Unit
Massachusetts General Hospital
Department of Pathology
Harvard Medical School
Boston, Massachusetts 02114

J. Donald Capra
Department of Microbiology
The University of Texas Health Science Center at Dallas
Southwestern Medical School
Dallas, Texas 75235

John J. Cebra
Mergenthaler Laboratory
Johns Hopkins University
Baltimore, Maryland 21218

Betty L. Chen
Department of Biophysics
The Johns Hopkins University School of Medicine
Baltimore, Maryland 21205

Joseph Davie
Department of Microbiology
Washington University School of Medicine
St. Louis, Missouri 63110

David R. Davies
Laboratory of Molecular Biology
National Institute of Arthritis, Metabolism and Digestive Diseases
National Institutes of Health
Bethesda, Maryland 20014

A. B. Edmundson
Division of Biological and Medical Research
Argonne National Laboratory
Argonne, Illinois 60439

Klaus Eichmann
Institut fur Genetic der Universität zu Köln
5 Köln 41
Weyertal 121, West Germany

Herman N. Eisen
Center for Cancer Research and Department of Biology
Massachusetts Institute of Technology
and
The Harvard-MIT Joint Program for Health Sciences and Technology
Cambridge, Massachusetts 02139

K. R. Ely
Division of Biological and Medical Research
Argonne National Laboratory
Argonne, Illinois 60439

Bruce W. Erickson
The Rockefeller University
New York, New York 10021

J. R. Firca
Division of Biological and Medical Research
Argonne National Laboratory
Argonne, Illinois 60439

B. A. Fraser
The Rockefeller University
New York, New York 10021

Moshe Gavish
Departments of Chemical Immunology and Biophysics
The Weizmann Institute of Science
Rehovot, Israel

Walter Gerhard
Department of Pathology
University of Pennsylvania Medical School
Philadelphia, Pennsylvania 19174

David Givol
Departments of Chemical Immunology and Biophysics
The Weizmann Institute of Science
Rehovot, Israel

Ronald Gold
Department of Pediatrics
University of Connecticut Health Center
Farmington, Connecticut 06032

Irving Goldschneider
Departments of Pathology and Pediatrics
University of Connecticut Health Center
Farmington, Connecticut 06032

Emil C. Gotschlich
The Rockefeller University
New York, New York 10021

Edgar Haber
Cardiac Unit
Massachusetts General Hospital
Department of Medicine
Harvard Medical School
Boston, Massachusetts 02114

Elvin A. Kabat
Departments of Microbiology, Human Genetics and Development, and Neurology
Columbia University;
The Neurological Institute
Presbyterian Hospital
New York, New York 10032
and
The National Cancer Institute
National Institutes of Health
Bethesda, Maryland 20014

T. J. Kindt
The Rockefeller University
New York, New York 10021

Norman R. Klinman
Department of Pathology
University of Pennsylvania Medical School
Philadelphia, Pennsylvania 19174

G. Kohler
Medical Research Council
Laboratory of Molecular Biology
Cambridge CB2 2 OH, England

Marion E. Koshland
Department of Bacteriology and Immunology
University of California, Berkeley
Berkeley, California 94720

David E. Krieger
The Rockefeller University
New York, New York 10021

Martha L. Lepow
Department of Pediatrics
University of Connecticut Health Center
Farmington, Connecticut 06032

Ronald Levy
Department of Pathology
University of Pennsylvania Medical School
Philadelphia, Pennsylvania 19174

Michael N. Margolies
Massachusetts General Hospital
Department of Surgery
Harvard Medical School
Boston, Massachusetts 02114

Johannes Meienhofer
Chemical Research Department
Hoffman-LaRoche Inc.
Nutley, New Jersey 07110

C. Milstein
Medical Research Council
Laboratory of Molecular Biology
Cambridge CB2 2QH, England

N. A. Mitchison
Zoology Department
University College
London W. C. 1, England

M. Mudgett
The Rockefeller University
New York, New York 10021

Eduardo A. Padlan
Laboratory of Molecular Biology
National Institute of Arthritis, Metabolism and Digestive Diseases
National Institutes of Health
Bethesda, Maryland 20014

N. C. Panagiotopoulos
Division of Biological and Medical Research
Argonne National Laboratory
Argonne, Illinois 60439

Charles W. Parker
Department of Internal Medicine
Washington University School of Medicine
St. Louis, Missouri 63110

R. Paul Phizackerley
Department of Biophysics
The Johns Hopkins University School of Medicine
Baltimore, Maryland 21205

Roberto J. Poljak
Department of Biophysics
The Johns Hopkins University School of Medicine
Baltimore, Maryland 21205

Michael Potter
Laboratory of Molecular Biology
National Cancer Institute
National Institutes of Health
Bethesda, Maryland 20014

Frank F. Richards
Department of Internal Medicine
Yale University–New Haven Medical Center
New Haven, Connecticut 06510

Mario S. Rosemblatt
Cardiac Unit
Massachusetts General Hospital
and
Department of Medicine
Harvard Medical School
Boston, Massachusetts 02114

Fred S. Rosen
Children's Hospital
Harvard Medical School
Boston, Massachusetts 02114

Stuart Rudikoff
National Cancer Institute
National Institute of Arthritis, Metabolism and Digestive Diseases
National Institutes of Health
Bethesda, Maryland 20014

Paul S. Russell
Transplantation Unit, the Surgical Services
Massachusetts General Hospital
and
Department of Surgery
Harvard Medical School
Boston, Massachusetts 02114

Frederick Saul
Department of Biophysics
The Johns Hopkins University School of Medicine
Baltimore, Maryland 21205

Matthew Scharff
Department of Cell Biology
Albert Einstein College of Medicine
Bronx, New York 10461

M. Schiffer
Division of Biological and Medical Research
Argonne National Laboratory
Argonne, Illinois 60439

Gary P. Segal
Department of Pathology
University of Pennsylvania Medical School
Philadelphia, Pennsylvania 19174

Jacqueline Sharon
Departments of Chemical Immunology and Biophysics
The Weizmann Institute of Science
Rehovot, Israel

Thomas W. Smith
Cardiovascular Division
Peter Bent Brigham Hospital;
Department of Medicine
Harvard Medical School
Boston, Massachusetts 02114

J. A. Sogn
The Rockefeller University
New York, New York 10021

Mark Vrana
National Cancer Institute
National Institute of Arthritis, Metabolism and Digestive Diseases
National Institutes of Health
Bethesda, Maryland 20014

F. A. Westholm
Division of Biological and Medical Research
Argonne National Laboratory
Argonne, Illinois 60439

Meir Wilchek
Departments of Chemical Immunology and Biophysics
The Weizmann Institute of Science
Rehovot, Israel

A. R. Williamson
Department of Biochemistry
University of Glasgow
Glasgow G12 8QQ Scotland

Rina Zakut
Departments of Chemical Immunology and Biophysics
The Weizmann Institute of Science
Rehovot, Israel

Vincent R. Zurawski, Jr.
Cardiac Unit
Massachusetts General Hospital
and
Department of Medicine
Harvard Medical School
Boston, Massachusetts 02114

Antibodies in Human Diagnosis and Therapy,
edited by Edgar Haber and Richard M. Krause.
Raven Press, New York © 1977.

Introduction

Edgar Haber

Cardiac Unit, Massachusetts General Hospital, Harvard Medical School, Boston, Massachusetts 02114

There are no simple substances and few biologic macromolecules that match the exquisite specificity of recognition of an antibody. This specificity has long been empirically utilized in clinical medicine through both active and passive immunization against infectious diseases and in the application of immunochemical tests for the identification of pathogens or the measurement of physiologic substances. But there are intrinsic limitations in the application of immunology as we have traditionally practiced it.

ACTIVE IMMUNIZATION

To achieve active immunization the injection of an antigen is required, and several weeks must elapse before adequate levels of antibody are achieved. Circulating antibody persists for a considerable period, and immunological memory for a number of years. These are acceptable or desirable characteristics when exposure to a pathogen can be anticipated and continued protection desired. Immunization can only be successful when undertaken after exposure to a pathogen in diseases characterized by a prolonged incubation period, such as rabies. Delay becomes a severe limitation when it is necessary to avert an infectious disease of short incubation period, to treat an ongoing pathophysiologic event, or to reverse the effects of a drug, toxin, or endogenous hormone.

When reversible effects are sought, such as infertility (see chapter by Mitchison) persistence of an immune response is undesirable. Active immunization may also not be possible when the immunogen is either infectious or toxic.

PASSIVE IMMUNIZATION

Where antibodies of human origin are available, passive immunization can be very useful. Present day applications include prophylaxis against tetanus in the nonimmune individual and the prevention of erythroblastosis fetalis in the newborn (see chapter by Rosen). The nature of these applications points out an important limitation of this approach. Sera from individuals immune to tetanus toxoid are available simply because active immunization to this material is an accepted and nearly universal procedure. Isoimmunization of Rh^- men with Rh^+ cells proved to be a benign procedure, and, consequently, volunteer donors could be found. But humans cannot be used as donors when immunization either modifies their own physiology, alters their response to a potentially useful drug such as digoxin (see chapter by Smith), necessitates the use of an immunogen that might be toxic, involves an antigen to which humans are normally tolerant, or causes undesirable allergic responses. While sufficient blood donors seem to be available, it is likely that if a specific immunization procedure must precede blood donation, the number of volunteers will necessarily diminish. For these reasons, humans cannot be realistic sources of antibodies for a wide scope of applications.

Animals have traditionally been employed as antibody donors. Even purified globulins from animal sources may result in serum sickness or anaphylaxis in the recipient, particulary on repeated use in the nonimmunosuppressed individual. Immune elimination of the foreign globulin limits their persistence.

CAN THE PRODUCTION OF HUMAN ANTIBODIES *IN VITRO* BE REALISTICALLY CONTEMPLATED?

Chemical Synthesis

Later in this volume, Bruce Merrifield writes: "it has become possible to obtain the variable regions of both heavy and light chains of certain immunoglobulins that can be reassociated to form functional binding sites. Therefore, we are now in a position to carry out meaningful synthetic experiments on protein chains of only 100 residues. I think this really is a feasible goal." He is, of course, discussing synthesis in the context of a laboratory exercise, the goal being a better understanding of combining site function. The feasibility of such an experiment is underlined by reported successful syntheses of a number of functional proteins of approximately the same size as the individual chains of an immunoglobulin domain. The question of whether or not large-scale production will be economically feasible by these methods has not yet been approached. The largest synthetic peptide presently produced in quantity as a drug is β-corticotropin, comprising 39 residues. While only three times as many residues need to be assembled to form the variable segment of the light or heavy chain, the complexity and cost of the operation probably exceeds three by an order of magnitude.

In Vitro Biological Synthesis

Allan Williamson's chapter provides a futuristic glimpse of cell-free synthesis of antibodies. While he does not hold out much hope for surpassing nature in the molecular biologist's laboratory, he does offer some attractive alternatives.

The antibody gene must first be isolated, then possibly amplified in a bacterial system, and finally attached to a viral vector. A mammalian cell line could then be infected by the virus which would introduce the new gene into the DNA of the cell. Hopefully, the gene product, a specific antibody, would now be secreted by the cells. At this time such proposals in genetic engineering have the ring of science fiction. Clearly, a great deal of fundamental research must be undertaken before such experiments can even be contemplated.

More practical in the light of available technology is the culture of lymphoid cells already producing antibody. Three kinds of problems must be overcome: the selection of an appropriate clone, encouraging its *in vitro* propagation, and devising means for mass culture so that useful quantities of antibody may be obtained.

Advances in cell fractionation now make it possible to select individual cells on the basis of antigenic markers on their surface, utilizing such methods as affinity chromatography (1) or a fluorescence-activated cell-sorting machine (2). In growing cultures, clones rather than single cells may be subcultured and then selected on the basis of the nature of the secreted antibody. The fragment culture technique suggested in the chapter by Norman Klinman and his colleagues provides an alternative that has already been proven to work.

Normal antibody-producing lymphoid cells do not proliferate indefinitely *in vitro*. Two rather different approaches have successfully overcome this problem: A normal clone can be transformed to a malignant clone either by a virus (see chapter by Black and Zurawski) or by cell fusion with a malignant lymphoid line (see chapter by Milstein and Kohler). Remarkably, after either of these maneuvers, the cells retain the capacity to produce specific antibody.

Large-scale culture of lymphoid cell lines has now been seriously addressed by a number of laboratories. Lynn and Acton (3) have devised procedures for operating on a most ambitious scale. In a 200-liter fermenter, as many as 10^{12} cells per week may be grown. While I know of no commercial facility that grows lymphoid cells in quantity, vaccine for hoof and mouth disease virus is grown commercially in a mammalian cell culture using 1,000-liter fermenters.

The elements of *in vitro* antibody production seem already at hand. Experiments must now be directed at choosing the best and most efficient approach and then refining it for production of optimal quantities.

APPLICATIONS OF ANTIBODIES

Antibodies as Reagents

The technique of radioimmunoassay has made a very large impact on physiologic research and the practice of clinical medicine. Successful measurement of very low concentrations of materials in biological fluids is very dependent on finding antibodies of high specificity and affinity. This is often an empirical process. Large numbers of animals are immunized and sera are screened during the development of the immune response. With some antigens, only a minority of sera contain useful antibodies. Reproducing an antiserum of given desirable properties is difficult. The standardization of assays among many laboratories and maintenance of standards for a period of years has not been possible. A continuous supply of a single homogeneous antibody of precisely the right properties would be most desirable; it would have a very large impact in the widespread application of these methods.

Antibodies as Therapeutic Agents

Aside from their classically recognized role in the control of pathogenic microorganisms or in the neutralization of toxins that these organisms produce, antibodies may serve a variety of other purposes that take advantage of their unique specificity. Some are discussed in depth in subsequent chapters; others are only hinted at.

Control of Tumor Growth

Herman Eisen's chapter reviews experiments in which antibodies specific for the unique antigens expressed on the surface of some tumors suppress the growth of the tumor. The importance of identifying and isolating the tumor-specific antigen is emphasized, while the scant progress in this field is acknowledged. Tumor-specific antibodies may also be envisioned as carriers of chemotherapeutic agents, resulting in an enhanced concentration of a toxic substance in tumor cells relative to normal cells that do not carry the antigen (4,5). The promise of this approach has not yet been realized. The most likely reason is the unavailability of purified tumor-specific antigens and, consequently, the lack of monospecific antibodies.

Modulation of the Immune Response

The most dramatic result of antibody in human therapy has been the virtual eradication of erythroblastosis fetalis by the prophylactic administration of human Rh^+ antibodies (see chapter by Rosen). Antilymphocyte globulin as a means of controlling rejection of an organ graft has not as yet been universally accepted. Paul Russell's chapter considers in detail problems inherent in identifying the proper antigen, standardization of the antibody, and the design of clinical trials in which variability in the origin of the immune globulin clouds the interpretation of results. Many of these problems could be overcome if an antibody of defined specificity were available in an indefinitely ample supply.

Klaus Eichmann, in the discussion following the chapter by Eisen, indicates that anti-idiotype antibodies of a given class are not only capable of suppressing a specific immune response but, under somewhat different circumstances, can substitute for an immunogen in stimulating a specific immune response. While certainly far in the future, the capacity to turn on or off a given immune response in man suggests many exciting therapeutic possibilities, ranging from the suppression of allergy and auto immune diseases to immunization using an idiotypic antibody rather than an antigen when the latter is either toxic or pathogenic.

Neutralization of Drug or Hormone Effects

Thomas Smith presents studies with digoxin as a model of drug which has a narrow toxic/therapeutic ratio. The effects of the drug may be neutralized *in vivo* with specific antibody, possibly because the antibody has a higher affinity for the drug than its physiologic receptor. Antibodies of animal origin have been used in initial human trials. The risk of allergic reactions to these foreign proteins will discourage their widespread application except under the most desperate circumstances. Antibodies of human origin would obviate this problem and provide the physician a therapeutic agent of considerable utility. Since man should not be immunized against a potentially useful drug, an *in vitro* method of human antibody production is the most desirable. The lessons learned from these experiments are readily extended to other drugs and toxins, particularly those potent at very low concentrations.

Fertility control may be effected by antibodies to chorionic gonadotropin (see chapter by Mitchison). Here, passively administered antibodies offer a great advantage over active immunization because the effect is reversed when the antibody has been metabolized.

Administered antibodies may also help in the diagnosis of certain endocrine disorders. Antibodies to renin have been shown to lower blood pressure in experimental renovascular hypertension (6). One can envision that in the practice of medicine of the future the availability of a panel of hormone-specific antibodies for both diagnostic or, in certain instances, therapeutic use.

Localization of a Healthy or Diseased Tissue

Cells that carry a unique antigen may be found within the body by radioactively tagged specific antibody. When an isotope emitting gamma radiation of the appropriate energy is used (such as ^{123}I), a scintillation camera can locate and size the region containing the antigen. Thus tagged antibody to myosin has been used to locate myocardial infarcts (7). While myosin is common to all cardiac cells, it is inaccessible to circulating antibody because of the impermeability of the cell membrane to macromolecules. The membrane of the necrotic cell, on the other hand, readily admits antibodies that concentrate within the cell by binding to the intracellular antigen.

When tumor-specific antigens can be identified, and purified antibodies are available, analogous *in vivo* localization of tumors may be possible.

At times an organ, such as the parathyroid gland, cannot be found readily because of an abnormal location within the body. This becomes especially important when a parathyroid adenoma must be found for surgical excision in the treatment of hyperparathyroidism. The discovery of an organ-specific antigenic determinant and the radioactive labeling of the complementary antibody could locate the gland and thereby circumvent extensive, and at times futile, surgical exploration.

The wanderings of a fertile imagination can readily speculate upon many more applications of the powerful drug or reagent which antibody represents, particularly when a specified and uniform product can be produced in indefinite supply. It is certainly a worthy goal for the expenditure of considerable investigative effort.

REFERENCES

1. Schlossman, S. F., and Hudson, L. (1973): *J. Immunol.,* 110:313.
2. Bonner, W. A., Hulett, H. R., Sweet, R. G., and Herzenberg, L. A. (1972): *Rev. Sci. Instrum.,* 43:404.
3. Lynn, J. D., and Acton, R. T. (1975): *Biotech. Bioengineer.,* 17:659.
4. Ghose, T., Norvell, S. T., Gluclu, A., Cameron, D., Bodurtha, A., and MacDonald, A. S. (1972): *Br. Med. J.,* 3:495.
5. Ghose, T., Path, M. R. C., and Nigam, S. P. (1972): *Cancer,* 29:1398.
6. Romero, J. C., Hoobler, S. W., and Kozak, T. J. (1973): *Am. J. Physiol.,* 225:810.
7. Khaw, B. A., Beller, G. A., Haber, E., and Smith, T. W. (1976): *J. Clin. Invest.,* 58:439.

The Covalent Structure of the Antibody-Combining Site

Antibodies in Human Diagnosis and Therapy,
edited by Edgar Haber and Richard M. Krause.
Raven Press, New York © 1977.

Covalent Structure of the Antigen Binding Site: Antigen-Binding Myeloma Proteins of the BALB/c Mouse

Michael Potter, Stuart Rudikoff, Eduardo A. Padlan, and Mark Vrana

National Cancer Institute, National Institute of Arthritis and Metabolic Diseases, National Institutes of Health, Bethesda, Maryland 20014

The susceptibility of the inbred BALB/c mouse to the induction of plasmacytomas by the intraperitoneal injection of mineral oils and related substances provides a virtually unlimited source of homogeneous immunoglobulins (1). These immunoglobulins are particularly useful for the analysis of genetic mechanisms in antibody diversity because they are all derived from the same autosomal haplotype. Furthermore, these proteins became highly useful in the structural analysis of antibodies when it was discovered that many of them had antigen-binding activity (Table 1)(2–4). Antigen-binding activity was first detected by interacting myeloma proteins with a battery of test antigens (2). While the method for detecting antigen binding activity is essentially a blind screening process, which would suggest that the antigen-binding proteins in the mouse are chance matchings of antigens with a binding protein, three lines of evidence have indicated that some of these proteins are close relatives of natural antibodies. First, close antigenic (i.e., idiotypic) relationships have been established between specific antigen binding myeloma proteins and induced or natural antibodies to the corresponding antigen (5–7). Antibodies induced to the R36A strain of pneumococci, for example, are directed predominantly in the mouse to phosphorylcholine (8), a constituent of the cell wall of this organism (9,10); and these antibodies are idiotypically identical to the TEPC15-S63 group of phosphorylcholine-binding myeloma proteins (6).

Similar relationships have been established between galactan (Mushinski and Potter, *unpublished*), levan (Lieberman and Potter, *unpublished*), and dextran-binding (5) myeloma proteins and induced or natural antibodies. In the latter situations, however, the antibody responses in the mouse to these

TABLE 1. *Antigen-binding myeloma proteins of BALB/c mice*

Phosphorylcholine	S63, S107, TEPC15, HOPC8, MOPC511, MOPC167, McPC603
β2→6 Fructosan	UPC10, ABPC48
β2→1 Fructosan	J606, W3082, UPC61, ABPC4, ABPC47, EPC109, AMPC1, TEPC803, TEPC957
β1→6 Galactan	J539, SAPC10, TEPC191, XRPC24, XRPC44, JPC1, TEPC601
α1→3 Dextran	MOPC10E, J558
α1→6 Dextran	QUPC52, W3129, W3434
N-Acetylglucosamine	S117
N-Acetylmannosamine	MOPC406
Flagellin	MOPC467
S. tranoroa lps	MOPC384, McPC870
Dinitrophenyl	MOPC315, MOPC460, XRPC25

antigens are not homogeneous, and the idiotypic relationships are shared cross-specific determinants rather than individual idiotypic identities as seen with TEPC15-S63 and phosphorylcholine-induced antibody. Second, there are groups of antigen-binding myeloma proteins that bind the same hapten (Table 1), indicating nonrandom detection of certain activities by the screening process. Third, many of the antigens bound by mouse myeloma proteins can be found in the environment of the mouse, indicating the relevancy of the antigen-binding activities to the antigenic experiences of the mouse (3,4).

ANTIGEN BINDING SITE

The antigen binding site is located on the surface of the immunoglobulin molecule that is most distal to the carboxylterminal domains (CH2, CH3) of the heavy chains. The binding site region faces the solvent environment and is in reality an entire surface approximately 30 Å by 40 Å in area. The antigen binding surface is formed by sections of the V_L and V_H domains and includes a cleftlike space between these domains (Fig. 1).

From the point of view of primary structure, the antigen binding site is not continuous, but rather is formed by a set of discontinuous sequences. In their classic analysis of amino acid sequence variations among V_L and V_H

FIG. 1. Drawing of the antigen binding site region of McPC603. The six complementarity regions are drawn from five different views. *Center view,* the binding site is seen in a direct frontal view. The phosphorylcholine antigen is depicted as two small spheres, the choline moiety has four spikes, and the phosphate has eight spikes. Note that the choline is bound more internally than the phosphate. The V_H CR are oriented around the choline as follows: H1 is at 1 o'clock, H2 from 2 to 5 o'clock, and H3 between 10 and 11 o'clock. The V_L CR are: L3 between 7 and 8 o'clock, L1 between 9 and 10 o'clock and L2 between 10 and 11 o'clock. H1 and L1 are solid lines, H3 and L3 are hatched lines, and H2 and L2 are open lines. The four additional binding sites are reviewed from 30° to the right (*left drawing*); 30° to the left (*right drawing*); from 30° below (*upper drawing*); and 30° above (*lower drawing*) of the position used in the center drawing. In the center composite the distance between the solid circles on H3 and L3 is 20 Å, on H2 and L2 is 25.5 Å, and H1 and L1 is 27 Å.

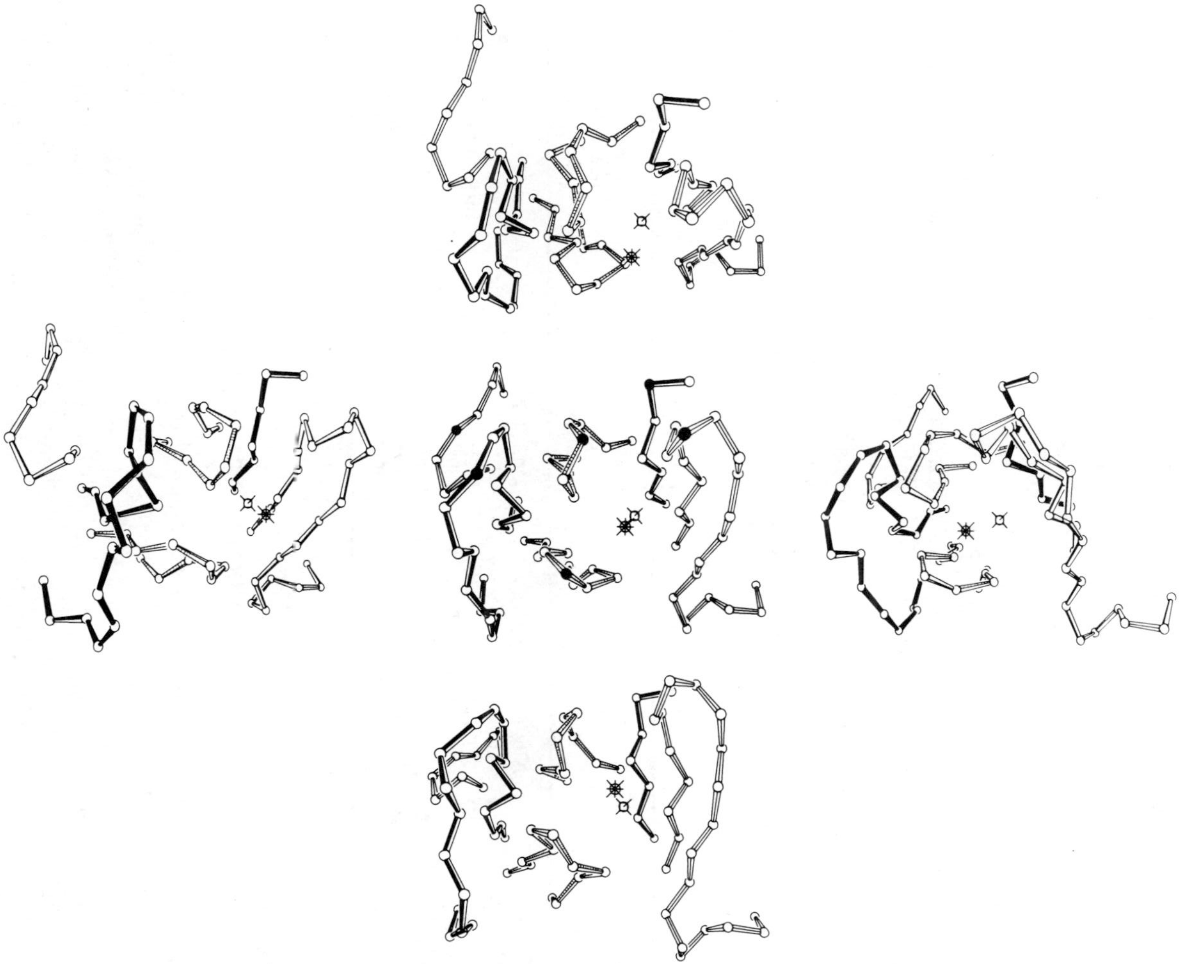

FIG. 1. Five views of the antigen-binding surface of McPC603. (See p. 10 for legend.)

polypeptides, Wu and Kabat (11,12) predicted that those regions which varied the most among homologous V polypeptides would, in the three-dimensional structure of the antibody molecule, form the antigen binding site. These regions were originally called complementarity regions (CR) and later hypervariable regions. We will adhere to the term complementarity region. The purpose of this chapter is to focus on the primary structures of CR of antigen-binding myeloma proteins of mice. Only five of the six CRs identified by primary sequences analysis have been shown by crystallographic analysis of two hapten-binding myeloma proteins to participate in the formation of the binding site. Two of these, L1 and L3, are from the light chain, and three, H1, H2, and H3, are from the heavy chains (13). As is well known, V polypeptides vary greatly in length; this has given rise to considerable confusion about the location of segments within V regions. There are, however, a number of invariant positions or homology positions that provide landmarks in the sequence; these can be used to define the location of the CR. The location and associated secondary structures of CRs in the mouse are given in Table 2.

TABLE 2. *Complementarity regions in mouse V_L and V_H*

CR	Lies between landmarks	Ref. protein[a]	McPC603 secondary structure
H1	Gly_{26}–Trp_{36}	McPC603	Stretch of extended loop
H2	Trp_{47}–Ser_{63}	McPC603	Hairpin loop
H3	Cys_{96}–Phe_{107}	McPC603	Tip of hairpin loop
L1	Cys_{23}–Trp_{35}	McPC603	Extended loop
L2	Ile_{48}–Gly_{57}	MOPC70	Tip of hairpin loop
L3	Cys_{88}–Gly_{102}	MOPC70	Tip of hairpin loop

[a] Landmark amino acids taken from McPC603 V_H sequence (14) and MOPC70 (15) sequence.

COMPLEMENTARITY REGIONS

CRs have different types of secondary structures: H3, L2, and L3 are tips of hairpin loops; L1 is a large part of an extended loop (containing the bend); H2 is a long hairpin loop; and H1 is a short stretch of beta structure (Table 2). Variations in primary structure fall into two categories: (i) variations in sequence of amino acids whose side chains do not play a role in maintaining the framework architecture of the domain; and (ii) variations in length. The first type of variation occurs chiefly along stretches, and the second involves bend portions of loops. Both forms of variation are permissive, in that the modifications do not affect the domain folding or interactions between segments.

The size variants play the dominant role in molding the contours of the binding surface. The interdomain cleft can be varied in depth by the presence or absence of space filling CRs. Grooves running in different directions across the antigen binding surface can be created, as a well as deep pockets which permit relatively close fitting of haptenic groups with amino acid side chains of CRs.

In the following section the primary structures of CRs of several available antigen-binding myeloma proteins in the mouse will be described. The sequences used have been taken from the literature and, in many cases, from studies which are presented here for the first time (Table 3).

TABLE 3. *Sources of myeloma proteins and sequences*

Myeloma protein	Hapten binding specificity	Source (ref.) of amino acid sequence V_L	V_H
McPC603	Phosphorylcholine	16	14
TEPC15	Phosphorylcholine	27	This chapter
MOPC167	Phosphorylcholine	This chapter	This chapter
MOPC315	DNP, TNP, menadione	28	29
MOPC460	DNP	30	31
EPC109	$\beta_2 \rightarrow 1$ Fructosan	This chapter	This chapter
MOPC104E	$\alpha 1 \rightarrow 3$ Dextran	23,24	
J539	$\beta 1 \rightarrow 6$ Galactan	25, this chapter	This chapter
UPC10	$\beta 2 \rightarrow 6$ Fructosan	This chapter	
EPC109	$\beta 2 \rightarrow 1$ Fructosan	This chapter	
XRPC25	DNP	This chapter	
MOPC70, MOPC41	None known	32	
MOPC221, TEPC124	None known	15	
MOPC63	None known		
MOPC21	None known	33	

H1

The H1 (Fig. 2) CR is a short stretch of amino acids (Fig. 1). This stretch corresponds to the carboxyl terminal end of L1. Remarkably, the bend portion of this loop, unlike the homologous section in L1, is not involved in size differences. Amino acid side chains of H1 interact with antigen in McPC603; Tyr_{33} forms a hydrogen bond with the phosphate of the phosphorylcholine hapten, and Glu_{35} interacts electrostatically with the choline group. The length of H1 is the same in all mouse proteins studied thus far, except MOPC315, which has an additional residue. TEPC15 and MOPC167 also have Tyr at 33 and Glu at 35. These residues appear to serve an important function in phosphorycholine binding.

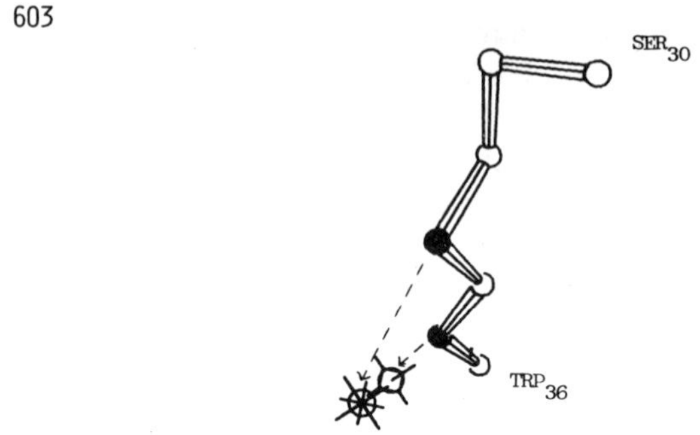

	26										36	
603	Gly	Phe	Thr	Phe	Ser	Asp	Phe	Tyr	Met	-	Glu	Trp	PC choline
T15	Gly	Phe	Thr	Phe	Ser	Asp	Phe	Tyr	Met	-	Glu	Trp	PC choline
M167	Gly	Phe	Thr	Phe	Ser	Asx	Phe	Tyr	Met	-	Glu	Trp	PC choline
M315	Gly	Tyr	Ser	Ile	Thr	Ser	Gly	Tyr	Phe	Lys	Asn	Trp	DNP
E109	Gly	Phe	Thr	Phe	Ser	Asn	Tyr	Trp	Met	-	Asn	Trp	β2→1 levan
M460	Gly	Ser	Asx	Ile	Thr	Asn	Gly	Tyr	Met	-	Asx	Trp	DNP

H-1

FIG. 2.

While the MOPC315 antigen binding site has not been determined by X-ray crystallography, a hypothetical structure has been proposed, based on plausible assumptions and the existence of a remarkable amount of structural homology in the framework of V_L and V_H chains from different species (21). In this hypothetical model the ring structure of the DNP lies close to L chain Trp_{91} in a planar fashion, and the nitro groups can be hydrogen-bonded to Tyr_{33} and Asn_{35} in H1 and Asn_{34} in L1. The types of variations in H1 are, in the available proteins, mostly amino acid substitutions.

H2

H2 (Fig. 3) is a long hairpin loop; it lies opposite L2 across the pseudodyad axis; and its long stretches form a large rim along the domain interface. The bend portion projects toward the solvent.

Six proteins have been sequenced through this region, and five of them appear to have a length similar to McPC603 (i.e., M167, TEPC15, EPC109, and MOPC173). Side chains emanating from the stretches interact with the phosphorylcholine hapten in McPC603. Arg_{52} interacts electrostatically with the phosphate group of the hapten, and Glu_{61} is very close to the choline group. Both of these residues are the same in homologous positions of M167 and T15, indicating their importance for the binding of the phosphorycholine hapten.

The H2 of MOPC315 is three amino acids shorter than the others. It is difficult to deduce the structure in MOPC315 without X-ray crystallographic data, but probably the structural variation affects the bend portion. It is most likely similar to the NEWM protein (13,14), which has a similar length.

The H2 regions of the different antigen binding proteins differ extensively from each other in sequence. More sequences are needed to determine if size variation will also occur frequently.

H3

H3 (Fig. 4) in McPC603, like L3, is the tip of a long hairpin loop. Like L1 in V_L (see below), H3 in V_H appears to be the most variable CR. In extreme forms, such as the EPC109 protein, the entire bend is eliminated; in this protein the bend probably forms nearer the interior where GLy_{99} permits the sharp turn and bridging of the two stretches of the loop.

In the three phosphorylcholine-binding proteins M603, T15, and M167, H3 also varies in length. T15 is one amino acid longer and M167 is three amino

603

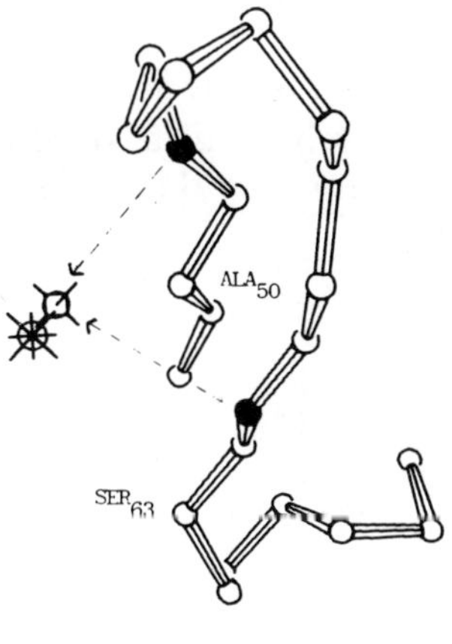

	50													63	
603	Ala	Ser	Arg	Asn	Lys	Gly	Asn	Lys	Tyr	Thr	Thr	Glu	Tyr	Ser	PC choline
T15	Ala	Ser	Arg	Asn	Lys	Ala	Asn	Asp	Tyr	Thr	Thr	Glu	Tyr	Ser	PC choline
M167	Ala	Ser	Arg	Ser	Lys	Ala	His	Asp	Tyr	Arg	Thr	Glu	Tyr	Ser	PC choline
M315	Phe	Ile	Lys	Tyr	Asp	Gly	Ser	Asx	Tyr	Gly	-	-	-	Asx	DNP
E109	Glu	Ile	Arg	Leu	Lys	Ser	His	Asn	Tyr	Ala	Thr	His	Tyr	Ala	β2→1 levan

H-2

FIG. 3.

acids longer than McPC603. These additions probably affect the size of the bend and also the interactions with L1. The three phosphorylcholine-binding myeloma proteins in the mouse have in common long L1 and H3 CRs.

The H3 region of McPC603 is an extended bend; however, the addition of three residues in the CR, as in the M167 protein, would favor extending the bend region into the solvent. In this environment a helical arrangement would be more stable. In fact, the rabbit anti-type III *Pneumococcus* polysaccharide homogeneous antibody BS-5 has a similar length in H3, and Davies et al. *(this volume)* hypothesize the H3 of this protein will be helical.

In the H3 CR of McPC603, T15 and M167 Tyr_{101}, Ser_{103}, Thr_{104}, and Tyr_{106}, or conservative replacements are found. These side chains are very close to the side chains of L1 and could form hydrogen bonds with L1 side chains. This is clearly seen in McPC603, where the L1 contains many Ser, Asx, and Glxs. Such interactions are less likely in T15 and M167 L1 that contain many charged groups in this region. In McPC603, only one of the side chains from H3, Trp_{105}, interacts with the hapten. This Trp also appears in T15, but is not found in M167. However, there are amino acids with bulky side chains in M167 that probably conserve the functions of this tryptophane. H3 appears to be the most structurally variable region in V_H, and much of this variation is attributable to size differences.

L1

L1 (Fig. 5) is a long open loop that begins at Cys_{23} and ends at Trp_{35}. The amino-terminal segment is a stretch of beta structure in McPC603 that projects outward toward the solvent, and then forms a complex bend that

603

	96													107	
603	Cys	Ala	Arg	Asn	Tyr	-	Tyr	Gly	Ser	Thr	-	-	Trp	Tyr	Phe	PC choline
T15	Cys	Ala	Arg	Asp	Tyr	-	Tyr	Gly	Ser	Ser	Tyr	-	Trp	Tyr	Phe	PC choline
M167	Cys	Thr	Arg	Asp	Ala	Asp	Tyr	Gly	(Asp)	Ser	Tyr	Phe	Gly	Tyr	Phe	PC choline
M315	Cys	Ala	Gly	Asp	Asn	Asp	-	-	-	-	-	His	Leu	Tyr	Phe	DNP
E109	Cys	Thr	Thr	Gly	-	-	-	-	-	-	-	-	-	-	Phe	β2→1 levan

H-3

FIG. 4.

ultimately projects toward the interdomain cleft. The carboxyl stretch of beta structure connects the two layers of the V_L domain (6). The L1 loop, then, is not hairpin-like where the straight stretches come close to each other over an extensive distance. In McPC603 the two stretches approach each other in a narrow area. The bend portion of L1 is long and is formed by six amino acids—31 a, b, c, d, e, and f. The side chains of these residues are close to and form close contacts with side chains of H3. The L1 bend, despite its length and bulk, apparently does not react with the phosphoryl-

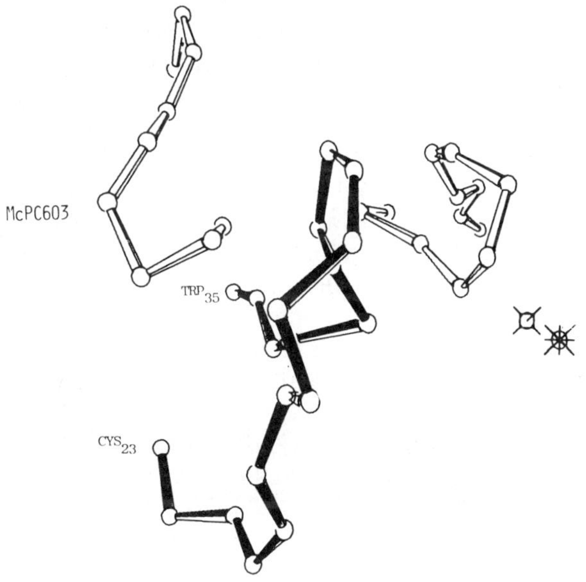

Kappa	23																		35	
TEPC15	Cys	Thr	Ala	Ser	Glx	Ser	Leu	Tyr	Ser	Ser	Lys	His	Lys	Val	His	Tyr	Leu	Ala	Trp	PC-choline
McPC603	Cys	Lys	Ser	Ser	Glx	Ser	Leu	Leu	Asx	Ser	Gly	Asx	Glx	Lys	Asx	Phe	Leu	Ala	Trp	PC-choline
MOPC167	Cys	Arg	Ser	Ser	Lys	Ser	Leu	Leu	Tyr	Lys	Asp	Gly	Lys	Thr		Tyr	Leu	Asn	Trp	PC-choline
MOPC460	Cys	Arg	Ser	Ser	Gln	Leu	Val	His	Ser	Thr	Asx	Gly	Asx			Tyr	Leu	His	Trp	DNP
EPC109	Cys	Gln	Ala	Ser	Gln	Gly	Thr	Asn	Ile	Asn							Leu	Asn	Trp	β2→1 Levan
UPC10	Cys	Arg	Ala	Ser	Glx	Asx	Ile	Ser	Asx							Tyr	Leu	Asx	Trp	β2→6 Levan
XRPC25	Cys	Arg	Ala	Ser	Gln	Ser	Ile	Ser	Asn	Asn							Leu	(His)	Trp	DNP
J539	Cys	Ser	Ala	Ser	Ser	Ser	Val	Ser	Ser								Leu	His	Trp	β1→6 Gal
Lambda																				
M104E	Cys	Arg	Ser	Ser	Thr	Gly	Ala	Val	Thr	Thr	Ser	Asn				Tyr	Ala	Asn	Trp	α1→3 Dex
M315	Cys	Arg	Ser	Ser	Thr	Gly	Ala	Val	Thr	Thr	Ser	Asn				Tyr	Ala	Asn	Trp	

Mouse L1

FIG. 5.

choline part of the antigen. Its structural function is related to the filling of the potential space between H3 and L2 and by its interactions with H3. H3 does contribute a residue, Trp_{105}, that does interact with the choline part of the hapten via van der Waal's forces.

The long amino-terminal stretch of L1, which projects into the solvent, probably plays a more direct role in antigen interaction. However, very little is known about the interactions from X-ray crystallography, because only hapten binding has been studied, and crystals have not been prepared with larger antigenic determinants. One antigen about which some structure is known is the *Pneumococcus* C polysaccharide (10). Although the exact structure of the repeating unit in this antigen has not been absolutely determined, enough is known to permit some educated guesses about possible interactions. The repeating unit in the *Pneumococcus* C polysaccharide antigen (with which McPC603 interacts) contains ribitol phosphate, *N*-acetylgalactosamine, glucose, *N*-acetylydiaminotrideoxyhexose, and choline phosphate. (See Watson and Baddiley, ref. 10, for discussion of structure.) The presence of sugars in the repeating unit provides many OH groups for hydrogen bond interactions with such side chains as serine, threonine, and tyrosine. The amino-terminal stretch of L1 in 603 contains eight residues that could participate in such interactions. It should be noted, however, that the *Pneumococcus* C polysaccharide is only one antigen with which McPC603 reacts. Others known to contain phosphorylcholine are derived from *Ascaris* integument, a *Lactobacillus* antigen, and, in addition, antigens from *Aspergillus* and *Trichoderma* species (3). The McPC603 protein also reacts with an antigen derived from a *Proteus morganii* species of mouse origin (3). The TEPC15 and MOPC167 proteins fail to react with this antigen, suggesting that the *Proteus morganii* antigen may not contain phosphorylcholine (L. Claflin, *personal communication*).

L1 Structural Variants

L1, based on present data, is the most structurally variable CR in the V_L. The drastic variations are reflected in great differences in polypeptide length. In different mouse kappa chains, length varies from 12 to 19 amino acids (17). In the shortest chains, which differ from McPC603 by seven amino acids, there is sufficient homology to pinpoint the "permissive" section, where amino acids can be deleted, to the region between position 30 and 33. [Note that this can be as long as eight amino acids or as short as one amino acid (Fig. 5).] Since only the McPC603 structure is known, one can only guess at the likely structures of the different L1 regions. The shortest L1 segments as in J606, UPC10, and XRPC25 (Table 4) resemble the human REI kappa chain in length (18). The structure of L1 in REI has been determined by X-ray crystallographic studies and is a simple bridge between

TABLE 4. *Amino acid sequences of BALB/c mouse V_K and V_λ to Trp_{35}*

Kappa	1 10 20 30 a b c d e f 35	No. of amino acids in L1	No. of amino acids to Trp_{35}	Ref
McPC603	D I V M T Q S P S S L S V S A G E R V T M S C K S S Z S L L B S G B Z K B F L A W	17	40	27
TEPC15 HOPC8 S107	—— T F – A – T – S K K — I — T A S Z S L Y S S K H K V H Y L A W	17	40	27
MOPC167	— I — B E L – D P – T S — S – S I T – R S S K S L L Y K D G K T Y L N W	16	39	17
MOPC321	— L —— A — A — L – Q – A – I — R A S K S V N T Y G N S F M Z W	15	38	15
TEPC124	— L ——. A — A — L – Q – A – I — R A S Q S V B W Y G B S F M Z W	15	38	15
BFPC32	— L —— A — A — L – Q – A – I — R A S Q V Z B S Y G I S F M Z W	15	38	15
MOPC70	— L —— A — A — L – Q – A – I — R A S Q S V B B S G I S F M N W	15	38	32
MOPC63	N — L —— A — A — L – Q – A – I — R A S Z S V B S Y G B S F M Z W	15	38	15
MOPC460	– V —— T – L — T — L – D – A S I — R S S Q L V H S T B G B Y L H W	15	38	31
MPC11	—— H K F M – T – V – B — S I T – K A S Q B V S T T V A W	11	34	34
MOPC21	N —— K – M – M – V —— L T – K A S E N V V T Y V S W	11	34	33
MOPC41	— Q —— A – L —— S L T – R A S Q B I G S L S B W	11	34	32
EPC109	– V Q – I —— A – L – D I —— T – Q A S Q G T N I N L N W	11	34	17
AMPC1	– V Q – I —— A – L – D I —— T – Z A S Z G T B I B L B W	11	34	17
MOPC46	— L —— A —— T P – D S -- S L — R A S Q B I S N N L H W	11	34	35
XRPC25	— L —— A T —— T P – D S – S L — R A S Q S I S N N L(H)W	11	34	17
UPC10	— Q —— T T —— A – L – D —— I — R A S Z B I S B Y L B W	11	34	17
J539	E — L —— A I T A A – L – Q K — I T – S A S S S V S S L H W	10	33	17
McPC773 MOPC265	E T T V —— A —— M A I — K —— – I T S T B I B G M B W	10	32	17
Lambda				
MOPC104 J558	p A V V T Q Q S A L T T S P G E T V T L T C R S S T G A V T T S N Y A N W	14	36	23,24
S178	—— Z Z – —— Z —— N —— B — B –			24
MOPC315	—— Z Z – —— G — I —— ——			28

the carboxyl and amino-terminal stretches of the L1 loop. The absence of the extended bend region creates a space between H3 and L2, which also may modify the secondary structure of H3.

The L1 lengths of V_L TEPC15 and MOPC167 that are also subunits of phosphorylcholine-binding proteins like the McPC603 L1 are unusually long, but the sequence in these chains differs greatly from McPC603. The L1 regions of two human chains, NEWM (19) and Mcg (20), have been determined by X-ray crystallography; in both cases the bend portion is partly helical. These proteins are similar in length to the M104E and M315 V_L L1 CR. While the sequences in the bend portion of human and mouse lambda chains differ, both species have amino acids with neutral side chains in these regions. Padlan et al. (21) postulate L1 in MOPC315 V_λ is also partly helical.

L2

The L2 (Fig. 6) region is in the lateral bounds of the antigen binding site and is structurally a bend in a relatively small loop between positions 49 (counting from Trp_{35}) and 56. The structure of the chain in this region is quite conserved. Large deletions have been observed only in one protein: the human NEWM (19). Aside from this, L2 sizes are remarkably constant, even between species. Amino acids in L2 could potentially provide side chains for interaction with antigens. However, in McPC603 (and probably also in T15 and M167), L2 is shielded from the hapten binding site, and in these cases its role in binding is doubtful.

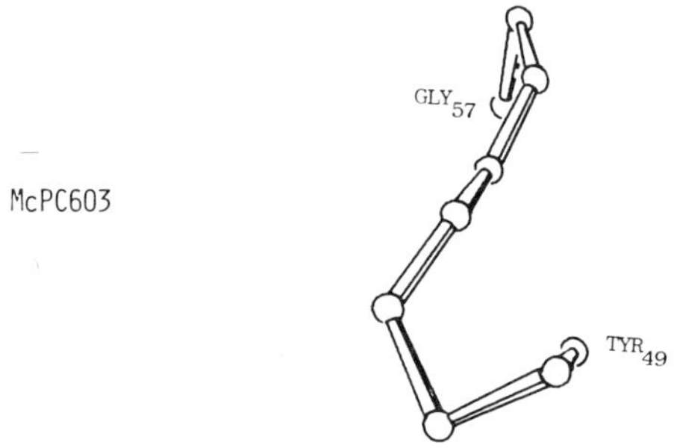

Kappa	48									57	
MOPC70	Ile	Tyr	Ala	Ala	Ser	Asn	Gln	Gly	Ser	Gly	
MOPC321	Ile	Tyr	Arg	Ala	Ser	Asn	Leu	Glx	Ser	Gly	
MOPC41	Ile	Tyr	Ala	Thr	Ser	Ser	Leu	Asx	Ser	Gly	
MOPC21	Ile	Tyr	Gly	Ala	Ser	Asn	Arg	Tyr	Thr	Gly	
Lambda											
MOPC104E	Ile	Gly	Gly	Thr	Asn	Asn	Arg	Ala	Pro	Gly	α1→3 Dex
MOPC315	Ile	Gly	Gly	Thr	Ser	Asx	Arg	Ala	Pro	Gly	DNP

L2

FIG. 6.

Hypothetical Construction of EPC109 Antigen Binding Site

The potential role of L2 in antigen binding is illustrated in a hypothetical construction of the binding site of the $\beta2\rightarrow1$ fructosan binding protein EPC109. H1, H2, H3, L1, and L3 from this protein have been sequenced. The L1 of EPC109 is 13 amino acids long and 6 amino acids shorter than McPC603. The length of L1 is the same as the L1 in the human REI protein, whose structure has been determined by X-ray crystallography (18) and was used in the hypothetical construction (Fig. 7). L3 of EPC109 has the same length as L3 in McPC603; this structure was used for the L3. The sequence of L2 has not been determined in EPC109, and the McPC603 was used. In EPC109 V_H H1 and H2 are similar in length to McPC603, and H3 is seven amino acids shorter. The H3 loop in EPC109 was constructed by closing the structure in McPC603 by a short bridge. The combination of these CRs drastically reshape the antigen binding surface. This is depicted in the computer printout of the binding site (Fig. 7). The great reduction of the lengths of L1 and H3 now creates a potential groove in which the antigen presumably lies. L2 forms one of the lateral boundaries of this groove.

L3

L3 (Fig. 8) is the bend portion of a long hairpin-like loop within the outer layer of the V_L domain. The stretches of this loop lie very close to a similar loop in the V_H domain and provide residues that are involved in interdomain

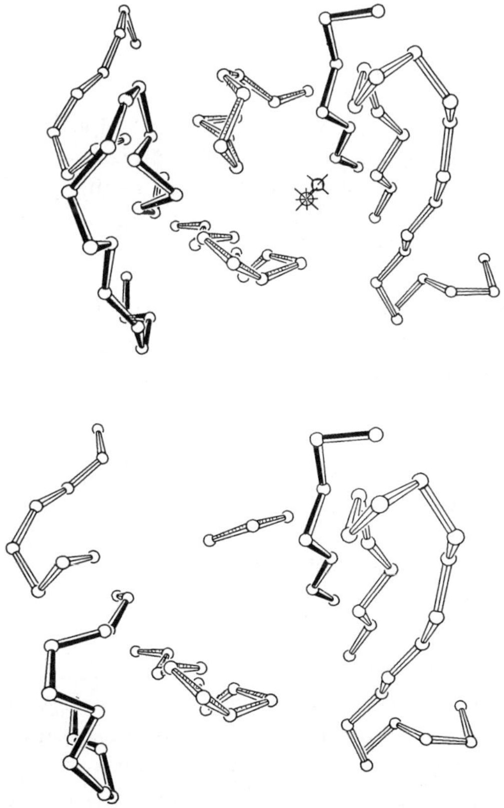

FIG. 7. Center view of the CR (complementarity regions) McPC603 (**upper**) and a hypothetical display of the EPC 109CR regions (**lower**). For orientation, compare with Fig. 1. In the McPC-603 the symbols indicate the location of the phosphorylcholine hapten. The EPC109 antigen binding region is characterized by very short H_3 and L1 CR.

603

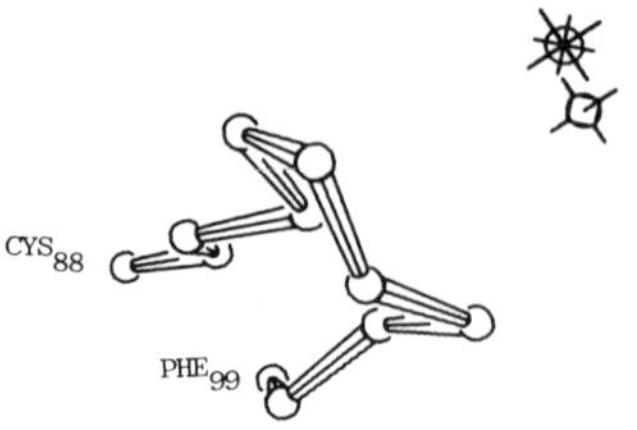

Kappa	88												101		
M70	Cys	Glx	Glx	Ser	Lys	Glu	Val	Pro	Trp	Thr	Phe	Gly	Gly	Gly	
M321	Cys	Glx	Glx	Ser	Asx	Glx	Asx	Pro	Trp	Thr	Phe	Gly	Ser	Gly	
T124	Cys	Glx	Glx	Ser	Asx	Glx	Ala	Pro	Trp	Thr	Phe	Gly	Ser	Gly	
M63	Cys	Glx	Glx	Asx	Asx	Glx	Asx	Pro	Trp	Thr	Phe	Gly	Gly	Gly	
M41	Cys	Leu	Gln	Tyr	Ala	Ser	Ser	Pro	Trp	Thr	Phe	Gly	Gly	Gly	
M21	Cys	Gly	Gln	Gly	Tyr	Ser	Tyr	Pro	—	Thr	Phe	Gly	Gly	Gly	
J539	Cys	Gln	Gln	Trp	Thr	Tyr	Pro	Leu	Leu	Thr	Phe	Gly	Ala	Gly	β1→6 Gal
E109	Cys	Leu	Gln	(Lys)	Ser	Tyr	Leu	Pro	Tyr	Thr	Phe	Gly	Gly	Gly	β2→1 Lev
Lambda															
M315	Cys	Ala	Leu	Trp	Phe	Arg	Asx	His	Phe	Val	Phe	Gly	Gly	Gly	DNP
M104	Cys	Ala	Leu	Trp	Tyr	Ser	Asn	His	Trp	Val	Phe	Gly	Gly	Gly	α1→3 Dex

L3

FIG. 8.

interactions. The potential antigen binding surface is confined to the bend portion of the loop. The sequence of L3 has not yet been determined for McPC603, and it is not known whether side chains derived from this bend are involved in binding the hapten itself or other parts of the antigen molecule. The proximity of L3 to the hapten, however, make this a likely possibility.

In a hypothetical model of the DNP binding protein MOPC315, Padlan et al. (21) have found that, by placing the DNP group in close proximity to Trp_{91}, the π-electron clouds of the two ring structures overlap, providing structural stability.

The available sequences in the mouse thus far indicate that the size of L3 is remarkably the same; however, in the rabbit L3 differs considerably in length in different antigen-binding proteins (22).

Genetic and Evolutionary Mechanisms in Producing a Variety of Antigen Binding Sites

The mouse has proven to be a particularly advantageous species for studying the genetic basis of antibody diversity. First, as mentioned above, the plasmacytoma-induction process appears to induce neoplasms in a wide range of functional plasma cell types, thus providing a broad representation of the immunoglobulin-producing potential of the mouse. However, this must be considered to be incomplete, since we have very few if any examples of immunoglobulins that react with protein antigens to which the mouse has been presumably naturally immunized. Second, the V_K amino-terminal segments in the mouse in contrast to man, for example, have a great variety of different sequences which can be easily analyzed. V_λ in the mouse occur less frequently and are less easily analyzed because the N terminus is blocked by a pyrrolidone carboxylic acid. However, the complete V region sequence of 12 of these proteins has been determined by Appella (23) and Weigert and colleagues (24).

Specific V_L sequences in the mouse in general appear to correlate with antigen-binding specificity. The α1→3 dextran binding proteins in the BALB/c mouse are all $V_{\lambda 1}$ (5); the β1→6 galactan (25), β2→6 fructosan, and β2→1 fructosan (26) binding proteins each thus far have V_K chains with the same amino-terminal sequence (Table 4). This suggests that certain V_L chains are specialized in the mouse to the extent that they must be activated exclusively if immunoglobulins of that specificity are to be produced. By contrast, the antibodies to phosphorylcholine and dinitrophenyl have different amino-terminal sequences. In these cases different structural genes can perform the same function.

Partial sequences of 55 V_K chains have been determined to Cys_{23} (see ref. 17 for references). Among these there are 24 sequences that differ from all the others by four to nine amino acids, i.e., 21 to 40% of the partial sequence (Table 5). In many cases there are one or more examples of a basic sequence, indicating that these have not arisen through some chance mutational process but rather are transcriptions of inherited germ-line structural genes. Furthermore, in addition to the 24 basic sequences, there are families of sequences that are more closely related, which have 4 or less differences (Tables 4 and 5).

The basic 24 V_K sequences in the mouse probably reflect, in a very conservative sense, the minimum number of V_K structural genes in the mouse. It is known that 17 differences exist between MOPC70 and MOPC321 proteins with the same amino-terminal sequence, indicating that these two are controlled by separate structural genes (15). Thus we can anticipate that more genes will be defined when the sequences are available. Hopefully, the possibilities will be exhausted when another 50 to 100 amino-terminal sequences are determined.

The available sequences do provide some insight into mechanisms of antibody variability. Twenty-four of the sequences have been completed through Trp_{35} and, hence, include the most variable CR of the V_L, L1 CR. As may be seen in Table 4, the length of L1 CR is the same in all proteins with the same amino-terminal sequence to Cys_{23}. This association suggests that the characteristic sequence and the length of L1 CR have coevolved in a single genetic unit. The structural importance of the L1 CR in forming the antigen binding site has been presented, and it may be surmised that the dominant structural feature selected in the evolution of new V_L genes is the L1 CR length. The association of a characteristic sequence to Cys_{23} with a characteristic length in L1 CR also makes it doubtful that DNA sequences

TABLE 5. *Number of different BALB/c V_κ sequences based on partial sequence to CYS_{23}*

Hapten Binding	V_k Chains	Minimal No. Amino Acid Differences to Cys_{23}	Length to Trp_{35}
PC Choline	[TEPC15 S107 HOPC8]	7	41
PC Choline	[McPC603] (bracket 1 with MOPC384 McPC870)	6	41
S. tel aviv lps	MOPC384 McPC870		
PC Choline	[MOPC167 MOPC511]	8	40
	McPC613 M321 MOPC70 TEPC124 BFPC32 (bracket 1 with MOPC63; bracket 2 with B61)	6	39
	MOPC63		
	B61		
Flagellin	[MOPC467] (bracket 4 with MOPC460)	8	
DNP	[MOPC460]	7	39
	Mc674 (bracket 1 with Mc843)	7	
	Mc843		
1→6 Dextran	[W3129 W3434]	9	
	MOPC41	6	35
2→1 Fructosan	[J606 W3082 ABPC4]	5	35
	LPC1	6	
	MOPC35	6	
2→6 Fructosan	[UPC10] MOPC173	6	35
	McPC600	6	
	MOPC21	6	35
	MOPC157 (bracket 4 with HOPC5)	8	
	HOPC5 (bracket 4 with MPC11)	7	
	MPC11	7	
	TEPC29	8	
DNP	S23 [XRPC25] MOPC46B MOPC172	8	35
	MOPC47B (bracket 2 with the RPC23 TEPC173 / MOPC31C MOPC178 group)	6	
	RPC23 TEPC173 (bracket 2 with MOPC31C MOPC178; bracket 3 with MOPC149)		
	McPC611 (bracket 2 with MOPC31C MOPC178)	6	
	MOPC31C MOPC178 (bracket 3 with MOPC149)	5	
	MOPC149	6	
1→6 Galactan	SAPC10 TEPC191 JPC1 XRPC24 XRPC44 J539	7	34
	MOPC29	8	
	MOPC265 McPC773	8	34

Numbers aside braquets = number of amino acid differences.

Enclosed entries have Hapten Binding activity.

Entries on same line have identical sequence.

controlling L1 CR are inserted in or are derived from an independent source and inserted into a V_L structural gene. The mutational mechanism involved in producing the different lengths of CR structures is not known but appears to involve a process that deletes or adds triplets. This type of mutation may occur more frequently in immunoglobulins than in other proteins. In most situations, triplet deletions or insertions would be selected against, because a triplet change could in many cases produce a nonfunctional protein, particularly when that triplet controls an amino acid that is essential for the folding and conformation of the protein. In the case of L1 CR, however, deletions or insertions of triplets are permissive in that they do not affect the conformation or folding of the framework of the V domain (13).

There is evidence in both man and rabbit that a similar process has occurred. Considering the length of the sequence from Cys_{23} to Trp_{35} as an index of this variability, the lengths given in the tabulation below have been described.

Length Between Cys_{23} and Trp_{35}

	V_κ	V_λ
Mouse	12, 13, 17, 18, 19	13
Man	13, 14, 17, 18, 19	13, 15, 16
Rabbit	13, 14, 15	

It could be argued the L1-CR length variants arise by deletion of a run of amino acid triplets as in a frameshift mutation. However, single-step deletions seem more plausible in many cases because, even though the changes affect a permissive section of the sequence that does not distort the folding of the domain, the amino acid side chains from CRs do interact with other CRs. Thus drastic changes in one CR will affect other CRs. If the evolutionary process is gradual, this would permit compensating mutuations in other CRs to develop coincidentally.

It would appear from the existing data that one of the dominant forces in immunoglobulin gene evolution is the selection of genes that provide drastic changes in the antigen binding surface through modifications in length. A second major process is the selection of forms with the same length, but different sequences in a given CR. In the latter cases the sequence differences are usually multiple involving much of the CR. The observation that diversity appears to be favored by extensive changes introduces a third and possibly most important factor, the duplication of immunoglobulin genes to provide the basis for evolving new structural mutants. These processes tend to relegate somatic changes to a more minor role, and indicate that major steps in the evolution of diversity are facilitated by a multiple gene system.

The BALB/c mouse system potentially provides the means for cataloguing the immunoglobulin-producing potential and provides a source for new information. It is hoped that the analysis of this system will ultimately define the number of different antibodies generated by a haplotype.

SUMMARY AND CONCLUSIONS

Crystallographic studies have revealed the potential antigen binding region of the immunoglobulin molecule to be an irregular surface approximately 30 × 40 Å that contains clefts, grooves, and outfoldings. This surface in the mouse McPC603 protein is lined by 6 CRs, 3 of which (L1, L2, L3) are from different regions of the V_L polypeptide, and 3 of which (H1, H2, H3) are from different parts of the V_H polypeptide. In the three-dimensional models the secondary structures of CR are straight segments of beta structure, bends (which include stretches) and tips of bends.

The primary structures of CR of antigen-binding myeloma proteins of the BALB/c mouse were compared. Two striking types of structural differences were observed in the few proteins available for study: (i) size differences particularly in L1 and H3, and (ii) extensive sequence variations (H1, H2, H3, L1, and L3). Very few sequences are currently available on L2 and L3 for comparison. L3 regions in the few available sequences have the same length.

Drastic restructuring of the antigen binding surface was illustrated in a hypothetical construction of the antigen binding site of the EPC109 $\beta_2 \rightarrow 1$ fructosan-binding myeloma protein.

The genetic origin of the diversity of mouse L chain structures was discussed on the basis of the relationship of the aminoterminal sequences to Cys_{23} (a framework sequence) and the L1 structure. In this species there is a great variety of different amino-terminal sequences that permit identification of different V_K structural genes. Thus far in the mouse, characteristic L1 structures are associated with a specific amino-terminal sequence, and there is no evidence that the same amino-terminal sequence is associated with a variety of different L1 structures.

It therefore appears that single immunoglobulin V-region structural genes as units are modified by mutations, so that the proteins synthesized from them: (i) maintain their ability to fold properly and interact with other domains, and (ii) generate diverse structural variations that permit binding to a variety of antigens. The fascinating aspect of the evolutionary process is that it exploits potential structural differences. Two major forms of variations are amino acid replacements that probably do not change the secondary structure of the segment in which they occur, and size differences that modify the secondary structure of bends. Apparently both forms are necessary for the evolution of an effective immunoglobulin-producing system.

REFERENCES

1. Potter, M. (1972): Immunoglobulin-producing tumors and myeloma proteins of mice. *Physiol. Rev.*, 52:631–719.

2. Cohn, M. (1967): Natural history of the myeloma. *Cold Spring Harbor Symp. Quant. Biol.*, 32:211–221.
3. Potter, M. (1971): Antigen binding myeloma proteins in mice. *Ann. NY Acad. Sci.*, 190:309–321.
4. Potter, M. (1970): Mouse IgA myeloma proteins that bind polysaccharide antigens of enterobacterial origin. *Fed. Proc.*, 29:85–91.
5. Blomberg, B., Carson, D., and Weigert, M. (1973): The immune response to dextran in mice. In: *Specific Receptors of Antibodies, Antigens, and Cells*, pp. 285–293. 3rd Int. Convoc. Immunol. Buffalo, N.Y. 1972, Karger, Basel.
6. Lieberman, R., Potter, M., Mushinski, E. B., Humphrey, W., Jr., and Rudikoff, S. (1974): Genetics of a new IgV_H (T15 idiotype) marker in the mouse regulating natural antibody to phosphorylcholine. *J. Exp. Med.*, 139:983–1001.
7. Claflin, J. L., and Davie, J. M. L. (1974): Clonal nature of the immune response to phosphorylcholine. IV. Idiotypic uniformity of binding site-associated antigenic determinants among mouse antiphosphorylcholine antibodies. *J. Exp. Med.*, 140:673–686.
8. Cosenza, H., and Kohler, H. (1972): Specific inhibition of plaque formation to phosphorylcholine by antibody against antibody. *Science*, 176:1027.
9. Tomasz, A. (1967): Choline in the cell wall of bacterium: A novel type of polymer linked in *Pneumococcus*. *Science*, 157:694–697.
10. Watson, M. J., and Baddiley, J. (1974): The action of nitrous acid on C-teichoic acid (C-substance) from the walls of *Diplococcus pneumonia*. *Biochem. J.*, 137:399–404, 1974.
11. Wu, T. T., and Kabat, E. A. (1970): An analysis of the sequences of the variable regions of Bence Jones proteins and myeloma light chains and their implications for antibody complementarity. *J. Exp. Med.*, 132:211–250.
12. Kabat, E. A. (1970): Heterogeneity and structure of antibody combining sites: Landsteiner Centennial Dec. 5–7, 1968. *Ann. NY Acad. Sci.*, 169:43–54.
13. Padlan, E. A., and Davies, D. R. (1975): Variability of three-dimensional structure in immunoglobulins. *Proc. Natl. Acad. Sci. USA*, 72:819–823.
14. Rudikoff, S., and Potter, M. (1974): Variable region sequence of the heavy chain from a phosphorylcholine binding myeloma protein. *Biochemistry,* 13:4033–4038.
15. McKean, D., Potter, M., and Hood, L. (1973): Mouse immunoglobulin chains. Pattern of sequence variation among K-chains with limited sequence differences. *Biochemistry,* 12: 760–771.
16. Segal, D. M., Padlan, E. A., Cohen, G. H., Rudikoff, S., Potter, M., and Davies, D. R. (1974): The three-dimensional structure of a phosphorylcholine-binding mouse immunoglobulin Fab and the nature of the antigen binding site. *Proc. Natl. Acad. Sci. USA* 71:4298–4302.
17. Potter, M., Padlan, E. A., and Rudikoff, S. (1976): Localized deletion-insertion mutations: A major factor in the evolution of immunoglobulin structural variability *(in preparation)*.
18. Epp. O., Colman, P., Fehlhammer, H., Bode, W., Schiffer, M., and Huber, R. (1974): Crystal and molecular structure of a dimer composed of the variable portions of the Bence Jones protein REI. *Eur. J. Biochem.*, 45:513–524.
19. Poljak, R. J., Amzel, L. M., Chen, B. L., Phizackerley, R. P., and Saul, F. (1974): The three-dimensional structure of the Fab fragment of a human myeloma immunoglobulin at 2.0-Å resolution. *Proc. Natl. Acad. Sci. USA,* 71:3440–3444.
20. Schiffer, M., Girling, R. L., Ely, K. R., and Edmundson, A. B. (1973): Structure of a λ-type Bence Jones protein at 3.5-Å resolution. *Biochemistry,* 12:4620–4631.
21. Padlan, E. A., Davies, D. R., Pecht, I., Givol, D., and Wright, C. (1976): Model building studies of antigen-binding site of MOPC315. *Cold Spring Harbor Symp. Quant. Biol. (In press.)*
22. Margolies, M. N., Cannon, L. E. III, Strosberg, A. D., and Haber, E. (1975): Diversity of light chain variable region sequences among rabbit antibodies elicited by the same antigens. *Proc. Natl. Acad. Sci. USA,* 72:2180–2184.
23. Appella, E. (1971): The amino acid sequences of two mouse immunoglobulin lambda chains. *Proc. Natl. Acad. Sci. USA,* 68:590–594.
24. Weigert, M., Cesari, I. M., Yonkovich, S. J., and Cohn, M. (1970): Antibody variability: Mouse λ-light chain sequences. *Nature,* 228:1045–1047.
25. Rudikoff, S., Mushinski, E. B., Potter, M., Glaudemans, C. P. J., and Jolley, M. E. (1973): Six BALB/c IgA myeloma proteins that bind B-1→6-D galactan. *J. Exp. Med.,* 138:1095–1106.

26. Lieberman, R., Potter, M., Humphrey, W. Jr., Mushinski, E. B., and Vrana, M. (1975): Multiple individual and cross-specific idiotypes on $\alpha 1 \rightarrow 3$ levan-binding myeloma proteins of BALB/c mice. *J. Exp. Med.,* 142:106–119.
27. Barstad, P., Rudikoff, S., Potter, M., Cohn, M., Konigsberg, W., and Hood, L. (1974): Immunoglobulin structure: Aminoterminal sequences of mouse myeloma proteins that bind phosphorylcholine. *Science,* 183:962–964.
28. Dugan, E. S., Bradshaw, R. A., Simms, E. S., and Eisen, H. N. (1973): Amino acid sequence of the light chain of a mouse myeloma protein (MOPC315). *Biochemistry,* 12:5400–5416.
29. Francis, S. H., Graham, R., Leslie, Q., Hood, L., and Eisen, H. N. (1974): Amino acid sequence of the variable region of the heavy (alpha) chain of a mouse myeloma protein with antihapten activity. *Proc. Natl. Acad. Sci. USA,* 71:1123–1127.
30. Lifter, J., Hew, C-L., Yoshioka, M., Richards, F. F., and Konigsberg, W. H. (1974): Affinity-labeled peptides obtained from the combining region of protein 460. I. Heavy-chain labeling patterns using dinitrophenylazide photoaffinity label. *Biochemistry,* 13:3567–3571.
31. Hew, C-L., Lifter, J., Yoshioka, M., Richards, F. F., and Konigsberg, W. H. (1973): Affinity-labeled peptides obtained from the combining region of protein 460. Light-chain labeling patterns using dinitrophenyl base photoaffinity labels. *Biochemistry,* 12:4685–4689.
32. Gray, W. R., Dreyer, W. J., and Hood, L. (1967): Mechanisms of antibody synthesis size differences between mouse kappa chains. *Science,* 155:461–467.
33. Svasti, J., and Milstein, C. (1972): The complete amino acid sequence of a mouse K-light chain. *Biochem. J.,* 128:427–441.
34. Smith, G. P. (1973): Mouse immunoglobulin kappa chain MPC-11 extraaminoterminal residues. *Science,* 181:941–943.
35. Melchers, F. (1969): The attachment site of carbohydrate in a mouse immunoglobulin light chain. *Biochemistry,* 8:938–947.

Antibodies in Human Diagnosis and Therapy,
edited by Edgar Haber and Richard M. Krause.
Raven Press, New York © 1977.

Allotype Expression and the Regulation of Immunoglobulin Synthesis

T. J. Kindt, M. Mudgett, J. A. Sogn,
B. A. Fraser, and B. Aasted

The Rockefeller University, New York, New York 10021

Studies on the serologically detected genetic markers of immunoglobulins have greatly aided understanding of the genetic mechanisms underlying antibody synthesis. Much of the basic information on the immunoglobulin genes has been obtained from studies on allotypes and idiotypes of the human, mouse, and rabbit. However, recent data on rabbit immunoglobulin allotypes suggest that certain concepts deduced from these studies require closer examination. This chapter concerns recent data obtained in serological and structural studies on the allotypes of rabbit immunoglobulins.

ANTIANTIBODIES IN IMMUNOTHERAPY

The immediate relevance of allotypes to immunotherapy has been demonstrated by the work of Jerne and Aasted (1) at the State Serum Institute in Copenhagen, Denmark. Their studies showed that administration of anti-RBC antibodies to rabbits prevented response to RBCs in animals challenged with RBCs on the previous day. More significantly, it was shown that all effects of this treatment were reversed when the injected rabbit was making antibodies directed against allotypes of the administered antibodies. Jerne and Aasted have suggested that effects similar to those shown in their model system could result in ineffective Rh prophylaxis. Indeed, testing of women for whom Rh prophylaxis was not successful revealed a higher frequency of antiantibodies in their sera than that found in the normal population. With more widespread use of antibodies in therapy, and especially with multiple courses of treatment, such interactions could assume major importance in antibody immunotherapy.

Control of autoimmune diseases and graft versus host reactions are other examples where immunological approaches to therapy stem from a consideration of immunoglobulin markers. McKearn (2) and Binz et al. (3) have shown the potential value of antiidiotype antibodies in suppression of certain

GVH reactions. The feasibility of long-term suppression of specific antibodies by low doses of certain heterologous antiantibodies has been demonstrated by Eichmann (4). Eisen (5) has reported on the use of antiidiotype antibodies to prevent growth of a plasmacytoma. In light of these findings, it is reasonable to consider the possibility of utilizing antibody–antiantibody interactions in the treatment of human disease.

Antibody–antiantibody interactions may not be limited solely to pathological states or to experimental systems but may be an intrinsic component of the normally functioning immune system. Indeed, Jerne (6) has proposed that antiantibodies are necessary for the constant regulation of the immune "network." Experimental evidence can be marshaled to support this network theory. Kluskens and Köhler (7) and McKearn et al. (8) have demonstrated spontaneous occurrence of antiidiotypes directed against various antibodies. Recent data by Adler and Noelle (9) showed persistence of antibodies directed against nonexpressed maternal allotypes in young rabbits (5–6 months of age).

GENETIC CONTROL OF ALLOTYPE SYNTHESIS

It has been widely accepted that immunoglobulin allotypes are encoded by allelic structural genes at three separate loci. Many data obtained from family studies support the representation depicted in Fig. 1. Recent data

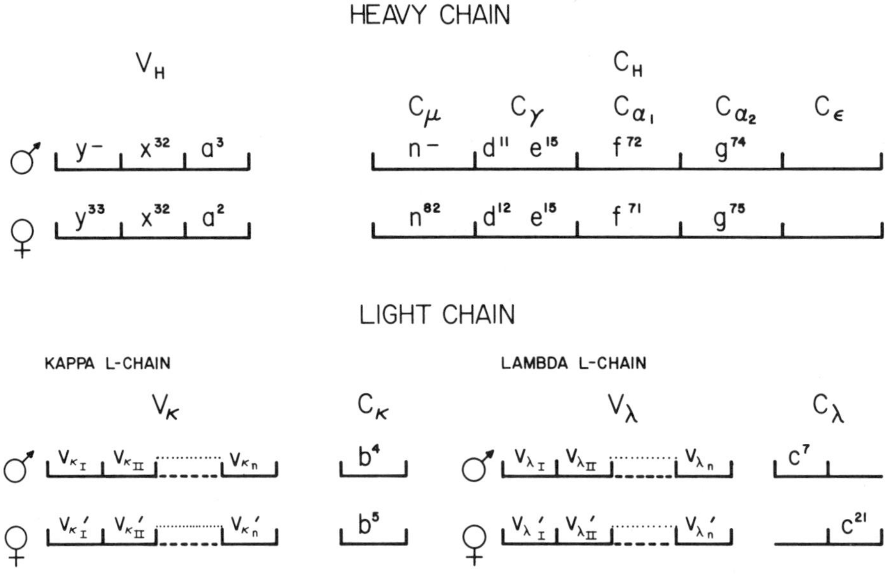

FIG. 1. Genetic loci involved in the synthesis of rabbit immunoglobulins as determined by allotype inheritance. It is not known whether V_H genes of groups a, c, and y exist in single or multiple copies.

obtained through structural and serological studies on rabbit immunoglobulin allotypes suggest that this description is incomplete. For example, recent evidence on the nature of rabbit allotype inheritance challenges the notion that these markers are encoded by allelic structural genes. There is evidence to suggest that their expression might be controlled by inherited regulator genes. In addition, structural and serological studies point to the existence of multiple forms among what were previously thought to be simple allotypes.

In this chapter, the questions listed below are considered:

1. Do group a allotype (V_H) genes exist as single or multiple copies? If multiple copies exist, are the allotypes they encode structurally and serologically identical?
2. Similarly, do group b allotype (C_K) genes exist as single or multiple copies?
3. Are immunoglobulin allotypes encoded by allelic structural genes?

COMPLEXITY OF THE GROUP a ALLOTYPES

The complexity of the group a allotypes is the first point to consider. If multiple forms of each allotype exist, then there is reason to suspect the idea that multiple copies of V_H genes exist for each. In this case, recombinations would be difficult to avoid, and allotypic inheritance would be quite different from that actually observed. That is to say, there would be no apparent pattern of inheritance of allotype (10).

An approach that has been used to investigate the nature and number of allotypic markers involves characterization of allotypically defined IgG pools and comparison of these to homogeneous antibodies. This approach has been made feasible through the observation of Osterland and co-workers in 1966 (11) that immunization of rabbits with bacterial vaccines often leads to the production of large amounts of antibody with remarkable homogeneity of structure. Early studies by Koshland et al. (12) and by Mole et al. (13,14) indicated that rabbit allotypes could be readily characterized by compositional and sequence difference. However, difficulties in similarly assigning these correlates to homogeneous antibodies have been pointed out by Jaton (15), Braun, and Haber and his colleagues (*this volume*).

A similar dichotomy between pooled material and homogeneous antibodies is observed when the serologic specificities of the group a allotypes are examined. Single homogeneous antibodies do not possess the full complement of allotypic specificities present in pooled IgG, even if the pooled IgG is isolated from those individuals that produced the antibody (16). This finding suggests that the full allotypic specificity is composed of discrete and possibly dissectable subspecificities.

An example of this component nature of allotypic specificity is illustrated in Fig. 2. Aasted (*unpublished results*) isolated two antibodies from a single

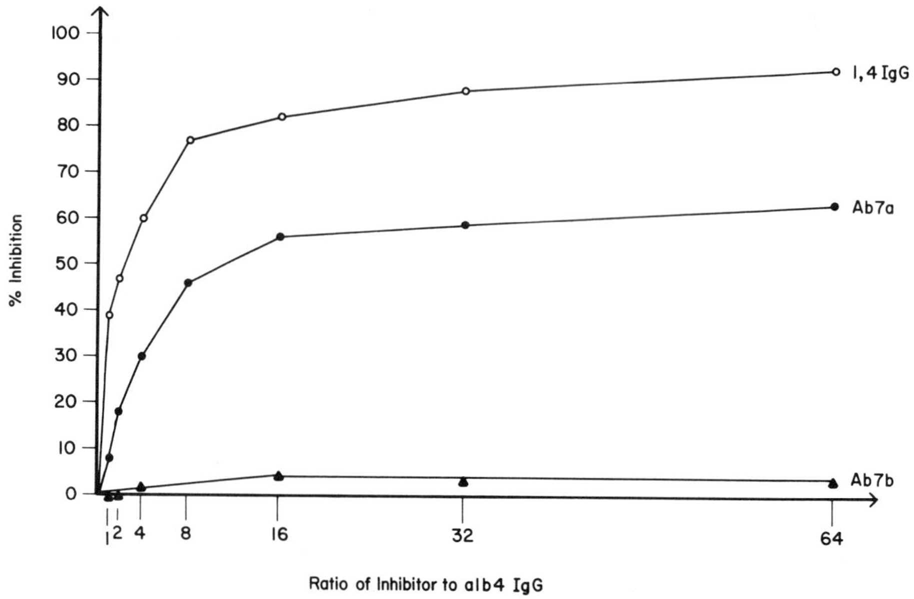

FIG. 2. Inhibition of binding of ^{125}I-labeled a1b4 IgG (0.25 μg) to solidified anti-a1 antiserum by a1b4 IgG, Ab7a, and Ab7b.

rabbit immunized with group C streptococci. The two antibodies, designated 7a and 7b, were shown to be similar in their binding properties for a group C immunoadsorbent column and in their light-chain electrophoretic mobility. Both light and heavy chains were blocked, thus precluding *N*-terminal sequence analysis. Data from inhibition of allotypic binding have shown that Ab7a has an allotypic subspecificity common to about 60% of the IgG a1 pool, while Ab7b has a rare subspecificity common to only a few percent of the pool. Both antibodies are a1 as shown by quantitative binding to anti-a1 antisera radioimmunoassays.

The precise structural basis of the group a allotypic subspecificities is at present unclear. Structural differences may reside in either the allotypic determinant in the constant part of the V_H region, or alternatively in hypervariable regions. While the latter explanation cannot be dismissed, available data best support the possibility that the differences exist in the constant parts of the V_H region. Complete V_H sequences from antibodies with known subspecificities will be required to accurately dissect the group a allotypes. In addition, structural and serologic comparisons must be made between pooled material and these homogeneous antibodies. These studies should not only explain the proper blending of subspecificities to achieve the entire allotypic site but should also help answer fundamental questions about the number of genes coding for antibody V_H regions.

COMPLEXITIES OF THE GROUP b ALLOTYPES

The group b allotypic markers, located in the constant region of kappa light chains, show less-pronounced evidence of serological subspecificities than the group a markers. No significant differences were found in the ability of any b4 homogeneous antibody tested to absorb the antibodies present in the anti-b4 antisera (17). The reaction between the anti-b4 and radiolabeled pooled b4 IgG was inhibited equally well by each of the homogeneous antibodies.

In contrast to the b4 system, Thunberg (10) has recently demonstrated an apparently simple serological difference within b9 allotypes. In a study of five homogeneous b9 antibodies, at least two different b9 allotypic subspecificities were observed. Of these, one was antigenically deficient relative to the other. Inhibition data for b9 IgG and two homogeneous antibodies of b9 allotype are shown in Table 1. The reaction between anti-b9 and any of the three ^{125}I-labeled b9 preparations is inhibited completely by the b9 pool or by homogeneous antibody 4182-I. The b9 allotype of antibody 4153-I, on the other hand, is deficient with respect to that of the pool or antibody 4182-I. These data suggest that b9 light chains of some homogeneous antibodies lack determinants present on other b9 light chains and in the b9 pool.

TABLE 1. *Relative inhibition of binding to b9 allotypic antisera*

	Unlabeled inhibitor[a]		
^{125}I-labeled antigen employed in test	b9 Pool (%)	4153-1 (%)	4182-1 (%)
b9 IgG pool	100	68	90
4153-1	99	100	100
4182-1	100	75	100

[a] All comparisons were carried out using 20-fold excess of unlabeled inhibitors. Numbers are relative percent inhibition using the inhibition given by the homologous inhibitor as 100%.

Recently, Sogn (*unpublished results*) identified two light-chain constant region differences at residues 121 and 124 between prototype b4 (4135) and a homogeneous antibody isolated from an offspring (4539) of rabbit 4135. It is interesting to note that one of these differences, a Ser instead of an Ala at residue 121, blurs an apparent distinction between b4, where Ala had always been found, and b9, where Ser was found. Pooled light chains from 4539 nonimmune IgG contained approximately equal amounts of the prototype and variant sequences. Structural studies within this extensive pedigreed family will elucidate the mode of inheritance of the b4 variant sequence and could provide evidence concerning the number of C_K genes.

Serological analysis has not yet shown any difference between the b4 determinant of the variant and prototype sequences, although this work is still in progress. This variation within b4, along with the previously reported differences found at residues 174 in b4 (18,19) and the single difference found within b9 pools (20,21) suggest that pool sequences should be examined very carefully for polymorphisms.

Studies indicate that C_L differences between L chains of varying allotype are much more extensive than would be expected for allelic variants. The first hint that this might be the case was the demonstration (22) that b9 IgG varies significantly on electrophoretic behavior from rabbit IgG of other group b allotypes. Several reports have appeared showing structural differences between light chains of different b allotypes (reviewed in ref. 23). The recently reported sequence of a complete b9 pool C_L region (20) differs at 34 of 102 positions (33%) from a prototype b4 sequence (24). It is difficult to imagine such a difference between two allelic forms of a single rabbit C_K gene, especially when one considers that human and rabbit kappa C regions differ in only about 50% of their residues.

REGULATION OF ALLOTYPE EXPRESSION

The data presented above give reason to question the notion that allotype genes, even those encoding constant region allotypes, exist as single copies.

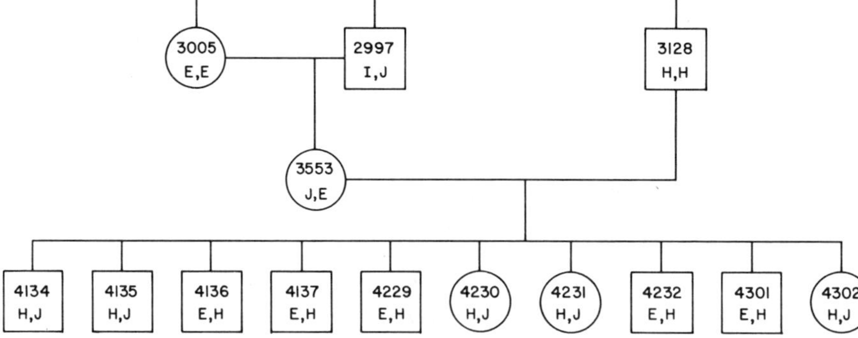

FIG. 3. R4136 pedigree: Complete allogroup typing of preimmune sera from members of rabbit 4136 family was done by W. Carey Hanly (University of Illinois at the Medical Center, Chicago, Ill.). The five H chain allogroups found in this family are indicated by single letter notations. A (a1, x^-, y^-, n81, f73, g74, d12, e15); E (a1, x^-, y^-, n81, f70, g76, d12, e15); H (a2, x^{32}, y^{33}, n82, f69, g77, d12, e15); I (a3, x^{32}, y^-, n^-, f71, g75, d^{12}, e^{15}), and J (a3, x^{32}, y^-, n^-, f72, g74, d^{11}, e^{15}). (From ref. 28.)

The finding of Strosberg et al. (25) that a single rabbit produced antibodies with three group a and three group b allotypes questions the interpretation of allotype genes as simple alleles. This and similar observations (26,27) have prompted the reexamination of allotype inheritance in our rabbit colony.

Observation of an unexpected allotype (termed latent allotype) was initially made in our laboratory in the immune sera of rabbit 4136, possessing nominal allotypes a1, a2, and b4. Elution of his streptococcal antiserum from a hapten column yielded a nonbound fraction (A) and two antibody fractions, one apparently homogeneous (B) and one a heterogeneous pool of higher affinity (C). These were pooled, concentrated, and their latent a3 allotypes quantitated by an inhibition of binding assay. The amount of inhibition corresponded to concentrations of the latent allotype a3 of 0.4, 6.6, and >10 μg/mg total protein for fractions A, B, and C, respectively.

Detection of this latent group a allotype in the serum from 4136 led to a detailed examination of this rabbit's pedigree (see Fig. 3). Complete allogroup typing, kindly done by W. Carey Hanly (University of Illinois at Chicago), demonstrated that in each case the complete allogroups segregated as expected. There were no indications of any unusual inheritance patterns in this family.

Method of Latent Allotype Detection

Allotypic markers in a variety of sera were determined by a sensitive radioimmune binding inhibition assay as previously described (29). A summary of this assay system is given below.

Binding Curve

A binding curve is constructed using antiallotype sera coupled to *N*-hydroxysuccinimide-activated Sepharose (HAS) (30) and ^{125}I-radiolabeled IgG of a given allotype. The concentration of HAS-antiallotype that binds 50% of the radiolabeled antigen is used in all subsequent inhibition assays.

Inhibition Assay

Each microfuge tube contains in a total volume of 200 μl:

a. 100 μl inhibitor solution
b. 50 μl ^{125}I-IgG antigen (0.25 μg) in phosphate buffered saline (0.05 M PO_4) which contains:
   1. Carrier protein
   2. Dithiothreitol (DTT) (0.04 M)
   3. ^{22}Na-volume marker

Solutions a and b are mixed and allowed to stand 30 min.

c. 50 μl of HAS-antiallotype at the concentration previously determined to bind 50% of the radiolabeled antigen (as in step 1).

Contents are mixed and agitated 3 to 4 hr at room temperature.

Dithiothreitol was added to the assays at a final concentration of 0.01 M to minimize binding interference by 19S rheumatoid factors in whole serum (31). At this concentration, DTT had a minimal effect on binding reactions. Carrier protein used to dilute antigens included 1% BSA and 10% (by volume) rabbit serum which contained from 50- to 100-fold excess of all group a and b allotypes, except that one being determined. The presence of an excess of all other allotypes in the inhibition system eliminates the possibility of allotypic cross reactions as an explanation for latent allotypes.

Frequency of Observation of the Latent Allotypes

Other members of the 4136 family (see Fig. 3) similarly showed latent allotypes in both their immune and preimmune sera. To test the generality of this phenomenon, preimmune sera were tested from rabbits of other related families and of a nonrelated family, as well as from random rabbits purchased from several different rabbitries. As shown by the data in Table 2,

TABLE 2. *Frequency of latent allotypes among rabbits with different combinations of group a allotypes*

	Latent allotype								
	a1			a2			a3		
	Nominal allotype								
	a2	a2,3	a3	a1	a1,3	a3	a1	a1,2	a2
No. tested	23	29	6	12	20	6	12	30	23
No. positive	15	20	3	1	3	2	8	14	14
% Positive	65	69	50	8	15	33	67	47	61

examples of latent allotypes were found for all nominal allotypic combinations. The frequency of latent allotype expression, however, was greater for latent a1 and a3 than for a2. Approximately 50% of nonimmune sera from 120 rabbits demonstrated latent allotypy levels ranging from 0.1 to 44 μg/ml.

Latent allotypes have similarly been detected by our system in nonimmune sera from certain rabbits of Drs. C. Hanly's and W. Mandy's laboratories. The results were also confirmed by testing for allotypes in their own laboratories.

Possible Explanations for Latent Allotypes

Several trivial explanations for the occurrence of latent allotypes have been ruled out. Cross reaction between allotypes was prevented by the experimental addition of rabbit sera to the binding system and by the production of the antisera in heterozygous animals (e.g., anti-a3 is produced by immunization with a3 IgG of an a^1a^2 rabbit). Nonspecific inhibition of binding is not a viable explanation, since nominal homozygotes expressing one latent allotype did not necessarily express the other. The possibility that latent allotypes are derived from maternal circulation is equally unlikely, because 10 rabbits in this study expressed detectable levels of latent allotypes absent as nominal allotypes in either parent.

A possible explanation for the occurrence of latent allotypes in only 50% of the rabbit sera is that allotype expression is under some form of regulatory control. In preliminary experiments aimed at testing this possibility, the expression of latent allotypy as a function of time was determined for 12 nonimmune rabbits. Bleedings were taken at 1-week intervals over a 6-week period. In seven of the 12 rabbits that were examined, latent allotype expression was sporadic and transitory.

Characterization of Molecules Bearing the Latent Allotypes

There is appreciable evidence that the molecules bearing the latent allotypes are IgG. Complete inhibition of binding of antiallotype antisera to radiolabeled IgG or radiolabeled homogeneous antibody has been demonstrated for (i) IgG fractions isolated from G200 Sephadex, (ii) IgG fractions isolated by DEAE-cellulose, (iii) antibodies eluted from hapten columns, and (iv) Fab fragments from IgG expressing a latent allotype. Moreover, in the case of latent allotype a3, these fractions have inhibited the binding of an a3 H_V piece (prepared by Dr. L. E. Mole).

Figure 4 depicts a double diffusion experiment that indicates immunologi-

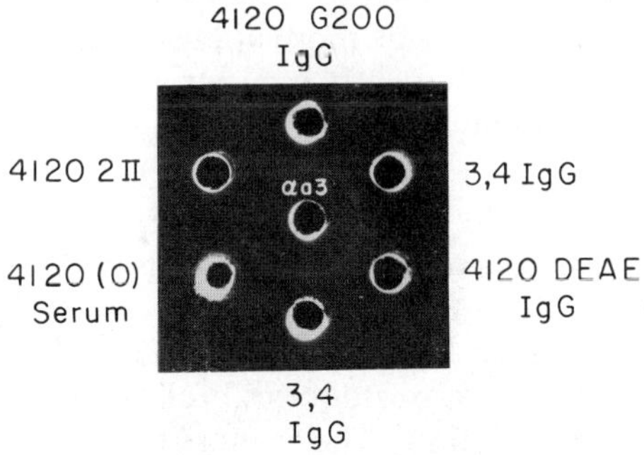

FIG. 4. Double diffusion showing identity of latent a3 allotype in serum and IgG fractions of rabbit 4120 (a^2a^2/b^4b^4) with an a3b4 IgG standard. Anti-a3 serum was placed in the center well.

cal identity of the a3 latent allotype from a homozygous a2 rabbit to a3b4 IgG. It is shown by this analysis that the standard a3 IgG is identical to the IgG contained in this serum and in IgG fractions from a rabbit with nominal allotype a2.

CONCLUSION

Data have been presented that indicate the need for closer examination of the conclusions drawn from studies on genetic markers of immunoglobulins. Three questions were posed initially. The data presented bear directly on these questions.

Returning to the first question, multiple forms of each group a allotype have been observed by structural and serological criteria. These data imply the existence of multiple V_H genes.

In regard to the second question, evidence has been presented for multiple serological forms of the constant region light chains allotype b9. This finding, along with the observation of structural variants of b4 C regions, implies the existence of multiple C_K genes. Whether these genes are alleles or pseudoalleles is not known.

Data on the occurrence of latent group a allotypes bear directly on the third question. These data are not easily compatible with the notion of allotypes as products of allelic structural genes and suggest that loci observed in breeding studies may be regulator gene loci. The new data do not invalidate previous descriptions of immunoglobulin genes based on allotype studies. Rather they do raise questions that must be answered with new and broader experimental approaches. It should be emphasized that information on the gene complex encoding immunoglobulins and governing their expression will have general biologic significance in regard to selective gene expression.

A possible conclusion to this presentation might include new genetic maps and revolutionary schemes for selective gene expression of immunoglobulin genes. On the other hand, it may be more appropriate at this time to quote an admonition of Sherlock Holmes: "It is a cardinal error to theorize in advance of the facts. Inevitably it biases the judgment" (33).

ACKNOWLEDGMENTS

This work was supported by PHS grants from the NIAID AI 08429, 11995 and 11439. M.M. is a James N. Jarvie Memorial Fellow of the New York Heart Association; J.A.S. is a postdoctoral fellow of the Arthritis Foundation; and B.A.F. is an NIH fellow. The authors thank Drs. R. M. Krause and J. E. Coligan for helpful suggestions, and Mss. D. Atherton, P. D. Hooke, and E. Ehlers for technical assistance.

REFERENCES

1. Jerne, D., and Aasted, B. In preparation.
2. McKearn, T. J. (1974): Antireceptor antiserum causes specific inhibition of reaction to rat histocompatibility antigens. *Science*, 183:94–96.
3. Binz, H., Lindenmann, I., and Wigzell, H. (1973): Inhibition of local graft-versus-host reaction by anti-alloantibodies. *Nature*, 246:146–148.
4. Eichmann, K. (1975): Idiotype suppression II. Amplification of a suppressor T cell with anti-idiotypic activity. *Eur. J. Immunol.*, 5:511–517.
5. Janeway, C. A., Sakado, N., and Eisen, H. (1975): Recognition of immunoglobulin idiotypes by thymus-derived lymphocytes. *Proc. Natl. Acad. Sci. USA*, 72:2357–2360.
6. Jerne, N. K. (1974): Toward a network theory of the immune system. *Ann. Immunol. (Paris)*, 125C:373–389.
7. Kluskens, L., and Köhler, H. (1974): Regulation of immune response by autogeneous antibody against receptor. *Proc. Natl. Acad. Sci. USA*, 71:5083–5087.
8. McKearn, T. J., Stuart, F. P., and Fitch, F. W. (1974): Anti-idiotypic antibody in rat transplantation immunity. I. Production of anti-idiotypic antibody in animals repeatedly immunized with alloantigens. *J. Immunol.*, 113:1876–1882.
9. Adler, F. L., and Noelle, R. J. (1975): Enduring antibody responses in "normal" rabbits to maternal immunoglobulin allotypes. *J. Immunol.*, 115:620–625.
10. Thunberg, A. L. (1974): Studies on the genetic control of antibody variable region synthesis. Ph.D. Thesis, The Rockefeller University. New York.
11. Osterland, C. K., Miller, E. J., Karakawa, W. W., and Krause, R. M. (1966): Characteristics of streptococcal group-specific antibody isolated from hyperimmune rabbits. *J. Exp. Med.*, 123:599–614.
12. Koshland, M. E., Davis, J. J., and Fugita, N. J. (1969): Evidence for multiple gene control of a single polypeptide chain: The heavy chain of rabbit immunoglobulin. *Proc. Natl. Acad. Sci. USA*, 63:1274–1278.
13. Mole, L. E., Geier, M. D., and Koshland, M. E. (1975): The isolation and characterization of the V_H domains from rabbit heavy chains of different a locus allotype. *J. Immunol.*, 114:1442–1448.
14. Mole, L. E., Jackson, S. A., Porter, R. R., and Wilkinson, J. M. (1971): Allotypically related sequences in the Fd fragment of rabbit immunoglobulin heavy chains. *Biochem J.*, 124:301–318.
15. Jaton, J. D. (1975): Comparison of the amino acid sequences of the variable domains of two homogeneous rabbit antibodies to Type III pneumococcal polysaccharide. *Biochem. J.*, 147:235–243.
16. Kindt, T. J., Seide, R. K., Tack, B. F., and Todd, C. W. (1973): Diverse expression of group a allotype specificities on the heavy chains of homogeneous rabbit antibodies. *J. Exp. Med.*, 138:33–43.
17. Kindt, T. J., Seide, R. K., Lackland, H., and Thunberg, A. L. (1972): Serologic identity of the b4 allotypic determinants present on homogeneous rabbit light chains with different *N*-terminal amino acid sequence. *J. Immunol.*, 109:735–741.
18. Appella, E., Roholt, O. A., Chersi, A., Radzimski, G., and Pressman, D. (1973): Amino acid sequence of the light chain derived from a rabbit *p*-azobenzoate antibody of restricted heterogeneity. *Biochem. Biophys. Res. Commun.*, 53:1122–1129.
19. Strosberg, A. D., Fraser, K. J., Margolies, M. N., and Haber, E. (1972): Amino acid sequence of rabbit pneumococcal antibody. I. Light chain cysteine-containing peptides. *Biochemistry*, 11:4978–4985.
20. Hood, L., Campbell, J. H., and Elgin, S. C. R. (1975): The organization, expression and evolution of antibodies and other multigene families. *Annu. Rev. Genet.*, 9:305–353.
21. Zeeuws, R., and Strosberg, A. D. (1975): Extensive sequence differences in rabbit light chains of allotype b4 and b9. *Arch. Int. Physiol. Biochim.*, 82:41.
22. Dubiski, S., and Muller, P. J. (1967): A "new" allotypic specificity (A9) of rabbit immunoglobulin. *Nature*, 214:696–697.
23. Kindt, T. J. (1975): Rabbit immunoglobulin allotypes: Structure, immunology and genetics. *Adv. Immunol.*, 21:35–86.

24. Chen, K. C. S., Kindt, T. J., and Krause, R. M. (1974): Amino acid sequence of an allotype b4 light chains from a rabbit antibody to streptococcal carbohydrate. *Proc. Natl. Acad. Sci. USA*, 71:1995–1998.
25. Strosberg, A. D., Hamer-Casterman, C., van der Loo, W., and Hamers, R. (1974): A rabbit with the allotypic phenotype: a1a2a3/b4b5b6. *J. Immunol.*, 113:1313–1318.
26. Bosma, M. J., and Bosma, G. C. (1974): Congenic mouse strains: the expression of a hidden immunoglobulin allotype in a congenic partner strain of Balb/c mice. *J. Exp. Med.*, 139:512–527.
27. Pothier, L., Borel, H., and Adams, R.A. (1974): Expression of IgG Allotypes in human lymphoid tumor lines serially transplantable in the neonatal Syrian hamster. *J. Immunol.*, 113:1984–1991.
28. Knight, K. L., and Hanly, W. C. (1975): Genetic control of L chains of rabbit IgA: allotypic specificities on the variable and the constant regions. In: *Contemporary Topics in Molecular Immunology,* Vol. 4, pp. 55–88, edited by F. P. Inman and W. J. Mandy. Plenum Press, New York.
29. Mudgett, M., Fraser, B. A., and Kindt, T. J. (1975): Nonallelic behavior of rabbit variable-region allotypes. *J. Exp. Med.*, 141:1448–1452.
30. Gottlieb, A. B., Seide, R. K., and Kindt, T. J. (1975): Quantitative determination of allotypic and idiotypic markers using antisera coupled to *N*-hydroxysuccinimide-activated sepharose. *J. Immunol.*, 114:51–54.
31. Kindt, T. J., Seide, R. K., Bokisch, V. A., and Krause, R. M. (1973): Detection of idiotypic cross-reactions among streptococcal antisera from related rabbits. *J. Exp. Med.*, 138:522–537.
32. Todd, C. W. (1972). Genetic control of H chain biosynthesis in the rabbit. *Fed. Proc.*, 31:188–192.
33. Watson, J. H. (1974): In: *The Seven-per-cent Solution*, p. 147, edited by N. Meyer. Random House, New York.

Antibodies in Human Diagnosis and Therapy,
edited by Edgar Haber and Richard M. Krause.
Raven Press, New York ©, 1977.

Recombination Among V Genes

Klaus Eichmann

Using idiotypic markers instead of allotypic markers for V genes, we have been able to show that recombinations exist between V genes as well as between V and C genes. Table 1 summarizes data from many laboratories. It shows a tentative map of

TABLE 1. *Recombinations between V gene and V and C gene Markers*

Mouse strain	Marker										
	A5A	–	S117	J558	ARS	(NIP)	T15	–	Ig1	–	Pre
Balb/c	–		+	+	–	–	+		a		a
CXBH	–		–	–	–	+	–		b	X	a
C57Bl	–		–	–	–	+	–		b		o
Balb/c	–		+	+	–	–	+		a		a
BAB14	–		+	+	–	–	+	X	b		o
C57B1	–		–	–	–	+	–		b		o
Balb/c	–		+	+	–	–	+		a		a
BB7	+	X	+	+	–	–	+		a		a
A/J	+		–	–	+	–	–		e		o

"X" denotes places at which crossovers are observed. "+" denotes the presence of a marker. Letters indicate alleles.

the variable region genes of the heavy chain of mouse immunoglobulin. The mapping has been done on the basis of various recombinants shown at the bottom of the table. The Ig-1 locus actually contains several genes that encode the C regions of the various classes of mouse heavy chains. On the left side of the Ig-1 locus there are several loci containing genes that encode variable regions of the mouse heavy chain; on the right side there is a locus that encodes prealbumin and that apparently has nothing to do with immunoglobulins. These V genes are marked either by idiotypes such as T-15 or J558 or A5A, or by a fine specificity such as NIP. The arrangement of these genes in this sequence is very tentative because the number of recombinants is limited. But what may be more important is that there are definitely recombinations that separate idiotypic markers from one another. For instance, the recombinant shown at the bottom of the table is a recombination that arose from a cross between BALB/c and A/J, and has A/J markers to the left and BALB/c markers to the right. Very clearly, BALB/c V gene markers and A/J V gene markers have been separated from one another in this crossover. A similar recombination is shown where C57/Bl and BALB/c markers have recombined such that here the recombination is between V gene and C gene markers. There are not many recombinations in that area, but those that have been observed have either been shown to separate all the V genes from the C genes or to separate V genes from other V genes.

Antibodies in Human Diagnosis and Therapy,
edited by Edgar Haber and Richard M. Krause.
Raven Press, New York © 1977.

A Problem in the Detection of Allotypic Variation

N. Avrion Mitchison

I would like to make a point about detection of allotypic variation. How efficient is alloimmunization likely to be as a method of detecting genetic variation in protein structure? I believe that it may be rather inefficient, because generally variants will be detected as alloantigens only if they introduce more than one antigenic determinant into the protein molecule. Consequently, most variants that introduce only a single new determinant—and these must surely include most single amino acid substitutions—will remain undetected as alloantigens or allotypes. This follows from the fact that most alloimmune responses are T-dependent, and thus need at least two distinct determinants on the antigen, one for T-cells and the other for B-cells to react against.

When the antigen carries only one new determinant, helper T-cells can be generated, and what happens then will presumably depend on the prevailing state of the B-cell population. If this population lacks cells reactive to the antigen, i.e., if the immune system is already tolerant at the level of B-cells, then nothing further will happen. This is presumably the state of affairs for proteins such as IgG to which B cells are exposed at high concentrations. Alternatively, if B cells are not tolerant, because the protein is either present in serum at low concentrations or sequestered from lymphocytes, the generation of T-helper cells by alloimmunization can lead to the production of autoantibody. An example of this is the production of autoantibody to the liver intracellular protein F as a result of alloimmunization (D. M. Silver and D. P. Lane, *J. Exp. Med.*, 142:1455, 1975).

Antibodies in Human Diagnosis and Therapy,
edited by Edgar Haber and Richard M. Krause.
Raven Press, New York © 1977.

Structure of the Framework and Complementarity Regions of Elicited Antibodies

Edgar Haber, Michael N. Margolies,
and L. Edward Cannon

*Massachusetts General Hospital and the Harvard Medical School,
Boston, Massachusetts 02114*

After many years of passionate controversy, the question of the origin of diversity among immunoglobulins is not yet settled. Quite central to the evidence which must be martialed in supporting any position is the enumeration of the number of variable (V) region genes. If the variety of phenotypes expressed is relatively small, plausible theories become far less convoluted than if the number is great. Recent studies on antigen-binding paraproteins and on some elicited antibodies suggest that variability among structures binding the same antigen is relatively limited. Considerable sequence data on V regions of a number of antigen-binding proteins has now been accumulated including: mouse myeloma proteins, particularly the set that manifests antigen-binding specificity for the phosphorylcholine determinant (1); human cold agglutinins specific for immunoglobulin G (2,3); guinea pig antihapten antibodies (4,5); and elicited antibodies from several strains of inbred mice that exhibit idiotypic cross-reactivity with either induced myeloma proteins (6) or elicited antibodies that occur commonly in other individuals of the same strain (7,8). Some of these data suggest that (a) certain framework sequences appear to be associated with antigen specificity, (b) very similar, if not identical, sets of complementarity regions are linked with a given antigen specificity, and (c) it follows from (a) and (b) that there is an association between the sequence within the V region framework and the sequence in complementarity regions (9). Are these conclusions based on data representative of all antibodies that can be elicited in the normal immune response or are they skewed by selection of the specific systems studied? Answers to these questions may be obtained from an examination of the structure of antibodies wherein the only selective factor was the antigen used for immunization.

The ability to obtain large amounts of antibodies that are monoclonal or greatly restricted in heterogeneity through hyperimmunization of rabbits with bacterial vaccines has provided material for detailed structural characterization of specific elicited antibodies (10,11). Although immunization with other antigens rarely yields sufficient amounts of homogeneous antibody for complete sequence analysis, improved methods of purification and advances in sequencing technology may now enable extended sequence studies of antibodies of a variety of specificities. In this chapter we examine the amino acid sequences of rabbit antibody V regions utilizing data obtained in our laboratory as well as that reported by others (see figures).

Type III or type VIII pneumococcal vaccines are employed as antigens in our iaboratory. Most of the antibodies produced are specific for the capsular polysaccharide of the organism. The structures of type III (S3) and type VIII (S8) polysaccharides have been determined (12) and found to be relatively simple. Both are unbranched linear polymers. S3 is comprised of a repeating disaccharide, while S8 is a repeating tetrasaccharide (Fig. 1).

Restricted responses are quite frequent among outbred rabbits (13). Of 40 animals immunized with type VIII vaccine, 38 produced a restricted response at some time during the course of a 1-year study (a restricted response was defined as one to four discrete bands in the gamma-globulin region on cellulose acetate electrophoresis of antiserum). Of 30 animals immunized with type III vaccine, 29 produced a restricted response. Considerable variability in the general patterns of response was noted. In certain rabbits a single antibody component was present at varying concentrations throughout the period of study. When more than one antibody was pro-

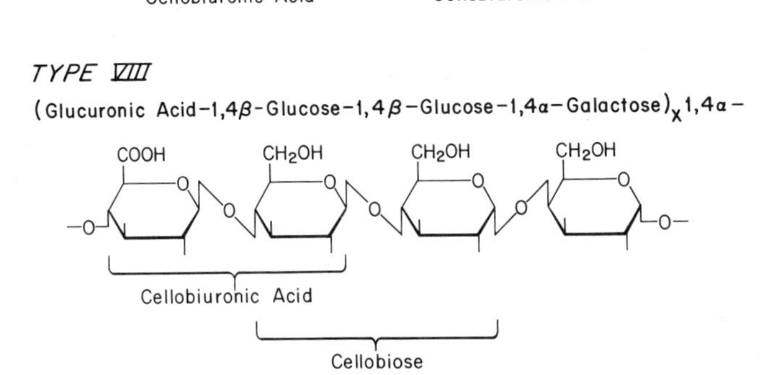

FIG. 1. Structure of the type III and type VIII pneumococcal polysaccharides.

duced, quantitative variations within the antibody population were often observed with time. In other instances, the quantitative relationship among a small number of antibody components was preserved throughout the immune response, although the overall concentration of all the antibodies would vary.

Both affinity chromatography utilizing polysaccharide-protein conjugates linked to sepharose (14) and ion-exchange chromatography on DEAE cellulose have been effective in isolating single antibody components, which may then be examined to determine amino acid sequence.

STRATEGY FOR SEQUENCE DETERMINATION

Classic methods of amino acid sequence determination are too time-consuming to permit comparisons among the primary structures of a large number of homogeneous Ig heavy (H) and light (L) chains. Successful automated Edman degradation in the liquid-phase spinning-cup sequencer results in extended degradations of polypeptides or peptides (15). Modifications of the programs employed, such as the use of dilute Quadrol and altered solvent extractions (16), allow for 35 to 63 cycles of Edman degradation on 5 to 100 nmoles of large peptides. Successful extended degradation of small amounts of protein or peptide require sensitive methods for identification of the PTH-amino acid recovered at each cycle of degradation. The identification methods used in the studies reported here include gas-liquid chromatography and two-dimensional thin-layer chromatography on polyamide sheets (17). In addition, for microsequence work, identification of amounts of PTH-amino acids as small as 20 pmoles is feasible by high-pressure liquid chromatography. The amides of aspartic and glutamic acid are identified directly, rather than inferred from electrophoretic methods. Both gas and liquid chromatography lend themselves to accurate quantitation. This is of considerable importance in the assessment of homogeneity in the hypervariable regions of the L and H chains. In the L-chain sequences reported here, only a single amino acid residue was found at each position including the hypervariable regions. This stands in contrast to the results for the pooled L chains of anti-arsonate antibodies elicited in inbred mice (18), or for the antihapten antibodies from inbred guinea pigs (5,19).

To provide large fragments suitable for automated amino acid sequencing, reproducible methods for cleaving the L and H chains into a limited number of peptides have been developed. Rabbit antibody b_4 L chains contain two to five arginine residues. The strategy for the sequence of an L chain containing three arginine residues is indicated in Fig. 2. The intact chain is first subjected to Edman degradation in the sequencer for 40 to 60 cycles. The identified PTH-amino acids at each cycle are examined for residues that permit specific cleavage of the chain. An arginine residue is found at position 50, and amino acid analysis reveals two other arginines. These two arginine

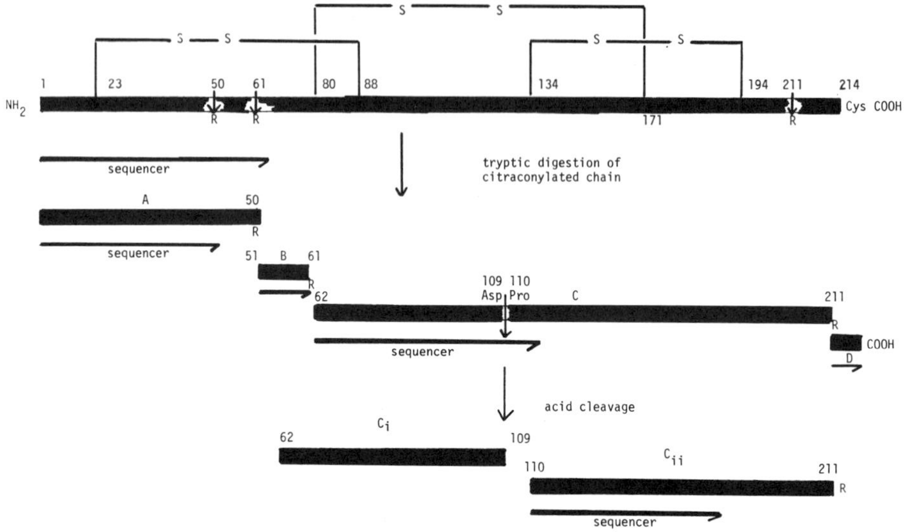

FIG. 2. Scheme for sequence determination of rabbit L chains by automated methods. The L chain is displayed linearly with the amino terminus at the left. The variable domain disulfide bridge (28–88), and the constant domain disulfide bridge (80–171) are indicated. The half-cystine residue at position 214 forms a disulfide bridge with the heavy chain. An R denotes the position of arginine residues where trypsin cleaves the citraconylated chain into the daughter peptides A-D. Each of the peptides A-C may be subjected to automated Edman degradation. Peptide C may be cleaved under mild acid conditions in b_4 L chains between aspartic acid 109 and proline 110.

residues at positions 61 and 211 are invariant in all rabbit b_4 L chains examined thus far. The arginine homologous to position 61 has been shown by X-ray crystallography of a human Fab′ fragment to be invariant because of steric constraints (20). Cleavage of the fully reduced and alkylated L chains with trypsin following modification of the lysine residues with citraconic anhydride results in the production of four fragments, A–D (see Fig. 2). The largest peptide, extending from residues 62 to 211 may be specifically cleaved in high yield at an acid-labile aspartyl–prolyl bond (109–110) (21) in the switch region between the V and constant (C) domains yielding two daughter peptides (62–109 and 110–211) which may be further examined in the sequencer. The amino acid sequence of the first hypervariable region (30–32 ff) is obtained from peptide A, the second hypervariable region from peptide B, and the third hypervariable region (89–97 ff) from peptide C.

In addition to economy in amounts of antibody and time required to obtain sequence data, extended automated Edman degradation reduces the probability of incorrect sequence alignment caused by inadequate "overlap" sequences. For example, L chain 3368 (see Fig. 3) contains the sequence Asn-Ser-Asn-Asn-Val-Val-Asn-Asn in the third hypervariable region. The correct assignment of this sequence by classic methods would require long peptides or multiple overlaps. Figure 3 reveals that there are large differences in chain length among L chains in this region that are best assessed by Edman degradation extending through this segment of the L chain. This is especially important, as results of X-ray crystallographic analyses of hapten-binding myelomas (20,22) reveal that variations in length of complementarity segments would be expected to result in marked changes in the size of the antigen-binding site.

Rabbit H chains have a blocked amino-terminus, pyrollidone carboxylic acid, rendering the intact chain inaccessible to Edman degradation. Cleavage with cyanogen bromide at one or two methionine residues in the V region permits isolation of an unblocked peptide (C_1) which may be examined in the sequencer. After complete reduction and alkylation, the amino terminal peptide may be isolated and sequenced following further enzymatic digestion. The remainder of the V region sequence is obtained from tryptic peptides of the citraconylated C_1 fragment.

TOPOLOGY OF THE RABBIT L CHAIN V REGION

The amino acid sequences for rabbit V_L sequences determined in this laboratory (23) are collated in Figs. 3, 5, and 6 with sequences reported elsewhere (24–29). In an examination of the L chain sequences from human and murine kappa and lambda chains, Wu and Kabat (30) found three regions of increased variability of sequence, the "hypervariable" regions. They predicted that these regions would fold together in the antibody with the H chain hypervariable regions to form the combining site. These predictions have been largely borne out by X-ray crystallographic studies of human and mouse hapten-binding myeloma Fab fragments (20,22).

When data for the rabbit L chains from Fig. 3 are used to calculate the variability at each position in the V region, three regions of increased variability are seen (see Fig. 4). Examination of the sequences in Fig. 3 and the variability plot in Fig. 4 reveals a variety of amino acid substitutions at the amino terminus as well as differences in chain length of up to two residues in this region. There is no apparent correlation here between the eliciting antigen and the found sequence. For example, the same sequence Ala-Phe-Glu-Leu may be found in an S3 antibody 3381 and in the antibody 120 specific for the carbohydrate or peptidoglycan cell wall of *M. lysodeikticus*. Moreover, several other amino terminal sequences may be associated with the same (S3) specificity. As X-ray crystallographic structures indicate that this region is not involved in antigen binding, it is probable that the hypervariability seen here is not related to antigen complementarity. A more detailed analysis of the amino terminus is given below.

The sequence data collated in Fig. 3 for the first and third rabbit L chain hypervariable regions have now been extended to include data from the S3

```
               0   1                                         10                                                  20                                                30  31  32  32A 32B 33
S3 3374        ALA ASP ILE VAL MET THR GLN THR PRO ALA SER VAL SER ALA ALA VAL GLY GLY THR VAL THR ILE ASN [CYS] GLN ALA SER GLN ASN ILE ASP SER TRP --- --- LEU ALA TRP TYR GLN GLN LYS
S3 3368        ———————————————————————————————— SER ———————————————————————————————————————— LYS —— ————————— GLU SER(——)GLY ASN GLU --- --- ————————————————————————————
S3 3381            ALA PHE GLU LEU ———————————————————— GLU ————————————————————————————————— ————————— GLU SER —— SER ASN —— --- --- ————————————————————————————
S3 BS-5            —— VAL ———————————————————————————————— GLU PRO ———————————————————— LYS —— ————————————— SER —— TYR ASN GLY --- --- ———————————————————X——X———
S3 BS-1            —— VAL ———————————————————————————————— GLU PRO ———————————————————— LYS —— ————————————— SER —— TYR —— GLY --- --- ———————————————————————X———
S8 3315        —— GLN ————————————————————————— SER ——————————————————————————————————————————————— —— SER ——— SER VAL TYR GLU ASN GLY ARG —— SER —— PHE ——————————
S8 3322A               ALA —— LEU ———————————————— SER ———————————————————————————————————— SER —— ————————————— SER VAL TYR ASN ASN ASN(   )———————————————————
S8 3322B           ALA TYR ASP ——————————————————— SER ———————————————————————————————————— LYS —— ————————————— SER LEU THR ASN GLN TYR --- ————————————————————————
S8 3322C           —— VAL ———————————————— ALA SER PRO ——————————————————————————————————————— ————————————— SER —— GLY ——(ASP)--- --- —— VAL ————————————————
Strep C 4135   ———————————————————————————————————————— GLU PRO ——————————————————————— LYS —— —— THR ——— SER ———— ASP TYR --- --- —— SER ————————————————
AZBGG 2717         VAL GLU —— LEU ———————————————— SER PRO ———————————————————————————————— SER —— —— SER THR LYS SER —— TYR ASX ASX ASX TYR ————————————————X———
M. lyso. 120       ALA PHE GLU LEU ———————————————— SER ——— GLU ————————————————————————————— LYS —— —— SER ———(SER)—— GLY THR TYR --- --- ———————————————X——X———

               40                                          50                                                60                                               70                                                80
S3 3374        PRO GLY GLN PRO  PRO LYS VAL LEU ILE TYR ARG THR SER THR LEU ALA SER GLY VAL PRO SER ARG PHE LYS GLY SER ARG SER GLY THR GLU PHE THR LEU THR ILE SER ASP LEU GLU [CYS] ALA
S3 3368        ———————————(——) ———— LEU ——————————— ALA —— LYS ——————————————————— SER ——— ——————————— GLY ————————————————————————————————————— GLY VAL GLN —— ASP
S3 3381        ———————————(——)(   )—— LEU ——————————— LYS ALA ——————————————————————— SER ——— ——————————— GLY ——————————— GLN ——————————————————————————————————
S3 BS-5        ————————————————————— LEU ——————————— LYS ALA ——————————— GLU ———————————————— ——————————— GLY ——————————— ASP ——————————————————————————————————
S3 BS-1        ————————————————————— LEU ——————————— LYS ALA ——————————————————————— SER ——— ——————————— GLY ————————————————————————————————————————————————————
S8 3315        ————————————————————— ARG ——————————————— ALA ——————————————————————— SER ——— —— THR ——— GLY ——————————— GLN ————————— SER ————————— VAL GLN —— ASP
S8 3322B                                                      LYS ——————————————————————— ——————————— GLY ——————————— GLN ——————————————————————————————————
Strep C 4135   ————————————————————— GLY ——————————————— ALA ——————————————————————————————————— —— ARG ——— GLY ——————————— ASP ——————————————————————————————————
AZBGG 2717     ————————————————————— ALA ——————————— THR ALA —— SER ——————————————————————————— —— THR ——— GLY ———————————X—————————————— LEU ————— VAL —X——— ASP
M. lyso. 120   ——————————————————————(   )——————————— ALA ——————————————————————— SER ——— ——————————— GLY ————————————————————————————————————— GLY VAL ——————————

                                   90                         97  97A 97B 97C 97D 98  100                                         110                                    120
S3 3374        ASP ALA ALA THR TYR TYR [CYS] GLN SER TYR TYR SER ILE(SER)SER ALA ---  --- --- --- PHE GLY GLY GLY THR GLU VAL VAL VAL LYS GLY ASP PRO VAL ALA PRO THR VAL LEU ILE PHE PRO PRO
S3 3368        ——— GLY ILE ———————————— GLN ASP TRP ASN SER ASN ASN VAL VAL  ASN ASN --- ———————————————————————————————————————
S3 3381        ——————————————————————————————— THR ASP
S3 BS-5        ——————————————— PHE ———X GLY SER ASX TYR THR GLY THR VAL ---  --- --- --- ————————————————————————————————————————————————————————————————————————
S3 BS-1        ——————————————— PHE ———X GLY SER THR TYR GLY GLY GLY TYR ---  --- --- --- ————————————————————————————————————————————————————————————————————————
S3 K7                                          ILE THR ALA THR THR ---  --- --- --- ——————————————————————————————(   )——————————————————————————————————————————(   )——
S8 3315        ——————————————————— LEU GLY ASN —— ASP [CYS] ———— GLY(ASP)(SER)PHE THR ————————————————————————————————————————————————————————————————————————————————
S8 2377                                                                                                         ————————————————————————————————————————————
Strep C 4135   —————————————————————————————— THR —— GLY VAL GLY --- --- ---  --- --- --- ————————————————————————————————————————————————————————————————————————
AZBGG 2717     ——————————————————— GLY GLY ALA ASP TYR THR GLY TYR SER ---  --- --- --- ————————————————————————————————————————X———————————————————————————————
M. lyso. 120   ——————————————————————————— GLY THR ——                                   ————————————————————————————————————————————————————————————————————————
pool                                                                                                                            ————————————————————————————————
```

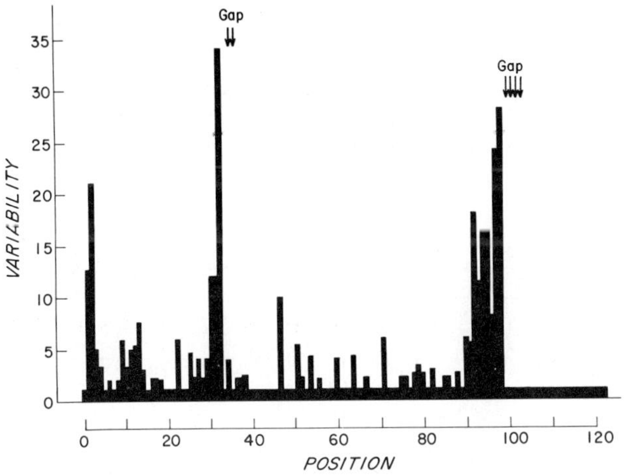

FIG. 4. Variability plot according to the method of Wu and Kabat based on the data from Fig. 3 except for the first 20 positions for which data from more than 50 rabbit L chains are included. Variability is defined as the number of different amino acids at a given position divided by the frequency of the most common amino acid at that position. The variability at the amino terminus is lessened when the sequences are aligned by an objective method (see Fig. 9). [Reprinted from *Proc. Natl. Acad. Sci. USA*, 72:2182, 1975.]

chains 3T70, 3T72, and 3T74 sequenced in our laboratory, the S3 chain K-25 reported by Jaton (28), and the streptococcal A-variant prototype L chain K9-106 reported by Braun (29). These data are summarized in Figs. 5 and 6. Inspection of the region homologous to the first hypervariable region of myeloma L chains (Fig. 5) reveals marked variability beginning at position 30 both in amino acid substitutions and in chain length. For the seemingly simple S3 antigen, nine unique hypervariable region sequences are found among nine chains examined. Similarly, among four S8 chains, including three obtained from the same rabbit (3322A, B, and C), four different sequences are seen in this region presumed to be complementarity-determining.

←

FIG. 3. V_L sequences from rabbit homogeneous antibodies are aligned by homologous invariant residues. The first five residues of K7 are aligned arbitrarily because this partial sequence begins in a hypervariable region. Numbering is according to the human V_{KI} Ag. A horizontal line indicates residue identity with sequence 3374; a broken line indicates a deletion. An X indicates that the amide or acid was not distinguished. The amino acid sequences of chains BS-1 (26), BS-5 (27), 4135 (25), and 2717 (24) were reported by others. The sequences of 3374, 3368, 3315, 3381, 3322A, B, and C, K7, 2377, 120, and nonimmune L chain are from Margolies et al. (23). [Reprinted from *Proc. Natl. Acad. Sci. USA,* 72:2181, 1975.]

```
                                                30  31  32  32A 32B 33
S3      3374     CYS GLN ALA SER GLN ASN ILE ASP SER TRP --- --- LEU ALA TRP TYR GLN GLN LYS PRO
S3      3368     ——————————————— GLU SER ——— GLY ASN GLU --- --- ———————————————————————————————
S3      3381     ——————————————— GLU SER ——— SER ASN ——— --- --- ———————————————————————————————
S3      3T70     ——————————————— GLU ——————— TYR ——— SER --- --- ——————————————————————————— ALA
S3      3T72     ——————————————————— SER ——— GLY ——————— --- --- ——— SER ———————————————————————
S3      3T74     ——————— SER ——————— SER VAL TYR ASN ASN ASN TYR ——— SER ——— PHE ———————————(———)
S3      BS-5     ——————————————————— SER ——— TYR ——— ASN --- --- —————————————————X———X———————
S3      BS-1     ——————————————————— SER ——— TYR ——— GLY --- --- —————————————————————X———————
S3      K25      ———————————————————————X——— TYR ——— TYR --- --- ——— SER ———————————————————————
S8      3315     ——————— SER ——————— SER VAL TYR GLU ASN GLY ARG ——— SER ——— PHE ———————————————
S8      3322A    ——————————————————— SER VAL TYR ASN ASN ASN ARG ———————————————————————
S8      3322B    ——————————————————— SER LEU THR ASN GLN TYR --- ———————————————————————————
S8      3322C    ——————————————————— SER ——— GLY ———(ASP)--- --- ——— VAL ———————————————————
Strep C 4135     ——————— THR ——————— SER ——————— ASP TYR --- --- ——— SER ———————————————————————
Strep A var.
        K9-106   ——————————————————— SER VAL TYR ——— ASN ASN ARG ——————————— PHE ———————————————
AZBGG   2717     ——————— SER THR LYS SER ——— TYR ASX ASX ASX TYR —————————————————————X———————
M. lyso. 120     ——————— SER ———————(SER)——— GLY THR TYR --- --- —————————————————X———X———————
```

FIG. 5. Amino acid sequence of rabbit L chains in the region homologous to the first hypervariable region of human kappa chains. The S3 chains 3374, 3368, 3381, 3T70, 3T72, 3T74, and the S8 chains 3315, 3322A, B, and C were determined in our laboratory. The sequences of the S3 chains BS-5 (27), BS-1 (26), and K-25 (28), the streptococcal C chain 4135 (25), the strep A variant chain K9-106 (29), and the azobenzoate-BGG chain 2717 (24) were reported by others; the *M. lysodeikticus* chain 120 is from ref. (23). The antigen specificity for each antibody L chain is listed on the left. Symbols and numbering are as in Fig. 2.

```
                                           90                          97  a   b   c   d  98      100
S3      3374     ALA THR TYR TYR CYS GLN SER TYR TYR SER ILE SER SER ALA --- --- --- --- PHE GLY GLY GLY
S3      3368     GLY ILE ——————————————— GLN ASP TRP ASN SER ASN ASN VAL VAL ASN ASN --- ———————————————
S3      3381     ——————————————————————————— THR ASP
S3      3T72     ——————————————————————— GLN ASP TRP THR SER ——— ASN VAL ASP ASN THR --- ———————————————
S3      3T74     ——————————————————— ALA GLY GLY ——————— SER ——————— HIS ASN ALA --- --- ———————————————
S3      BS-5     ——————————— PHE ———————X GLY SER ASX TYR THR GLY THR VAL --- --- --- --- ———————————————
S3      BS-1     ——————————— PHE ———————X GLY SER THR TYR GLY GLY GLY TYR --- --- --- --- ———————————————
S3      K25      ——————————————————————— THR ——— SER TYR SER ——— THR TYR --- --- --- --- ———————————————
S3      K7                                           ILE THR ALA THR THR --- --- --- --- ———————————————
S8      3315     ——————————————————— LEU GLY ASN ——— ASP CYS ——————— GLY ASP SER PHE THR ———————————————
S8      3322B    ———————————————————
Strep C 4135     ——————————————————————————— THR ——— GLY VAL GLY --- --- --- --- --- --- ———————————————
AZBGG   2717     ——————————————————— GLY GLY ALA ASP TYR THR GLY TYR SER --- --- --- --- ———————————
M. lyso. 120     ——————————————————————— GLY THR ——— TYR GLX(———)ALA SER --- --- --- --- ———————————
         pool
```

FIG. 6. Amino acid sequence of rabbit L chains in the region homologous to the third hypervariable region of human kappa chains. References, symbols, and numbering are as in Figs. 3 and 5.

Rabbit K_B L chains demonstrate little variation in sequence in the region (50–56) homologous to the second L chain hypervariable region of human and murine kappa chains (Fig. 3). An identical sequence in this region is found among an S3 antibody 3368, an S8 antibody 3315, a streptococcal C antibody 4135, and an *M. lysodeikticus* antibody 120. If this region contributes to the binding site, it may represent a common feature of antipolysaccharide antibodies; a corollary of this idea must be that an S3 antibody is distinguished from an S8 antibody by virtue of residues contributed by other hypervariable regions of L and/or H chains. An alternative explanation is that this region of the rabbit L chain is not involved in the antigen-binding site. An examination of the binding-site structure of hapten-binding myeloma Fab fragments as determined by X-ray crystallography supports this contention. The second L-chain hypervariable region of the phosphorylcholine-binding murine myeloma McPC 603 does not participate in the binding site; in this protein there is a large first L chain hypervariable region (22). In the human lambda New Fab′ which binds vitamin K, this region is also not involved in binding, perhaps because there is an adjacent seven-residue deletion in the polypeptide chain (20). When the position of the alpha-carbons of the V_L kappa dimer REI are compared with those of McPC 603, this region is superimposable between the two structures (31), suggesting that it is part of the "framework." Rabbit kappa chain framework sequences are more homologous to the human V_{KI} framework than to human V_{KII} or V_{KIII}. When a variability plot of human V_{KI} proteins is inspected (32), significant variability is not seen at positions 50 to 56. Further generalizations concerning the role of this segment of V_L in the binding site require the determination of additional sequences from antibodies of other than carbohydrate specificity and additional three-dimensional structures.

The portion of the rabbit L chain V region homologous to the third hypervariable region in other species is also hypervariable in rabbits (see Fig. 6). This presumed complementarity-determining region encompasses residues 89–97 ff. Marked variations in amino acid substitutions are seen, even among eight L chains from antibodies directed against the S3 polysaccharide. In addition, differences in chain length of up to six residues have been identified. The homologous region in human myeloma kappa chains shows no variation in length. The binding specificity for the majority of these myelomas is unknown; they may represent only a narrow window of the entire spectrum of antigen specificities. The insertions in the hypervariable regions in rabbit L chain antipolysaccharide antibodies may reflect a larger combining site in certain antibodies, with consequent differences in affinity or specificity.

The amino acid sequence of the nonbinding site portion ("framework") of rabbit V_L is relatively conserved as are the homologous regions in other species. Rabbit K_B L chains have seven half-cystine residues (see Fig. 2) in contrast to five found in K_A and in L chains of other species. The intradomain disulfide bridges (23–88 and 134–194) are homologous to those of human and murine kappa chains. The observations that the molecular size of K_B L chains is not decreased following specific acid cleavage in the switch region until the chain was fully reduced led us to postulate (33) that an extra disulfide bridge spans the V and C regions. The location of the rabbit L chain disulfide bridge between position 80 in the V region and 171 in the C region (see Fig. 2) was established by the placement of the bridge peptides in the complete sequence of the S8 L chain 3315 (34). The X-ray crystallographic structure of Fab′ lambda New (20) and of the human L chain dimer Mcg (35) reveals that the homologous residues in these L chains are located at a distance from each other consistent with disulfide bridge formation. Thus the interdomain disulfide bridge does not introduce structural constraints interfering with domain folding. All residues identified by Wu and Kabat (30) that are invariant in myeloma L chains, except for a variation in proline 59, are present in rabbit L chains (glutamine 6, tryptophan 35, arginine 61, aspartic acid 82, tyrosine 86, phenylalanine 98, and glycines at 16, 41, 57, 64, 68, 99, and 101) (Fig. 3). Examination of the X-ray structures reveals that these glycines are located at hairpin turns of the V_L polypeptide chain, allowing antiparallel folding of adjacent segments.

The rabbit third L chain hypervariable region (Fig. 6) ends at an invariant phenylalanine 98. Phenylalanine 98, glycine 99, and glycine 101 are invariant because of steric constraints as described by Poljak for Fab′ lambda New (20). All rabbit L chains sequenced to date are invariant at these positions. Residue 100 is often a glycine in human kappa chains, but a threonine is also seen.

Residue 46 in rabbit L chains exhibits five different amino acid replacements among eight chains examined (see Fig. 3). This may be an example of a position in the framework, in which, because of a lack of absolute steric constraint, changes in amino acid side chains may occur without disturbing the chain folding necessary for Ig function.

Residues at positions 61, 62, and 64 in rabbit L chains are found to be invariant in accordance with steric requirements as seen in the lambda New crystal structure. The constancy of the arginine residue at position 61 accounts for the ability to isolate a tryptic peptide beginning at position 62 in a variety of species (36). Among the rabbit L chains reported (Fig. 3), certain chains exhibit variations at position 63 and L chain 3374 has an arginine substitution at position 66 consistent with the evidence from the crystal structure that variations in side chains at these positions might be permitted. Such variations would be difficult to detect in sequence analysis of pooled Ig L chains where only the dominant residue might be identified. These variations in sequence are thus best evaluated by sequence analysis on several homogeneous antibodies. The conservation of framework residues in the rabbit K_B L chain that are thought necessary for proper V_L domain folding in other species suggests that rabbit K_B L chains fold in a homologous fashion.

ANALYSIS OF FRAMEWORK AND COMPLEMENTARITY REGION SEQUENCES

The amino terminal region (positions 0–20) is that portion of the rabbit V_L framework for which the greatest amount of sequence data is available. In Fig. 7 we have recorded 66 amino terminal sequences of rabbit antibody L chain. Antigen specificities represented in this set of sequences are the pneumococcal type II, type III, and type VIII polysaccharides, the strep tococcal group A, group A-variant, and group C carbohydrates, the carbohydrate or peptidoglycan cell wall constituents of *M. lysodeikticus* and the protein conjugates of the benzenearsonate, aminophenyltrimethylammonium, and the benzoate haptens.

The sequences are arranged in Fig. 7 to illustrate the occurrence of three different amino terminal chain lengths in rabbit kappa chains. Light chains having the longest chain length (e.g., 3368) are one amino acid residue longer than L chains having the most common chain length (e.g., 7331-G), and are two residues longer than L chains that have the shortest chain length (e.g., 3322-A). In this figure, the sequences have been aligned by arbitrarily placing deletions relative to the longest chain length at the amino terminal end (positions 0 and 1) of the L chains.

In addition to the occurrence of different chain lengths, this portion of the V_L framework is characterized by increased variability at position 1 (six amino acid residue alternatives) and at position 2 (eight amino acid residue alternatives). The observation that there are 45 unique sequences among the 66 amino acid terminal sequences is an additional indication of the diversity among the frameworks of rabbit V_L. Even among the amino terminal sequences of L chains associated with the same antigen specificity, considerable diversity is noted. For example, the 18 L chains that are derived from antibodies elicited by the type III pneumococcal vaccine demonstrate 14 unique sequences. In addition to the occurrence of diverse amino terminal sequences among L chains associated with the same antigen specificity, there are also numerous instances of identical amino terminal sequences for L chains associated with very different specificities (e.g., 3413, K9-106, 4140-4, and 161; in Fig. 7). These findings preclude a definitive correlation between antigen specificity and V_L framework sequence.

Because L chain amino terminal sequences are frequently used to assign subgroup designations (38,40,44) and in gene counting calculations (52), it is important to determine if relationships inferred by analysis of the amino terminal 21 positions of the L chain may be extrapolated to the complete framework of the V_L. The sequences of rabbit antibody L chains that have been determined for the complete framework V_L are presented in Fig. 3 [except K-25 (28)]. The amino acid residue differences among the framework sequences of these L chains are recorded in Table 1. If residue differences in the amino terminal portion of these L chains (Fig. 7) are predictive of

Antigen[a]	Antibody Designation	Allo-type	0 5 10 15 20	Refer-ence
Pneu III	3368	b4	A D I V M T Q T P S S V S A A V G G T V T	37
Strep C	K19	b4	— A —	38
Pneu III	3374	b4	— A —	37
Pneu VIII	3315	b4	– Q —	37
Pneu III	3T70	b4	— A — E —	37
Pneu VIII	722369	b4	— A — E —	39
Strep Av	K4820	b4	— A — E —	38
Strep Av	K429-1	b4	— A — E —	38
M. Lyso.	1966-1	b6	– E — E —	37
Pneu III	3380-3	b4	— A – E P —	37
Strep C	4135	b4	— A — E P —	25
Strep C	2711	b4	— V — A — E P —	40
Strep C	3521	b4	– N — P — G —	41
Ap-BGG	7331-G	b4	A ? —	42
Strep C	K20	b4	A — Z — A —	38
Strep C	4182-1	b9	A L — T ? —	43
Strep Av	K429-7S	b4	A L — E — A —	38
Strep C	2461-Ib	b4	A L – L – Z — A —	40
Strep Av	K6-89	b4	A L – L – Z — A —	38
M. lyso.	153B	b4	A L — P – E V — B —	37
Pneu III	3T75	b4	– V —	37
Pneu VIII	2388-S	b4	– V — A —	37
Pneu III	722714-T	b4	– V — A — E —	39
Pneu III	722714-G	b4	– V — A — Q —	39
Strep C	4136	b4	– V — A ? – E —	44
Pneu VIII	3322-C	b5	– V — A – P —	37
Pneu III	3T76	b4	– V — A — E P —	37
Pneu VIII	2348-3	b4	– V — A — E P —	14
Pneu III	K17	b4	– V — A — E P —	45
Pneu III	BS-1	b4	– V — A — E P —	38
Pneu III	BS-5	b4	– V — A — E P —	38
Pneu VIII	325	b4	– V — Z — A — Z P —	46
Pneu III	724812-P	b4	— A — E P —	39
Pneu VIII	3322-B	b5	A Y D —	37
Pneu VIII	2388-C	b4	A Y D —	37
Strep A	3547	b4	A Y D —	47
Ap-BGG	7331-F2	b4	A Y ? —	42
Ap-BGG	7331-E	b4	A Y — A — ? —	42
Pneu III	3382	b4	A Y D — A — E V —	37
Pneu III	3T72	b4	A Y D — A — E V —	37
Strep A	3013-III	b4	A Y D — Z — A — T V —	48
Pneu VIII	2388-2Aa	b4	– P – L — P —	14
M. lyso.	1966-II	b6	– P – L — E —	37
Pneu III	K25	b4	A V E L — A — E —	38
Pneu III	3381-2	b4	A F E L — A — E —	37
M. lyso.	120	b4	A F E L — E —	37
Strep C	2869-2	b9	A V – L — ? —	43
Rp-Ed	134	b5	A V – L – Z – A – P — P —	49
Strep C	2461-1	b4	A F Z — Z — A — T – P —	40
Strep C	2461-1a	b4	A A Z L – Z — A — Z P —	40
Ab-BGG	2717	b4	V E – L — P —	24
Strep C	2990	b4	— P T A – Z P —	44
Pneu III	724812-G	b4	A V — V — A V – Q E V —	39
Pneu VIII	3322-A	b4	A – L —	37
Pneu III	3T74	b4	A – L —	37
Pneu II	BS-6	b4	A – L —	38
Strep C	4153-I	b9	V — T — E —	43
Strep C	4153-II	b9	V — T K — E —	43
Strep C	2869-1	b9	– L — T ? — E —	43
M. lyso.	161	b4	— K – V P — B —	37
M. lyso.	166	b4	— K – V P — B —	37
Rp-BGG	4140-4	b4	— K – V P — B —	42
Strep Av	K9-106	b4	— K – V P — B —	38
Strep C	3413	b4	— ? – K – V P — D —	43
Strep A	3113	b4	— ? – V — D —	47
Strep C	2690	b4	— Z — ? – V P — P —	40

FIG. 7. Amino terminal sequences of rabbit antibody kappa chains. Amino acid residues are expressed in the single letter code (50): A = Ala, B = Asx, C = Cys, D = Asp, E = Glu, F = Phe, G = Gly, H = His, I = Ile, K = Lys, L = Leu, M = Met, N = Asn, P = Pro, Q = Gln, R = Arg, S = Ser, T = Thr, V = Val, W = Trp, Y = Tyr, Z = Glx. A? designates an undetermined residue. A line indicates identity with the top sequence (3368). Position numbering is based on the human V_{κ}I prototype Ag (51). [a]Pneu II, Pneu III, and Pneu VIII designate the pneumococcal type II, type III, and type VIII capsular polysaccharides, respectively. M. lyso. designates the carbohydrate and peptidoglycan cell wall constituents of *M. lysodeikticus.* Strep C, Strep A, and Strep Av designate the streptococcal group C, group A, and group A-variant carbohyrates. Rp-BGG, Ap-BGG, and Ab-BGG designate the bovine Ig conjugates of the benzenearsonate, aminophenyltrimethylammonium, and benzoate haptens, respectively. Rp-Ed is the edestin conjugate of the benzenearsonate hapten.

TABLE 1. *Number of residue differences among the framework sequences (100 or 101 residues) of rabbit antibody kappa chains*

Ref. no.	L chain										
(23)	3374	0									
(23)	3368	16	0								
(23)	3315	19	17	0							
(25)	4135	11	18	18	0						
(27)	BS-1	13	16	21	11	0					
(27)	BS-5	14	19	23	10	3	0				
(27)	K25	14	17	20	15	10	13	0			
(23)	3381-2	13	17	20	16	12	14	7	0		
(23)	120	16	13	17	17	13	16	9	9	0	
(24)	2717	22	20	18	23	21	22	22	20	17	0
		3374	3368	3315	4135	BS-1	BS-5	K25	3381-2	120	2717

differences in their complete frameworks, then for a given pair of L chains the total number of framework differences (Table 1) should be about five times greater than the number of differences in the amino terminal 21 positions of the two L chains. This clearly is not the case for many of the L chains. For example, based on their amino terminal sequences (Fig. 7), the 3368, 3374, and 3315 L chains should differ by only 5 to 10 residues in their complete framework sequences; in fact, they differ by two to three times this number. Other L chains with substantial differences in amino terminal sequence (BS-1, BS-5, or 4135 compared to K-25, 3381, or 120) have very similar sequences in the remainder of their frameworks. For example, BS-1 and 3381 differ at 7 positions in the amino terminal 20 residues (Fig. 7), but differ at only 5 positions in the remaining 80 positions of the framework. These discrepancies between amino terminal sequence and complete framework sequence place an important constraint on the use of amino terminal sequences as markers for germ-line V region genes in the rabbit, and suggest that attempts to delineate subgroups among rabbit L chains by analysis of amino terminal sequences is at best misleading.

A definitive analysis of a large volume of amino acid sequence data must rely on objective methods for aligning sequences and for detecting relationships among sequences. We have elected to use a minimum nucleotide base-change criterion for these purposes and have developed computer methods to analyze the sequence data presented. The algorithm employed for aligning sequences determines for each sequence the alignment that minimizes the theoretical number of nucleotide base changes required to convert that sequence to every other sequence in the set being aligned. Relationships among sequences are determined by analysis of similarity measures that are calculated from the theoretical minimum number of nucleotide base changes required to interconvert individual sequences. From the distribution of these similarity measures criteria are computed for assigning sequences to clusters. The results of the clustering process are displayed in phenograms that indicate hierarchical relationships between clusters and are analogous to genealogic trees. Both alignment and clustering procedures have been exhaustively tested with: (a) sequences generated by the artificial "mutation" of a precursor sequence, (b) randomly generated sequences that have amino acid compositions similar to the rabbit L chain sequences being analyzed, and (c) human kappa, lambda, and H chain V region sequences. Analyses of these "model" sequences have demonstrated the following. (a) Alignment of artifically mutated sequences provides an exact indication of the events separating them from the precursor sequence. (b) The level (nucleotide base changes) at which clusters are detected among randomly generated sequences is significantly higher than the level at which clusters are defined among the rabbit L chain sequences ($p < 10^{-7}$, using a double-tailed t-test). (c) Cluster analysis of human kappa chains yields the expected subgroup division including the delineation of subsubgroups (53). (d) Alignment of amino terminal sequences of human kappa, lambda, and H chains locates a deletion in lambda and H chain V regions at position 9 corresponding to the "kappa kink."

As previously noted, rabbit kappa chains demonstrate three different amino terminal chain lengths and hypervariability at positions 1 and 2. In the alignment shown in Fig. 7, deletions caused by differences in amino terminal chain length are arbitrarily placed at positions 0 and 1. However, if these 66 amino terminal sequences are aligned by a minimum nucleotide base-change criterion, a different distribution of deletions is obtained. As shown in Fig. 8, the sequences can be grouped into six subsets based on the occurence of homologously positioned deletions. Subset 1 contains the longest chain-length sequences, and subset 2 consists of sequences having a single deletion placed at position 0. The members of subset 3 have a single deletion placed at position 1 and those of subset 4 have the deletion at position 2. Although the sequences in subsets 2, 3, and 4 have the same amino terminal chain length, each subset has a single deletion located at different positions. This is also the case for those L chains that have the shortest amino terminal chain length; deletions may occur either at positions 1 and 2 as in subset 5 or at positions 0 and 1 as in subset 6.

An important result of the alignment obtained by the minimum base change criterion is that the variability at positions 1 and 2 is reduced (Fig. 9). The variability plot shown in Fig. 9B was calculated from the alignment shown in Fig. 8 and reveals a 26% decrease in the variability at position 1 and a 50% decrease in the variability at position 2 when compared with the plot constructed from the alignment in Fig. 7 (Fig. 9A). Thus, the variability at these two positions is reduced to a level comparable to that at other positions in the V_L framework.

The placement of internal deletions near the amino terminus for many of the L chains (Fig. 8) suggests that differences in amino terminal chain length

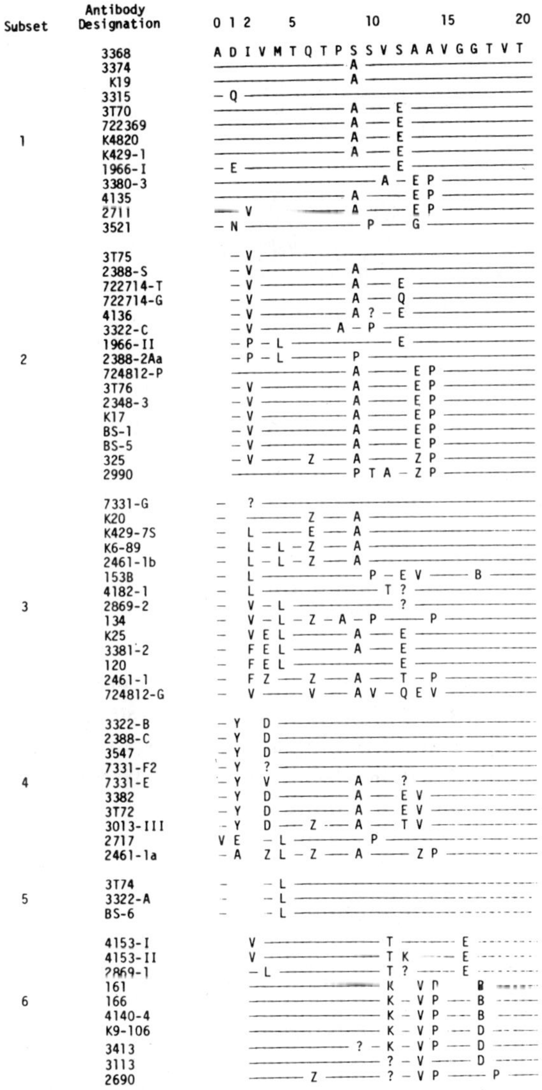

FIG. 8. Alignment of rabbit antibody kappa chains on the basis of a minimum base-change criterion. Deletions are indicated by blanks. A line indicates identity with the 3368 L chain. Sequences are grouped into subsets on the basis of deletions in homologous positions.

are not the result of posttranslational modification of an L chain precursor. Close examination of the sequences in Fig. 7 also indicates that such an explanation is unlikely. Since alanine does not occur at position 1 or 2 in L chains that have the longest chain length, it is not possible to obtain the 26 L chains that have both a shorter chain length and an alanine amino terminus by degrading the amino terminus of the longest chain length L chains. Similarly, the group of L chains having the shortest chain length and the sequence Ile-Val-Met (IVM) at positions 2, 3, and 4 (e.g., 161) cannot be

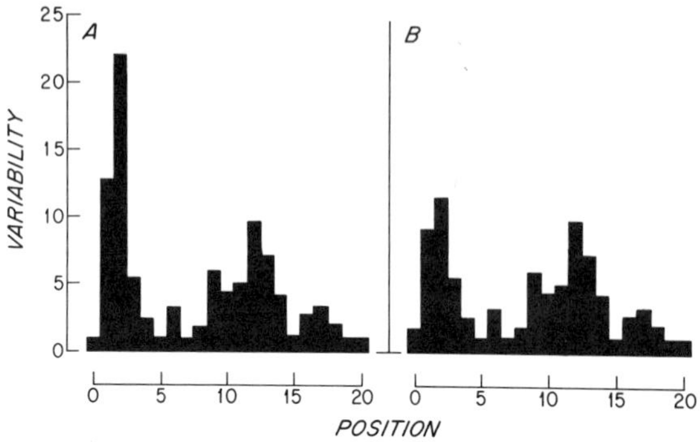

FIG. 9. Variability plot constructed from the amino terminal 21 positions of rabbit antibody L chains. Variability equals the number of different amino acids at a position divided by the frequency of the most common amino acid at that position (30). **A,** calculated from alignment given in Fig. 7. **B,** calculated from alignment based on a minimum base-change criterion (Fig. 8).

derived simply by amino terminal degradation of longer L chains but also require mutations at positions 11, 13, 14, and 17. Although the aspartic acid at position 1 in many of the L chains that have the longest chain length suggests that L chains one residue shorter and having an aspartyl amino terminus (e.g., 3T75) originate by removal of the amino terminal alanine from the longer L chains, the residue alternatives at position 2—isoleucine in 12 of 13 of the longest chain length L chains and valine in 12 of 16 L chains that have an aspartyl amino terminus—make this unlikely. It is probable, therefore, that the different amino terminal chain lengths in these rabbit L chains are due to deletions of genetic material rather than posttranslational degradation.

When the 66 amino terminal sequences are subjected to the clustering procedure, well-defined clusters of sequences can be identified. However, antigen specificities are found to be randomly distributed among the clusters (Fig. 10). For example, cluster number contains five different carbohydrate or polysaccharide specificities and aminophenyltrimethylammonium hapten specificity.

As expected from the data presented in Table 1, cluster analysis of the entire framework sequences results in the definition of two clusters (Fig. 11). The BS-1 and BS-5 L chains, which differ by only four base changes in their frameworks, form one cluster, and the 3381 and K-25 L chains, differing by

FIG. 10. Phenogram constructed from cluster analysis of the amino terminal (positions 0 to 20) sequences of rabbit L chains (Fig. 7). Distances, measured as nucleotide base changes, are indicated by the horizontal and vertical linkages. In the interest of clarity, only the intercluster linkages are shown. Clusters containing two or more L chains are indicated in brackets (clusters 1 through 11). L chains in the same cluster link at 3.5 base changes or less.

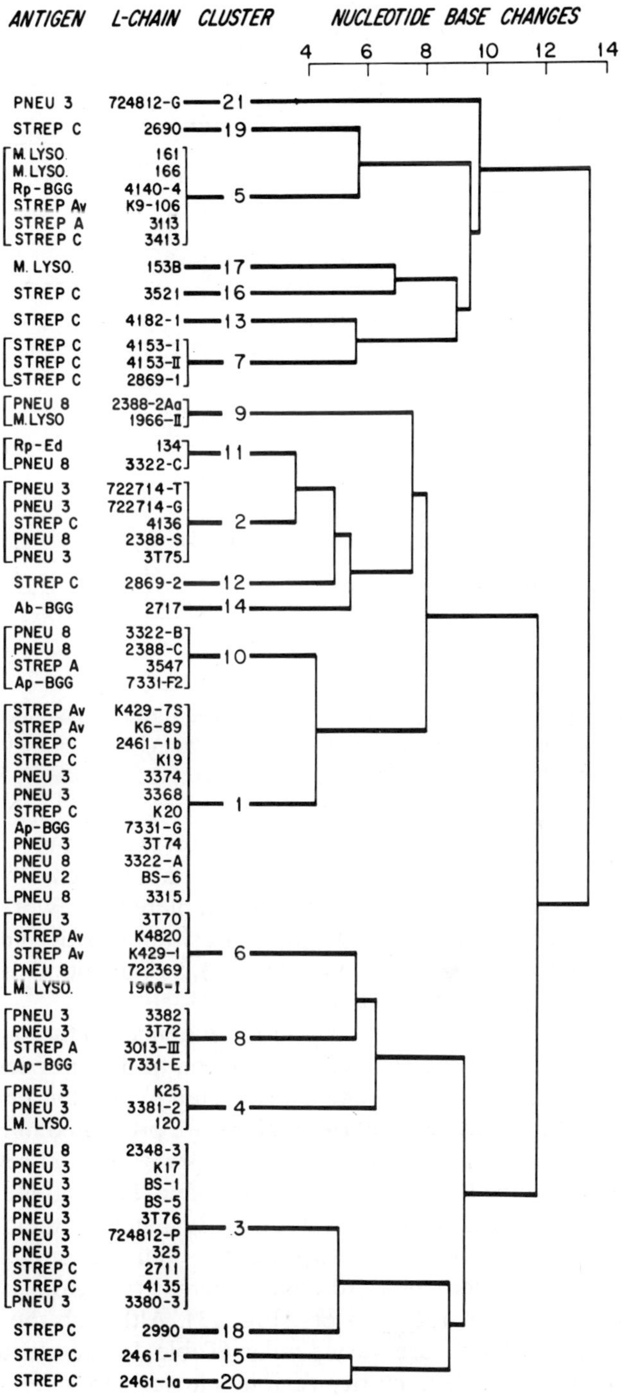
ANTIGEN L-CHAIN CLUSTER NUCLEOTIDE BASE CHANGES
4 6 8 10 12 14
PNEU 3 724812-G 21
STREP C 2690 19
M. LYSO. 161
M. LYSO. 166
Rp-BGG 4140-4
STREP Av K9-106
STREP A 3113
STREP C 3413
5
M. LYSO. 153B 17
STREP C 3521 16
STREP C 4182-1 13
STREP C 4153-I
STREP C 4153-II
STREP C 2869-1
7
PNEU 8 2388-2Aa
M. LYSO 1966-II
9
Rp-Ed 134
PNEU 8 3322-C
11
PNEU 3 722714-T
PNEU 3 722714-G
STREP C 4136
PNEU 8 2388-S
PNEU 3 3T75
2
STREP C 2869-2 12
Ab-BGG 2717 14
PNEU 8 3322-B
PNEU 8 2388-C
STREP A 3547
Ap-BGG 7331-F2
10
STREP Av K429-7S
STREP Av K6-89
STREP C 2461-1b
STREP C K19
PNEU 3 3374
PNEU 3 3368
STREP C K20
Ap-BGG 7331-G
PNEU 3 3T74
PNEU 8 3322-A
PNEU 2 BS-6
PNEU 8 3315
1
PNEU 3 3T70
STREP Av K4820
STREP Av K429-1
PNEU 8 722369
M. LYSO. 1966-I
6
PNEU 3 3382
PNEU 3 3T72
STREP A 3013-III
Ap-BGG 7331-E
8
PNEU 3 K25
PNEU 3 3381-2
M. LYSO. 120
4
PNEU 8 2348-3
PNEU 3 K17
PNEU 3 BS-1
PNEU 3 BS-5
PNEU 3 3T76
PNEU 3 724812-P
PNEU 3 325
STREP C 2711
STREP C 4135
PNEU 3 3380-3
3
STREP C 2990 18
STREP C 2461-1 15
STREP C 2461-1a 20

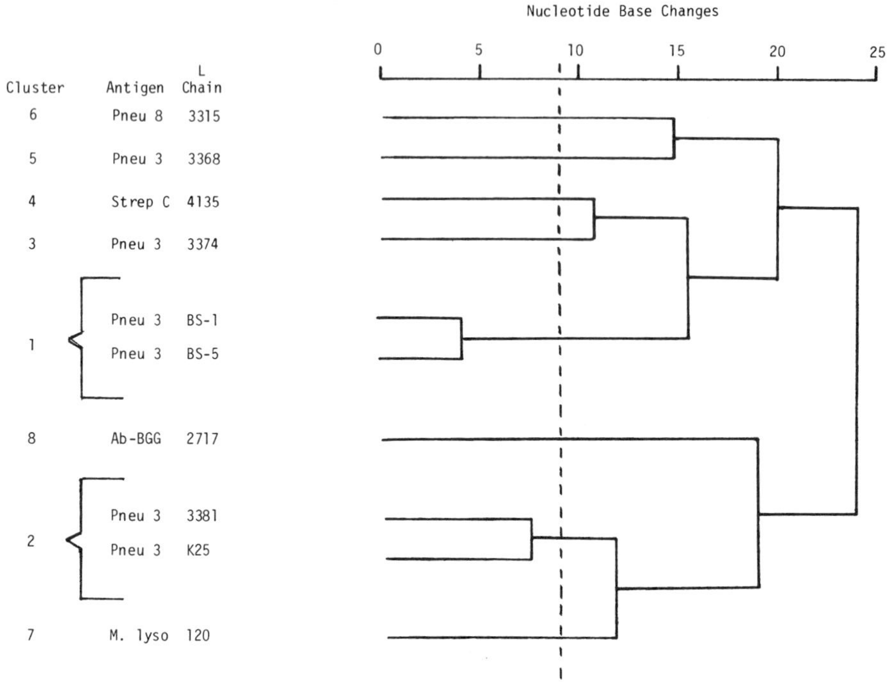

FIG. 11. Phenogram constructed from cluster analysis of the framework sequences of rabbit antibody L chains. Clusters were detected at linkages below nine nucleotide base changes (indicated by the dashed line).

seven base changes, constitute the other cluster. The other six L chains differ by 11 to 25 base changes and link with each other above the clustering thresholds (Fig. 11). Alignment of this set of complete framework sequences results in the placement of deletions only at position 0 (BS-1 and BS-5), at position 1 (K25,3381,120), or at position 2 (2717), as was found for the amino terminal portion of the L chain (Fig. 8).

The availability of complementarity region sequences in the same set of 10 L chains permits a test of the correlation between framework sequence and combining site sequence. Phenograms constructed by cluster analysis of the L chain first and third complementarity regions are presented in Figs. 12 and 13, respectively. Comparison of these two figures with Fig. 11 reveals a lack of association between framework sequences and binding-site sequence. The BS-1 and BS-5 L chains form a well-defined cluster for their framework sequences (Fig. 11) and also cluster together for their first complementarity regions (Fig. 12). However, the third complementarity regions of these two L chains distribute in separate clusters (Fig. 13). Although the framework sequences of K25 and 3381 are very similar (Table 1) and cluster together (Fig. 11), their first complementarity regions cluster separately (Fig. 12).

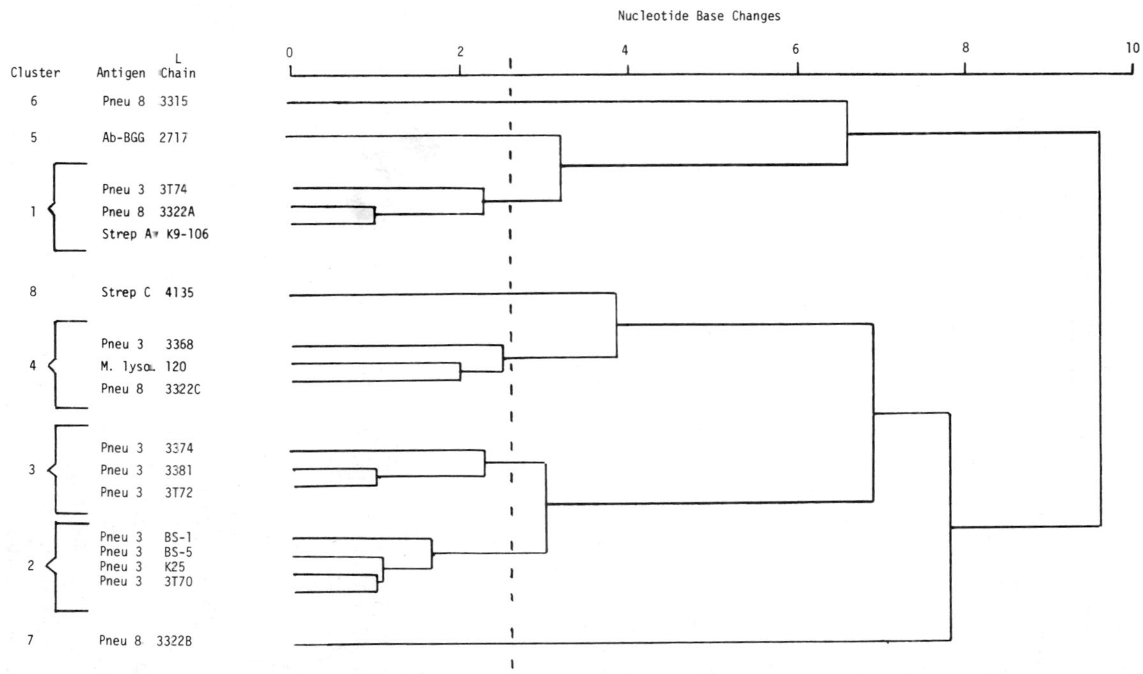

FIG. 12. Phenogram constructed by cluster analysis of the L chain first complementarity region sequences (positions 30–32 f⁻). Clusters were defined at linkages below 2.6 nucleotide base changes.

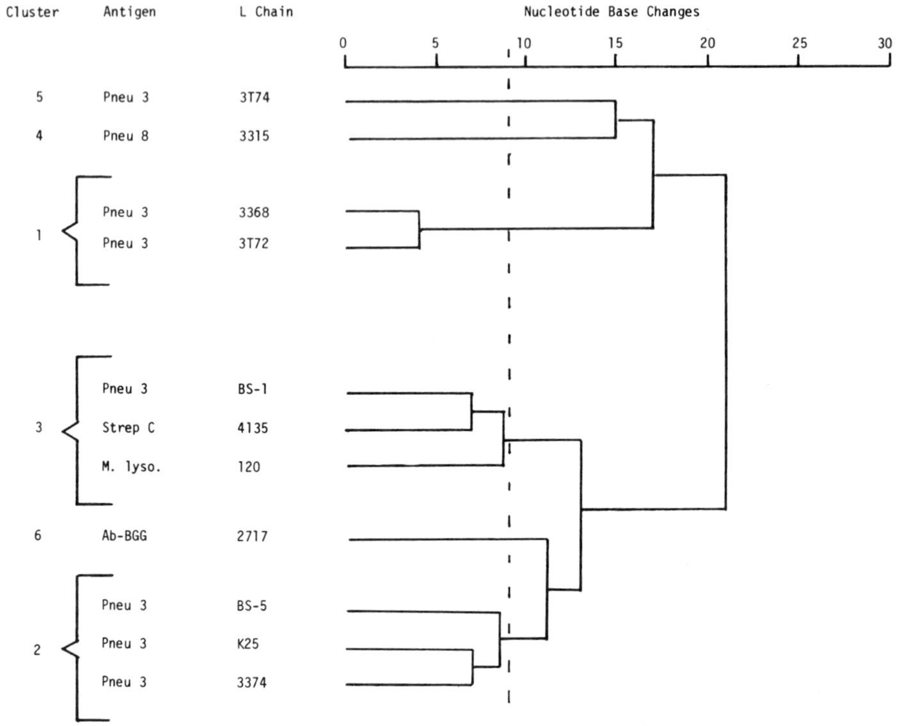

FIG. 13. Phenogram constructed by cluster analysis of the complementarity region at positions 89–97 ff of rabbit antibody L chains. Clusters were detected at linkages below nine nucleotide base changes.

In contrast to the apparent hypervariability at positions 1 and 2, the variability in the L chain complementarity regions (30–32 ff and 89–97 ff) is not significantly reduced by alignment using the minimum nucleotide base-change criterion. The alignments obtained for the complementarity regions are shown in Figs. 14 and 15. Note that deletions are not placed solely at the C-terminal end of the complementarity regions, as they were arbitrarily placed in Figs. 5 and 6. Comparative variability plots are shown in Fig. 16 for the first complementarity region and in Fig. 17 for the third complementarity region. Although the variability in these regions is reduced by alignment using the minimum base-change method, the reduction is less marked than was obtained for the amino terminus (Fig. 9), and these complementarity regions remain hypervariable. Thus, for rabbit L chains there are only two hypervariable regions (positions 30–32 ff and 89–97 ff), and each corresponds to an antigen complementarity region as delineated by X-ray crystallographic analyses.

In comparing framework sequences with complementarity region sequences, we have noted that two features distinguish these two regions. The

L Chain	Antigen	29	30	31	32	a	b	33
3T74	Pneu 3	VAL	TYR	ASN	ASN	ASN	TYR	LEU
2717	Ab-BGG	ILE	—	X	X	X	—	—
3322A	Pneu 8	—	—	—	—	—	ARG	—
K9-106	Strep Av	—	—	SER	—	—	ARG	—
3315	Pneu 8	—	—	GLU	—	GLY	ARG	—
3322B	Pneu 8	LEU	---	THR	—	GLN	—	—
K25	Pneu 3	ILE	—		SER	---	—	—
3T70	Pneu 3	ILE	—	---	SER	---	SER	—
BS-5	Pneu 3	ILE	—	SER	—	---	---	—
BS-1	Pneu 3	ILE	—	---	SER	---	GLY	—
3381	Pneu 3	ILE	SER	---	SER	---	TRP	—
3T72	Pneu 3	ILE	SER	---	SER	---	TRP	—
3374	Pneu 3	ILE	ASP	---	SER	---	TRP	—
3322C	Pneu 8	ILE	GLY	---	SER	---	ASP	—
120	M. lyso	ILE	GLY	---	THR	---	—	—
4135	Strep C	ILE	ASP	---	ASP	---	—	—
3368	Pneu 3	ILE	GLY	---	—	---	GLU	—

FIG. 14. Alignment of the first complementarity region sequences of rabbit L chains on the basis of a minimum nucleotide base-change criterion. A line indicates identity with the top sequence (3T74). A broken line denotes a deletion relative to the top sequence.

first, as shown in Table 2, is the significant difference in the frequency of one, two, and three base-change events and of deletions or insertions that occur in these two regions. Note that amino acid differences that occur in the framework are due predominately to single base-change events. In the complementarity regions, double mutations and deletions or insertions account for the majority of residue differences, suggesting that the mechanism that generates the variability in these two regions may, indeed, be different. This is also supported by the failure to find an association between framework sequence and complementarity region sequence.

A second feature that distinguishes these two regions is the relative occurrence of certain amino acid residues as shown in Table 3. As expected, cysteine is found primarily in the framework where it participates in the formation of an invariant intradomain disulfide bridge and also an interdo-

L Chain	Antigen	88		90					95		97	a	b	c	d	98
3315	Pneu 8	CYS	LEU	GLY	ASN	TYR	ASP	CYS	SER	SER	GLY	ASP	SER	PHE	THR	PHE
3T72	Pneu 3	—	GLN	GLN	ASP	TRP	THR	SER	—	ASN	VAL	—	ASN	---	—	—
3368	Pneu 3	—	GLN	GLN	ASP	TRP	ASN	SER	ASN	ASN	VAL	VAL	ASN	---	ASN	—
3T74	Pneu 3	—	ALA	—	GLY	—	SER	SER	—	—	---	HIS	ASN	---	ALA	—
BS-1	Pneu 3	—	GLN	---	GLY	SER	THR	TYR	GLY	---	—	GLY	TYR	---	---	—
BS-5	Pneu 3	—	GLN	---	GLY	SER	X	TYR	THR	---	---	GLY	THR	---	VAL	—
120	M. lyso.	—	GLN	---	GLY	THR	TYR	TYR	GLX	—	---	ALA	—	---	---	—
3374	Pneu 3	—	GLN	---	---	SER	TYR	TYR	—	ILE	---	SER	—	---	ALA	—
K25	Pneu 3	—	GLN	---	THR	—	SER	TYR	—	—	---	THR	TYR	---	---	—
2717	Ab-BGG	—	GLY	---	GLY	ALA	—	TYR	THR	---	---	GLY	TYR	---	SER	—
4135	Strep C	—	GLN	---	---	SER	THR	TYR	GLY	---	---	VAL	GLY	---	---	—

FIG. 15. Alignment of the "third" complementarity region sequences of rabbit antibody L chains on the basis of a minimum nucleotide base change criterion.

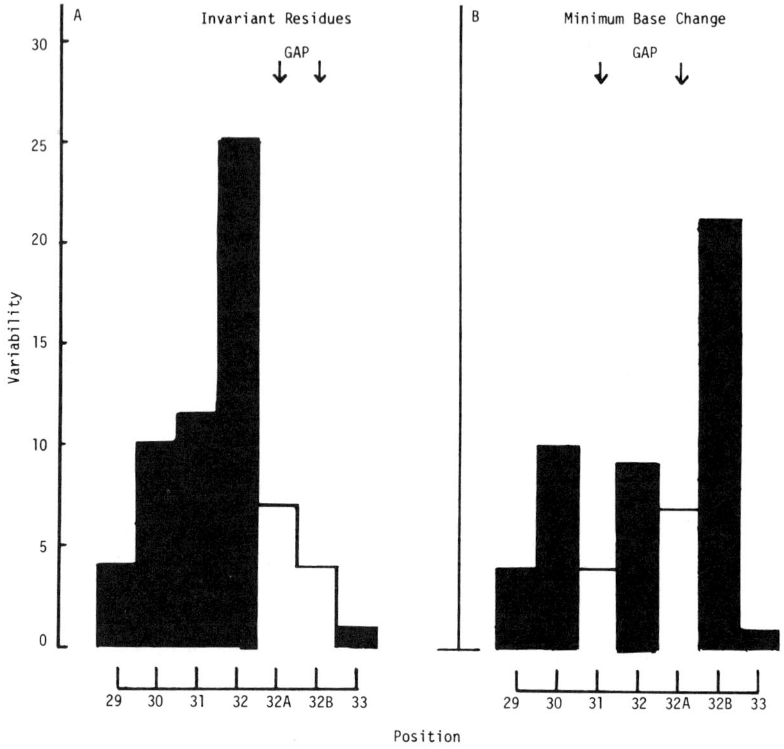

FIG. 16. Variability plot for the first complementarity region of rabbit antibody L chains. **A,** calculated from the alignment of Fig. 5. **B,** calculated from the alignment of Fig. 14. Nonblackened positions indicate positions where deletions occur.

main disulfide bond that is unique to rabbit subtype B kappa chains. To date, proline, lysine, and methionine have not been identified in a rabbit L-chain complementarity region. The finding of asparagine and, to a lesser extent, tyrosine primarily in the complementarity regions could reflect antigen selection for binding-site complementarity, since most of these L chains are derived from antipolysaccharide antibodies.

DIVERSITY OF ANTIBODIES IN A SINGLE RABBIT

The diversity of framework and complementarity sequences noted in the previous discussion could simply be attributed to a great degree of genetic polymorphism in an outbred species. A comparison of antibodies raised in a single individual in response to a single relatively simple antigenic stimulus might give considerable insight into this question.

A rabbit hyperimmunized with type VIII pneumococcal vaccine produced an antiserum that indicated the presence of multiple antibody components

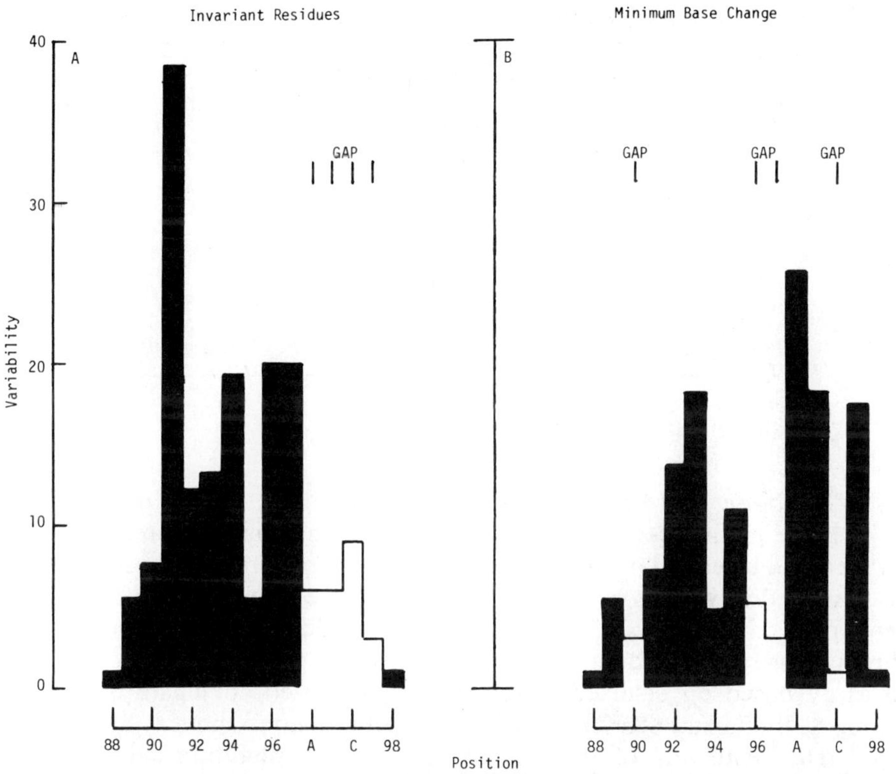

FIG. 17. Variability plot for the "third" complementarity region of rabbit antibody light chains. **A,** calculated from the alignment in Fig. 6. **B,** calculated from the alignment in Fig. 15. Nonblackened positions indicate positions where deletions occur.

TABLE 2. *Frequency of base changes and deletions*

	Region	
Base changes	Complementarity	Framework
1	0.42	0.88
2	0.22	0.07
3	0.01	0.00
Deletion/Insertion	0.35	0.05

TABLE 3. *Relative frequency of residue occurrence*

Residue	Ratio of frequencies: Complementarity		Framework
CYS	1	:	7.2
TYR	3.5	:	1
GLU	1	:	4
ASN	30	:	1
ALA	1	:	4
ILE	1	:	9.6
LEU	1	:	12.6
VAL	1	:	4.2

PRO, LYS, MET found only in framework.
Ratio for all other residues is less than 2.

(54). Cellobiose (Fig. 1) played an immunodominant role in the antigen-binding site of all these antibodies since 90% of the antibody present in the serum could be eluted from an S8 immunoadsorbant with cellobiose alone or in combination with 3 M NaCl (3 M NaCl alone eluted only 3% of total antibody). Utilizing NaCl and cellobiose gradient elution in affinity chromatography, four homogeneous antibody components were isolated. In an assay based on displacement of a radioiodinated conjugate of bovine gamma-globulin and S8 (^{125}I-S8-BGG), the heterologous antigen, S3, did not cross react with any of these antibodies. When inhibition studies were performed with cellobiose, considerable differences among these antibodies were observed (Fig. 18). Antibody 3322A was displaced from ^{125}I-S8-BGG by cellobiose at the lowest concentration of the disaccharide. Antibodies 3322B and C both required a greater concentration of cellobiose, while 3322D required a much higher concentration. As also shown in Fig. 18, an

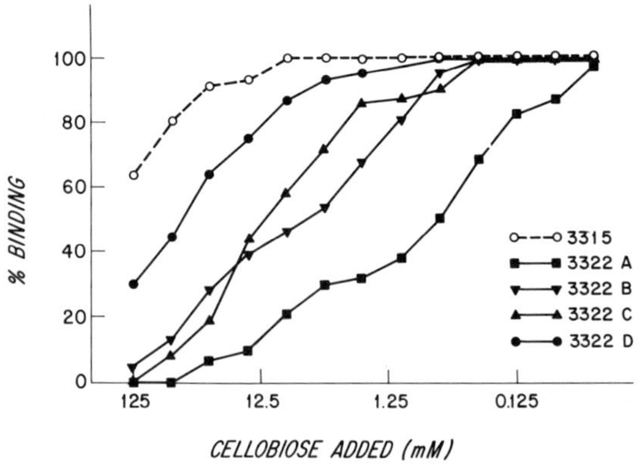

FIG. 18. Inhibition of binding of anti-S8 antibodies to ^{125}I-BGG-S8 by cellobiose.

S8-specific antibody from another rabbit, 3315, appeared to have very little affinity for cellobiose.

The amino acid sequences through position 40 of the L chains of 3322A, B, and C are shown in comparison to 3315 in Fig. 19. The L chain of 3322D could not be sequenced because of an apparent blocked amino terminal residue. The other three antibodies from rabbit 3322 are characterized by variability in chain length and sequence at positions 0, 1, and 2. The L chain from 3322A is characteristic of the shortest class of L chains discussed previously; 3322B and C are of intermediate length. The amino acid sequences for the first three residues demonstrate little homology among these chains. Relatively greater sequence homology is apparent, however, between residues 5 and 29. Between residues 30 and 32B, considerable sequence diversity and differences in chain length are seen. Light chain 3322A has the longest sequence in this region, 3322B is one residue shorter, and 3322C is two residues shorter than 3322A. A leucine residue which is invariant at position 33 begins a relatively conserved region.

Thus, in the serum from a single rabbit, four antibodies can be isolated manifesting differing binding properties for antigen as well as different sequences in both framework and hypervariable regions of their L chains. It would not be surprising that antibodies with different binding properties would have different sequences in their complementarity region, yet 3322B and C are essentially indistinguishable in their binding to cellobiose and have different sequences and lengths in the L chain first complementarity region. As a measure of possible diversity among antibodies, rabbit 3322 was able to synthesize four different antibodies to accommodate the same antigen, each with a unique binding site and a unique framework. Since this animal was

	0	1									10										20	
3315	ALA	GLN	ILE	VAL	MET	THR	GLN	THR	PRO	SER	SER	VAL	SER	ALA	ALA	VAL	GLY	GLY	THR	VAL	THR	ILE
3322A			ALA	—	LEU	—	—	—	—	—	—	—	—	—	—	—	—	—	—	—	—	—
3322B		ALA	TYR	ASP	—	—	—	—	—	—	—	—	—	—	—	—	—	—	—	—	—	—
3322C		ASP	VAL	—	—	—	—	—	ALA	—	PRO	—	—	—	—	—	—	—	—	—	—	—
3322D	BLOCKED AMINO-TERMINUS																					

									30	31	32	32A	32B	33							40
3315	ASN	CYS	GLN	SER	SER	GLN	SER	VAL	TYR	GLU	ASN	GLY	ARG	LEU	SER	TRP	PHE	GLN	GLN	LYS	PRO
3322A	SER	—	—	ALA	—	—	—	—	—	ASN	—	ASN	—	—	ALA	—	TYR	—	—	—	
3322B	LYS	—	—	ALA	—	—	—	LEU	THR	ASN	GLN	TYR	---	—	ALA	—	TYR	—	—	—	
3322C	—	—	—	ALA	—	—	—	ILE	GLY	SER	ASP	---	---	—	VAL	—	TYR	—	—	—	—

FIG. 19. Amino acid sequences for positions 0–40 of anti-S8 antibody L chains.

heterozygous at the L chain allotypic locus (b4b5) and three of the antibodies were of the b5 allotype (3322B, C, and D), genetic polymorphism as an explanation for diversity is effectively excluded.

DIVERSITY IN HEAVY CHAIN COMPLEMENTARITY REGIONS

Some evidence has now accumulated in several systems indicating that diversity of both framework and complementarity regions is greater among L chains than among H chains. Phosphorylcholine binding myeloma proteins possessing identical or very similar H chains may have different L chains (55,56). H chains of idiotypically cross-reactive antibodies to the arsonilic acid determinant in A/J mice appear identical in framework and complementarity regions but differ in their L chains (18,57). The hapten-specific antibodies examined by Cebra in inbred guinea pigs appear to have considerably greater sequence restriction in H chains than in L chains (4,5,19). It would be reasonable to determine if the diversity seen among rabbit antibodies follows a similar pattern.

Although available rabbit H chain data are still limited, an examination of Fig. 20 reveals considerable diversity in the region homologous to the H chain second complementarity-determining region in other species (20,22). The amino acid sequences of six antibody H chains between residues 45 and 63 are shown. Five are from antibodies specific for S3; three of these sequences were obtained in our laboratory (3374, 3381, 3T72) and one is specific for group C streptococcus polysaccharide (58). Four of the S3 antibodies are of the same V_H allotype, a1. The region 49-60 appears considerably more variable than the polypeptide chain at either end. There is little similarity in sequence within this region, even among chains of the same specificity with the exception of K-25 and BS-5 which differ by only one residue. It is apparent that considerable diversity exists among complementarity regions of rabbit H chains directed against the same antigen.

Allotype	Ab	Antigen	45					50					55					60			
a1	BS-5	Pneu III	GLU	TRP	ILE	GLY	ILE	ILE	TYR	ALA	SER	GLY	SER	THR	TYR	TYR	ALA	SER	TRP	ALA	LYS
a1	3381	Pneu III	—	TYR	—	—	VAL	—	SER	SER	GLY	—	—	ALA	—	—	—	—	—	—	—
a1	3374	Pneu III	—	—	—	—	THR	—	ASN	ASN	VAL	—	THR	—	ASN	—	—	—	—	—	—
a1	3T72	Pneu III	—	—	—	—	VAL	—	SER	SER	—	ALA	ASP	—	—	—	—	ASN	—	—	—
a2	K25	Pneu III	—	—	—	—	VAL	—	—	—	—	—	—	—	—	—	—	THR	—	—	—
a2	2690	Strep C	—	—	—	—	ALA	—	ASP	GLY	TYR	—	THR	—	—	—	—	—	—	—	—

FIG. 20. Amino acid sequences of rabbit antibody H chains in the region homologous to the H chain second complementarity region. H chains 3374, 3381, and 3T72 are from our laboratory, K25 and BS-5 were reported by Jaton (28), and 2690 was reported by Fleischman (58).

DISCUSSION

We have demonstrated the following facts in elicited antibodies in outbred rabbits.

(i) The light chain contains three apparently hypervariable segments, located at the amino terminus, at positions 30–32 ff and 89–97 ff. The region around position 50 identified as hypervariable by Kabat and Wu in myeloma L chains is quite conserved among rabbit antibodies. Neither the amino terminal hypervariable segment nor the segment about position 50 contributes to the structure of the antigen-binding region in two hapten binding myeloma Fab fragments from other species studied by X-ray crystallography.

(ii) Each of the three apparent hypervariable regions is characterized by insertions or deletions. In comparison to the longest chain, length may differ by one or two residues both at the amino terminus and at the first hypervariable region and by up to six residues in the third hypervariable region.

(iii) Alignment of amino acid sequences by a method that maximizes nucleotide base sequence homology diminishes the apparent variability at the amino terminal region of the L chain. Variability in this region then becomes comparable to variability elsewhere within the framework. Such alignment still permits identification of the other two hypervariable regions that correspond to complementarity-determining regions.

(iv) Alignment by minimal base-change criteria places amino terminal deletions at any of the first three positions of the chain, suggesting that differences in L chain length cannot be explained by posttranslational degradation of the L chain. Examination of sequence data also supports this view because the great majority of shorter chains cannot be generated by removal of the amino terminal residues of the longer chains. Alignment by minimal base-change criteria may result in noncontiguous deletions in the other hypervariable regions.

(v) When objective analysis of nucleotide base sequences is carried out, an absolute correlation between framework and complementarity region sequences is not seen. A correlation between framework and complementarity sequences is seen among myeloma proteins, however.

Within complementarity regions, apparent mutation frequency and the occurrence of deletions are greater than in the framework. These observations can be taken as support for a gene integration model for generation of diversity. However, the need to postulate such a model is diminished by the large numbers of both framework and complementarity region sequences seen in rabbit antibodies.

(vi) Cluster analysis also does not show a correlation between framework sequence and antigen specificity when either complete framework sequences or sequences between positions 0 and 20 are analyzed.

(vii) Homology among the complete framework sequences cannot be predicted by examination of only the first 20 positions of rabbit L chains.

(viii) Among all the complementarity regions of S3- and S8-specific antibodies examined, no two have been found that are identical.

(ix) Although cluster analysis of complementarity region sequences demonstrates some correlations with antigen specificity, there is more than one cluster of sequences associated with this specificity among S3 antibodies where the data are greatest.

(x) Diversity is also seen among L chain complementarity and framework sequences when antibodies of the same specificity are compared in a single animal. Antibodies possessing a similar affinity for antigen may have different complementarity and framework sequences. Since three of these antibodies carried the same allotypic marker in a heterozygous individual, genetic polymorphism as an explanation for diversity seems unlikely.

(xi) Preliminary data on an H chain complementarity region indicates that the degree of diversity in complementarity sequences of H chains is not likely to be less than that of L chains.

Accumulated evidence from examination of other sets of antigen-binding immunoglobulins gives a different picture from the one presented here. The degree of complexity and diversity apparent from studies of elicited rabbit antibodies does not seem to be reflected in other systems. The specific manner in which antigen-binding immunoglobulins are selected for study needs to be examined critically in order to determine if selection of a specific subset is the primary determinant of the result observed.

Elicited antibodies subjected to the most intense study have been either those that exhibit idiotypic cross-reactivity with antigen-binding myeloma proteins (6) or those that occur or can be elicited in all members of a given strain of animals (7,8). In each instance, the phenotypes appear heritable and seem to segregate by simple Mendelian principles (see ref. 59). A very close relationship exists between framework and complementarity region sequences, both associated with antigen specificity. Cebra and colleagues (4) suggest that all elicited antibodies follow this rule. Pooled hapten-specific antibodies not selected on the basis of idiotypic cross-reactivity were examined in inbred guinea pigs. The results are somewhat difficult to interpret because it is now apparent that the antibodies studied were heterogeneous. While a major sequence seemed to be present in the framework and first two hypervariable regions of the H chain, the third hypervariable region was so complex that a sequence could not be determined (5). Multiple L chain spectrotypes were also present (19). For these reasons it is possible that the sequence of the H chain up to the third hypervariable region represents only the dominant residue at each position derived from a mixture of sequences. Linkage between specific framework and complementarity residues cannot be inferred under these circumstances since it is difficult to know whether one is relating residues within the same polypeptide chain or among different chains.

Are antibodies elicited in inbred animals and selected on the basis of shared idiotypy representative?

The phosphorylcholine binding myeloma protein T15 represents a dominant clone occurring in all members of a given strain, even prior to immunization (6). However, even in the same animal, when antibodies other than the common idiotype but of the same antigen-binding specificity are sought, different results are obtained. A variable fraction of antibodies specific for arsonilic acid in A/J mice share a common idiotype (8). These antibodies are characterized by a common amino acid sequence in the amino terminal segment of the framework and in the first hypervariable region of the H chain (57). Their L chains, however, are heterogeneous (18). Arsonate-specific antibodies which do not share a common idiotype may share H chain framework sequences, but differ in their first hypervariable region (57). Thus an immune response in a member of an inbred strain elicits antibodies of both common V region structure ("public phenotype") and antibodies unique to an individual ("private phenotype").

Public phenotypes are characterized by (a) heritability, (b) clonal dominance, at times prior to immunization, and (c) linkage of framework and complementarity region structures. Private phenotypes are characterized by a multiplicity of unique complementarity and framework structures that may not be shared among individuals. Among the set of private phenotypes, an association between complementarity and framework sequences is not found. It is likely that public phenotypes are the products of germ-line genes. If private phenotypes are also products of a very large pool of germ-line genes, then they must be subjected to selective control in their expression. Alternatively, these phenotypes may be generated by somatic variation of a smaller number of germ-line genes. The rabbit antibodies we have discussed in this chapter fall into the group of private phenotypes. Which of these groups is more representative of the normal immune response?

This question cannot be answered until both groups of phenotypes are examined simultaneously in several systems so that quantitative relationships may be better appreciated. It is evident that if selection is made on the basis of heritable characteristics, heritable or public phenotypes will be revealed, but if antigen is the only selector, both private and public phenotypes will be seen and their relationships can be assessed. By sequence analysis of antibodies elicited in outbred rabbits, public phenotypes have not yet been found.

STRUCTURAL BASIS OF DIVERSITY

The reasons for complementarity site diversity among antibodies specific for a single antigen need to be considered. First, in examining crystallographic models, it becomes apparent that not all of the amino acid side chains within a complementarity region participate in antigen contact. Indeed, with smaller antigens, such as phosphorylcholine and menadione, large surfaces of the apparent combining site are unoccupied. Could the lack of correlation between hypervariable region sequences and antigen specificity be related simply to free variation among residues not involved in antigen contact? An inspection of Figs. 14 and 15, where homology has been maximized, fails to reveal even a single position where the same residue is found in all L chains of a given specificity.

Other explanations for diversity which must be listed include the complexity of even an apparently simple antigen such as S3. For example, the number of sugar residues participating in binding to the site may vary. It is of interest that insertions or deletions within L chain complementarity regions of these rabbit antibodies are more marked than in human kappa chains. The crystallographic models seem to indicate that the size of the combining site is governed by the length of polypeptide chain available within complementarity regions. Another difference in specificity may relate to the location of the "reading frame," i.e., the beginning and end of the oligosaccharide determinant recognized may differ among antibodies.

Functionally equivalent binding sites associated with different structures have long been suggested by the intrinsic diversity of a normal immune response. One can point to comparisons between proteins of vertebrates and bacteria, which often perform similar functions and bind the same ligands, yet may have very different residues within their binding sites (60). There is little reason not to extend such observations to antibodies, where more than one constellation of amino acid side chains could bind an antigen with similar affinity and specificity. Structural variation allowed in contact residues can only be delineated by X-ray crystallographic studies.

ACKNOWLEDGMENT

This work was supported by Grant AI04967 from the National Institutes of Health, and by a Grant-in-Aid from the American Heart Association. M.N.M. is an Established Investigator of the American Heart Association.

REFERENCES

1. Potter, M. J. This volume.
2. Capra, J. D., and Kehoe, J. M. (1974): Structure of antibodies with shared idiotypy: The complete sequence of the heavy chain variable regions of two immunoglobulin M anti-gamma globulins. *Proc. Natl. Acad. Sci. USA*, 71:4032.
3. Capra, J. D. This volume.
4. Cebra, J. J., Koo, P. H., and Ray, A. (1974): Specificity of antibodies: Primary structural basis of hapten binding. *Science*, 186:263.
5. Cebra, J. J. This volume.
6. Claflin, J. L., and Davis, J. M. (1974): Clonal nature of the immune response to phosphorylcholine. *J. Immunol.*, 113:1678.
7. Eichmann, K. (1973): Idiotype expression and the inheritance of mouse antibody clones. *J. Exp. Med.*, 137:603.

8. Nisonoff, A., Tung, A. S., and Capra, J. D. (1974): Factors determining immune responsiveness at the level of B-cells. In: *Progress in Immunology, Vol. 2,* edited by L. Brent and J. Holborow, p. 17. North-Holland Publishing Co., Amsterdam.
9. Cohn, M. (1974): A rationale for ordering the data on antibody diversification. In: *Progress in Immunology, Vol. 2*, edited by L. Brent and J. Holborow, p. 261. North-Holland Publishing Co., Amsterdam.
10. Krause, R. M. (1970): The search for antibodies with molecular uniformity. *Adv. Immunol.*, 12:1.
11. Haber, E. (1970): Antibodies of restricted heterogeneity for structural study. *Fed. Proc.*, 29:66.
12. Jones, J. K. N., and Perry, M. B. (1957): The structure of the type VIII pneumococcus specific polysaccharide. *J. Am. Chem. Soc.*, 79:2787.
13. Chen, F. W., Strosberg, A. D., and Haber, E. (1973): Evolution of the immune response to type III and VIII pneumococcal polysaccharides. *J. Immunol.,* 110:98.
14. Cheng, W. C., Fraser, K. J., and Haber, E. (1973): Fractionation of antibodies to the pneumococcal polysaccharides by affinity chromatography. *J. Immunol.*, 111:1677.
15. Edman, P., and Begg, G. (1967): A protein sequenator. *Eur. J. Biochem.*, 1:80.
16. Brauer, A. W., Margolies, M. N., and Haber, E. (1975): The application of 0.1 M Quadrol to the microsequence of proteins and the sequence of tryptic peptides. *Biochemistry*, 14:3029.
17. Summers, M., Smythers, G., and Oroszlan, S. (1973): Thin-layer chromatography of sub-nanomole amounts of phenylthiohydantoin (PTH) amino acids on polyamide sheets. *Anal. Biochem.*, 53:624.
18. Capra, J. D., Tung, A. S., and Nisonoff, A. (1975): Structural studies on induced antibodies with defined idiotypic specificities. *J. Immunol.*, 115:414.
19. Trischmann, I., Dugan, E., and Cebra, J. (1975): A comparison of the heavy chain variable regions of different anti-dinitrophenyl (DNP) antibodies raised in inbred guineas pigs. *Fed. Proc.*, 34:970.
20. Poljak, R. J., Amzel, L. M., Chen, B. L., et al. (1974): The three-dimensional structure of the Fab′ fragment of a human myeloma immunoglobulin at 2.0 Å resolution. *Proc. Natl. Acad. Sci. USA*, 71:3440.
21. Fraser, K. J., Poulsen, K., and Haber, E. (1972): Specific cleavage between variable and constant domains of rabbit antibody light chains by dilute acid hydrolysis. *Biochemistry*, 11:4974.
22. Segal, D. M., Padlan, S. A., Cohen, G. H., Silverton, E. W., Davies, D. R., Rudikoff, S., and Potter, M. (1974): The structure of McPC 603 Fab and its hapten complex. In: *Progress in Immunology, Vol. 2*, edited by L. Brent and J. Holborow, pp. 93–102. North-Holland Publishing Co., Amsterdam.
23. Margolies, M. N., Cannon, L. E., Strosberg, A. D., and Haber, E. (1975): Diversity of light chain variable region sequences among rabbit antibodies elicited by the same antigens. *Proc. Natl. Acad. Sci. USA*, 72:2180.
24. Appella, E., Roholt, O. A., Chersi, A., Radzimski, G., and Pressman, D. (1973): Amino acid sequence of the light chain derived from a rabbit p-azobenzoate antibody of restricted heterogeneity. *Biochem. Biophys. Res. Commun.,* 53:1122.
25. Chen, K. C. S., Kindt, T. J., and Krause, R. M. (1974): Amino acid sequence of an allotype b4 light chain from a rabbit antibody to streptococcal carbohydrate. *Proc. Natl. Acad. Sci. USA*, 71:1995.
26. Jaton, J.-C. (1974): Amino acid sequence of the N-terminal 139 residues of light chain derived from a homogeneous antibody. *Biochem. J.*, 141:1.
27. Jaton, J.-C. (1974): Comparison of the amino acid sequences of the variable regions of light chains derived from two homogeneous rabbit antipneumococcal antibodies. *Biochem. J.*, 141:15.
28. Jaton, J.-C. (1975): Comparison of the amino acid sequences of the variable domains of two homogeneous rabbit antibodies to type III pneumococcal polysaccharide. *Biochem. J.*, 147:235.
29. Braun, D. G., Huser, H., and Jaton, J.-C. (1975): Identical sequence of light chains from rabbit anti-streptococcal antibodies. *Nature*, 258:368.
30. Wu, T. T., and Kabat, E. A. (1970): An analysis of the sequences of the variable regions of

Bence Jones proteins and myeloma light chains and their implications for antibody complementarity. *J. Exp. Med.*, 132:211.
31. Padlan, E. A., and Davies, D. R. (1975): Variability of three-dimensional structure in immunoglobulins. *Proc. Natl. Acad. Sci. USA*, 72:819.
32. Kabat, E. A., and Wu, T. T. (1971): Attempts to locate complementarity determining residues in the variable positions of light and heavy chains. *Ann. NY Acad. Sci.*, 190:382.
33. Strosberg, A. D., Fraser, K. J., Margolies, M. N., and Haber, E. (1972): Amino acid sequence of rabbit pneumococcal antibody. I. Light chain cysteine-containing peptides. *Biochemistry*, 11:4978.
34. Strosberg, A. D., Margolies, M. N., and Haber, E. (1975): The interdomain disulfide bond of a homogeneous rabbit pneumococcal antibody light chain. *J. Immunol.*, 115:1422.
35. Edmundson et al. This volume.
36. Stanton, T., Sledge, L., Capra, J. D., Woods, R., Clem, W., and Hood, L. (1974): A sequence restriction in the variable region of immunoglobulin light chains from sharks, birds, and mammals. *J. Immunol.*, 112:633.
37. Cannon, L. E., Margolies, M. N., Strosberg, A. D., Chen, F. W., Newell, J., and Haber, E. (1976): Diversity among rabbit antibody light chain amino terminal sequences. *J. Immunol.*, 117:160.
38. Braun, D. G., and Jaton, J.-C. (1973): The amino terminal sequence of antibody light chains: Evidence for possible inheritance of structural genes. *Immunochemistry*, 10:387.
39. Freedman, M., Hofmann, T., Thee, S., Yeger, H., Slaughter, M., and Milandre, J. (1975): Studies of hyperimmune restricted and partially restricted anti-pneumococcal polysaccharide antibodies from allotype-defined pedigreed rabbits. IV. Amino-terminal light chain sequence analysis of restricted homozygous b4 rabbit anti-SIII and SVIII antibodies from partially inbred rabbits. *Immunochemistry,* 12:263.
40. Hood, L., Eichmann, K., Lackland, H., Krause, R. M., and Ohms, J. J. (1970): Rabbit antibody light chains and gene evolution. *Nature (Lond.)*, 228:1040.
41. Waterfield, M. D., Prahl, J. W., Hood, L. E., Kindt, T. J., and Krause, R. M. (1972): Restricted structural heterogeneity in antibodies: Might different heavy chains have a common light chain. *Nature* [*New Biol.*], 240:215.
42. Freedman, M. H., Guyer, R. B., and Terry, W. D. (1972): Isolation and characterization of electrophoretically homogeneous rabbit antihapten antibody populations. *J. Biol. Chem.*, 247:7051.
43. Thunberg, A. L., Lackland, H., and Kindt, T. J. (1973): Sequence variations in b9 light chains as potential V region genetic markers. *J. Immunol.*, 111:1755.
44. Kindt, T. J., Thunberg, A. L., Mudgett, M., and Klapper, D. G. (1974): A study of V region genes using allotypic and idiotypic markers. In: *The Immune System—Genes, Receptors, Signals,* edited by E. E. Sercarz, A. R. Williamson, and C. F. Fox, p. 69. Academic Press, New York.
45. Strosberg, A. D., Fraser, K. J., Margolies, M. N., and Haber, E. (1972): Amino acid sequence of rabbit pneumococcal antibody. I. Light-chain cysteine-containing peptides. *Biochemistry*, 11:4978.
46. Jaton, J.-C., Waterfield, M. D., Margolies, M. N., Bloch, K. J., and Haber, E. (1971): Variation in the primary structure of antibodies during the course of immunization. *Biochemistry*, 10:1583.
47. Kindt, T. J., Seide, R. K., Lackland, H., and Thunberg, A. L. (1972): Serologic identity of the b4 allotypic determinants present on homogeneous rabbit light chains with different N-terminal amino acid sequences. *J. Immunol.*, 109:735.
48. Hood, L., Waterfield, M. D., Morris, J., and Todd, C. W. (1971): Light chain structure and theories of antibody diversity. *Ann. NY Acad. Sci.*, 190:26.
49. Mage, R. G., Appela, E., Young-Cooper, G. O., Winfield, J. B., Pincus, J. H., and Chersi, A. (1972): A rabbit anti-p-azobenenearsonate antibody of restricted heterogeneity with light chains of b5 allotype. *J. Immunol.*, 110:722.
50. IUPAC-IUB Commission on Biochemical Nomenclature. (1968): A one letter notation for amino acid sequences, tentative rules. *J. Biol. Chem.*, 243:3557.
51. Putnam, F. W., Titani, K., and Whitley, E. (1966): Chemical structure of light chains: Sequence of type κ. *Proc. R. Soc. Lond.* [*Biol.*], 166:124.

52. Cohn, M., Blomberg, B., Geckeler, W., Raschke, W., Riblet, R., and Weigert, M. (1974): First order considerations in analyzing the generator of diversity. In: *The Immune System—Genes, Receptors, Signals*, edited by E. E. Sercarz, A. R. Williamson, and C. F. Fox, p. 89. Academic Press, New York.
53. Smith, G. P., Hood, L., and Fitch, W. M. (1971): Antibody diversity. *Ann. Rev. Biochem.*, 40:969.
54. Chen, F. W., Cannon, L. E., Margolies, M. N., Strosberg, A. D., and Haber, E.: Purification, specificity, and hypervariable region sequence of antipneumococcal polysaccharide antibodies elicited in a single rabbit. *J. Immunol. (In press.)*
55. Barstad, P., Rudikoff, S., Potter, M., Cohn, M., and Hood, L. (1974): Immunoglobulin structure: Amino terminal sequences of mouse myeloma proteins that bind phosphorylcholine. *Science*, 183:962.
56. Barstad, P., Farnsworth, V., Weigert, M., Cohn, M., and Hood, L. (1974): Mouse immunoglobulin heavy chains are coded by multiple germ line variable region genes. *Proc. Natl. Acad. Sci. USA*, 71:4096.
57. Capra, J. D., Tung, A. S., and Nisonoff, A. (1975): Structural studies on induced antibodies with defined idiotypic specificities. I. The heavy chains of anti-p-azophenylarsonate antibodies from A/J mice bearing a cross-reactive idiotype. *J. Immunol.*, 114:1548.
58. Fleischman, J. B. (1971): A partial amino acid sequence of the heavy chain of a rabbit antibody to group C streptococcal carbohydrate. *Biochemistry*, 10:2753.
59. Riblet, R., Weigert, M., and Mäkelä, O. (1975): Recombination between V_H genes and allotype. *Eur. J. Immunol.*, 5:778.
60. Nolan, C., and Margoliash, E. (1968): Comparative aspects of primary structure of proteins. *Ann. Rev. Biochem.*, 37:727.

Antibodies in Human Diagnosis and Therapy,
edited by Edgar Haber and Richard M. Krause.
Raven Press, New York © 1977.

Relationship Among Sequences of Hypervariable Regions, V_H Framework, and Ligand-Binding Specificities of Antihapten Antibodies from Inbred Guinea Pigs

John Cebra

I would like to summarize some of our findings concerning the relationship among the sequences of hypervariable (complementarity) regions, the sequences of V_H "frameworks," and ligand-binding specificities of anti-hapten antibodies from inbred guinea pigs. Some years ago we found that the primary structure of V_H from guinea pig IgG-2 antibodies of three general specificities—anti-dinitrophenyl (DNP), anti-*p*-azobenzenearsonate (ARS), and anti-*p*-azo-benzenetrimethylammonium (TMA)—was markedly restricted in its heterogeneity relative to V_H from pooled "normal" IgG-2 (1). Moreover, we found that at least two of the three regions of V_H (Hv1, N31-N35; Hv2, N48-N59), which accounted for much of the sequence heterogeneity found in normal V_H, each had apparently either a single sequence or one of markedly restricted heterogeneity if derived from the specific anti-hapten antibodies. Further, the primary structures of Hvl and Hv2 were different for each of the three antibody specificities and distinctive in that Hv1 and Hv2 from different preparations of antibodies of the same specificity always had approximately the same primary structures (1). The primary structure of the framework of V_H, comprising those residue positions located outside the complementarity regions (N1-N30, N36-N48, N60-N98), was remarkably similar whether V_H was derived from normal IgG-2 or from antibodies of any of the three anti-hapten specificities. Heterogeneity at framework residue positions had only been firmly established for N2, N16, N79 (1), and, more recently, for N84 and N88. We had found that the pattern of residues at N2, N16, and N79 was both unique and distinctive for each of the three anti-hapten antibody specificities examined. It now appears that the residues present at N84 and N88 will also fit into this pattern (see Table 1). We have suggested that the pattern of amino acids at the five distinctive framework residue positions may be taken as a marker for the expression of a particular V_H-structural gene (1,2).

We have recently completed an examination of the above-mentioned primary structural features of V_H from anti-DNP and anti-ARS antibodies raised in both available inbred strains (2 and 13) and in strain 13 animals that had been subjected to treatments known to alter the normal immune response. Our aim was to determine whether antibodies of the same general specificity but of different qualities—of different isotypes or average ligand-binding affinity—would display different patterns of distinctive framework residues and/or hypervariable regions with different primary structures. By assessing whether the relationships among framework markers, structure of Hv regions, and specificity previously established could be altered, we hoped to infer the extent of linkage of framework and Hv regions and thus something about V_H structural genes.

Thus we have analyzed V_H from IgG2 anti-DNP and anti-ARS antibodies raised in strains 2 and 13 by direct immunization with DNP- or ARS-keyhole limpet

TABLE 1. *Distinctive "framework" residue positions and hypervariable regions Hv1 and Hv2 of V_H from specific antibodies (6)*

Residue position(s):	2	16	Hv1[a]	Hv2[b]	79	84	88
Ab type							
Anti-DNP							
St.2 IgG-2	Glu	Ala	+	+	Leu	ND[c]	ND
Conventional IgG-2	Glu	Ala	+	+	Leu	Asn	Ala
Conventional IgG-1	Glu	Ala	+	+	Leu	Asn	ND
Deviated IgG-2	Glu	Ala	+	+	Leu	ND	ND
Deviated IgG-1	Glu	Ala	+	+	Leu	ND	ND
Suppressed IgG-2	Glu	Ala	+	+	Leu	ND	ND
Anti-ARS							
St. 2 IgG-2	Glu	Ser	+	+	Val	( )	Pro
Conventional IgG-2	Glu	Ser	+	+	Val	( )	Pro
Conventional IgG-1	Glu	Ser	+	+	Val	ND	ND
Deviated IgG-2	Glu	Ser	+	+	Val	ND	ND
Deviated IgG-1	Glu	Ser	+	+	Val	ND	ND
Anti-TMA							
St. 13 IgG-2	Val	Lys	+	+	Leu	Tyr	ND

[a] The sole or major sequence of Hv1 was: anti-DNP, Ser Tyr Tyr Met Ala; anti-ARS, Ser Tyr Thr Met Tyr; anti-TMA, Ser Tyr Trp Met Asn. The + signifies that this sequence was found to be the sole or major one of Hv1 from the particular anti-hapten antibody.

[b] The sole or major sequence of Hv2 was: anti-DNP, Val Thr Trp Ile Gly Asn Thr Gly Gly Ser Ile Gly; anti-ARS, Ile Ser Tyr Ile Ser Ser Ser Ser Ser Tyr Ile Lys; anti-TMA, Ile Ser Ala Ile Asn Ser Asp Gly Ser Ser Tyr. The + signifies that this sequence was found to be the sole or major one of Hv2 from the particular anti-hapten antibody.

[c] ND = not determined.

hemocyanin (KLH) (conventional IgG2) and also from conventional IgG-1 antibodies raised in strain 13 guinea pigs. In addition, we have also analyzed V_H from both IgG-2 and IgG-1 obtained in strain 13 animals carrier-deviated (3) with KLH to increase markedly the proportion of IgG-1 antibodies when challenged with either DNP- or ARS-KLH (deviated IgG-1 or deviated IgG-2). Finally, we have analyzed V_H from IgG-2 anti-DNP antibodies raised in strain 13 animals that had been suppressed with DNP-poly-D-(glutamyl, lysyl) copolymer (2,4). These suppressed IgG-2 antibodies had an average affinity for ϵ-DNP-lysine that was one to two orders of magnitude lower than that shown by conventional anti-DNP (4,5). Our general finding, based on extensive automatic sequential degradation of large CNBr fragments of V_H and on detailed analysis of enzymatically generated small peptides of V_H containing the pertinent residue positions, is that all anti-DNP or anti-ARS antibodies examined had the same pattern of distinctive framework residues as did their corresponding prototype conventional anti-DNP or anti-ARS antibodies raised in strain 13 animals (see Table 1). Furthermore, the primary structures of Hv1 and Hv2 from all of these antibodies of the two specificities were approximately the same as those from the corresponding conventional IgG-2 antibodies both in sequence and in degree of restriction of their heterogeneity. Thus we inferred that either the same structural gene for V_H or family of structural genes defined by a distinctive pattern of framework residues was, in general, being expressed by both strains 2 and 13 inbred animals responding to a given haptenic determinant. In particular, the same V_H gene (or genes) was apparently expressed by different plasma cells making one or another isotype (IgG-1 or IgG-2) of that anti-hapten antibody as well as by different cells making anti-haptens with different ligand-binding affinities.

To examine further the nature and bases for the heterogeneity that seemed to occur in anti-DNP antibodies with respect to affinity, we have compared some thermodynamic quantities reflective of the interaction of ligands with conventional and suppressed anti-DNP antibodies (4). Moreover, we have also attempted to evaluate Hv3 (N99-~N112) chemically (5). Briefly, although low-affinity antibodies from suppressed animals differed from conventional antibodies by about two orders of magnitude in average association constant for ε-DNP-lysine ($\sim 8 \times 10^6$ versus $\sim 7 \times 10^8$ liters/mole), they had the same affinity for DNP-aniline ($\sim 5 \times 10^6$ liters/mole) (4). Microcalorimetric measurements of both sorts of anti-DNP antibodies reacting with ε-DNP-lysine showed an exothermic reaction yielding about 9 Kcal/mole in each case. It was concluded that the affinity differences for ε-DNP-lysine reacting with conventional and suppressed antibodies were due to differences in entropy change upon ligand binding in the two cases. These differences could possibly be attributed to differences in interaction with the lysyl side-chain by two populations of antibodies having similar recognition subsites for DNP, possibly contributed in part by Hv1 and Hv2. Of course, Hv3 and hypervariable regions of light chain are candidates for contributors to the subsite reactive with the lysyl side chains. Finally, analysis of γ_2- chain from "conventional" anti-DNP antibodies shows it to be oligodisperse (five to seven major components) despite the apparent homogeneity of its primary structure from N1 to N98 (5).

An analysis of the CNBr fragment from γ_2-chain-bearing Hv3 was carried out to assess its size heterogeneity. The tryptic peptides from the succinylated CNBr fragment were sized by gel filtration in 8 M urea in dilute HCl. That peptide containing Hv3 and a radiolabeled lysyl residue was estimated to have a size of 20 ± 2 amino acids. Nevertheless, analyses of a variety of enzymic digests of this CNBr fragment from anti-DNP antibody show it to have a heterogeneous primary structure. We propose that the same structural V_H gene—defined by its pattern of distinctive framework residues—is often expressed by plasma cells in guinea pigs responding to DNP antigens, but that this gene may be modified during somatic cell division and that those modifications most often detected are expressed in the Hv3 of its antibody product. Such modifications of Hv3 probably do not alter the essential specificity for the DNP determinant and so may still be expressed in a response to DNP antigens.

REFERENCES

1. Cebra, J. J., Koo, P. H., and Ray, A. (1974): Specificity of antibodies: Primary structural basis of hapten binding. *Science*, 186:263.
2. Cebra, J. J., Cordle, C. T., Dugan, E. S., Graziano, S. L., Liu, S. H., Massey, G. R., Ricardo, M. J., Tracey, D. E., and Trischmann, T. M. (1974): Anti-hapten antibodies from inbred guinea pigs: Relationships of their structure to specificity and to other biologic activities. Proceedings of the Robert A. Welch Foundation Conferences on Chemical Research XVIII. *Immunochemistry*, pp. 67–96.
3. Liu, S. H., Koo, P. H., and Cebra, J. J. (1974): Effect of carrier priming on the distribution of anti-hapten antibodies between IgG1 and IgG2 isotypes in the hyperimmune guinea pig. *J. Immunol.*, 113:677.
4. Halsey, J. F., Cebra, J. J., and Biltonen, R. L. (1975): Structural and thermodynamic basis of affinity in anti-dinitrophenyl antibody. *Biochemistry*, 14:5221.
5. Trischmann, T. M. (1975): A structural basis for the heterogeneity of anti-dinitrophenyl antibodies raised in inbred strains of guinea pigs. Doctoral Dissertation, Johns Hopkins University.
6. Ricardo, M. J., Liu, S. H., Dugan, E. S., and Trischmann, T. M.: Unpublished observations.

Antibodies in Human Diagnosis and Therapy,
edited by Edgar Haber and Richard M. Krause.
Raven Press, New York © 1977.

Light-Chain Hypervariable-Region Sequences of Rabbit Antibodies Elicited by Group A-Variant Streptococcal Vaccines

Dietmar G. Braun

Special breeding programs have led to considerable advance in the selection of rabbits with amplified immune responses to streptococcal group polysaccharides (1,2). Many of these antisera contain monoclonal antibody suitable for the analysis of the amino acid sequence of heavy and light chains. These antibodies, obtained from closely related rabbits and directed to the streptococcal groups A, A-variant, and C polysaccharides, have stimulated a search for variable region markers common to antibodies specific for the same polysaccharide antigen (3–7).

In the past 5 years, we have mainly concentrated our efforts toward accumulating data on antibodies induced by immunization with streptococcal group A-variant vaccines (5–7) where the relevant antigen is a branched homopolymer of rhamnose (8,9). Given below are the amino acid sequences of the first (V_LI) and third (V_LIII) hypervariable regions of light chains, and these are compared with amino acid sequences of rabbit light chains published by other laboratories (10–13). Because the second (V_LII) hypervariable region of rabbit κ light chains does not seem to show a comparable degree of variability as in human κ chains (10–13), a discussion of this sequence stretch of the variable region is not included.

In κ light chains from rabbit antistreptococcal group A-variant polysaccharide antibodies, the current sequence information concerning the V_LI region is summarized in Table 1. In principle, three different V_LI regions were found, represented by: (i) the sequences of K4820 and K30-267; (ii) the sequence of K16-167; and (iii) the sequence of K9-106. Another V_LI (K27-489) contained an additional asparaginyl residue. It was interesting that the two allotype b5 chains (K87-529 and K87-551) analyzed in this region differed in length from the above chains of allotype b4; however, they also showed substantial similarities due to two asparaginyl residues in contiguous positions. When compared to each other, these allotype b5 chains differed in the two homologous positions 30 and 31.

V_LI region sequences from anti-group A-variant polysaccharide antibodies were significantly different from V_LI regions of anti-group C streptococcal polysaccharide antibodies (Table 1). Most importantly, these latter three light chains shared identical V_LI sequences that were highly charged, owing to two contiguous aspartic acid residues.

Although asparaginyl residues are also a frequent constituent of V_LI regions in rabbit antipneumococcal type 3 (S3) and type 8 (S8) antibodies, sequence variations in V_LI among these light chains (12,13) were far greater than in V_LI regions of antistreptococcal group A-variant antibodies. The fact that sequence differences in this area were of the same order among both anti-S3 and anti-S8 antibodies as compared to V_LI regions of one anti-*Micrococcus leisodeikticus* antibody and one anti-*p*-azobenzoate antibody emphasizes the rather unique structural characteristics of antistreptococcal group A-variant and group C antibodies with regard to V_LI.

V_LIII regions of antistreptococcal group A-variant antibodies presented a sequence

TABLE 1. *Light-chain hypervariable region L_I from rabbit antibodies*

			23		25					30	31	a	b	c	32			35					40
Av-CHO	K 4820	b4	Cys	Gln	Ala	Ser	Gln	Ser	Ile	Gly	·	·	Asn	·	Phe	Leu	Ser	Trp	Tyr	Gln	Gln	Lys	Pro
″	K 30-267	″						Asn			·	·		·	Arg		Ala						
″	K 16-167	″							Val	Tyr	Ser	·		Asn	·				Phe				
″	K 27-489	″						Asn	Leu	Tyr	Ser	Asn		Asn	·								
″	K 6-139	″							Val	Tyr	Ser	·		Asn	·		Ala						
″	K 33-380	″							Val	Tyr	Ser	·		Asn	·		Ala						
″	K 31-147	″							Val	Tyr	( )	·		Asn	·		( )						
″	K 9-106	″							Val	Tyr	Ser	·		Asn	Arg		Ala		Phe				
″	K 29-213	″							Val	Tyr	Ser	·		Asn	Arg		Ala		Phe				
″	K 27-306	″							Val	Tyr	Ser	·		Asn	(Arg)		Ala		Phe				
″	K 9-335	″							Val	Tyr	Ser	·		Asn	Arg		Ala		Phe				
″	K 9-338	″							Val	Tyr	Ser	·		Asn	Arg		Ala		Phe				
″	K 6-141 II	″							Val	Tyr	Ser	·		Asn	Arg		Ala		Phe				
″	K 732-50	″							Val	Tyr	Ser	·		Asn	Arg		Ala		Phe				
″	K 87-529	b5								·	·	·		Asn	Arg		Ala						
″	K 87-551	b5								·	Ser	·		Asn	·								
C-CHO	R 4135	b4			Thr					·	·	·	Asp	Asp	Tyr								
″	K 19	″								·	·	·	Asp	Asp	Tyr								
″	K 20	″								·	·	·	Asx	Asx	Tyr								
S3	BS-1	b4								Tyr	Ser	Gly	·	·	·		Ala						
″	BS-5	″								Tyr	Ser	·		·	·		Ala						
″	K 25	″								Tyr	Ser	·	·	·	Tyr								
″	3374	″						Asn		Asp	Ser	·	·	·	Trp		Ala						
″	3368	″					Glu		(–)	Gly	·	·		Glu	·		Ala						
″	3381	″					Glu			·	Ser	·		·	Trp		Ala						
S8	3315	″			Ser				Val	Tyr	·	Glu		Gly	Arg		Ser		Phe				
″	3322A	b5							Val	Tyr	·	Asn		Asn	( )		Ala						
″	3322B	″							Leu	·	Thr	·		Gln	Tyr		Ala						
″	3322C	″									Ser	·	Asp	·	·		Val						
M.lyso.	120	b4			Ser			(–)			Thr	·	·	·	Tyr		Ala			Glx	Glx		
A-p-B	2717	b4			Ser	Thr	Lys			Tyr	·	Asx	Asx	Asx	Tyr		Ala				Glx		

Av-CHO and C-CHO, light chains from antistreptococcal group A-variant and group C-specific antibodies, respectively. S3 and S8, light chains from antipneumococcal types 3 and 8 (12,13). M.lyso., light chain from an antibody to an antibody specific for *M. lysodeikticus* (13). A-p-B, light chain from antibody to *p*-azobenzoate (10). Rabbit chain R4135 was taken from (11). b4 and b5 indicate the allotype specificity. Solid lines indicate sequence identity to the heading sequence K4820. Dots mark deletions in order to maximize the homology; (   ) indicate unidentified residues, (–) indicate uncertain residues.

TABLE 2. *Light-chain hypervariable region L_{III} from rabbit antibodies*

						90				93		94	a	b	c	95			98		100					
Av-CHO	K 4820	b4	Tyr	Tyr	Cys	Gln	Gln	Gly	Asn	Leu	Gly	Asn	·	·	·	Ile	Val	·	Phe	Gly	Gly	Gly	Thr	Glu	Val	Val
″	K 30-267	″	—	—	—	—	Leu	Ala	—	(–)	·	—	·	·	Gly	(—	—)	·	—	—	—	—	—	—	—	—
″	K 16-167	″	—	—	—	—	Gly	Thr	—	Thr	—	—	Asn	·	·	—	—	·	—	—	Thr	—	—	—	—	—
″	K 27-489	″	—	—	—	Leu	Gly	Tyr	(–)	Thr	Tyr	—	Asn	Thr	Gly	·	—	Thr	—	—	—	—	—	—	—	—
″	K 29-213	″	—	—	—	Arg	Val	Ala	Ser	Thr	·	—	Asn	·	·	—	—	·	—	—	—	—	—	—	—	—
″	K 27-306	″	—	—	—	—	Gly	Thr	—	(Thr	Gly)	—	Asn	·	·	—	—	·	—	—	—	—	—	—	—	—
″	K 9-335	″	—	—	—	Lys	Val	Ala	—	Ile	·	—	·	Thr	·	—	—	·	—	—	—	—	—	—	—	—
″	K 9-338	″	—	—	—	Lys	Val	Ala	—	Ile	·	—	·	Thr	·	—	—	·	—	—	—	—	—	—	—	—
C-CHO	R 4135	″	—	—	—	—	Ser	·	·	Thr	Tyr	·	Gly	Val	Gly	·	·	·	—	—	—	—	—	—	—	—
S3	3374	″	—	—	—	—	Ser	Tyr	Tyr	Ser	Ile	Ser	(Ser)	Ala	·	·	·	·	—	—	—	—	—	—	—	—
″	3368	″	—	—	—	—	—	Asp	Trp	Asn	Ser	—	Asn	Val	Val	Asn	Asn	·	—	—	—	—	—	—	—	—
″	3381	″	—	—	—	—	—	Thr	Asp																	
″	BS-1	″	—	Phe	—	—	Gly	Ser	Thr	Tyr	Gly	Gly	Gly	Tyr	·	·	·	·	—	—	—	—	—	—	—	—
″	BS-5	″	—	Phe	—	—	Gly	Ser	Asx	Tyr	Thr	Gly	Thr	Val	·	·	·	·	—	—	—	—	—	—	—	—
″	K 25	″	—	—	—	—	Thr	Tyr	Ser	Tyr	Ser	Ser	Thr	Tyr	·	·	·	·	—	—	—	—	—	—	—	—
S8	3315	″	—	—	—	Leu	Gly	Asn	Tyr	Asp	Cys	Ser	Ser	Gly	(Asp	Ser)	Phe	Thr	—	—	—	—	—	—	—	—
M.lyso.	120	″	—	—	—	—	Gly	Thr	Tyr																	
pAB	2717	″	—	—	—	Gly	Gly	Ala	Asp	Tyr	Thr	Gly	Tyr	Ser	·	·	·	·	—	—	—	—	—	—	—	—

Designations, symbols, and references as in legend for Table 1.

pattern of greater complexity than was observed for V_LI (Table 2). Although this region was equally long, only in five out of eight chains, the degree of homology is remarkable in comparison to V_LIII regions of antibodies with unrelated specificities. Furthermore, V_LIII of antibodies K9-335 and K9-338, produced by two littermates, differed only in residue position 92 where an isoleucine substitution was found for an alanyl residue. As in V_LI, asparaginyl residues were also prevalent in V_LIII; from this the relative importance of this residue may be deduced in hypervariable light-chain regions of antistreptococcal group A-variant antibodies. Different from V_LIII regions of antibodies with other specificities, residue positions 95 and 96 were predominantly occupied by an Ile-Val sequence in anti-group A-variant antibodies.

While there might soon be a straightforward explanation at hand for the variability pattern of V_LI in antistreptococcal group A-variant rabbit antibodies, this explanation is certainily not necessarily relevant for V_LIII; a reasonable explanation of V_LIII for antibodies of this specificity produced by the rabbit will require further sequence work of selected light-chain probes.

REFERENCES

1. Krause, R. M. (1970): The search for antibodies with molecular uniformity. *Adv. Immunol.*, 12:1–56.
2. Braun, D. G., and Jaton, J.-C. (1974): Homogeneous antibodies: Induction and value as probe for the antibody problem. *Curr. Topics Microbiol. Immunol.*, 66:29–76.
3. Eichmann, K., and Kindt, T. J. (1971): The inheritance of individual antigenic specificities of rabbit antibodies to streptococcal carbohydrates. *J. Exp. Med.*, 134:532–552.
4. Kindt, T. J., Seide, R. K., Bokisch, V. A., and Krause, R. M. (1973): Detection of idiotypic cross-reactions among streptococcal antisera of related rabbits. *J. Exp. Med.*, 138:522–537.
5. Braun, D. G., Kjems, E., and Cramer, M. (1973): A rabbit family of restricted high responders to the streptococcal group A-variant polysaccharide. Selective breeding narrows the isoelectric focusing spectra of dominant clones. *J. Exp. Med.*, 138:645–658.
6. Braun, D. G., and Kelus, A. S. (1973): Idiotypic specificity of rabbit antibodies to streptococcal group polysaccharides. *J. Exp. Med.*, 138:1248–1265.
7. Braun, D. G., Huser, H., and Jaton, J.-C. (1975): Identical sequence of light chains from rabbit anti-streptococcal antibodies. *Nature,* 258:363–365.
8. Krause, R. M. (1963): Symposium on relationship of structure of microorganisms to their immunological properties. IV. Antigenic and biochemical composition of hemolytic streptococcal cell walls. *Bact. Rev.*, 27:369–380.
9. Coligan, J. E., Schnute, W. C., and Kindt, T. J. (1975): Immunochemical and chemical studies on streptococcal group-specific carbohydrates. *J. Immunol.*, 114:1654–1658.
10. Appella, E., Roholt, O. A., Chersi, A., Radzinski, G., and Pressman, D. (1973): Amino acid sequence of the light chain derived from a rabbit anti-p-azobenzoate antibody of restricted heterogeneity. *Biochem. Biophys. Res. Commun.*, 53:1122–1129.
11. Chen, K. C. S., Kindt, T. J., and Krause, R. M. (1974): Amino-acid sequence of an allotype b4 light chain from a rabbit antibody to streptococcal carbohydrate. *Proc. Natl. Acad. Sci. USA*, 71:1995–1998.
12. Jaton, J.-C. (1975): Comparison of the amino acid sequences of the variable domains of two homogeneous rabbit antibodies to type III pneumococcal polysaccharide. *Biochem. J.*, 147:235–247.
13. Margolies, M. N., Cannon, L. E., III, Strosberg, A. D., and Haber, E. (1975): Diversity of light chain variable region sequences among rabbit antibodies elicited by the same antigens. *Proc. Natl. Acad. Sci. USA*, 72:2180–2184.

Antibodies in Human Diagnosis and Therapy,
edited by Edgar Haber and Richard M. Krause.
Raven Press, New York © 1977.

Patterns of Variability in the Variable Regions of Immunoglobulin Heavy Chains

J. Donald Capra

Department of Microbiology, The University of Texas Health Science Center at Dallas, Southwestern Medical School, Dallas, Texas 75235

The chemical definition of the idiotype and the precise location of the antibody combining site on immunoglobulin molecules constitutes one of the major themes which has underscored much of the work our laboratory has been involved with over the past six years. Fundamental to both of these concepts is the nature of antibody variability itself. Much of this work was begun as a fellow in Dr. Henry Kunkel's laboratory here at Rockefeller University in 1967. At that time it was demonstrated that human IgM cold agglutinins shared certain antigenic determinants which were termed "cross-idiotypic specificities." The experimental evidence indicated that these cross-idiotypic determinants were located on the heavy chain and they were somehow related to the antibody combining site (1). This led to a focus on human immunoglobulin heavy chains as a system to test the relationship of idiotypy and the antibody combining site, and early amino acid sequence studies led to the delineation of heavy chain hypervariable regions in 1971 (2). This chapter will develop the concept that, in the main, these three concepts, hypervariable regions, idiotypy, and the antibody combining site, by and large represent one and the same entity (3).

SUBGROUPS OF HUMAN IMMUNOGLOBULIN HEAVY CHAINS

As early as 1967, it was recognized that the variable regions of immunoglobulin light chains could be easily classified into subsets or families of proteins on the basis of similarities in amino acid sequence (4–6). This type of classification has probably reached its fullest extension in the human kappa chain subgroups, but is a much more difficult construct in certain systems such as the mouse kappa chains and human lambda chains. Human heavy chains, like human kappa chains, can be easily divided into three variable region subgroups (7). These are shown in Fig. 1. Leaving aside for

											10										20
$V_H 1$																					
Eu	(γ1)	PCA	VAL	GLN	LEU	VAL	GLU	SER	GLY	ALA	GLU	VAL	LYS	LYS	PRO	GLY	SER	SER	VAL	LYS	VAL
Ca	(γ1)	PCA	—	—	—	—	—	—	—	—	—	ARG	—	—	—	—	ALA	—	—	—	ILE
$V_H 11$																					
Daw	(γ1)	PCA	VAL	THR	LEU	ARG	GLU	SER	GLY	PRO	ALA	LEU	VAL	ARG	PRO	THR	GLN	THR	LEU	THR	LEU
Ou	(μ)	PCA	—	—	—	THR	—	—	—	—	—	—	—	LYS	—	PRO	—	—	—	—	—
He	(γ1)	PCA	—	—	—	LYS	—	ASN	—	—	THR	—	—	LYS	—	—	—	—	—	—	—
Cor	(γ1)	PCA	—	—	—	—	—	—	—	—	—	—	—	LYS	—	—	—	—	—	—	—
$V_H 111$																					
Tei	(γ1)	GLU	VAL	GLN	LEU	VAL	GLU	SER	GLY	GLY	GLY	LEU	VAL	GLN	PRO	GLY	GLY	SER	LEU	ARG	LEU
Was	(γ1)	—	—	—	—	LEU	—	—	—	—	—	—	—	—	—	—	—	—	—	—	—
Jon	(γ3)	ASP	—	—	—	—	—	—	—	—	—	—	—	LYS	—	—	—	—	—	—	—
Zap	(α1)	—	—	—	—	—	—	—	—	—	ALA	—	—	—	—	—	—	—	—	GLY	—
Tur	(α1)	—	—	—	—	LEU	—	—	—	—	—	—	—	—	—	—	—	—	—	—	—
Nie	(γ1)	PCA	—	—	—	—	GLN	—	—	—	—	VAL	—	—	—	—	ARG	—	—	—	—
Gal	(μ)	—	—	—	—	—	—	—	—	—	ASP	—	—	—	—	—	ARG	—	—	—	—

FIG. 1. The variable region subgroups of the human immunoglobulin heavy chains. Only those amino acids which differ from the first sequence in each group are indicated. Otherwise, a line indicates identity with the first sequence in each group.

the moment the genetic meaning of a subgroup, it is obvious that heavy-chain variable regions can be grouped such that their sequences are far more similar to each other than they are to members of another V region subgroup. In general, intermediate sequences are not encountered, and the grouping is simple and straightforward.

The precise distribution of normal immunoglobulins and myeloma proteins into the three V_H subgroups in man has never been fully defined. Far more sequences are available for the V_HIII subgroup primarily because the unblocked nature of the *N*-terminal amino acid allows for a more easily approachable variable region sequence. It is likely, however, that over 80% of human immunoglobulin heavy chains belong to V_HI or V_HII.

DISTRIBUTION OF V_H SUBGROUPS AMONG THE IMMUNOGLOBULIN CLASSES

While it is known that immunoglobulins of virtually all classes can be found in each of the V_H subgroups, a clear bias has been found in myeloma proteins and pooled immunoglobulins in regard to the distribution of the V_H subgroups among the human immunoglobulin classes. For example (as shown in Table 1), while randomly chosen IgG myeloma proteins, as well as pooled IgG, contain approximately 20% of proteins of the V_HIII subgroup, in a study of 30 randomly chosen human IgA myeloma proteins, 66% were V_HIII (8). In addition, 5 of 7 human IgE myeloma proteins belong to the V_HIII subgroup (9), while relatively few IgM and IgD myelomas can be so classified (10). Thus, the human immunoglobulin classes can be roughly divided into two groups—those in which the V_HIII subgroup is a major component (IgA and IgE), and those which utilize either the V_HI or V_HII subgroup predominantly (IgG, IgM, and IgD).

The biologic significance of this observation is not clear; but in our limited studies it is not an artifactual bias because of the use of myeloma proteins, for heavy chains of normal IgG, IgA, and IgM gave similar results. The most likely explanation probably relates to antigenic selection, although more complex interpretations should be kept in mind (discussed in ref. 8). This

TABLE 1. *Distribution of the V_HIII subgroup among immunoglobulin classes (human)*

	From myelomas		From pooled sera
IgG	18%	(5/28)	20%
IgA	67%	(20/30)	75%
IgM	27%	(9/33)	24%
IgD	14%	(1/7)	ND[a]
IgE	71%	(5/7)	ND

[a] ND, not done.

V_H/class distribution may explain why the BALB/c IgG pool contains only 27% V_HIII, while virtually all BALB/c myelomas (which are primarily IgA) are of the V_HIII subgroup (11,12). Recently, it has been shown that the NZB myelomas, which are mainly IgG, are predominantly "blocked" (13).

STRUCTURAL STUDIES ON HUMAN HEAVY-CHAIN VARIABLE REGIONS: IDENTIFICATION OF FOUR HYPERVARIABLE REGIONS

The variable regions of 16 human immunoglobulin heavy chains have now been sequenced. Six of these sequences became available in 1969 and 1970 (14–18). An additional variable region was reported by Watanabe et al. in 1973 (19). Seven complete human V_H sequences have been previously reported from our laboratory (20,21) and, two additional complete sequences will be reported at this time (22–23). Since our work has exclusively centered on human proteins of the V_HIII subgroup, the bulk of the following discussion will focus on human V_HIII sequences. Figure 2 displays the amino acid sequence depicted in the one letter code of the 11 complete human V_HIII heavy-chain variable regions. The first nine sequences were done in our laboratory, the latter two by Hilschmann's group (18,19).

Vast regions of the immunoglobulin heavy-chain variable region are conserved, and other areas are extremely variable. Within a subgroup, variability is largely confined to these so-called hypervariable regions. Some years ago Wu and Kabat applied a statistical analysis to the light-chain variable regions sequenced at that time and developed the concept of hypervariable regions (24). Both Franêk (25) and Milstein (6), utilizing different approaches, independently developed a similar construct.

Figure 3 depicts the variability of human heavy chain variable regions plotted according to the method of Wu and Kabat. The figure is somewhat out of date and does not take into consideration several additional sequences. However, the general pattern is obvious and is not changed by more recent data. Four broad regions of sequence hypervariability are seen, and within each area of hypervariability, sequence gaps are required to align all the proteins.

THE "FRAMEWORK" OF V_HIII

Figures 4 and 5 are a restatement of Fig. 2, except that in this construct, only the "framework" residues are depicted in Fig. 4 in a linear array, while in Fig. 5, the hypervariable regions are similarly depicted. In this way one can look at the heavy chain as two separate structures—a framework structure and a hypervariable region structure. We can now ask certain questions concerning each of these "separate" portions of the human heavy chain variable region. In Fig. 4, in which 11 human "frameworks" are depicted, each compared to protein Tie, one can see the remarkable lack of variability within a specific V region subgroup. Thus, some proteins differ by only a

```
                        10                  20                  30                  40
Tie    E V Q L V E S G G G L V Q P G G S L R L S C A A S G F T F S T S A V Y ( ) W V R
Was    ______ L __________________________________________ S _____ D _ M _ ( ) _____
Jon    D ______________________ K __________________________________ A W M K ( ) _____
Zap    _______________ A _______________ G ______________________ T S R F ( ) _____
Tur    ______ L ____________________________________________ R V L S S ( ) _____
Pom    ______ L ______________________________________ S ___ M S ( ) _____
Lay    A ____ L ____________________________________________ A ___ M S ( ) _____
BroM   _________________________________________________________ Y Y M N ( ) _____
Gra    _______________________________________________________ I _ K T V Y E ( ) _____
Nie    Z ______ Q ______ V ______ R ______________________ R Y T I H ( ) _____
Gal    _______________ D __________ R __________________(B V L B B F)M T ( ) _____

                       50                  60                  70                  80
Tie    Q A P G K G L E W V G W R Y E G S S L T H Y A V S V Q G R F T I S R N D S K N T
Was    ___________________ A _ K _ Q E A _ N S _ F _ D T _ N ________________________
Jon    ___________________ V ___ V _ Q V V E K A F _ N ___ N ________________________
Zap    ___________________ E F _ V Q ___ A I S ____ D ____ A ________________________
Tur    ___________________ S G _ L N A ___ N L _ F ____ A ___________________________
Pom    ___________________ A _ K ___ N G N D K ____ D ___ N _________________________
Lay    ___________________ A _ K ___ N G N D K ____ D ___ N _________________________
BroM   _ V E ____________ S A I G _ A G D Q()Y Y _ D ___ K __________________________
Gra    ___________________ T Y V G Q V _ K F S()_____________________________________
Nie    ___________________ A V M S Y B G B B K ____ D ___ N _________________________
Gal    ___________________ A N I K Z B G _ Z Z B _ V D ___ K ___________ D N A ___ S

                        90                  100                  110
Tie    L Y L Q M L S L E P E D T A V Y Y C A R V T P A A A S L T F S A V W G Q G T L
Was    _________ N R ___ A ____________________ F R Q P F V Q ( ) _ F D _ F _________
Jon    _________ I _ V T _______________________ V V S T (     ) S M D ___________ P
Zap    _________ N T G _ A _____________________ T R _ G G Y (   ) ___ D ____________
Tur    _______________ Q A _____________________ L S V T _ V (   ) A F D __________ K
Pom    _____ L _ N ___ Q A ______ L ___________ D A G P Y V _ P ___ F _ H Y ________
Lay    _________ N G _ Q A _ V S _ I __________ D A G P Y V _ P ___ F _ H __________
BroM   _________ N ___ R A _____________________ S P V S L V D G W L Y*S ___________
Gra    _____ E _ K T G _________________________ H I Y V T L Y Y Y*_ M ____________
Nie    _____ N _ N ___ R _______________________ I R D T _ M (   ) _ F _ H __________
Gal    _________ N ___ R V _____________________ G W G (         ) G G D Y __________
```

FIG. 2. The amino acid sequence depicted on one letter code of the 11 complete human V_HIII heavy chain variable regions. All are compared to protein Tie, and only substitutions from the Tie sequence are indicated. *, Additional residues.

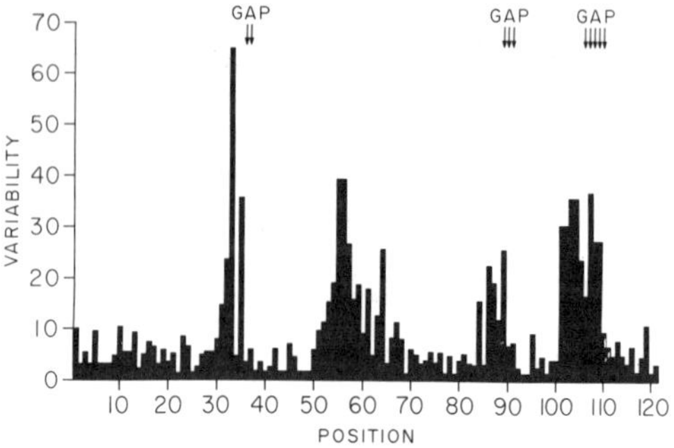

FIG. 3. Variability of the human heavy chain variable regions plotted according to the method of Wu and Kabat.

single amino acid in their framework from the Tie protein (Gra), while others differ by as many as nine residues (Gal). The average protein differs by only three residues. The exact number of differences will, of course, depend upon the precise definition of the hypervariable regions, but a variation of between 3 and 5 seems reasonable.

These framework residues display several important characteristics: (i) They are largely invariant among all human V_H sequences or they are subgroup specific; (ii) as mentioned above, they vary from a prototypic framework sequence by an average of only three to five amino acids; (iii) the variations are largely conservative; thus, of the 35 sequence differences in the 11 proteins, only 4 require more than a one base change in the genetic code; (iv) there is no obvious linear pattern to the substitutions noted in these framework residues. There are, in fact, only two positions in which the same amino acid substitution has been found in the framework of two proteins. Proteins Was, Tur, Pom, and Lay have a leucine in position 5 rather than the more common valine. However, among the other substitutions noted in the framework, none correlate with the substitution of leucine for valine in position 5. Similarly, in the two proteins Nie and Gal, arginine has been found in position 16 rather than the more common glycine; among the 11 other substitutions noted between these two proteins, none are the same. Thus, the variations display no obvious linear pattern, and a subdivision of the V_HIII subgroup in man is impossible at present (26).

THE HYPERVARIABLE REGIONS

Figure 5 depicts (in a linear fashion) the hypervariable regions of the V_HIII subgroup of human immunoglobulin heavy chains. All these proteins have been selected at random with the exception of the Pom/Lay proteins. I would like to make several points concerning these hypervariable region sequences: (i) They constitute approximately 25% of the V_H region. (ii) Deletions that are not related to subgroup have been found in three of the four hypervariable regions. This is best noted in Fig. 2 where proteins BroM and Gra require single amino acid deletions in the second hypervariable

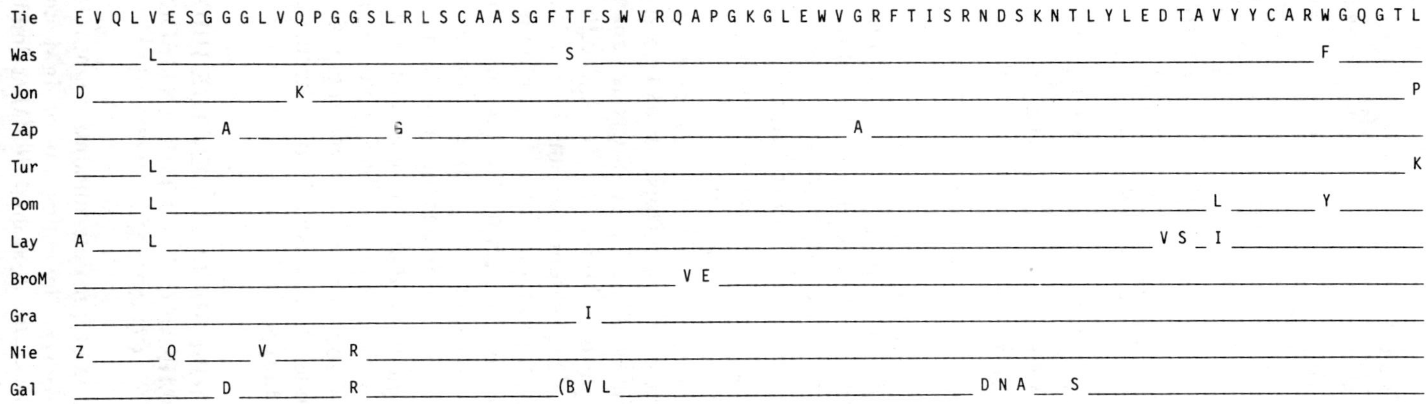

FIG. 4. A linear array of the framework residues of the 11 complete human V_HIII sequences.

```
Tie    T S A V Y G W R Y E G S S L T H Y A V S V Q Q M L S L E P V T P A A A S L T F S A V
Was    _ D _ M _ A _ K _ Q E A _ N S _ F _ D T _ N ___ N R ___ A F R Q P F V Q ( ) _ F D _
Jon    _ A W M K V ___ V _ Q V V E K A F _ N ___ N ___ I _ V T ___ V V S T (     ) S M D _
Zap    _ T S R F E F _ V Q ___ A I S _____ D _________ N T G _ A T R _ G G Y (   ) ___ D _
Tur    R V L S S S G _ L N A ___ N L _ F _____ A ___________ Q A L S V T _ V (   _ A F D _
Pom    S ___ M S A _ K ___ N G N D K _____ D ___ N L _ N ___ Q A D A G P Y V _ P ___ F _ H
Lay    A ___ M S A _ K ___ N G N D K _____ D ___ N ___ N G _ Q A D A G P Y V _ P ___ F _ H
BroM   _ Y Y M N S A I G _ A G D Q ( ) Y Y _ D ___ K ___ N ___ R A S P V S L V D G W L Y*S _
Gra    K T V Y E T Y V G Q V _ K F S ( ) ___________ E _ K T G ___ H I Y V T L Y Y Y*_ M ___
Nie    R Y T I H A V M S Y B G B B K _____ D ___ N N _ N ___ R _ I R D T _ M (   ) _ F _ H
Gal    B B F)M T A N I K Z B G _ Z Z B _ V D ___ K ___ N ___ R V G W G (         ) G G D Y
```

FIG. 5. A linear array of the hypervariable regions of the human V_HIII subgroup.

region. Many proteins, of course, require deletions in the fourth hypervariable region. In the third hypervariable region, the only deletions noted are required in protein Cor (17). No deletions have been noted within a subgroup in the first hypervariable region. (iii) Substitutions in hypervariable regions are not conservative in regards to the genetic code. Thus, at virtually every position, the substitutions are multiple and cannot be generated from a single residue by single base changes. (iv) Substitutions are not conservative in regards to amino acid charge or size. Thus, for example, in the first residue shown in Fig. 5 (position 31 of the V_HIII subgroup) many proteins contain threonine but arginine, serine, alanine, lysine, and either aspartic acid or asparagine have been noted. In four proteins of the V_HII subgroup three different amino acids have been found: glycine, threonine and serine. Thus, in the first residue of the first hypervariable region, seven different amino acids have been detected, most of which differ dramatically both in size and charge. (v) Most importantly, no obvious linear pattern is noted within these sequences. Thus, while many proteins share a particular residue in the first hypervariable region, there is no obvious sharing among these same proteins in the second, third, and fourth hypervariable regions. Proteins Pom and Lay are, of course, excluded from this particular discussion because of their similar specificity and idiotype.

THE RELATIONSHIP BETWEEN THE IDIOTYPIC DETERMINANTS AND THE ANTIBODY COMBINING SITE

The first evidence that idiotypic determinants might relate to the antibody combining site came from an investigation of IgM cold agglutinins by Williams et al. (1). This study demonstrated that immunoglobulins with similar specificities shared antigenic determinants (termed "cross-idiotypic specificity" or "shared idiotypy") and it was postulated that the structures shared by these proteins included the antibody combining site. The inhibition of the reactivity of idiotypic determinants when the cold agglutinins were bound to their red cell antigens provided further evidence for this relationship. These studies were extended to the IgM anti-γ-globulins, and it was demonstrated that these proteins could be classified into two cross-idiotypic groups (27). Again, evidence was presented which suggested that the antibody combining site was involved in the idiotypic cross reactions. The elegant studies of Brient and Nisonoff (28) showed that the reaction between idiotypic antibodies could be inhibited by homologous haptens by as much as 69%. Similar results were obtained by Sher and Cohn (29) who showed that phosphorylcholine inhibited the reaction between a mouse antiphosphorylcholine myeloma protein and its idiotypic antiserum.

More recent studies linking idiotypic determinants to the combining site come from the work of Weigert et al. (30). These authors studied the relationship between the idiotype and the combining site of mouse myeloma proteins with anti-α(1 → 6)-dextran activity, and anti-α(1 → 3)-dextran and antilevan specificities. The authors introduced the term *ligand-modifiable idiotype* to indicate an idiotypic reaction that is inhibited by the ligand. They stress that, although proteins demonstrating idiotypic cross-reactions generally have similar combining specificities, the converse is not true; that is, all antibodies that bind a specific ligand need not show cross-idiotypic specificity. The concepts of ligand-modifiable idiotypes and idiotypes not modifiable by ligand have become increasingly useful in immunochemical and immunogenetic studies.

Thus, while it is clear that most idiotypic determinants are related to the antibody combining site, not all can be modified by ligand. Idiotypic determinants not modifiable with ligand may represent those portions of the combining site not occupied by hapten, since it is evident from the X-ray crystallographic studies that the combining site can accommodate larger molecular weight molecules (31–33).

THE RELATIONSHIP BETWEEN THE HYPERVARIABLE REGIONS AND THE IDIOTYPIC DETERMINANTS

Two anti-γ-globulin antibodies with cross-idiotypic specificity were selected for complete amino acid sequence analysis of the variable regions of the heavy and light polypeptide chains. These two proteins (Lay and Pom) have only eight amino acid differences in their entire heavy-chain variable regions (21). Three of the interchanges occur within hypervariable regions and five in the "framework" residues. When one compares all previously studied V_HIII proteins, it is seen that these proteins differ in approximately three to five residues in their framework—a very similar number to the differences between the Pom and Lay proteins. However, as noted previously, randomly chosen proteins are dramatically different in their hypervariable regions, while these two proteins are virtually identical. These data strongly imply that the major idiotypic determinants on immunoglobulin molecules are related to the hypervariable regions.

More recent studies demonstrating the relationship between the idiotype and the hypervariable regions come from deliberately induced antibodies to the azophenylarsonate group. These studies have been done in collaboration with Tung and Nisonoff (34–36). Anti-*p*-azophenylarsonate antibodies have been isolated from individual mice and separated into two subpopulations—those bearing the cross-reacting idiotype and those devoid of the cross-reacting idiotype. Amino acid sequence analysis of the heavy chains of the idiotypically related proteins has demonstrated that the first and second hypervariable regions of the heavy chain are homogeneous, while those anti-*p*-azophenylarsonate antibodies not bearing the cross-reacting idiotype have a heterogeneous amino acid sequence in the first and second hypervariable regions. One must reserve judgment on the homogeneity of the entire V_H region since these sequences are not complete, especially since Cebra's group has noted heterogeneity in the fourth hypervariable region of guinea pig antibodies despite homogeneity in the first, second, and third hypervariable regions (37). Nonetheless, these studies suggest that antibodies selected on the basis of idiotypic specificity demonstrate a more homogeneous sequence in at least some of their hypervariable regions and provide additional evidence relating idiotypy and hypervariable regions.

THE RELATIONSHIP BETWEEN THE HYPERVARIABLE REGIONS AND THE ANTIBODY COMBINING SITE

Conclusions concerning the relationship between the hypervariable regions and the combining site come from several different areas of investigation. In light chains, affinity labels have been localized near or within the hypervariable regions by several authors, and most of the recent work on heavy chains supports this general view. Ray and Cebra (38) localized affinity labels to the first and fourth hypervariable regions of guinea pig antibodies, while Haimovich et al. (39) affinity labeled residue 54 of mouse myeloma protein 315. Thus, the general idea has emerged that the same regions of the molecule which show the highest degree of sequence variation are near or part of those exact regions of the heavy chain where affinity labels have been localized.

Cebra's group (37) has elegantly demonstrated that when guinea pigs are immunized with defined haptens and the antibodies isolated, the first three hypervariable regions are restricted in their sequence and differ between antibodies of differing specificities. This clearly relates the specificity of antibodies to the sequence in their hypervariable regions. Much of the work done on the mouse myeloma proteins, particularly antiphosphorylcholine myeloma proteins, supports this view.

Finally, the X-ray crystallographic evidence indicates that there is a clustering of five of the seven hypervariable regions near one end of the antibody molecule. Here they form a cavity, the depth of which varies as insertions or deletions are made in hypervariable regions. The third heavy chain hypervariable region clearly plays no role in the combining site, and the second hypervariable region of the light chain may not (31–33).

IS THE V_H SUBGROUP LINKED TO THE HYPERVARIABLE REGION–IDIOTYPE SPECIFICITY?

One of the important issues now revolves around whether there is a direct colinear relationship between the V region subgroup and the hypervariable region—idiotype specificity of a myeloma protein and/or an antibody. If no correlation exists between the framework and the hypervariable region sequences, then the possibility must be strongly entertained that they are encoded separately in the genome, although other explanations could deal with this finding.

Four lines of evidence suggest a direct correlation between the framework and the hypervariable regions. Limited sequence analysis of anti-γ-globulin antibodies showed that proteins belonging to certain cross-idiotypic specificity groupings had marked restrictions of the V region subgroups of their heavy and/or light chains (40). All members of one particular cross-idiotypic group contained light chains of the rare human V_KIIIB sub-subgroup. While studies have not been completed on the hypervariable regions of these proteins, if, as suspected, idiotype and hypervariable regions are related, this study demonstrates a direct relationship between framework sequences and hypervariable region sequences.

Work from Cebra's group (37) provides further support for this relationship. In their studies certain residues in the framework of the V_H region were always correlated with the sequences found in the hypervariable regions and were related to antibody specificity.

Among deliberately induced anti-*p*-azophenylarsonate antibodies (34–36) all were found to belong to a rare mouse V_H subgroup, thereby linking specificity to V_H subgroup. No anti-*p*-azophenylarsonate antibodies which did not belong to the V_HII subgroup were found in the A/J mouse despite the fact that the vast majority of the unblocked A/J mouse heavy chains are of the V_HIII subgroup. Finally, the studies of Barstad et al. (12) in the mouse myeloma system clearly indicate a direct relationship between the genealogy of the V_H framework and the idiotype and/or specificity of the myeloma proteins. Since many of these proteins have been completely sequenced, these data provide the strongest evidence for the direct relationship between framework and hypervariable regions.

There is, however, a considerable amount of evidence against this point of view. Kindt et al. (41) demonstrated an idiotypic cross-reaction between two different rabbit antibodies to the streptococcal Group C carbohydrate isolated from the same rabbit serum. One antibody was of allotype a3, while the other was y0. Consistent with the allotypic difference, preliminary amino acid sequence data on the heavy chains have shown that they differ at the amino terminus (42). If a3 and y0 molecules are as different as H chains belonging to different V region subgroups, they would have over 30% amino acid sequence differences in the framework portions of the V region; yet they are idiotypically identical. This suggests that there is not an obligatory association between allotypic and idiotypic markers.

Recent data from Haber's group, which has been reported at this symposium, indicates that there is no direct relationship between the light chain subgroup, the light chain hypervariable regions, and the specificities of homogeneous rabbit antibodies (43). Finally, Hopper et al. (44) have evidence that two proteins with shared idiotypic determinants were produced by the same cell and have different light chain types (IgM lambda and IgG kappa). Careful idiotypic analysis of the heavy chains of these proteins indicates that they have identical idiotypic determinants. We have collaborated with Hopper on a study of these proteins, and one of the V_H sequences shown earlier (BroM) is that of the IgM heavy chain from this pair. We are presently studying the IgG heavy chain. The first 30 residues of this protein, which shares complete idiotypic identity with the IgM protein, has seventeen amino acid differences from its IgM counterpart. Complete amino acid sequence analysis on this protein should be enlightening.

Thus, while much of the experimental evidence seems to indicate that there is a direct relationship between the framework and the hypervariable regions, there are certain puzzling inconsistencies which suggest that there may not be such a direct relationship. If a direct relationship does not exist, it could be accommodated within the two general theories of antibody diversity which have been proposed to date (germline versus somatic mutation). However, recently an alternative to this has been proposed which suggests that there may be some form of gene interaction accounting for the diversity in immunoglobulin chains (45–46).

THE RELATIONSHIP BETWEEN THE V_H SUBGROUP, THE IDIOTYPE, AND THE MOUSE V_H MAP

In recent years a number of idiotypic systems have been developed in inbred mouse strains which have allowed the construction of a tentative genetic map of the V_H region of the mouse (Fig. 6). What has emerged from

FIG. 6. The tentative mouse V_H map. On the basis of several recombinants together with those observed between IgG-1 and 3, the chromosomal order of the genes is tentatively established.

A5A S117	ARS J558 NP NBrP	T15	Ig-1 Ig-2 Ig-3 Ig-4	PRE

these studies is that seven specific idiotypes tentatively placed in the map occur in three "clusters" of genes. Those coding for the A5A and S117 idiotypes, which are antibodies to streptococcal A polysaccharide, are at a considerable distance from the CH (Ig) locus. At an intermediate distance there are at least four idiotypic markers, those antibodies against dextran (the so-called J558 idiotype) (47), the anti-Ars idiotype (48), and the NIP idiotypes (49). Closer to the CH loci is the T15 idiotype representing proteins with antiphosphorylcholine activity (50). Since in each of these instances the idiotype has been linked to the CH locus; and since there is no linkage between light chains and the CH locus, it has been assumed that mapping the idiotypes represented a map of the V_H locus of murine immunoglobulin heavy chain genes. However, it must be borne in mind that, with the exception of the T15 idiotype, each of these markers may represent the mapping of only a portion of the V_H genome.

A study was undertaken of the *N*-terminal amino acid sequence of the heavy chains of many of these proteins in order to ask the question as to whether there was a correlation between "framework" residues (V_H subgroup) and the genetic map. Such information should be informative concerning the origin and evolution of heavy chain V_H regions. The results of our preliminary study are presented in Fig. 7 which allows for the first comparison between the V_H subgroup of proteins represented in all three gene clusters (51). The data on the J558 and T15 myeloma proteins have been taken from Barstad et al. (12). The striking finding is that there appears to be no collinearity between the sequence and the genetic map. On the one hand, in the gene cluster represented by the A5A and S117 proteins, antibodies exist which are clearly related to the V_HIII subgroup. On the other hand, the T15 idiotype, which maps closest to the CH locus and is clearly

A5A E V K L E E S G G G V_HIIIC	ARS E V Q L Q Q S G A G V_HII	T15 E V K L V E S G G G V_HIIIB
S117 E V K L L E S G G G V_HIIIA	J558 E V Q L Q E S G P E V_HII	

FIG. 7. Amino acid sequence analysis and subgroup designation of the proteins previously placed in the mouse V_H map for which sequence data are available.

separated from S117 and A5A by the Ars and J558 idiotype, also includes proteins of the V_HIII subgroup. In the central regions, the genes coding for the Ars and J558 idiotype had very similar V_H subgroup structures (V_HII) and differ significantly from the flanking S117, A5A and TEPC-15 heavy chains.

Whatever mechanisms are involved in the generation of antibody diversity, it is anticipated that a continued correlation between the V_H genetic map and the sequences of the proteins encoded by that region of the murine chromosome will continue to provide insights and put constraints on our present conceptualizations concerning the generation of antibody diversity. These data present evidence which suggest that a simple collinear relationship between the two does not exist.

ACKNOWLEDGMENTS

The author wishes to express his gratitude to his many associates and collaborators over the past seven years, in particular, Drs. Henry Kunkel, J. Michael Kehoe, and more recently Al Nisonoff and Klaus Eichmann. The work in the author's laboratory would not have been possible without the dedicated and skilled technical assistance of Vicki Ecker, Bonnie Gerber, Donna Atherton, Ellen Bogner, Suzanne Astor, Jackie Rosenthal, and Margaret Cason. The skilled and patient secretarial assistance of Betty Bechtel is graciously appreciated. This work was done under the auspice of grants from United States Public Health Service, AI-12127, AI-12796, and the National Science Foundation. Their generous support over the past seven years is acknowledged.

REFERENCES

1. Williams, R. C., Kunkel, H. G., and Capra, J. D. (1968): Antigenic specificities related to the cold agglutinin activity of IgM globulins. *Science,* 161:379.
2. Capra, J. D. (1971): A hypervariable region of human immunoglobulin heavy chains. *Nature,* 230:61.
3. Capra, J. D., and Kehoe, J. M. (1974): Hypervariable regions, idiotypes and the antibody combining site. *Adv. Immunol.,* 20:1.
4. Niall, H. D., and Edman, P. (1967): Two structurally distinct classes of kappa-chains in human immunoglobulins. *Nature,* 216:262.
5. Gray, W. R., Dreyer, W. J., and Hood, L. (1967): Mechanism of antibody synthesis: Size differences between mouse kappa chains. *Science,* 155:467.
6. Milstein, C. (1967): Linked groups of residues in immunoglobulin κ chains. *Nature,* 216:330.
7. Kohler, J., Shimizu, A., Paul, C., Moore, V., and Putnam, F. W. (1970): Three variable region pools common to IgM, IgG and IgA immunoglobulins. *Nature,* 277:1318.
8. Capra, J. D., and Kehoe, J. M. (1975): Distribution of heavy and light chain variable region subgroups among human IgA immunoglobulins. *J. Immunol.,* 114:678.
9. Capra, J. D., and Penn, G. Unpublished observations.
10. Capra, J. D. Unpublished observations.
11. Barstad, P., Rudikoff, S., Potter, M., Cohn, M., Konigsberg, W., and Hood, L. (1967):

Immunoglobulin structure: Amino terminal sequences of mouse myeloma proteins that bind phosphorylcholine. *Science,* 183:962.
12. Barstad, P., Weigert, M., Cohn, M., and Hood, L. (1974): Mouse immunoglobulin heavy chains are coded by multiple germline variable genes. *Proc. Natl. Acad. Sci. USA,* 10:4096.
13. Weigert, M. Personal communication.
14. Edelman, G. M., Cunningham, B. A., Gall, W. E., Gottieb, P. D., Rutishauser, U., and Waxdal, M. (1969): The covalent structure of an entire IgG immunoglobulin molecule. *Proc. Natl. Acad. Sci. USA,* 63:78.
15. Cunningham, B. A., Pflumm, M. N., Rutishauser, U., and Edelman, G. M. (1969): Subgroups of amino acid sequences in the variable regions of immunoglobulin heavy chains. *Proc. Natl. Acad. Sci. USA,* 64:997.
16. Wikler, M., Kohler, H., Shinoda, T., and Putnam, F. W. (1969): Homology of mu and gamma heavy chains of human immunoglobulins. *Science,* 163:75.
17. Press, E. M., and Hogg, N. M. (1970): The amino acid sequences of the Fd fragments of two human gamma-1 heavy chains. *Biochem. J.,* 117:641.
18. Ponstingl, H., Schwartz, J., Reichel, W., and Hilschmann, N. (1970): Die Primarstruktur eines Monoklonalen Gamma-1-Immunoglobulins (Myelomprotein Nie). *Z. Physiol. Chem.,* 351:1591.
19. Watanabe, S., Barnikol, H. U., Horn, J., Bertram, J., and Hilschmann, N. (1973): In lab. *Hoppe Seylers Z. Physiol. Chem.,* 354:1505.
20. Capra, J. D., and Kehoe, J. M. (1974): Variable region sequences of five human immunoglobulin heavy chains of the V_HIII subgroup: Definitive identification of four heavy chain hypervariable regions. *Proc. Natl. Acad. Sci. USA,* 71:845.
21. Capra, J. D., and Kehoe, J. M. (1974): Structure of antibodies with shared idiotypy: The complete sequence of the heavy chain variable regions of two IgM anti-gammaglobulins. *Proc. Natl. Acad. Sci. USA,* 71:4032.
22. Capra, J. D. Unpublished observations.
23. Capra, J. D., and Hopper, J. E. *Immunochemistry* (*In press.*)
24. Wu, T. T., and Kabat, E. A. (1970): An analysis of the sequences of the variable regions of Bence-Jones proteins and myeloma light chains and their implications for antibody complementarity. *J. Exp. Med.,* 132:211.
25. Franêk, F. (1969): *Symposium on Developmental Aspects of Antibody Formation and Structure,* p. 311. Prague.
26. Capra, J. D. (1976): The implications of phylogenetically associated residues and idiotypes on theories of antibody diversity. Academic Press, New York. (In press.)
27. Kunkel, H. G., Agnello, V., Joslin, F. G., Winchester, R. J., and Capra, J. D. (1973): Cross-idiotypic specificity among monoclonal IgM proteins with anti-γ-globulin activity. *J. Exp. Med.,* 137:331.
28. Brient, B. W., and Nisonoff, A. (1970): Quantitative investigations of idiotypic antibodies. IV. Inhibition by specific haptens of the reaction of anti-hapten antibody with its anti-idiotypic antibody. *J. Exp. Med.,* 132:951.
29. Sher, A., and Cohn, M. (1972): Effect of haptens on the reaction of anti-idiotype antibody with a mouse anti-phosphorylcholine plasmacytoma protein. *J. Immunol.,* 109:176.
30. Weigert, M., Raschke, W. C., Carson, D., and Cohn, M. (1974): Immunochemical analysis of the idiotypes of mouse myeloma proteins with specificity for levan or dextran. *J. Exp. Med.,* 139:137.
31. Schiffer, M., Girling, R. L., Ely, K. R., and Edmundson, A. B. (1973): Structure of a λ-type Bence-Jones protein at 3.5-Å resolution. *Biochemistry,* 12:4620.
32. Segal, D. M., Padlan, E. A., Cohen, G. H., Rudikoff, S., Potter, M., and Davies, D. R. (1974): The three-dimensional structure of a phosphorylcholine-binding mouse immunoglobulin Fab and the nature of the antigen binding site. *Proc. Natl. Acad. Sci. USA,* 71:4298.
33. Poljak, R. J., Amzel, L. M., Chen, B. L., Phizackerley, R. P., and Saul, F. (1974): The three-dimensional structure of the Fab' fragment of a human myeloma immunoglobulin at 2-0-Å resolution. *Proc. Natl. Acad. Sci. USA,* 71:3440.
34. Nisonoff, A., Tung, A. S., and Capra, J. D. (1974): Factors determining immune responsiveness at the level of B-cells. Second International Congress of Immunology, Brighton, England, *Prog. Immunol.,* II, 2:17.

35. Capra, J. D., Tung, A. S., and Nisonoff, A. (1974): Structural studies on induced antibodies with defined idiotypic specificities. I. The heavy chains of anti-*p*-azophenylarsonate antibodies from A/J mice bearing a cross-reactive idiotype. *J. Immunol.,* 114:1548.
36. Capra, J. D., Tung, A. S., and Nisonoff, A. (1975): Structural studies on induced antibodies with defined idiotypic specificities. II. The light chains of anti-*p*-azophenylarsonate antibodies from A/J mice bearing a cross-reactive idiotype. *J. Immunol.,* 115:414.
37. Cebra, J. J., Koo, P. H., and Ray, A. (1975): Specificity of antibodies: Primary structural basis of hapten binding. *Science,* 186:263.
38. Ray, A., and Cebra, J. J. (1972): Localization of affinity-labeled residues in the primary structure of anti-dinitrophenyl antibody raised in strain 13 guinea pigs. *Biochemistry,* 11:3647.
39. Haimovich, J., Eisen, H. N., Hurwitz, E., and Givol, D. (1972): Localization of affinity—labeled residues on the heavy and light chain of two myeloma proteins with anti-hapten activity. *Biochemistry,* 11:2389.
40. Kunkel, H. G., Winchester, R. J., Joslin, F. G., and Capra, J. D. (1974): Similarities in the light chains of anti-gammaglobulins showing cross-idiotypic specificities. *J. Exp. Med.,* 139:128.
41. Kindt, T. J., Seide, R. K., Bokisch, V. A., and Krause, R. M. (1973): Detection of idiotypic cross-reactions among streptococcal antisera from related rabbits. *J. Exp. Med.,* 138:522.
42. Kindt, T. J., Klapper, D. G., and Waterfield, M. D. (1973): An idiotypic cross-reaction between allotype a3 and allotype a negative rabbit antibodies to streptococcal carbohydrates. *J. Exp. Med.,* 137:636.
43. Haber, E. This volume.
44. Hopper, J. E. (1975): Comparative studies on monotypic IgM lambda and IgG kappa from an individual patient. I. Evidence for shared V_H idiotypic determinants. *J. Immunol.,* 115:1101.
45. Capra, J. D., and Kehoe, J. M. (1974): Antibody diversity. *Scand. J. Immunol.,* 3:1.
46. Capra, J. D., and Kindt, T. J. (1975): Antibody diversity: Can more than one gene encode each variable region? *Immunogenetics,* 1:417.
47. Lundblad, S., Stellar, R., Kabat, E. A., Hirst, J. W., Weigert, M., and Cohn, M. (1972): Immunochemical studies on mouse myeloma proteins with specificity for dextran or for levan. *Immunochemistry,* 9:535.
48. Kuettner, M. G., Wang, A. L., and Nisonoff, A. (1972): Quantitative investigations of idiotypic antibodies. VI. Idiotypic specificity as a potential genetic marker for the variable regions of mouse immunoglobulin polypeptide chains. *J. Exp. Med.,* 135:579.
49. Imanishi, T., and Mäkelä, O. (1974): Inheritance of antibody specificity. I. Anti-(4-hydroxy-3-nitrophenyl)acetyl of mouse primary response. *J. Exp. Med.,* 140:1498.
50. Lieberman, R., Potter, M., Mushinski, E. B., Humphrey, W., and Rudikoff, S. (1974): Genetics of a new IgV_H (T15 Idiotype) marker in the mouse regulating natural antibody to phosphorylcholine. *J. Exp. Med.,* 139:983.
51. Capra, J. D., and Eichman, K. Unpublished observation.

The Three-Dimensional Structure of the Antibody Combining Site

Antibodies in Human Diagnosis and Therapy,
edited by Edgar Haber and Richard M. Krause.
Raven Press, New York © 1977.

Three-Dimensional Structure of the Antibody-Combining Site: Introduction

Elvin A. Kabat

Departments of Microbiology, Human Genetics and Development, and Neurology, Columbia University, The Neurological Institute, Presbyterian Hospital, New York, New York 10032; and the National Cancer Institute, National Institutes of Health, Bethesda, Maryland 20014

X-Ray crystallographic studies at high resolution (1–6; *cf.* 7) on the variable regions of the light and heavy chains of immunoglobulins, as well as affinity-labeling studies (see 8), have confirmed the prediction from a statistical analysis of variability (9,10) that the three hypervariable segments of the light chain (10–12) and the three corresponding hypervariable segments of the heavy chain (13,14) are complementarity-determining and form the antibody-combining site while a fourth segment (14,15) is remote from the site. The structure of each variable region is made up of two twisted antiparallel β-pleated sheets, one of three strands and one of four strands forming a sandwich-type structure with an additional loop composed largely of hypervariable residues in between. The three complementarity-determining segments are at one end of the domain, and the two domains are so positioned that they intersect at an angle of about 120°. One may thus distinguish two types of segments in each domain, those involved in three-dimensional structure and those involved in site complementarity. (Information about the detailed structures of the site is given elsewhere in this volume.) I wish in this introduction to pose certain questions about the directions in which studies should go in order best to elucidate further the structural basis for antibody specificity and diversity.

Since it would be manifestly impossible to make detailed crystallographic and sequence studies on homogeneous antibodies or myeloma antibodies of all specificities, one must make certain choices to determine how effectively we can use any information gained about the structure of a given site to predict structures of other kinds of sites.

I should like briefly to mention two relevant types of study from my laboratories. The first (16,17) involves a study of two mouse myelomas with α1→6 anti-dextran specificity, W3129 and QUPC52. In studies on the sizes of their combining sites, we found that the former was complementary to a terminal nonreducing chain of five $\alpha 1 \rightarrow 6$ linked glucoses whereas the latter was complementary to six $\alpha 1 \rightarrow 6$ glucoses (16). In all antibody systems previously studied, a larger site always had had a higher association constant (K^a) than a smaller site. When the K^a of each of these proteins was measured (17), we were astonished to find that W3129 with the smaller site had a K^a of 1×10^5 while QUPC52 with the larger site had a K^a of 8×10^3. Equilibrium dialysis displacement measurements of the contributions of each residue to the site showed that with W3129, methyl α-D-glucoside contributed 55% and the disaccharide isomaltose 60%; in contrast, with QUPC52 they contributed less than 5% of the binding energy and the trisaccharide isomaltotriose was required for good binding and accounted for 60 to 70% of the total. These findings indicate that W3129 has a site which is directed toward a terminal nonreducing chain in which at least the terminal one or two glucoses are firmly held in a cavity with interactions from all sides while QUPC52 has a site which is a shallow groove specific for internal stretches of $\alpha 1 \rightarrow 6$ linked glucoses. These findings were strongly supported when a synthetic linear polymer with about 200 $\alpha 1 \rightarrow 6$ linked units prepared by Ruckel and Schuerch (18,19) was tested. This polymer reacted with W3129 to inhibit precipitation on a molar basis equivalent to the pentasaccharide but it actually precipitated with QUPC52.

These findings indicate substantial differences in the shapes of these two sites. The human anti-dextrans we had previously studied were found to be mixtures of both types of molecules. These shape differences may well represent the extremes of structural variation among antibody sites. Since they are specific for the same kind of matter, if one knew their sequences and three-dimensional structures, one would learn a great deal about antibody specificity and perhaps be able to predict the shapes of sites directed toward other kinds of determinants.

Thus antibody sites specific for the linear SIII molecule (20,21) may be like QUPC52 while those directed toward the streptococcal C carbohydrate (22) with its branches may be like W3129 or perhaps intermediate between the two.

A second approach which T. T. Wu (Northwestern University), Howard Bilofsky (Bolt Beranek and Newman, Inc.), and I have begun is to attempt to analyze the complementarity-determining segments of the light and heavy chains, attempting to correlate pairs of residues in these segments and looking for unusual frequencies of occurrence as compared with those calculated from random probabilities (29). This is being done in the PROPHET system into which we have compiled all of the published light- and heavy-chain variable-region sequences. One computes for any two residues in the complementarity-determining segments the probability that two amino acids will be associated by chance as compared with the observed frequency. Such associations may suggest patterns of residues with a given specificity. Thus residues selected from all the heavy-chain sequences, considering them as one population, showed Phe 32, Tyr 33, and Glu 35 to occur 3.5 σ more frequently than expected in 6 proteins and these all turned out to be phosphorylcholine-binding mouse myelomas. All of these also have Met at position 34, but this was not selected as unique because Met occurs in 36 other myeloma proteins at that position. Thus one would postulate that Met is not a contacting residue but that it plays essentially a structural role whereas Phe, Tyr, and Glu either are complementarity-determining residues or may have a conformational influence on a complementarity-determining residue. In the phosphorylcholine-binding myeloma McPC603, Tyr 33 has been found to be a contacting residue, as is Glu 35, while Met 34 is not (4,7). Phe 32 might have a conformational influence on Tyr 33.

Another finding (23) from our tabulation of sequences of variable regions is that Bence Jones protein Vil (24) (a λII) and Mcg (25) (a λV), despite 21 differences in the variable region, are identical in all 14 positions of their first complementarity-determining region. Accepting the inference that different subgroups involve different genes, the finding of identical hypervariable regions in a λII and a λV light chain is most readily explained by insertion (10,26,27) into DNA of the nucleotides responsible for the complementarity-determining segments. Since the Mcg light-chain dimer is known to bind many low molecular weight substances, it would be of great interest to determine if Vil has similar binding properties and cross idiotypic specificities and also to study Vil–Mcg hybrid molecules.

I will not dwell on the many other relationships of evolutionary and structural significance that can be made by correlating sequences and structure in the constant regions (see, for example, 28). I believe that an understanding of the problems of antibody complementarity and of generation of diversity will require many more sequences and a number of additional X-ray crystallographic studies on myeloma proteins with antibody activity and on antibodies in which the dimensions of the antigenic determinants to which they are complementary are known.

ACKNOWLEDGMENTS

The studies with the PROPHET system are supported by the National Cancer Institute, National Institute of Allergy and Infectious Diseases, National Institute of Arthritis, Metabolic and Digestive Diseases, National Institute of General Medical Sciences, Division of Research Resources (contract no. NO1-RR-4-2147) National Institutes of Health, and also by a grant from the National Science Foundation (BMS-72-02219-A02, A03) to the author.

REFERENCES

1. Schiffer, M., Girling, R. L., Ely, K. R., and Edmundson, A. B. (1973): Structure of a λ-type Bence Jones protein at 3.5 Å resolution. *Biochemistry,* 12:4620–4631.

2. Poljak, R. J., Amzel, L. M., Avey, H. P., Chen, B. L., Phizackerly, R. P., and Saul, F. (1973): Three dimensional structure of the Fab′ fragment of a human immunoglobulin at 2.8 Å resolution. *Proc. Natl. Acad. Sci. USA,* 70:3305–3310.
3. Poljak, R. J., Amzel, L. M., Chen, B. L., Phizackerly, R. P., and Saul, F. (1974): The three dimensional structure of the Fab′ fragment of a human myeloma immunoglobulin at 2.0Å resolution. *Proc. Natl. Acad. Sci. USA,* 71:3440–3444.
4. Segal, D. M., Padlan, E. A., Cohen, G. H., Rudikoff, S., Potter, M., and Davies, D. R. (1974): The three dimensional structure of a phosphorylcholine-binding mouse immunoglobulin Fab and the nature of the antigen-binding site. *Proc. Natl. Acad. Sci. USA,* 71:4298–4302.
5. Epp, O., Colman, P., Fehlhammer, H., Bode, W., Schiffer, M., Huber, R., and Palm, W. (1974): Crystal and molecular structure of a dimer composed of the variable portions of the Bence Jones protein REI. *Eur. J. Biochem.,* 45:513–524.
6. Edmundson, A. B., Ely, K. R., Girling, R. L., Abola, E. E., Schiffer, M., Westholm, F. A., Fausch, M. D., and Deutsch, H. F. (1974): Binding of 2,4-dinitrophenyl compounds and other small molecules to a crystalline λ-type Bence Jones dimer. *Biochemistry,* 13:3816–3827.
7. Davies, D. R., Padlan, E. A., and Segal, D. M. (1975): Three-dimensional structure of immunoglobulins. *Annu. Rev. Biochem.,* 44:639–667.
8. Givol, D. (1974): Affinity labelling and topology of the antibody combining site. *Essays Biochem.,* 10:1–31.
9. Kabat, E. A. (1970): Heterogeneity and structure of antibody combining sites. Landsteiner Centennial Dec. 5–7, 1968. *Ann. NY Acad. Sci.,* 169:43–54.
10. Wu, T. T., and Kabat, E. A. (1970): An analysis of the sequences of the variable regions of Bence Jones proteins and myeloma light chains and their implications for antibody complementarity. *J. Exp. Med.,* 132:211–250.
11. Franêk, F. (1970): The character of variable sequences in immunoglobulins and its evolutionary origin. In: *Developmental Aspects of Antibody Formation and Structure,* Vol. I, edited by J. Šterzl and I. Řiha, pp. 311–313. Academia Publishing House of the Czechoslovak Academy of Sciences, Prague, Czechoslovakia.
12. Weigert, M. G., Cesari, I. M., Yonkovich, S. J., and Cohn, M. (1970): Variability in the λ-light chain sequences of mouse antibody. *Nature,* 228:1045–1047.
13. Kabat, E. A., and Wu, T. T. (1971): Attempts to locate complementarity-determining residues in the variable positions of light and heavy chains. *Ann. NY Acad. Sci.,* 190:382–393.
14. Kehoe, J. M., and Capra, J. D. (1971): Localization of two additional hypervariable regions in immunoglobulin heavy chains. *Proc. Natl. Acad. Sci. USA,* 68:2019–2021.
15. Capra, J. D., and Kehoe, J. M. (1974): Variable region sequences of five human immunoglobulin heavy chains of the V_HIII subgroup: Definitive identification of four heavy chain hypervariable regions. *Proc. Natl. Acad. Sci. USA,* 71:845–848.
16. Cisar, J. O., Kabat, E. A., Liao, J., and Potter, M. (1974): Immunochemical studies on mouse myeloma proteins reactive with dextrans or with fructosans and on human antilevans. *J. Exp. Med.,* 139:159–179.
17. Cisar, J. Kabat, E. A., Dorner, M., and Liao, J. (1975): Binding properties of immunoglobulin combining sites specific for terminal or non-terminal antigenic determinants in dextran. *J. Exp. Med.,* 142:435–459.
18. Ruckel, E. R., and Schuerch, C. (1966): Chemical synthesis of a stereoregular linear polysaccharide. *J. Am. Chem. Soc.,* 88:2605.
19. Ruckel, E. R. and Schuerch, C. (1967): Chemical synthesis of a dextran model, poly α (1→6)-D-glucan. *Biopolymers,* 5:515–523.
20. Margolies, M. N., Cannon III, L. E., Strosberg, A. D., and Haber, E. (1975): Diversity of light chain variable region sequences among rabbit antibodies elicited by the same antigens. *Proc. Natl. Acad. Sci. USA,* 72:2180–2184.
21. Jaton, J.-C. (1975): Comparison of the amino acid sequences of the variable domains of two homogeneous rabbit antibodies to type III pneumococcal polysaccharide. *Biochem. J.,* 147:235–247.
22. Chen, K. C. S., Kindt, T. J., and Krause, R. M. (1974): Amino acid sequence of an allotype b4 light chain from a rabbit antibody to streptococcal carbohydrate. *Proc. Natl. Acad. Sci. USA,* 71:1995–1998.

23. Wu, T. T., Kabat, E. A., and Bilofsky, H. (1975): Similarities among hypervariable segments of immunoglobulin chains. *Proc. Natl. Acad. Sci. USA*, 72:5107–5110.
24. Ponstingl, H., and Hilschmann, N. (1971): Die vollstandige Primärstruktur einer monoklonalen Immunglobulin-L-Kette vom λ-Typ, Subgruppe II (Bence Jones protein Vil) *Hoppe-Seyler's Z. Physiol. Chem.*, 352:859–877.
25. Fett, J. W., and Deutsch, H. F. (1974): Primary structure of the Mcg λ chain. *Biochemistry*, 13-4102-4114.
26. Kindt, T. D., and Capra, J. D. (1975): Antibody diversity: Can more than one gene encode each variable region. *Immunogenetics*, 1:417–427.
27. Capra, J. D., and Kehoe, J. M. (1975): Hypervariable regions, idiotypy and the antibody combining site. *Adv. Immunol.*, 20.1–40.
28. Kabat, E. A., Padlan, E. A., and Davies, D. R. (1975): Evolutionary and structural influences on light chain constant (C_L) region of human and mouse immunoglobulins. *Proc. Natl. Acad. Sci. USA*, 7:2785–2788.
29. Kabat, E. A., Wu, T. T., and Bilofsky, H. (1976): Attempts to locate residues in complementarity determining regions which make contact with antigen. *Proc. Nat. Acad. Sci. USA*, 73:617–619.

Antibodies in Human Diagnosis and Therapy,
edited by Edgar Haber and Richard M. Krause.
Raven Press, New York © 1977.

Three-Dimensional Structure of the Fab Fragment of a Human Myeloma Immunoglobulin

Roberto J. Poljak, L. Mario Amzel, Betty L. Chen, R. Paul Phizackerley, and Frederick Saul

Department of Biophysics, The Johns Hopkins University School of Medicine, Baltimore, Maryland 21205

The three-dimensional structure of the Fab fragment of the human myeloma protein IgG New has been determined by single crystal X-ray diffraction analyses to a resolution of 6 Å (1), 2.8 Å (2), and 2.0 Å (3). A number of physicochemical and immunological properties of immunoglobulins and the patterns of variations in their amino acid sequences have been correlated with the structural models obtained in these studies. The high-resolution studies (2,3) provide a detailed structural characterization of the antigen combining region.

THE IMMUNOGLOBULIN-FOLD

The overall dimensions of the Fab fragment, as determined by our X-ray diffraction studies, are 80 × 50 × 40 Å. Fab consists of two globular domains, V and C_1, each containing two homology subunits, V_H and V_L, and C_H1 and C_L, respectively. These homology subunits share a basic pattern of polypeptide chain folding, which we have called the immunoglobulin-fold (2). A characteristic of this polypeptide chain folding is the presence, in each homology subunit, of two irregular, roughly parallel β-pleated sheets surrounding a tightly packed interior of hydrophobic amino acid side chains. The invariant intrachain disulfide bond, at the center of this hydrophobic interior, links the two β sheets in a direction approximately perpendicular to their planes. The immunoglobulin-fold describes the overall three-dimensional structure of the C_H1 and C_L homology subunits. In V_L and V_H an additional loop of polypeptide chain is added to this basic structure (2). No major helical regions are present in any of the subunits.

In each structural domain, V and C_1, the homology subunits, are related by an approximate local twofold axis of symmetry. In the L chain the major axes of the C_L and V_L subunits make an angle greater than 90° (100–110°), whereas in the Fd chain the corresponding axes of C_H1 and V_H make an angle smaller than 90° (80–85°) (2).

The high-resolution model of Fab New has been used to analyze different arrangements of disulfide bonds which occur in IgM, IgA, and IgG (2). On the basis of this analysis, we postulated (2) that the immunoglobulin-fold is common to all classes of immunoglobulins of different animal species. This postulate has been supported by the results of the crystallographic studies of Davies and colleagues (4) on the Fab fragment from a mouse IgA, and by the results of Huber and colleagues (5), who studied a human V_κ dimer.

Analysis of the three-dimensional structure of the homology subunits of Fab New indicates the presence of stereochemical constraints which restrict the possibility of amino acid sequence variations in many parts of the molecule (3,5). These constraints frequently correlate with the pattern of variability observed by comparison of the amino acid sequences of different immunoglobulins. For example, the tightly packed, hydrophobic amino acid side chains, which occur between the two β-pleated sheets in V_H and V_L, are generally invariant or semiinvariant, replaced in the latter case by similar hydrophobic residues. Another example is that of Gly residues, which occur at hairpin bends of the polypeptide chain, preceded by semiinvariant Pro, Ser, or Thr residues. Also, the constant sequence Gly-X-Gly at positions 100–102 in V_L and the homologous sequence Gly-X-Gly at positions 108–110 in V_H, is explained by the close intrachain contacts of these Gly residues with other residues of V_L and V_H. As a last example, positions 35, 37, 42, 43, 86, and 99 in human V_L and positions 37, 39, 43, 45, 47, 95, and 108 in V_H, which provide the closer interactions between V_L and V_H, are invariant or nearly invariant. This observation explains the frequently observed reassociation of heterologous light (L) and heavy (H) immunoglobulin chains (6).

THE COMBINING SITE

The regions of hypervariable sequences of V_L and V_H occur at one end of Fab, fully exposed to solvent. The structural model can be described as one in which the exposed hypervariable positions occur at adjacent bends of tightly packed, hydrogen-bonded strands of polypeptide chain. Different immunoglobulins will each present a unique pattern of amino acid side chains attached to the compact, constant structural framework of the V_L and V_H subunits. Since the hypervariable segments of both H and L chains occur at one end of Fab, we concluded that this region of the molecule contains the antigen combining site (2). Thus, the combining site of IgG New consists of a relatively shallow groove or pocket delineated by the main polypeptide chain and side chains which occur at positions 27a–30 and 90–95 in V_L, 30–34, 50–58, and 99–106 in V_H. A comparison of the amino acid sequences at these positions in different H and L chains indicates that the structure of the combining site can vary in chemical character and in dimensions due to amino acid substitutions and deletions (or insertions) which occur at these regions. The antigen complementarity region around positions 100–105 in V_H is the most variable in length, and will consequently be a major determinant of the conformational variability of the combining site. The first and third hypervariable regions of L chains also exhibit variation in number of amino acids, and thus they also contribute significant variability to the conformation of the combining site. A wide range of recognition specificities can be incorporated in this design without altering the constant immunoglobulin-fold. In addition, assorted pairing or recombination of H and L chains could generate a large number of specificities without the need to introduce new elements into the set of V_H and V_L structures.

Crystallographic studies of ligand–Fab complexes (4,7) have added to the characterization of antibody combining sites. The ligand vit. K_1OH (a hydroxyl derivative of vitamin K_1) was found to bind to IgG New with an affinity constant of 1.7×10^5 L/M (8). An X-ray diffraction study of the vit. K_1OH–Fab complex to a resolution of 3.5 Å showed that vit. K_1OH binds to the combining site of Fab′ (New). The quinone moiety of vit. K_1OH makes close contact with the phenolic ring of L chain Tyr 90, with the backbone and side chain of H chain residue 101 and with the backbone of L chain residues 29 and 30. The phytyl tail of vit. K_1OH is bent at positions 3′ and 4′, making close contacts with L chain residues Gly 29 and Asn 30. From that point it proceeds in an extended form, making close contacts with L chain residues 93 and 94 and H chain residue 101. Toward its end it interacts with the side chain of H chain Trp 47 and with H chain residues 50 and 58. The extensive interaction of vit. K_1OH with the active site of Fab′ (New) gives a general indication of the type of contacts that could be involved in the formation of antigen–antibody complexes.

Using the combining site of Fab New as a basic structural model, a striking correlation can be obtained between amino acid sequence (9) and function in the antidinitrophenyl (DNP) mouse myeloma protein IgA MOPC 315 (3). The L (λ) chains of IgG New and MOPC 315 are highly homologous and contain the same number of amino acid residues in the first and third hypervariable regions. Also, the antigen complementarity regions of V_H New and V_H MOPC 315 are very similar in number of amino acid residues. Based on this similarity we were able to fit the MOPC 315 sequence to the basic V_L and V_H structures obtained in the 2.0 Å resolution Fourier map of Fab New. The model of the MOPC 315 combining site that emerges from this comparison includes a crevice similar to that shown in Fig. 1, in which L chain Tyr 34 forms the "upper" limit, H chain Trp 47 and Phe 50 form the "lower" limit, L chain Phe 99, H chain Tyr 104 and Phe 34 contribute to the "sides," and L chain Trp 98 and Phe 103 and H chain Phe 105 form the 6–10-Å deep "floor." The high density of adjacent aromatic side chains that line this crevice, at the center of the active region, is striking and correlates

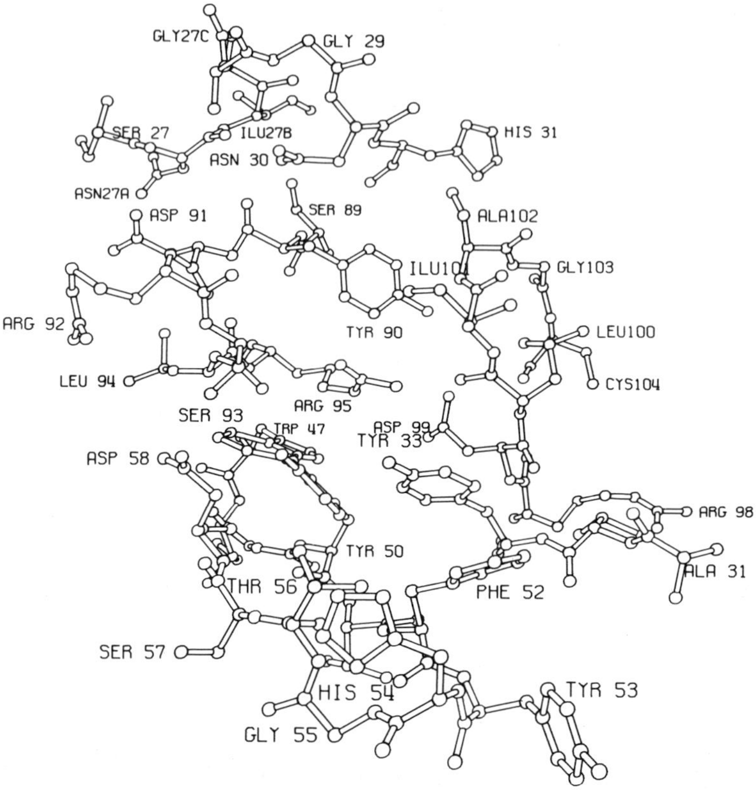

FIG. 1. A view of the combining site of Fab New. [Reproduced from Poljak et al. (3).]

with the observed specificity of MOPC IgA immunoglobulin for DNP and other haptens that include benzene and naphthalene aromatic rings. The relatively shallow depth of the active center is in agreement with the electron microscopy study of a complex between MOPC 315 IgA and the bifunctional hapten bis (DNP-β-alanyl)diaminosuccinate (10) in which the DNP groups that join the Fab arms of different IgA molecules end to end are only 15 Å apart.

HAPTEN BINDING AND COMBINING SUBSITES

In parallel with current work on the refinement of the atomic coordinates of Fab New we have recently reexamined its combining site. This analysis and evidence obtained by others to be discussed below, leads us to propose a schematic subdivision of the combining site into two main subsites.

Referring to Fig. 1, the first subsite is located in the region surrounding the L chain residue Tyr 90 (upper middle in Fig. 1). Since this subsite is surrounded by hypervariable regions from both the L and H chains we will refer to it as the "L–H subsite." It is bounded by residues in the first L chain hypervariable region (Ala 28, Gly 29, Asn 30, and His 31), by residues in the third L chain hypervariable region (Tyr 90, Asp 91, Arg 92, Ser 93, Leu 94, and Arg 95) and by H chain residues 99–103 (Asp 99, Leu 100, Ile 101, Ala 102, and Gly 103). Binding of menadione, uridine, orceine, and other ligands including the naphtoquinone moiety of vit. K_1OH occurs at this subsite of Fab New (7).

The second subsite is located in the region surrounding the H chain residue Phe 52 (lower right in Fig. 1). This subsite is formed by residues from three H chain hypervariable regions and with the possible exception of L chain residue 95 is not influenced by the L chain. We therefore refer to it as the "H subsite." Its boundaries are determined by H chain residues 31–35 in the first hypervariable region (the only remaining residues of V_H New to be sequenced), by H chain residues in the second hypervariable segment (Tyr 50, Val 51, Phe 52, Tyr 53, His 54, Gly 55, and Thr 56) and by H chain residues in the third antigen complementarity region (Asp 99, Leu 100, and Ile 101). Thus, this second subsite is entirely delineated by the H chain which, as indicated before (3), makes a larger contribution to the combining site. Binding of phosphorylcholine, involving specific interactions with Tyr 33 and Arg 52 in the crystalline phosphorylcholine–McPC 603 Fab complex, takes place mostly at this subsite (4).

The preceding description of the combining site stresses the fact that the antigen complementarity region around H chain residues 99–105 has a very important role: (i) it contributes to the conformation and specificity of the L–H subsite, complementing L chain residues; (ii) it divides the combining site into the two main subsites described above; and (iii) it contributes to the conformation and specificity of the H subsite, complementing H chain residues 31–35 and 50–56. It is more difficult to visualize a similar role for the homologous third hypervariable region of the L chain. As can be seen in Fig. 1 this region contributes Arg 95, placed as a dividing line between the two subsites; however, the contribution of this region to the combining site is clearly more limited (see Fig. 1). Also, the H chain region around residues 100–105 is much more variable in length than the homologous region of the L chain (either κ or λ) and consequently it makes a larger contribution to the variability of combining sites of different antibodies.

Amino acid sequences of some antibodies and immunoglobulins of known ligand specificity can be analyzed in terms of these subsites. For example, phosphorylcholine binds to the H subsite in McPC 603; therefore, we would expect that antiphosphorylcholine activity will be correlated mainly with an H chain sequence and will be much less dependent on L chain type of sequence (11). Another example is that of the guinea pig antidinitrophenyl (DNP), anti-*p*-azobenzene arsonate (ARS), and anti-*p*-azobenzenetri-methylammonium (TMA) antibodies. Due to their relatively small size, the

DNP, ARS, and TMA haptens can be expected to bind within one subsite. Inspection of the partial amino acid sequences of H chains from these antibodies (12) indicates that the important positions 33, 50, and 52 of the H subsite have a unique sequence for each specificity. We therefore conclude that the H subsite is the locus of binding of the DNP, ARS, and TMA haptens. Consistent with this idea, affinity labeling reagents modify amino acid side chains in the first and second H chain hypervariable regions of the anti-DNP and anti-ARS guinea pig antibodies (12). Also, the L chains from these antibodies show more heterogeneity than the H chains (J. J. Cebra, *personal communication*) in agreement with a subsite model in which L chains play a less important role since binding of the DNP, ARS, and TMA haptens occurs mainly at the H subsite.

A two-subsite model may be less appropriate for antibodies against a periodic, regularly repeating antigen such as carbohydrate polymers. However, in spite of its schematic nature the model is appealing to us because it assigns a functional role to both L and H chains, emphasizes the role of H chains and suggests that V_H by itself could have a receptor function, recognizing antigens in cells which may not have complete antibody molecules at their surface.

Although encouraging progress has recently been made in our knowledge about the structure of immunoglobulins and antibodies and their combining sites, it still seems premature to attempt to predict the detailed conformation of a combining site based solely on the primary structure or the binding specificity of an immunoglobulin. Further analysis of antibody molecules and correlation of the primary and tertiary structures with their specificities will be required before this goal is reached.

ACKNOWLEDGMENTS

Our research was supported by grants AI 08202 from the National Institutes of Health, NP-141A from the American Cancer Society and by an NIH Research Career Development Award AI 70091 to R.J.P.

REFERENCES

1. Poljak, R. J., Amzel, L. M., Avey, H. P., Becka, L. N., and Nisonoff, A. (1972): Structure of Fab′ New at 6 Å Resolution. *Nature* [*New Biol.*], 235:137–140.
2. Poljak, R. J., Amzel, L. M., Avey, H. P., Chen, B. L., Phizackerley, R. P., and Saul, F. (1973): The three-dimensional structure of the Fab′ fragment of a human immunoglobulin at 2.8 Å resolution. *Proc. Natl. Acad. Sci. USA*, 70:3305–3310.
3. Poljak, R. J., Amzel, L. M., Chen, B. L., Phizackerley, R. P., and Saul, F. (1974): The three-dimensional structure of the Fab′ fragment of a human myeloma immunoglobulin at 2.0 Å resolution. *Proc. Natl. Acad. Sci. USA*, 71:3440–3444.
4. Segal, D. M., Padlan, E. A., Cohen, G. H., Rudikoff, S., Potter, M., and Davies, D. R. (1974): The three-dimensional structure of a phosphorylcholine-binding mouse myeloma immunoglobulin Fab and the nature of the antigen binding site. *Proc. Natl. Acad. Sci. USA*, 71:4298–4301.

5. Epp, O., Colman, P., Fehlhammer, H., Bode, W., Schiffer, M., and Huber, R. (1974): Crystal and molecular structure of a dimer composed of the variable portions of the Bence Jones protein Rei. *Eur. J. Biochem.*, 45:513–524.
6. Poljak, R. J., Amzel, L. M., Chen, B. L., Phizackerley, R. P., and Saul, F. (1975): Structural basis for the association of heavy and light chains and the relation of subgroups to the conformation of the active site of immunoglobulins. *Immunogenetics*, 2:393–394.
7. Amzel, L. M., Poljak, R. J., Saul, F., Varga, J., and Richards, F. F. (1974): The three-dimensional structure of a combining region–ligand complex of immunoglobulin New at 3.5 Å resolution. *Proc. Natl. Acad. Sci. USA*, 71:1427–1430.
8. Varga, J. M., Lande, S., and Richards, F. F. (1974): Multiple binding functions. II. The use of nylon–polyserine whisker discs in screening myeloma immunoglobulins for binding activity. *J. Immunol.*, 112:1565–1570.
9. Francis, S. H., Leslie, R. G. Q., Hood, L., and Eisen, H. N. (1974): Amino acid sequence of the variable region of the heavy (α) chain of a mouse myeloma protein with anti-hapten activity. *Proc. Natl. Acad. Sci. USA*, 71:1123–1127.
10. Green, N. M., Dourmashkin, R. R., and Parkhouse, R. M. E. (1971): Electron miscroscopy of complexes between IgA (MOPC 315) and a bifunctional hapten. *J. Mol. Biol.*, 56:203–206.
11. Barstad, P., Rudikoff, S., Potter, M., Cohn, M., Konigsberg, W., and Hood, L. (1974): Immunoglobulin structure: amino terminal sequences of mouse myeloma proteins. *Science,* 183:962–964.
12. Cebra, J. J., Koo, P. H., and Ray, A. (1974): Specificity of antibodies: primary structural basis of hapten binding. *Science*, 186:263–265.

Antibodies in Human Diagnosis and Therapy,
edited by Edgar Haber and Richard M. Krause.
Raven Press, New York © 1977.

Correlations Between Antigen-Binding Specificity and the Three-Dimensional Structure of the Antibody Combining Sites

David R. Davies and Eduardo A. Padlan

Laboratory of Molecular Biology, National Institute of Arthritis, Metabolism and Digestive Diseases, National Institutes of Health, Bethesda, Maryland 20014

From the X-ray diffraction data on immunoglobulin fragments that have been acquired to date (1-5), we can now begin to understand the structural nature of the diversity of antibody specificities. Further, the crystallographic studies on the Fab of mouse McPC 603 immunoglobulin, which binds natural antigens, permit us to visualize the general appearance of an antibody combining site and the nature of its interaction with antigen (6,7). Here we shall describe the structure of McPC 603, discuss its relationship to other immunoglobulin fragments, and from this discuss the basis of our present understanding of the antibody combining site.

McPC 603 FAB, STRUCTURE AND COMBINING SITE

The three-dimensional structure of McPC 603 Fab has been reported in a number of papers and reviews (4,6,8–10). This IgA(k) protein has been shown to precipitate with phosphorylcholine-containing natural antigens from a variety of organisms including *Pneumococcus, Lactobacillus,* and *Proteus morganii* (11). The Fab binds phosphorylcholine in solution and also in the crystal where the binding constant is 4×10^3 liters/mole, or about 45-fold lower than in free solution (12), a result that was assumed to be due to the ammonium sulfate of crystallization. The phosphorylcholine binding site was located at the amino tip of the molecule (4,6) in a cavity formed by hypervariable regions (13–15).

The overall appearance of the McPC 603 Fab is very similar to the Fab New structure (2) as well as to the L chain Mcg dimer (1). The individual domains exhibit the sandwich-like structure observed to be a common feature of the immunoglobulin fragments (1–5).

A quantitative comparison between the McPC 603 V_L fold and that of the V_L of the human K chain Bence-Jones protein REI (3) is shown in Fig. 1 (16) where it can be immediately seen that there is close homology of the two domains except in the vicinity of the first hypervariable region. Similar invariance of the fold has been observed for V_L and V_H domains of McPC 603 (16). It is thus reasonable to regard each variable domain structure as consisting of a relatively invariant framework at one end of which the hypervariable regions are looped.

Figure 2 illustrates the binding site for phosphorylcholine in McPC 603. The site consists of a cavity formed principally by the segments 24–34 (L1) and 89–97 (L3) of the light chain and 31–35 (H1), 49–66 (H2), and 99–104b (H3) of the heavy chain. The hypervariable region 50–56 (L2) of the light chain does not participate in the formation of the cavity, nor does the hypervariable region 82–89 found in heavy chains (15). Phosphorylcholine binds in the deep cleft with the choline moiety interacting with both light and heavy chains, whereas the phosphate interacts exclusively with heavy chain residues. The choline is buried with the phosphate exposed to solvent and making hydrogen bonds to the phenolic hydroxyl of Tyr 33 and the guanidinium group of Arg 52, both of the heavy chain. The charge on the phosphate will be partially neutralized by the positively charged guanidinium group of Arg 52. The choline is in van der Waals contact with residues from L3 and the side groups of Tyr 33 and Trp 104a of the heavy chain. Electrostatic interactions may occur between the acid groups of Glu 35 and Glu 59 of the heavy chain and the positively charged trimethylammonium moiety. It should be noted that the binding of phosphorylcholine to the Fab in the crystal produced no significant changes in the overall conformation of the Fab (6).

The heavy chain sequences of six other phosphorylcholine-binding immunoglobulins are available (17,26). For T15, S107, H8, M511, and possibly also W3207, there is no significant change in the hypervariable residues actively involved in the binding of phosphorylcholine. In all cases, the phosphate-locating residues Tyr 33 and Arg 52 are unchanged. Similarly, the choline-contacting residues Glu 35 and Glu 59 are also unchanged. Sequence as well as length differences occur in H3, although it would appear that these differences might be accommodated without in any way affecting phosphorylcholine binding; except in M167 where the replacement of Trp 108 by a Gly could leave a large hole in the general region of the H3-choline contact. The next residue in M 167 is Phe 107 which could, with rearrangement, be made to occupy some of the space occupied by Trp 108 in McPC 603. Other differences are observed elsewhere, e.g., at position 99. Thus it is apparent that although the phosphorylcholine binding site in all three proteins could be basically the same, some of the contacting residues from H3 will be different, which could account for the observed differences in their hapten binding properties (18). Since there is a priori no limit on the number of

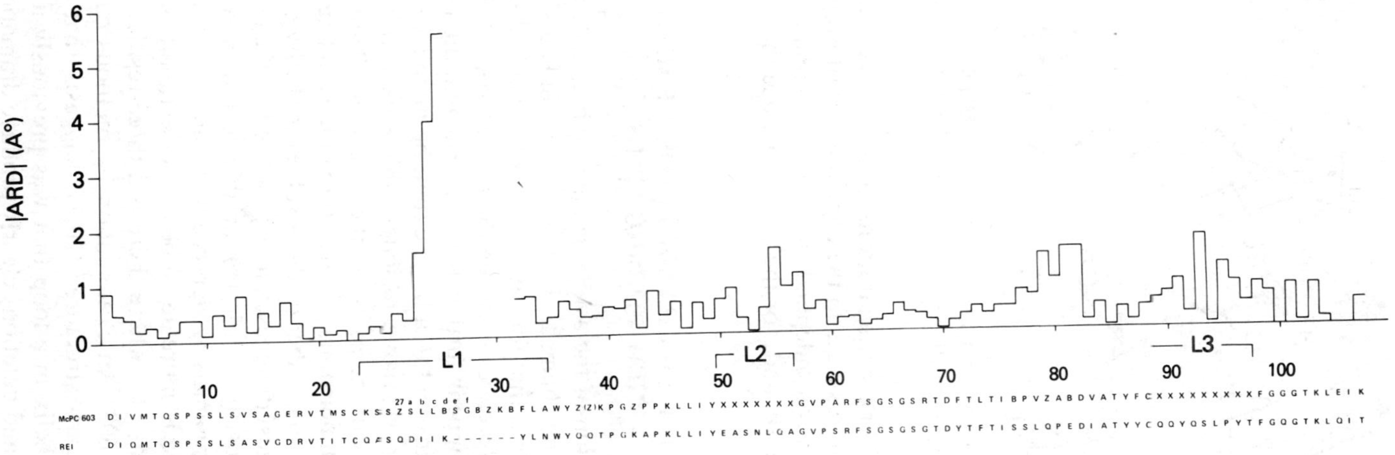

FIG. 1. Plot of the average relative displacement (16) of individual alpha carbons against sequence position in McPC 603 and REI V_L domains.

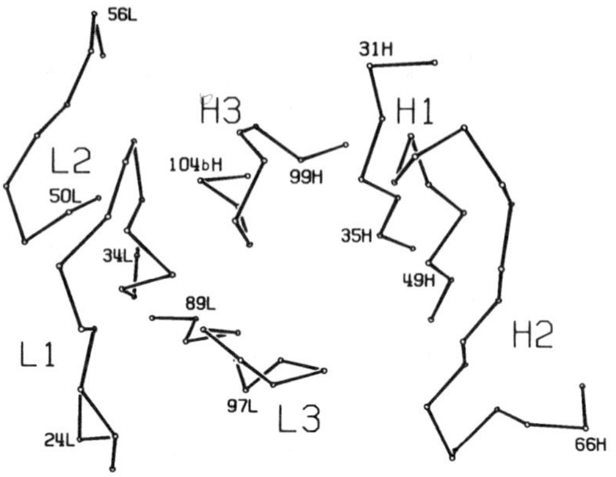

FIG. 2. Schematic drawing of the hypervariable loops of McPC 603 protein. The first and last hypervariable positions in each segment are indicated.

phosphorylcholine-binding configurations that could conceivably be constructed by varying the residues in the complementarity regions, it is rather surprising that these seven independently derived immunoglobulins appear to have the same location and, for the most part, use the same residues for hapten binding.

STRUCTURAL MECHANISMS FOR GENERATING NEW BINDING SITES

A general examination of the available data on immunoglobulin structure immediately suggests mechanisms by which new binding sites may be produced. It is by now clear that for the individual variable domains there will be relatively small differences between the frameworks of the chains of the same class. It would appear further that the principal interactions between light and heavy domains will to a large extent be maintained since they are determined by nonhypervariable residues which are highly conserved and occur in relatively invariant parts of the molecule. Thus, the source of diversity in the three-dimensional architecture of the binding site will depend mainly on the nature and number of the residues in the hypervariable loops. Here, perhaps the biggest changes can be produced by the effects of multiple insertions and deletions in each hypervariable region, particularly if these changes are also accompanied by significant changes in secondary structure. An example considered in more detail below is the rabbit BS-5 anti-type III polysaccharide antibody (19) where there is a three-residue insertion in the H3 relative to McPC 603. An examination by methods currently used to predict protein secondary structures (20,21) suggests that there is now a significant amount of helix in a loop that was previously a simple bend.

Although insertions and deletions clearly produce dramatic changes in the binding site, one should not underestimate the role of simple amino acid substitutions in modifying the nature of the binding site, e.g., by substituting for a charged residue a neutral or oppositely charged residue. A well-known example of the effects of such a simple change occurs in the serine proteases, chymotrypsin and trypsin, where the substitution of an Asp for a Ser in the pocket changes the binding specificity of the enzyme. Similar effects might be produced by changing the phosphate-bonding Arg 52 in McPC 603 to a glutamic acid.

The location of residue positions that have been found hypervariable in one species or another is shown in Fig. 3. It can be seen that, for the most part, the hypervariable residues form a continuous surface. However, not all residues in hypervariable positions could be directly involved in the binding of antigen. Some hypervariable residues are in the domain interiors and some are involved in the contact between the two variable domains. Changes in the size of the residues at the interior of the cavity could produce significant changes in the size and shape of the cavity. Similarly, changes in the size and shape of the contacting residues, both within a chain and between chains can produce significant changes in the configurations of the hypervariable loops which should not be regarded as completely rigid. In the Bence-Jones protein Mcg it has been observed that there are significant differences in the configurations of the loops even when the amino acid sequence is unchanged (22). These differences presumably arise from different interactions with the neighboring molecules. In McPC 603, it was found that the complementarity regions L1 and H3 are in close contact. The fact that insertions and deletions are frequently found in these two regions makes this close contact an apparently very important interaction that could have a modulating effect on the size and shape of the binding cavity.

CORRELATION BETWEEN BINDING SPECIFICITY AND COMBINING SITE STRUCTURE

It is to be expected that the factor determining the high specificity of antibody–antigen reactions is the complementarity in the physicochemical natures of the antibody combining site and the antigenic determinant. This is clearly illustrated by the specific interactions observed in the binding of phosphorylcholine to McPC 603 protein. In McPC 603 the existence of a deep binding cavity ensures maximal envelopment of the hapten. In addition the presence of the negatively charged carboxylic groups of Glu 35 and Glu 59 deep in the binding pocket and the location of the positively charged Arg 52 at the entrance of the cavity determines the directionality in the binding of phosphorylcholine. This is obviously the most logical binding orientation for phosphorylcholine since the hapten is linked to a polysaccharide carrier by an ester linkage involving the phosphate group (23).

Further evidence for the structural complementarity of ligand and binding

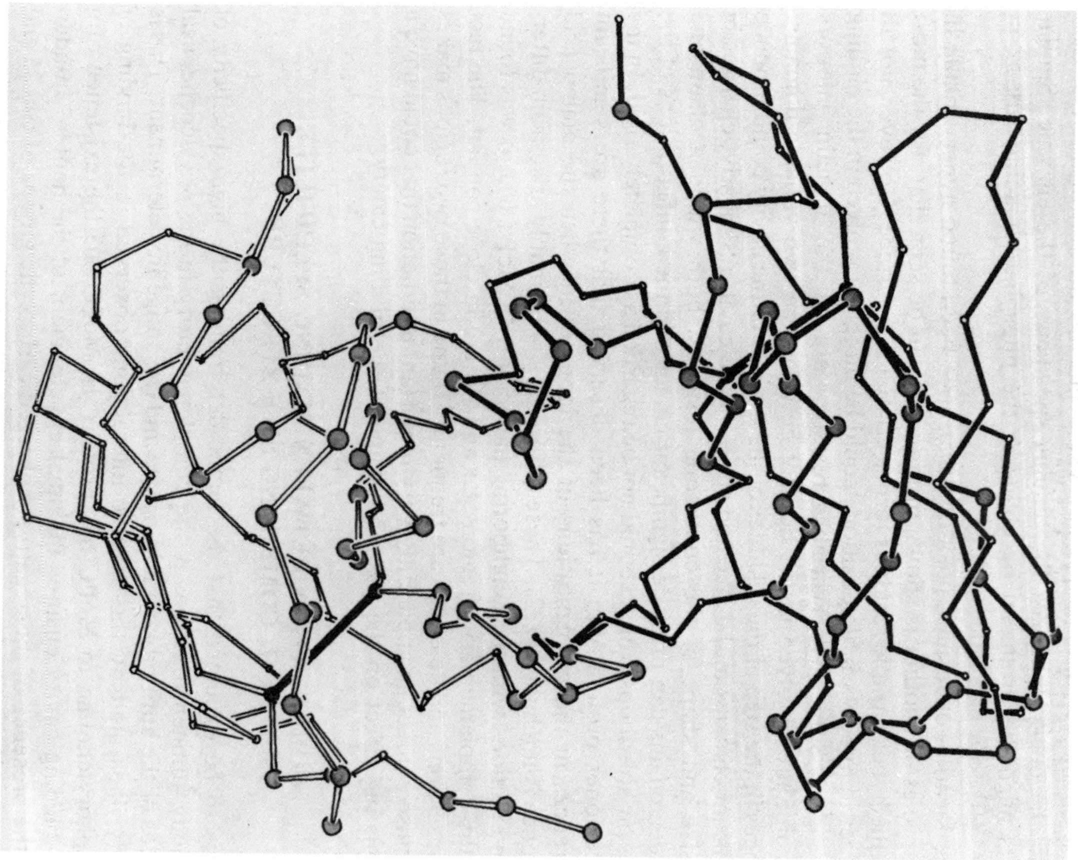

FIG. 3. Schematic drawing of the variable region of McPC 603. The positions that have been found to be hypervariable in one species or another are indicated by large filled circles.

site is the interaction that probably occurs in the binding of dinitrophenyl haptens to MOPC 315 protein. While the three-dimensional structure of this protein has not been elucidated, examination of the binding site residues (24) and model building studies (25) indicate that the MOPC 315 binding cavity would be lined by a number of large aromatic side chains as would be expected from the aromatic nature of the hapten ligand. A binding configuration was found feasible (25) whereby the dinitrophenyl moiety is parallel to a Trp ring and the nitro groups of the hapten are involved in hydrogen-bonding to various protein groups.

In a discussion of this nature it is also important to keep in mind that the potential combining surface is much larger than haptens such as phosphorylcholine. It is reasonable to assume that portions of the hypervariable surface that are not involved in hapten binding are involved in the interaction with the rest of the antigenic determinant or are concerned with carrier specifications. The availability of sequence information for rabbit antibodies against type III polysaccharide (19,27) affords an opportunity to investigate the more extensive interaction between antibody and antigen. Ultimately, it is to be hoped that a crystallographic study will be possible on one of these systems. However, until this happens, considerable information may be obtained from a comparative study of the sequence and from model building.

We have made a preliminary model building study on one of these proteins (BS-5). The V_L sequences of the two rabbit antibodies, BS-5 and K-25, are listed in Table 1 together with those of McPC 603 and REI, a human V_K of known structure (3). The V_H sequences of BS-5, K-25, McPC 603 and New proteins are listed in Table 2.

It can be seen in Tables 1 and 2 that there is considerable homology between the chains from these three different species. Comparing the BS-5 and McPC 603 framework sequences we find that, in V_L, 77% of the interior residues are identical, as are 58% of the residues exposed to solvent. In V_H, 75% of the interior residues and 51% of the exposed residues are identical. Moreover, among the residues that are not identical, there are many Ser for Thr, Tyr for Phe, Arg for Lys and similar conserved interchanges. Even more conserved are the residues involved in the contact between the V_L and V_H domains. Of the 8 nonhypervariable interface residues in the heavy chain, 6 are identical in BS-5 and McPC 603; in the light chain, all are identical. Comparable figures are found on comparing K-25 with McPC 603.

The high degree of similarity observed in the sequences of the rabbit, mouse and human V_L domains (Table 1) indicates that the structure of the rabbit domains will be very similar to that found for McPC 603 and REI proteins. The rabbit V_H domains will probably also be similar to those of the mouse and human proteins although to a lesser degree. There is a deletion of 2 residues occurring at around position 72 and a one-residue deletion at around position 79 in the rabbit chains. Since, however, these changes in length occur at bends, they can be readily accommodated by minor restructur-

TABLE 1. *Alignment of the V_L sequences of McPC 603, REI, BS-5, and K-25 proteins*

		0	3	0	4	0	4	1	0	0	0
McPC603	1	Asp	- Ile	- Val	- Met	- Thr	- Gln	- Ser	- Pro	- Ser	- Ser
REI	1	Asp	- Ile	- Gln	- Met	- Thr	- Gln	- Ser	- Pro	- Ser	- Ser
BS-5	1	Asp	- Val	- Val	- Met	- Thr	- Gln	- Thr	- Pro	- Ala	- Ser
K-25	1	Ala	- Val	- Glu	- Leu	- Thr	- Gln	- Thr	- Pro	- Ala	- Ser
		4	0	4	0	0	0	0	0	4	0
	11	Leu	- Ser	- Val	- Ser	- Ala	- Gly	- Glu	- Arg	- Val	- Thr
	11	Leu	- Ser	- Ala	- Ser	- Val	- Gly	- Asp	- Arg	- Val	- Thr
	11	Val	- Ser	- Glu	- Pro	- Val	- Gly	- Gly	- Thr	- Val	- Thr
	11	Val	- Glu	- Ala	- Ala	- Val	- Gly	- Gly	- Thr	- Val	- Thr
		4	0	4	0	4	0	0	0	3	0
	21	Met	- Ser	- Cys	- Lys	- Ser	- Ser	- Glx	- Ser	- Leu	- Leu
	21	Ile	- Thr	- Cys	- Gln	- Ala	- Ser	- Gln	- Asp	- Ile	- Ile
	21	Ile	- Lys	- Cys	- Gln	- Ala	- Ser	- Gln	- Ser	- Ile	- Tyr
	21	Ile	- Lys	- Cys	- Gln	- Ala	- Ser	- Gln	- Asx	- Ile	- Tyr
		0	0	C	C	C	1	C	3	4	c
	27d	Asx	- Ser	- Gly	- Asx	- Glx	- Lys	- Asx	- Phe	- Leu	- Ala
	31		-	-	-	-	-	- Lys	- Tyr	- Leu	- Asn
	31		-	-	-	-	-	- Ser	- Asn	- Leu	- Ala
	31		-	-	-	-	-	- Ser	- Tyr	- Leu	- Ser
		4	C	3	C	0	0	0	1	C	C
	35	Trp	- Tyr	- Glx	-(Glx)	Lys	- Pro	- Gly	- Glx	- Pro	- Pro
	35	Trp	- Tyr	- Gln	- Gln	- Thr	- Pro	- Gly	- Lys	- Ala	- Pro
	35	Trp	- Tyr	- Gln	- Gln	- Lys	- Pro	- Gly	- Gln	- Pro	- Pro
	35	Trp	- Tyr	- Gln	- Gln	- Lys	- Pro	- Gly	- Gln	- Pro	- Pro
		0	C	4	4	C	1	1	0	1	0
	45	Lys	- Leu	- Leu	- Ile	- Tyr	- X	- X	- X	- X	- X
	45	Lys	- Leu	- Leu	- Ile	- Tyr	- Glu	- Ala	- Ser	- Asn	- Leu
	45	Lys	- Leu	- Leu	- Ile	- Tyr	- Lys	- Ala	- Ser	- Thr	- Leu
	45	Lys	- Leu	- Leu	- Ile	- Tyr	- Lys	- Ala	- Ser	- Thr	- Leu
		0	0	0	4	1	0	1	4	0	4
	55	X	- X	- Gly	- Val	- Pro	- Ala	- Arg	- Phe	- Ser	- Gly
	55	Gln	- Ala	- Gly	- Val	- Pro	- Ser	- Arg	- Phe	- Ser	- Gly
	55	Glu	- Ser	- Gly	- Val	- Pro	- Ser	- Arg	- Phe	- Lys	- Gly
	55	Ala	- Ser	- Gly	- Val	- Ser	- Ser	- Arg	- Phe	- Lys	- Gly
		0	4	0	0	2	0	4	0	4	0
	65	Ser	- Gly	- Ser	- Arg	- Thr	- Asp	- Phe	- Thr	- Leu	- Thr
	65	Ser	- Gly	- Ser	- Gly	- Thr	- Asp	- Tyr	- Thr	- Phe	- Thr
	65	Ser	- Gly	- Ser	- Gly	- Thr	- Asp	- Phe	- Thr	- Leu	- Thr
	65	Ser	- Gly	- Ser	- Gly	- Thr	- Glu	- Phe	- Thr	- Leu	- Thr
		4	0	0	2	2	0	0	1	2	4
	75	Ile	- Asx	- Pro	- Val	- Glx	- Ala	- Asx	- Asp	- Val	- Ala
	75	Ile	- Ser	- Ser	- Leu	- Gln	- Pro	- Glu	- Asp	- Ile	- Ala
	75	Ile	- Ser	- Asp	- Leu	- Glx	- Cys	- Ala	- Asp	- Ala	- Ala
	75	Ile	- Ser	- Asp	- Leu	- Glx	- Cys	- Ala	- Asp	- Ala	- Ala

TABLE 1. (*Continued*)

	c	4	C	4	C	4	C	3	0	3
85	Thr	- Tyr	- Phe	- Cys	- X	- X	- X	- X	- X	- X
85	Thr	- Tyr	- Tyr	- Cys	- Gln	- Gln	- Tyr	- Gln	- Ser	- Leu
85	Thr	- Tyr	- Phe	- Cys	- Gln	- Gly	- Ser	- Asx	- Tyr	- Thr
85	Thr	- Tyr	- Tyr	- Cys	- Gln	- Thr	- Tyr	- Ser	- Tyr	- Ser
	0	C	4	C	4	c	0	4	0	4
95	X	- X	- X	- Phe	- Gly	- Gly	- Gly	- Thr	- Lys	- Leu
95	Pro	- Tyr	- Thr	- Phe	- Gly	- Gln	- Gly	- Thr	- Lys	- Leu
95	Gly	- Thr	- Val	- Phe	- Gly	- Gly	- Gly	- Thr	- Glu	- Val
95	Ser	- Thr	- Tyr	- Phe	- Gly	- Gly	- Gly	- Thr	- Glu	- Val
	0	3	0	0						
105	Glu	- Ile	- Lys	- Arg						
105	Gln	- Ile	- Thr	- Gly						
105	Val	- Val	- Lys	- Gly						
105	Val	- Val	- Lys	- Gly						

The McPC 603 sequence is not known beyond position 49 and the residues given are those most frequently found in other mouse κ chains (28). The location of each residue as it occurs in McPC 603 is indicated: 0, completely exposed to solvent; 1, mainly exposed; 2, partly buried, partly exposed; 3, mainly buried in the domain interior; 4, completely buried; or C, involved in the V_L:V_H contact. The numbers on the left correspond to the sequence numbers of the first residues in any given line according to the original reports.

ing of these bends and nearby regions. It is also clear that the mode of association of the V_L and V_H domains of the rabbit proteins will be similar to that found in McPC 603 Fab.

A most interesting result of our model building study of the BS-5 binding site is the distinctive chemical nature of the continuous hypervariable surface. The entire potential binding surface displays a large number of Ser, Thr, Tyr, Asn, and other residues capable of forming hydrogen bonds (Fig. 4). Among the 43 hypervariable residues from the complementarity regions L1, L2, L3, H1, H2, and H3 in BS-5 that would be at least partly exposed to solvent and that in theory could interact with antigen, there are 10 Ser, 5 each of Thr and Tyr, 3 Gln, 1 each of Asn, Asp, Glu, His, and an undetermined Asx. In K-25, 31 of the 44 hypervariable residues that are probably exposed are capable of hydrogen bonding. A preliminary examination of the light chain sequences of other rabbit antibodies against polysaccharides (19, 27) also reveals a large number of such residues in potential binding positions. Thus it appears that the combining sites of these antibodies are distinctly suited to interact by hydrogen bond information with the many hydroxyl groups of the polysaccharide antigens.

TABLE 2. *Alignment of the V_H sequences of McPC 603, New, BS-5, and K-25 proteins*

		0	2	0	4	0	2	0	4	0	0
McPC603	1	Glu	- Val	- Lys	- Leu	- Val	- Glu	- Ser	- Gly	- Gly	- Gly
New	1	Pca	- Val	- Gln	- Leu	- Pro	- Glu	- Ser	- Gly	- Pro	- Glu
BS-5	1		- Glu	- Ser	-(Val	, Glu)	- Glu	- Ser	- Gly	- Gly	- Arg
K-25	1		- Glu	- Ser	- Val	- Lys	- Glu	- Ser	- Glu	- Gly	- Gly
		0	4	0	0	0	0	0	4	0	4
	11	Leu	- Val	- Gln	- Pro	- Gly	- Gly	- Ser	- Leu	- Arg	- Leu
	11	Leu	- Val	- Ser	- Pro	- Gly	- Glx	- Thr	- Leu	- Ser	- Leu
	10	Leu	- Val	- Thr	- Pro	- Thr	- Pro	- Gly	- Leu	- Thr	- Leu
	10	Leu	- Phe	- Lys	- Pro	- Thr	- Asp	- Thr	- Leu	- Thr	- Leu
		0	4	0	4	0	1	4	0	4	0
	21	Ser	- Cys	- Ala	- Thr	- Ser	- Gly	- Phe	- Thr	- Phe	- Ser
	21	Thr	- Cys	- Thr	- Gly	- Ser	- Thr	- Val	- Ser	- Thr	- Phe
	20	Thr	- Cys	- Thr	- Val	- Ser	- Gly	- Phe	- Ser	- Leu	- Ser
	20	Thr	- Cys	- Thr	- Val	- Ser	- Gly	- Phe	- Ser	- Leu	- Ser
		0	4	2	4	2	4	C	3	C	2
	31	Asp	- Phe	- Tyr	- Met	- Glu	- Trp	- Val	- Arg	- Gln	- Pro
	31	Ala	- Val	- Tyr	- Ile	- Val	- Trp	- Val	- Arg	- Gln	- Pro
	30	Ser	- Tyr	- Asp	- Met	- Gly	- Trp	- Val	- Arg	- Gln	- Ala
	30	Gly	- Tyr	- Asp	- Met	- Ser	- Trp	- Val	- Arg	- Gln	- Ala
		0	0	0	0	C	2	C	4	4	1
	41	Pro	- Gly	- Lys	- Arg	- Leu	- Glu	- Trp	- Ile	- Ala	- Ala
	41	Pro	- Gly	- Arg	- Gly	- Leu	- Glu	- Trp	- Ile	- Ala	- Tyr
	40	Pro	- Gly	- Lys	- Gly	- Leu	- Glu	- Trp	- Ile	- Gly	- Ile
	40	Pro	- Gly	- Lys	- Gly	- Leu	- Glu	- Trp	- Ile	- Gly	- Val
		4	2	0	0	0	0	0	2	0	2
	51	Ser	- Arg	- Asn	- Lys	- Gly	- Asn	- Lys	- Tyr	- Thr	- Thr
	51	Val	- Phe	- Tyr	- His	- Gly	-	-	-	- Thr	- Ser
	50	Ile	- Tyr	- Ala	- Ser	- Gly	-	-	-	- Ser	- Thr
	50	Ile	- Tyr	- Ala	- Ser	- Gly	-	-	-	- Ser	- Thr
		0	3	1	0	0	4	0	0	1	4
	59	Glu	- Tyr	- Ser	- Ala	- Ser	- Val	- Lys	- Gly	- Arg	- Phe
	58	Asp	- Thr	- Asp	- Thr	- Pro	- Leu	- Arg	- Ser	- Arg	- Val
	57	Tyr	- Tyr	- Ala	- Ser	- Trp	- Ala	- Lys	- Gly	- Arg	- Phe
	57	Tyr	- Tyr	- Ala	- Thr	- Trp	- Ala	- Lys	- Ser	- Arg	- Ser
		0	4	0	2	0	0	0	0	3	0
	69	Ile	- Val	- Ser	- Arg	- Asp	- Thr	- Ser	- Gln	- Ser	- Ile
	69	Thr	- Met	- Leu	- Val	- Asn	- Thr	- Ser	- Lys	- Asn	- Gln
	67	Thr	- Ile	- Ser	- Lys	-	- Thr	- Ser	-	- Thr	- Thr
	67	Thr	- Ile	- Thr	- Arg	-	- Thr	- Ser	-	- Asx	- Thr
		4	1	4	0	4	0	0	4	0	0
	79	Leu	- Tyr	- Leu	- Gln	- Met	- Asn	- Ala	- Leu	- Arg	- Ala
	79	Phe	- Ser	- Leu	- Arg	- Leu	- Ser	- Ser	- Val	- Thr	- Ala
	75	Val	- Asp	- Leu	- Lys	-	- Thr	- Ser	- Leu	- Pro	- Thr
	75	Val	- Asx	- Leu	-	- Met	- Asp	- Ser	- Leu	- Thr	- Ala

TABLE 2. *(Continued)*

	0	2	3	4	2	4	C	4	c	3
89	Glu	- Asp	- Thr	- Ala	- Ile	- Tyr	- Tyr	- Cys	- Ala	- Arg
88	Ala	- Asp	- Thr	- Ala	- Val	- Tyr	- Tyr	- Cys	- Ala	- Arg
84	Glu	- Asp	- Thr	- Ala	- Thr	- Tyr	- Phe	- Cys	- Ala	- Arg
84	Gln	- Asp	- Thr	- Ala	- Thr	- Tyr	- Phe	- Cys	- Ala	- Arg
	3	2	C	0	0	C				C
99	Asn	- Tyr	- Tyr	- Gly	- Ser	- Thr	-	-	-	- Trp
99	Asx	- Leu	- Ile	- Ala	-	-	-	-	-	- Gly
94	Gln	- Gly	- Thr	- Gly	- Leu	- Val	- His	- Leu	-	- Ala
94	Gly	- His	- Thr	- Gly	- Leu	- Ser	- Tyr	- Leu	- Lys	- Ser
	C	C	1	4	C	4	0	4	2	0
107	Tyr	- Phe	- Asp	- Val	- Trp	- Gly	- Ala	- Gly	- Thr	- Thr
107	Cys	- Ile	- Asx	- Val	- Trp	- Gly	- Gln	- Gly	- Ser	- Leu
103	Phe	- Val	- Asp	- Val	- Trp	- Gly	- Pro	- Gly	- Thr	- Leu
104	Ser	- Val	- Asp	- Val	- Trp	- Gly	- Pro	- Gly	- Thr	- Leu
	4	0	4	0	0					
117	Val	- Thr	- Val	- Ser	- Ser					
117	Val	- Thr	- Val	- Ser	- Ser					
113	Val	- Thr	- Val	- Ser	- Ser					
114	Val	- Thr	- Val	- Ser	- Ser					

See footnote of Table 1 for further description.

CONCLUSION

In order to understand the nature of antibody interaction with antigen it has not been necessary to invoke any new binding mechanisms. The crystallographic and model building studies described above, although limited, clearly indicate that the interaction involves dispersion forces, hydrogen bonds, and charge interactions, the customary forces of the protein physical chemist. It is the distribution of these forces that results in the total interaction with antigen and defines the specificity of the antigen–antibody reaction.

It should also be emphasized that the surface of the antibody molecule that is lined by hypervariable residues is rather extensive and provides a large potential area of interaction with antigen. Thus the continuous hypervariable surface extends from a bend near the carboxyl end of V_L, around the amino tip of the Fab over to bends near the carboxyl end of V_H. The outer limits of the hypervariable surface are then only a few angstroms removed from the switch region. It is not inconceivable, therefore, that antigen binding, that involves the far edges of the combining site, could cause significant perturbations in V region–C region interactions. The interactions between V and C domains are different for the light and heavy

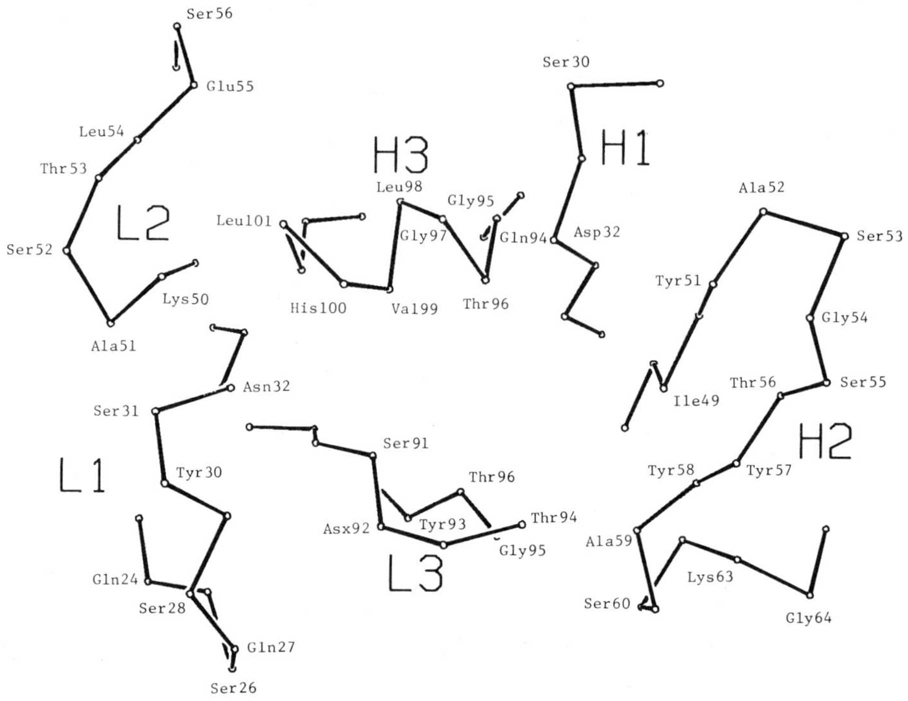

FIG. 4. Schematic drawing of the hypervariable loops of BS-5 antibody based on model-building studies. The hypervariable residues that would be at least partly exposed to solvent are shown.

chains in the Fab causing a bend of about 135° in the fragment. This is intriguing, since the bend causes the binding cavity to be asymmetric relative to the long axis of the Fab.

The model building studies on new immunoglobulins based on the sequence alone do provide considerable information. However, at a precise level they are clearly quite tentative. In order to place them on a firmer footing much more X-ray diffraction data is needed of Fabs with different length hypervariable loops. With this information it should be possible to make accurate predictions regarding the three-dimensional structure of antigen-binding sites.

REFERENCES

1. Schiffer, M., Girling, R. L., Ely, K. R., and Edmundson, A. B. (1973): Structure of a λ-type Bence-Jones protein at 3.5 Å resolution. *Biochemistry,* 12:4620–4631.
2. Poljak, R. J., Amzel, L. M., Avery, H. P., Chen, B. L., Phizackerley, R. P., and Saul, F. (1973): Three-dimensional structure of the Fab′ fragment of a human immunoglobulin at 2.8 Å resolution. *Proc. Natl. Acad. Sci. USA,* 70:3305–3310.

3. Epp, O., Colman, P., Fehlhammer, H., Bode, W., Schiffer, M., Huber, R., and Palm, W. (1974): Crystal and molecular structure of a dimer composed of the variable portions of the Bence-Jones protein REI. *Eur. J. Biochem.,* 45:513–524.
4. Segal, D. M., Padlan, E. A., Cohen, G. H., Rudikoff, S., Potter, M., and Davies, D. R. (1974): The Three-dimensional structure of a phosphorylcholine-binding mouse immunoglobulin Fab and the nature of the antigen binding site. *Proc. Natl. Acad. Sci. USA,* 71:4298–4302.
5. Davies, D. R., Padlan, E. A., and Segal, D. M. (1975): Immunoglobulin structures at high resolution. In: *Contemporary Topics in Molecular Immunology,* Vol. 4, pp. 127–155, edited by F. P. Inman and W. J. Mandy. Plenum Press, New York.
6. Padlan, E. A., Segal, D. M., Rudikoff, S., Potter, M., Spande, T., and Davies, D. R. (1973): Structure at 4.5 Å resolution of a phosphorylcholine-binding Fab. *Nature* [*New Biol.*], 245:165–167.
7. Amzel, L. M., Poljak, R. J., Saul, F., Varga, J. M., and Richards, F. F. (1974): The three-dimensional structure of a combining region—ligand complex of immunoglobulin New at 3.5 Å resolution. *Proc. Natl. Acad. Sci. USA,* 71:1427–1430.
8. Padlan, E. A., Segal, D. M., Cohen, G. H., Davies, D. R., Rudikoff, S., and Potter, M. (1974): The three-dimensional structure of the antigen binding site of McPC 603 protein. In: *The Immune System,* pp. 7–14, edited by E. E. Sercarz, A. R. Williamson, and C. F. Cox. Academic Press, New York.
9. Segal, D. M., Padlan, E. A., Cohen, G. H., Silverton, E. W., Davies, D. R., Rudikoff, S., and Potter, M. (1974): The structure of McPC 603 Fab and its hapten complex. In: *Progress in Immunology II,* Vol. 1: *Immunochemical Aspects,* pp. 93–102, edited by L. Brent and J. Holborow. North-Holland, Amsterdam.
10. Davies, D. R., Padlan, E. A., and Segal, D. M. (1975): Three-dimensional structure of immunoglobulins. *Annu. Rev. Biochem.,* 44:639–667.
11. Potter, M. (1971): Antigen-binding myeloma proteins in mice. *Ann. NY Acad. Sci.,* 190:306–321.
12. Rudikoff, S., Potter, M., Segal, D. M., Padlan, E. A., and Davies, D. R. (1972): Crystals of phosphorylcholine-binding Fab-fragments from mouse myeloma proteins: Preparation and X-ray analysis, *Proc. Natl. Acad. Sci. USA,* 69:3689–3692.
13. Wu, T. T., and Kabat, E. A. (1970): An analysis of the sequences of the variable regions of Bence-Jones proteins and myeloma light chains and their implications for antibody complementarity. *J. Exp. Med.,* 132:211–250.
14. Kabat, E. A., and Wu, T. T. (1971): Attempts to locate complementarity-determining residues in the variable positions of light and heavy chains. *Ann. NY Acad. Sci.,* 190:382–392.
15. Capra, J. D., and Kehoe, J. M. (1974): Variable region sequences of five human immunoglobulin heavy chains of the V_H III subgroups: Definitive identification of four heavy chain hypervariable regions. *Proc. Natl. Acad. Sci. USA,* 71:845–848.
16. Padlan, E. A., and Davies, D. R. (1975): Variability of three-dimensional structure in immunoglobulins. *Proc. Natl. Acad. Sci. USA,* 72:819–823.
17. Potter, M., Rudikoff, S., Padlan, E. A., and Vrana, M. (1975): The covalent structure of the antigen-binding site: Antigen-binding myeloma proteins of the BALB-c mouse. *This volume.*
18. Claflin, J. L., Lieberman, R., and Davie, J. M. (1974): Clonal nature of the immune response to phosphorylcholine. II. Idiotypic specifity and binding characteristics of anti-phosphorylcholine antibodies. *J. Immunol.,* 112:1747–1756.
19. Jaton, J.-C. (1975): Comparison of the amino acid sequences of the variable domains of two homogeneous rabbit antibodies to type III pneumococcal polysaccharide. *Biochem. J.,* 147:235–247.
20. Chou, P. Y., and Fasman, G. D. (1974): Prediction of protein conformation. *Biochemistry,* 13:222–245.
21. Wu, T. T., and Kabat, E. A. (1971): An attempt to locate the non-helical and permissively helical sequences of proteins:Application to the variable regions of immunoglobulin light and heavy chains. *Proc. Natl. Acad. Sci. USA,* 68:1501–1506.
22. Edmundson, A. B., Ely, K. R., Girling, R. L., Abola, E. E., Schiffer, M., and Westholm, F. A. (1974): Structure and binding properties of a λ-type Bence-Jones dimer. In: *Progress*

in Immunology. II. Vol. 1, pp. 103–113, edited by L. Brent and J. Holborow, North-Holland, Amsterdam.
23. Watson, M. J., and Baddiley, J. (1974): The action of nitrous acid on C-teichoic acid (C-substance) from the walls of *Diplococcus pneumoniae. Biochem. J.,* 137:399–404.
24. Poljak, R., Amzel, L. M., Chen, B. L., Phizackerley, R. P. and Saul, F. (1974): The three-dimensional structure of the Fab′ fragment of a human myeloma immunoglobulin at 2.0 Å resolution. *Proc. Natl. Acad. Sci. USA,* 71:3440–3444.
25. Padlan, E. A., Davies, D. R., Givol, D., Pecht, I., and Wright, C., (1976): Model building studies of antigen binding sites: The hapten binding site of MOPC 315. *Cold Spring Harbor Symp. Quant. Biol. (in press).*
26. Hood, L., Campbell, J. H., and Elgin, S. C. R. (1975): The organization, expression and evolution of antibody gene families. *Annu. Rev. Genetics,* 9:305–353.
27. Margolies, M. N., Cannon, III, L. E., Strosberg, A. D., and Haber, E. (1975): Diversity of light chain variable region sequences among rabbit antibodies elicited by the same antigens. *Proc. Natl. Acad. Sci. USA,* 72:2180–2184.

Antibodies in Human Diagnosis and Therapy,
edited by Edgar Haber and Richard M. Krause.
Raven Press, New York © 1977.

Conformational Changes in IgM on Hapten Binding

Marion Koshland

Dr. John Brown in my laboratory has recently obtained evidence for antigen-induced conformational changes in IgM antibody. He isolated the IgM fraction from rabbit antibody raised against Lac-BGG and measured its capacity to fix complement in the presence of the Lac hapten and in the presence of mono-Lac-substituted ribonuclease. The hapten had no effect, but the monovalent antigen induced a response equivalent to that observed with the homologous multivalent antigen. When Dr. Brown repeated these experiments with anti-Lac IgG antibody, he obtained quite different results. The mono-Lac ribonuclease did not induce complement fixation although under similar conditions the homologous multivalent Lac-BGG elicited a maximum response. These results suggested to us that both aggregation and antigen-induced conformational changes are essential for the expression of Fc function. We suspect that aggregation serves to stabilize the Ig structure so that energy from antigen binding can be transferred to the Fc effector sites. In IgM antibodies, stabilization is accomplished by the intersubunit S–S bonds, and thus monovalent antigen interaction provides a sufficient signal. In IgG antibodies, on the other hand, stabilization must be effected by antigen cross-linking, and thus a multivalent antigen is required for effective signaling.

Dr. Brown then took advantage of the J-chain component in IgM antibody to obtain an independent measure of Fc conformational changes. The J polypeptide is known to be attached in the C_H4 domain and to be folded within the free IgM molecule so that very few determinants are available for reaction with anti-J-chain antibody. Using radioimmune assay, Dr. Brown showed that the interaction of anti-Lac IgM with its specific antigen increased by fourfold the number of available J-chain determinants. Again, Lac hapten was found to be totally ineffective. In view of these results, I wonder if Dr. Davies would be able to put a carrier tail on the phosphocholine hapten and still get it into the combining site.

Antibodies in Human Diagnosis and Therapy,
edited by Edgar Haber and Richard M. Krause.
Raven Press, New York © 1977.

Implications of Conformational Isomerism and Rotational Allomerism to the Binding of Small Molecules by the Mcg Bence-Jones Dimer

A. B. Edmundson, E. E. Abola, K. R. Ely, J. R. Firca, N. C. Panagiotopoulos, M. Schiffer, and F. A. Westholm

Division of Biological and Medical Research, Argonne National Laboratory, Argonne, Illinois 60439

The serum IgG1 protein and the urinary Bence-Jones λ chain dimer from a patient (Mcg) with multiple myeloma and amyloidosis are being characterized by crystallographic techniques at Argonne (1–8) and by chemical methods (9–13) at the University of Wisconsin. Diffraction data to 3.5-Å resolution have been collected for the IgG1 molecule and are currently being evaluated. An electron density map has been calculated for the Bence-Jones dimer at a nominal resolution of 2.3 Å, and an atomic model has been constructed (6–8). The amino acid sequence of the heavy chain of the IgG1 molecule is presently being determined in H. F. Deutsch's laboratory, and Fett and Deutsch (12) previously completed the sequence of the Bence-Jones protein.

The γ_1 heavy chain and the λ light chain in the myeloma protein represent classes of constituents found in normal antibodies, although the Mcg type of heavy chain is relatively rare (11–13). It has a deletion of 15 residues in the "hinge" region, including the half-cystine residues which usually participate in heavy–light and heavy–heavy interchain disulfide bonds (11). The light chain is of normal length and is covalently linked through its penultimate half-cystine residue to a second light chain molecule in both the IgG1 protein and Bence-Jones dimer (9,10).

In the carboxyl ("constant" or C) domain of the light chain, there are three substitutions which indicate coding by a previously undetected gene (13). These Mcg "factors" are associated with the Oz (−) and Kern (+) substitutions (14,15) in the C domain and an interchange in position 103 in the amino ("variable" or V) domain (13). The isotypic nature of the Mcg λ chain is shown by the finding of similar molecules in light chain pools from normal individuals. The Mcg type of light chain can associate with γ_1 heavy chains with intact hinge regions, as well as with α and μ chains in IgA and IgM immunoglobulins (13). In these molecules the light chains presumably are covalently linked to the heavy chains by disulfide bonds. The Mcg-type light chains thus appear to be sufficiently versatile to meet the structural requirements for formation of a covalently linked Bence-Jones dimer and larger immunoglobulins containing either heavy–light or light–light disulfide bonds.

In this article we shall discuss the principal features of the three-dimensional structure of the prototype of this class of light chains. Evolutionary changes in the domains will also be considered, particularly in regard to the binding sites for hapten-like molecules, the interactions between domains, and the surface properties of the dimeric modules of domains.

CONFORMATIONAL ISOMERISM OF THE LIGHT CHAINS IN THE BENCE-JONES AND IgG1 PROTEINS

Differences in the Switch Regions

When associated with heavy chains, the Mcg light chains have identical conformations, as evidenced by the crystallographic twofold axis of rotation between halves of the IgG1 molecule [(2); see Fig. 1]. In contrast, light chains with the same amino acid sequences in the Bence-Jones dimer behave as conformational isomers [(5, 6); see Fig. 2]. One monomer is similar to the heavy chain component of a human (16–20) or murine antigen-binding (Fab) fragment (21–24), while the second resembles the light chain component. The bend in the switch region between the V and C domains in the heavy chain type of conformation (see monomer 1 in Fig. 2) is more acute than in the light chain conformation (monomer 2). The V–C interdomain disulfide bond in rabbit light chains (25) can only be accommodated in molecules with conformations similar to that of monomer 2 (5).

Conformational isomerism is important for the binding of hapten-like molecules in the Bence-Jones dimer (6,7). In its absence, the binding sites would not exist as we observe them in the three-dimensional structure (for comparisons with V domain dimers from other light chains, see refs. 26–28).

Differences in Pairs of Like Domains

Like domains in the Bence-Jones protein are paired in modules of different size and structure (3,5). The V dimeric module is larger because the domains are separated by a solvent channel in which hapten-like molecules can be bound. An approximate pseudotwofold axis of rotation passes through the middle of this solvent channel (see Fig. 2). The C domains are related by a pseudodiad through a solvent-free zone in which side chains interact to stabilize the dimeric module.

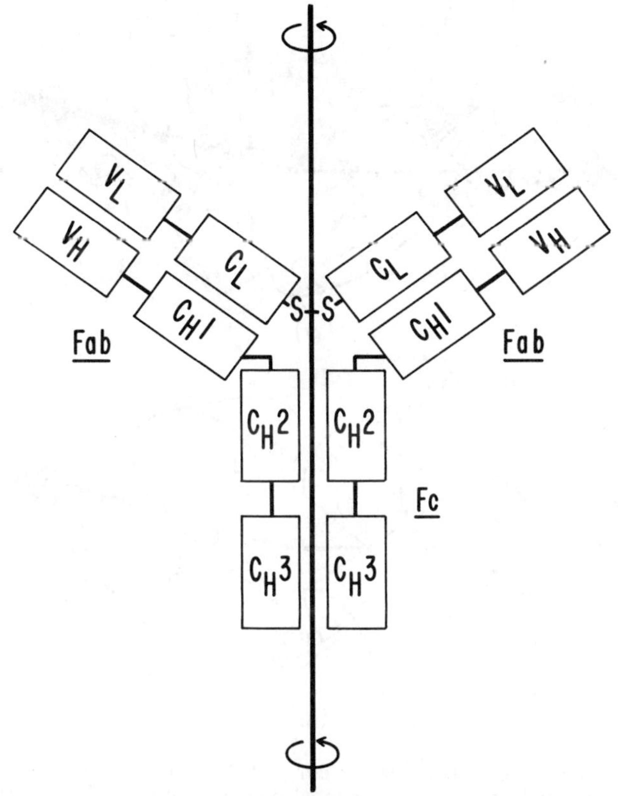

FIG. 1. Schematic drawing of the domain structure of the serum IgG1 protein from a patient (Mcg) with multiple myeloma and amyloidosis. Domains of the light and heavy chains are denoted by the subscripts L and H. The amino-terminal ("variable") domains are designated V and the remaining domains, which are relatively "constant" in immunoglobulins of the same class, are labeled C. Parts of the molecule involved in antigen-binding and in effector functions are marked Fab and Fc. The crystallographic twofold axis of rotation is indicated by the line between halves of the molecule. The interchain disulfide bond connecting penultimate residues of the λ-type light chains is represented by an —S—S—. Since the Mcg heavy chain has a deletion of 15 amino acid residues in the "hinge" region between C_H1 and C_H2, the normal interchain disulfide bonds are missing between heavy and light chains and between the two heavy chains.

Conformational Differences in the V Domains

The pseudotwofold symmetry is not strictly maintained in the V domain dimer. Some of the conformational variations are attributable to differences in the crystal packing forces on the two domains (see Fig. 3). For example, the V domain of monomer 1, dimer I, is closely approached by monomer 2, dimer II, in the crystal lattice. As a result of the interactions, part of the binding cavity of dimer I is capped and somewhat distorted by the complementary surfaces of dimer II (see Fig. 4).

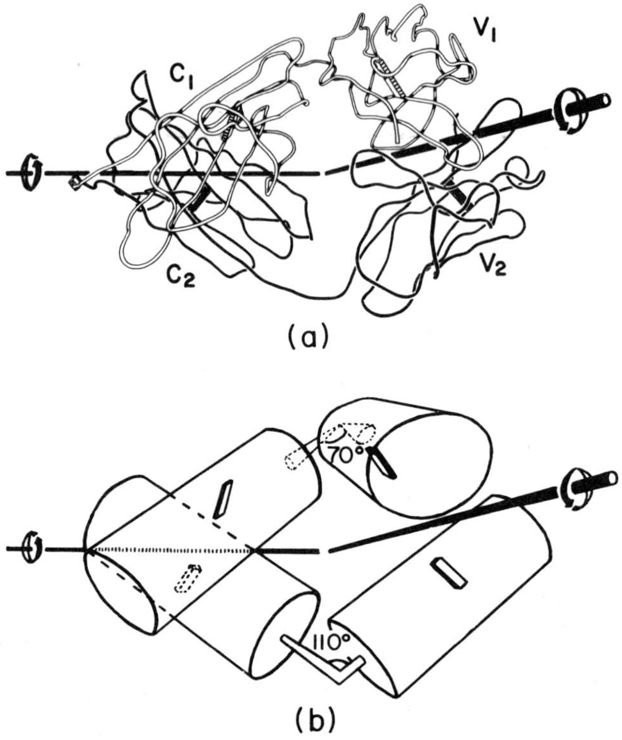

FIG. 2. a: Photograph of the model (2 cm = 1 Å) of the polypeptide chains of the Bence-Jones dimer, with superimposed local pseudotwofold axes of rotation between pairs of like domains. The pseudodiad between the V domains passes through the solvent channel in which hapten-like molecules are bound. The pseudotwofold axis between the C domains passes through a solvent-free zone in which side chains interact to stabilize the C_1–C_2 dimer. Disulfide bonds are represented by bars, with the interchain bond on the extreme left. Note the differences in the spatial relations between the V and C domains in the two monomers. **b:** Same view, but a schematic drawing in which the domains are represented by cylinders. The long axes of the cylinders of monomer 1 intersect in the "switch" region at an angle of ~70°, while the corresponding angle in monomer 2 is ~110°.

The switch region and adjoining segments of monomer 2, dimer II, pass over the cavity and the side chain of Leu 110 protrudes into the cavity to occupy a position close to Tyr 34 of monomer 1, dimer I (see site A′ in Fig. 4a,b). Lys 114 from the switch region of monomer 2, dimer II, is in contact with Asp 97 of monomer 2, dimer I. Glu 202 from the C-terminal segment of monomer 2, dimer II, interacts with Tyr 93 of monomer 2, dimer I. Tyr 93 is also in close proximity to the polypeptide backbone at positions 95 and 96. The interactions in the crystal lattice thus result in the blocking of site A′ along the solvent front of monomer 1 and a decrease in the accessibility of two side chains of monomer 2.

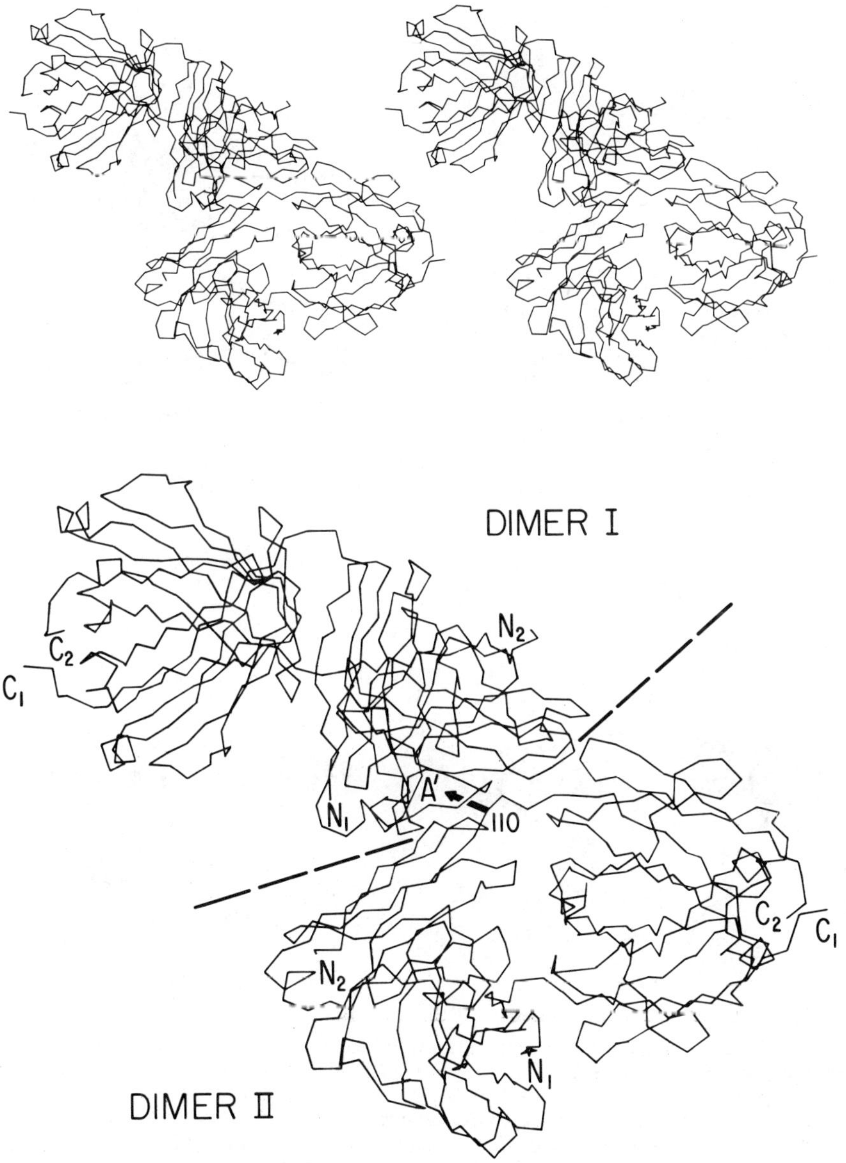

FIG. 3. Stereo view of two Bence-Jones dimers in the crystal lattice to show that part of the binding region of one molecule is blocked by interactions with a second dimer. Only the αC atoms of the polypeptide backbones are shown. A labeled monocular view is given below the stereo plot to indicate that the side chain of leucine 110 of monomer 2, dimer II, penetrates into site A′ of the binding cavity of dimer I (see Fig. 4a,b). The amino- and carboxyl-terminal residues of the monomers are marked with N and C, and the interface between dimers is indicated by a dashed line.

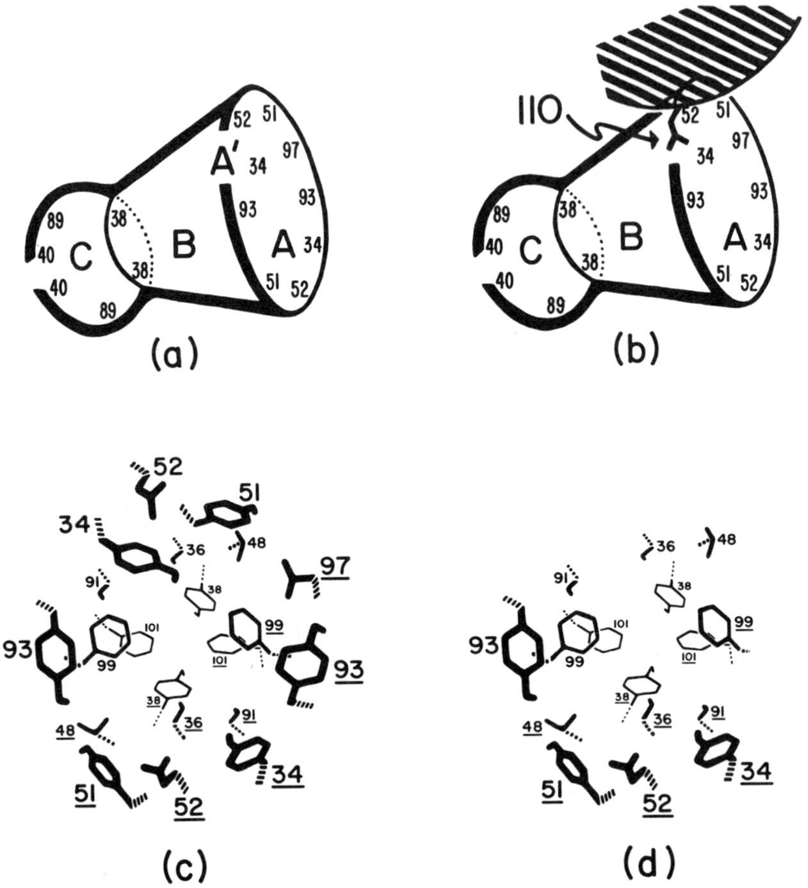

FIG. 4. a: Schematic drawing of the binding regions in the solvent channel between the V domains of the Bence-Jones dimer. Key side chains in the binding regions are numbered in appropriate positions. Sites A, B, and C bind various hapten-like molecules in crystals of the Bence-Jones dimer. **b:** Schematic drawing to illustrate the blocking of site A′ in the crystals, as shown in Fig. 3. **c:** Perspective drawing of the side chains lining the main binding cavity of the Bence-Jones dimer. The residue numbers of the constituents of monomer 2 are underlined. The bonds between the α and β carbon atoms of the side chains are indicated by dashed lines. All of these side chains are available for binding in solution. **d:** Drawing of the cavity lining available for binding in the crystal after the interactions shown in **b**.

Fortunately for the binding studies with the crystals, the remainder of the cavity is accessible to hapten-like molecules, and three binding sites have been identified along the solvent channel between the V domains (see sites A, B, and C in Fig. 4a). The expected differences between the accessible binding regions in crystals and in solution are illustrated in Fig. 4c,d.

The Binding of Hapten-like Molecules in Crystals and in Solution

Crystal sites A and B are located at successive levels in the conical main cavity, while the deepest site C is reached through a small opening in the floor of the main cavity (see Fig. 4). Twelve of the 21 side chains lining the main cavity are aromatic (see Fig. 4c), and it is apparent from their distribution that aromatic hapten-like molecules such as dinitrophenyl (Dnp) compounds should readily be bound.

The general properties of sites A and B were first defined by specific and covalent labeling of tyrosines 34 (monomer 2) and 38 (monomer 2) by diffusion of iodinated 1-fluoro-2, 4-dinitrobenzene (FIDNB) into the crystals (7). The locations of the Dnp groups were determined by difference Fourier analysis (see Fig. 5). Then Dnp and other aromatic compounds were diffused into crystals for noncovalent binding. It was found that sites A and B can be simultaneously occupied by two molecules of Dnp leucine. A and B act as a cooperative site for the binding of one molecule of di-Dnp lysine. While the Dnp leucine can be quantitatively removed from a crystal by washing with the crystallizing solution of 1.9 M ammonium sulfate, di-Dnp lysine remains tightly bound even after 2 weeks of perfusion. The contact residues in the binding of Dnp and other compounds are mainly aromatic in nature (Tyr 34 of monomer 2 in site A; Tyr 38 and Phe 101 of both monomers in site B; Phe 99 of the two monomers can contribute to both sites A and B; see Fig. 5e). Other residues are in close proximity and can probably influence the binding (e.g., Glu 52 and Tyr 51, monomer 2, near site A; Ser 36 and 91 of both monomers near site B). In solution the residues bordering site A′, as well as Tyr 93 and Asp 97 of monomer 2, should be available for binding (see Fig. 4c).

The results for the binding of di-Dnp lysine in solution are consistent with the conclusions based on the crystal binding studies. The binding constant for di-Dnp lysine is 7.6×10^4 liters/mole, as determined by equilibrium dialysis (J. R. Firca et al., *unpublished work*). Two molecules are bound per dimer, as expected if site A′ is available. However, sites A and A′ are both more spatious than site B, and two molecules of di-Dnp lysine cannot penetrate as deeply as one molecule does in the crystal (i.e., without distortion of site B).

Antibodies and myeloma proteins which bind Dnp groups often show cross-reactivity with menadione (vitamin K_3), and with purine and

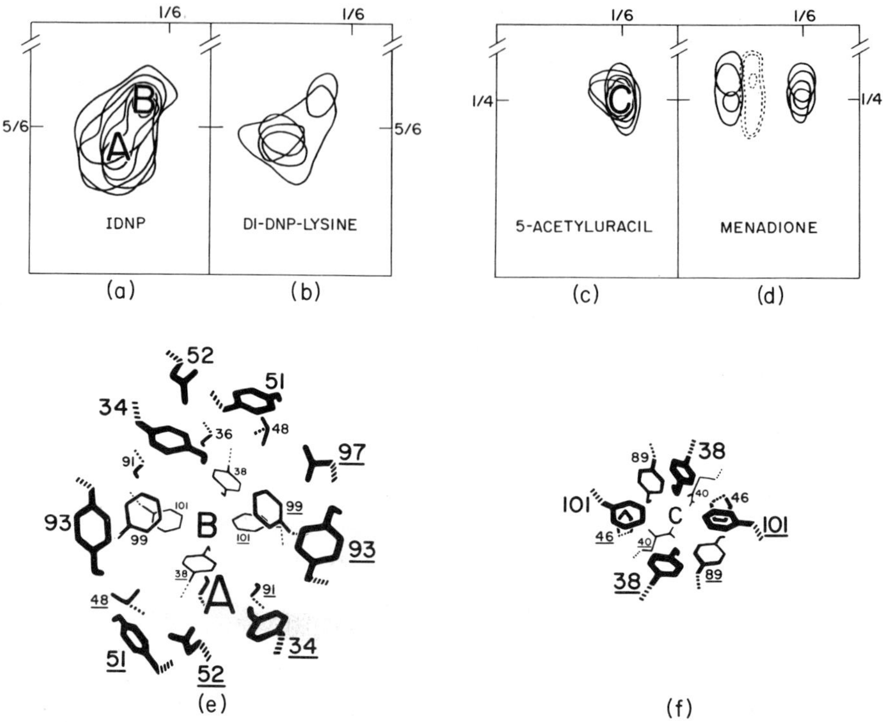

FIG. 5. a: Superposed sections of a 6.5-Å difference Fourier map for the Bence-Jones dimer after reaction with 1-fluoro-5-iodo-2,4-dinitrobenzene. The 5-iodo-2,4-dinitrophenyl (IDnp) groups are covalently attached to tyrosines 34 and 38 of monomer 2 (see Fig. 5e). These specific labeling reactions define sites A and B. **b:** Difference Fourier map for the noncovalent binding of di-Dnp-lysine in sites A and B. **c:** Difference Fourier map for the binding of 5-acetyluracil in site C. **d:** Difference Fourier map for the binding of menadione in site C. The negative trough and positive peak to the left of site C indicate that the polypeptide chain at positions 44 and 45 of monomer 2 were displaced from one position (trough) to the other (peak) by menadione; this is an example of induced fit in the binding of small molecules. **e:** Lining of the main binding cavity, showing the locations of sites A and B. **f:** Lining of the deep pocket, reached through the floor of the main cavity and containing site C.

pyrimidine derivatives (29–34). Crystals of the Mcg Bence-Jones dimer also bind such compounds, but not in site A. Menadione and 5-acetyluracil (5-AU) are found in site C, while caffeine and theophylline are bound in site B.

Local Conformational Changes Accompanying Binding of Hapten-like Molecules

The covalent labeling of Tyr 34 with FIDNB is accompanied by local conformational changes. The side chain of Tyr 34 is rotated about its α–β bond and the polypeptide chain in positions 34–36 is shifted ~2.5 Å. There is less space for side chain movements in site B, and the binding of purine derivatives does not cause detectable conformational changes in this part of the cavity.

Colchicine occupies sites A and B, and causes such extensive damage to the remainder of the main cavity that the crystals are destroyed within 2 days. Phenanthroline, which is bound with full occupancy in site A and about 20% occupancy in site C, also destroys the crystals. A similar pattern of crystal damage and dual binding in sites A and C is observed with ϵ-dansyl lysine. Both phenanthroline and dansyl lysine are too large to enter or remain in site C without distortion of the dimer. Entry of even smaller molecules like 5-AU through the floor of the main cavity can only be explained if the opening is expanded, presumably by rotation of the side chains of Tyr 38 from both monomers (see Fig. 5f). 5-AU is bound isomorphously in site C, but the binding of menadione provides an example of induced fit. The difference Fourier maps indicate that the polypeptide chain at residues 44 and 45 of monomer 2 is displaced to make room for the menadione [(7); see Fig. 5d].

Conclusions from the Binding Studies

Collectively, the studies indicate that the Bence-Jones dimer has multiple binding sites capable of acting alone or in combination to accommodate hapten-like molecules of various sizes and chemical structures. The solvent channel can be expanded to permit binding of an oversized compound with the proper affinity. Local conformational changes and examples of induced fit can be studied by crystallographic techniques. Destruction of the crystals indicates more extensive changes that may have to be assessed by solution methods.

Conformational Differences in the C Domains

The pseudotwofold symmetry between the C domains is generally maintained except in the C-terminal sequences preceding the interchain disulfide bond. In contrast to monomer 1, the polypeptide chain in monomer 2 makes an approximate right-angled turn at proline 212 (see Fig. 6a). After reduction of this bond either in the crystal or in solution, a mercury atom can be inserted between the sulfur atoms without major perturbations of the conformations in this region (4). The Bence-Jones dimer can also be reversibly reduced and alkylated, dissociated in acid, and reassembled to the original disulfide-linked form (J. R. Firca et al., *unpublished work*).

The flexibility permitting conformational isomerism in both the C-terminal and switch regions markedly influences this reassembly process, and appears to be a fundamental feature of the light chain structure. Flexibility and conformational isomerism are particularly important in understanding the

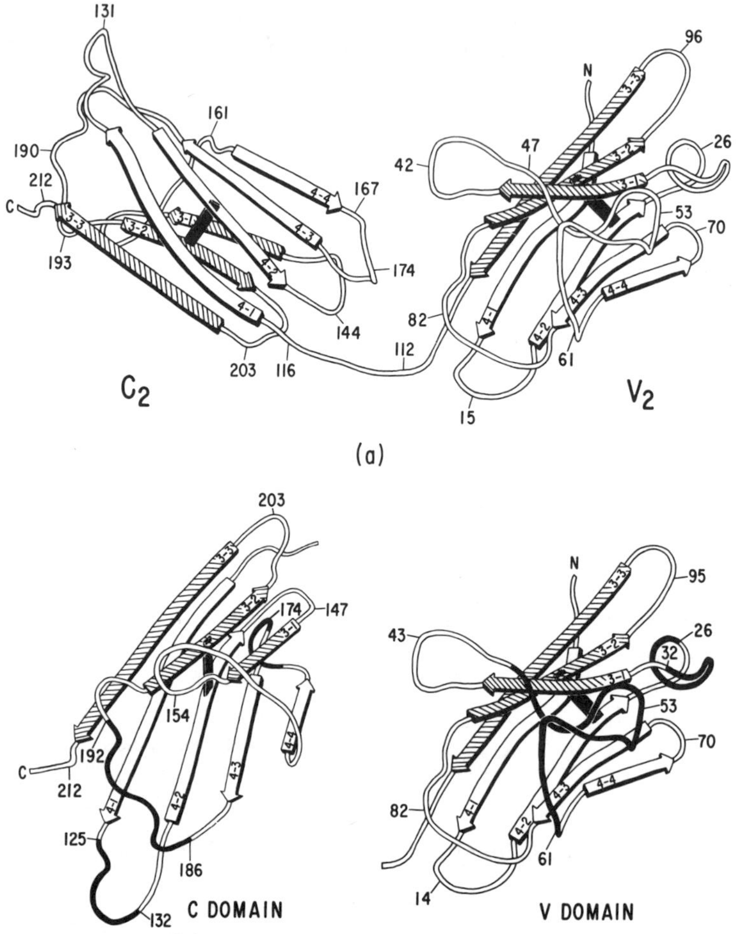

FIG. 6. a: Drawing of the polypeptide chain of monomer 2, with directional arrows superimposed to indicate strands participating in antiparallel β-pleated sheets. Three-chain layers are represented by numbered striated arrows, and four-chain layers by white arrows. Residue positions are numbered to aid in the correlation of sequence and structure. **b** and **c**: For comparison of their structures, the C domain is rotated and placed in a position approximating that of the V domain. Chain segments present in only one domain or having different structures in the two domains are shown as solid black lines. Since the V and C domains have different sequences, but similar structures after rotation, we refer to them as rotational allomers.

diversity in the kinds of complexes formed by the Mcg type of light chain: i.e., the light–heavy covalent complexes with α, γ, and μ chains; the noncovalent light–heavy complex with a light–light interchain disulfide bond in the Mcg IgGl protein; and the covalent complex of conformational isomers in the Bence-Jones dimer. Since the dimer resembles an Fab fragment in both structure and function, we have proposed that it be considered a model for a primitive antibody (6,7).

ROTATIONAL ALLOMERISM AND DIVERGENT EVOLUTION OF V AND C DOMAINS

Basic Immunoglobulin Fold

The similarities in the structures of the V and C domains in the Bence-Jones dimer (5,6) and Fab fragments (16–24) support the hypothesis of a common ancestral gene (35,36). The basic immunoglobulin fold in each domain is a layered structure. One layer is a β-pleated sheet with four antiparallel segments, and the second is a three-stranded pleated sheet (see Fig. 6). The layers are linked by the intrachain disulfide bond, which is shielded from solvent by hydrophobic side chains. For comparative purposes the domains can be considered as cylinders of β-pleated sheets enclosing hydrophobic interiors.

As shown in Fig. 6, the layers are less distinct in the V domain. For example, strand 1 in the four-chain layer of the V domain closely parallels segment 3 of the three-chain layer for part of its length. One of the other notable differences involves the loop containing the second hypervariable region (residues 51–55), which is not present in the C domain (see Fig. 6). In general the segments present in only one domain are located at the ends of the cylinders in positions appropriate for their functions but not deleterious to the maintenance of the basic layered fold (see segments indicated by solid black lines in Fig. 6b,c).

With the completion of the atomic model of the dimer at Argonne and the amino acid sequence of the light chain by Fett and Deutsch (12), we attempted to define the structural basis for the retention of this common fold. At the same time we sought an explanation for what C. C. F. Blake (37) calls the "functional differentiation" of the V and C domains.

Rotational Allomerism of the V and C Domains

Divergence of the V and C domains during evolution can be explained in terms of rotational allomerism (8). The cylinders of β-pleated sheets have rotated in such a way that the homologous three- and four-chain layers in the two domains perform different functions in the association of pairs of light chains or complexes of light and heavy chains (see Figs. 6 and 7). In the

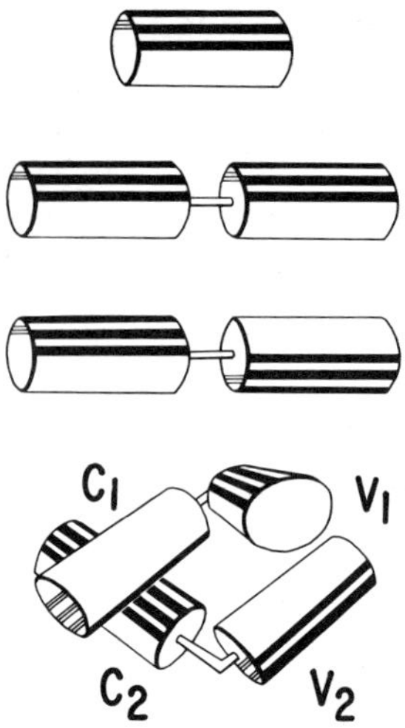

FIG. 7. Possible evolutionary pathway from an ancestral domain to the present Bence-Jones dimer. Four-chain layers are indicated by stripes, and the three-chain layers occupy the blank parts of the cylinders.

Bence-Jones dimer, V domain dimers (26–28), or in Fab fragments (16–24), the three-chain layers of the V domains face each other across the solvent channel in which hapten-like molecules are bound (see earlier section and Fig. 6). The corresponding layers provide most of the external surfaces of the dimeric module of C domains. In contrast, the important interactions stabilizing the compact arrangement of the C domains occur across the solvent-free zone between the four-chain layers. These layers form the external surfaces in the V_1–V_2 dimer. Rotational allomerism thus markedly influences not only antigen binding by the V domain dimer and stabilization of the C domain dimer, but also the surface properties which are involved in the formation of amyloid fibrils and in the thermal behavior of Bence-Jones proteins.

The patterns of amino acid sequences in the β-pleated sheets are consistent both with the observed similarities in the V and C domains and with the changes resulting from rotational allomerism. In pleated sheets the side chains emerge on alternating sides of the polypeptide backbone. With only two layers of pleated sheets in a domain, one set of side chains in each layer is directed toward the interior and the other set is oriented toward the surface of the cylinder. To make functional changes in a cylinder by rotation, it is necessary only to change the set of surface residues. The internal set can be left relatively undisturbed.

The pleated sheets providing the external surfaces of the Bence-Jones dimer generally have an alternating pattern of external polar and internal hydrophobic residues (see sequences of four-chain layers of the V domains and three-chain layers of the C domains in Fig. 8). The preservation of the hydrophobic character of these key internal sites largely explains the marked similarities in the overall structures of the V and C domains.

The alternating pattern of polar and apolar residues seen in the surface layers is broken in the remaining pleated sheets of the dimer. Along the expected "polar" side of the three-chain layer in the V domains, for example, tyrosyl and phenylalanyl residues are substituted for aliphatic polar residues in positions important for the maintenance of the general architecture of the hapten-binding sites (see Fig. 8). These and other side chains along the three-stranded layer also act as contact residues in the binding. The combined observations lead to the conclusion that binding specificity in immunoglobulins can be changed by a limited number of substitutions along only one side of a pleated sheet.

The alternating sequence pattern is also interrupted in the four-chain layers of the C domains by the substitution of hydrophobic for polar residues at sites of interactions in the C_1–C_2 interface (see Fig. 8). Again, stabilization of the allomer after rotation is achieved by the replacement of a small number of residues along one side of a pleated sheet.

DIFFERENCES IN SURFACE PROPERTIES OF THE V AND C DIMERIC MODULES RESULTING FROM ROTATIONAL ALLOMERISM

External Surfaces of the V and C Dimers

Rotational allomerism results in substantial differences in the side chains exposed to solvent on the outer surfaces of the V_1–V_2 and C_1–C_2 dimeric modules. In the four-chain layers of the V domains the external polar constituents in the alternating sequences are mainly seryl and threonyl residues (see Fig. 8). Surface residues on the polar side of the three-chain pleated sheets in the C domains are more diversified. In particular, there is a higher proportion of residues with accessible, large side chains. Such residues can serve as recognition sites for proteolytic enzymes, and the C domains of Bence-Jones proteins are generally more susceptible to extensive enzymatic cleavage than V domains at physiological temperatures (38).

It should be emphasized that even the reduced and alkylated Mcg Bence-Jones protein remains a dimer in solution until the pH is lowered to about 2.5. For most light chains, therefore, the dimeric form can be assumed to be the initial molecular species in enzymatic or thermal reactions, with different surface features on the V and C dimeric modules.

Formation of Amyloid Fibrils

The preceding observations are pertinent to studies of the type of amyloidosis in which the characteristic fibrils are produced by proteolytic cleavage of immunoglobulin light chains (39–43). In 10–20% of Bence-Jones proteins, fibril formation can be simulated *in vitro* by hydrolysis with pepsin (41–43). These fibrils are composed predominantly of large fragments of V

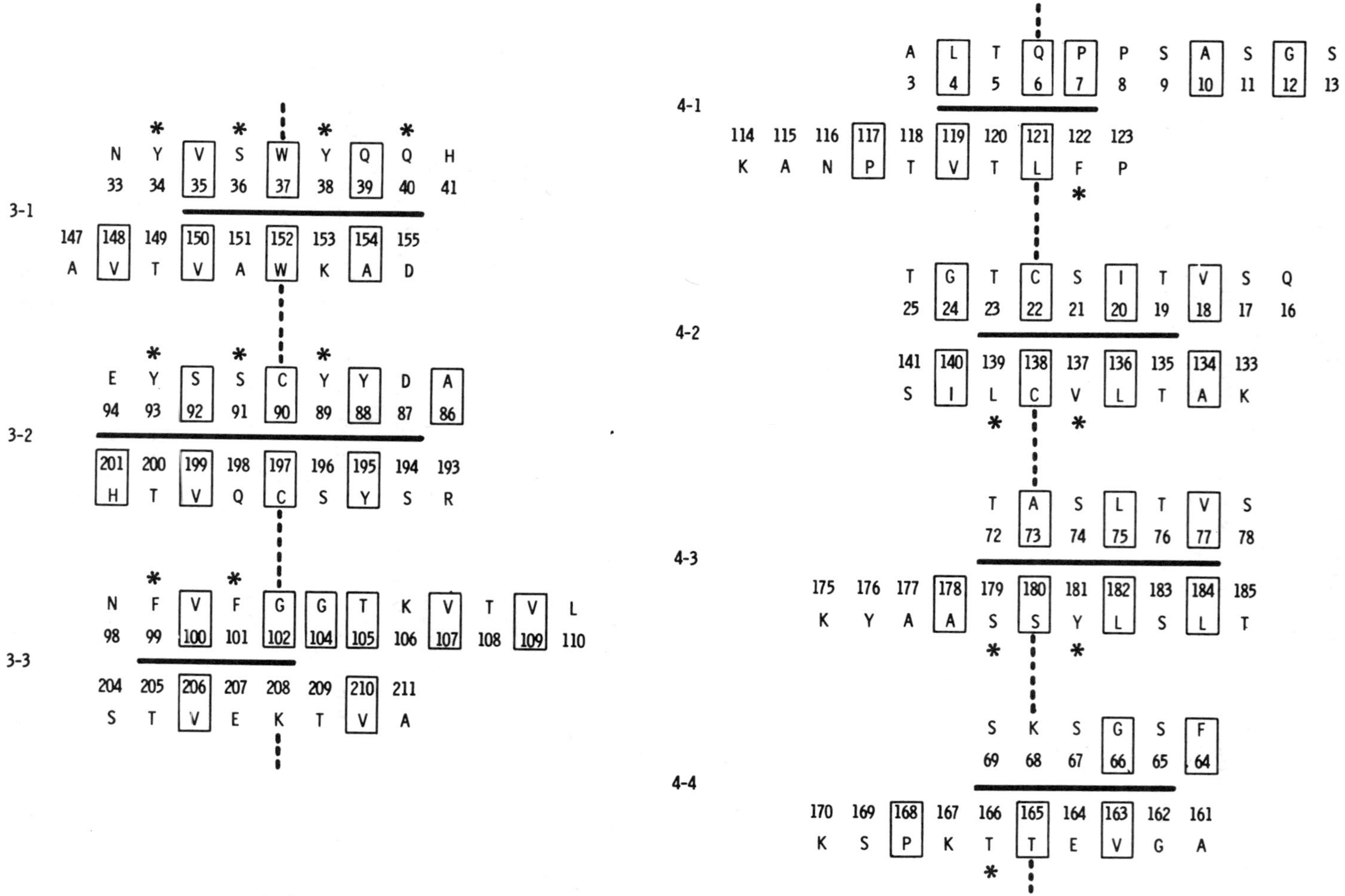

3-1
N Y V S W Y Q Q H
33 34 35 36 37 38 39 40 41
147 148 149 150 151 152 153 154 155
A V T V A W K A D
3-2
E Y S S C Y Y D A
94 93 92 91 90 89 88 87 86
201 200 199 198 197 196 195 194 193
H T V Q C S Y S R
3-3
N F V F G G T K V T V L
98 99 100 101 102 104 105 106 107 108 109 110
204 205 206 207 208 209 210 211
S T V E K T V A
4-1
A L T Q P P S A S G S
3 4 5 6 7 8 9 10 11 12 13
114 115 116 117 118 119 120 121 122 123
K A N P T V T L F P
4-2
T G T C S I T V S Q
25 24 23 22 21 20 19 18 17 16
141 140 139 138 137 136 135 134 133
S I L C V L T A K
4-3
T A S L T V S
72 73 74 75 76 77 78
175 176 177 178 179 180 181 182 183 184 185
K Y A A S S Y L S L T
4-4
S K S G S F
69 68 67 66 65 64
170 169 168 167 166 165 164 163 162 161
K S P K T T E V G A

(a) (b)

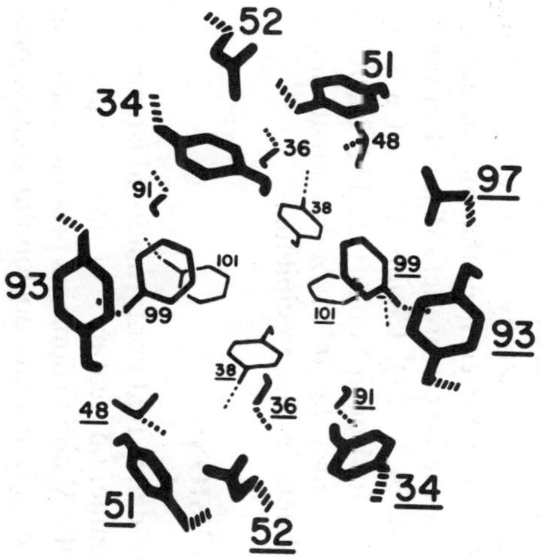

(c)

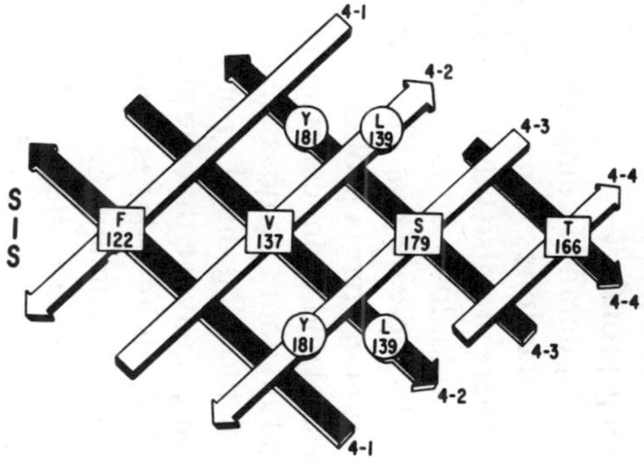

(d)

FIG. 8. Amino acid sequences, as determined by Fett and Deutsch (18), of the segments in the three-chain (**a**) and four-chain (**b**) layers of β-pleated sheets (see Fig. 6). Equivalent pairs of residues in the two domains are listed, with the V components on top. Abbreviations of residues with side chains oriented toward the interior of each domain are enclosed in boxes. In the three-chain layer of the V domain, the residues marked with asterisks are constituents of the binding regions shown in (**c**) (also see Fig. 4). Note the contrast between the V and C domains in these positions. In the four-chain layer of the C domain, the residues designated by asterisks occupy the interdomain space and help to stabilize the C_1–C_2 dimer (**d**). The four-chain layers of the C domains of monomer 1 and 2 are schematically represented by white and black arrows. Abbreviations in squares indicate the approximate positions of the αC atoms of residues facing each other in the interdomain space. Other important residues contributing to the stabilization of the C_1–C_2 dimer are denoted by circles.

domains, a finding consistent with the distribution of external bulky side chains in the V and C domains of the Mcg protein.

Although the patient Mcg had amyloidosis, detailed comparisons have to be made with caution because the Mcg Bence-Jones dimer does not form amyloid fibrils when treated with pepsin. There are also other cases in which Bence-Jones proteins from patients with amyloidosis do not produce fibrils after peptic hydrolysis (42,43). These results indicate that the specificity of the enzyme(s) responsible for fibril formation *in vivo* is probably not identical with that of pepsin. However, the physical and chemical properties of the natural and induced fibrils are similar, and it should be possible to correlate these properties with the atomic model of the Mcg dimer. For example, it can be concluded from the sites of enzymatic cleavage that the characteristic β-pleated sheet conformation in amyloid fibrils (44–46) is derived mainly from the four-chain layer and chain 3-1 of the V domain. Moreover, there is a preponderance of λ chains among amyloidogenic proteins (47), an indication that key surface features may be different in V dimeric modules of κ chains.

Thermal Properties of Bence-Jones Proteins

Solomon and McLaughlin (48) found that the characteristic thermal behavior of Bence-Jones proteins was associated only with the V domains. Again, the differences in the surface structures of the V and C dimers probably have a strong influence on the thermal properties. In particular, segments of the four-chain layers of the V domains have the most extensive alternating hydroxy-apolar sequences in the Mcg protein, and are well-suited for the formation of micelles of highly ordered β-pleated sheets. The pleated sheet bands in infrared patterns are generally enhanced after heat treatment of Bence-Jones proteins (46). Other differences that may affect the thermal properties include the fact that the layers are less distinct in the V domains of both λ and κ chains (5,19,24,26–28). Chains 4-1 and 3-3 are sufficiently close in one region to form hydrogen bonds (see Fig. 6). After heat treatment, therefore, both layers may be in favorable positions to contribute to the observed pleated sheets in the precipitates.

CONCLUSIONS

In this article we have presented evidence that conformational isomerism and rotational allomerism are fundamental features of immunoglobulin structures. The differences between the V and C domains provide a striking example of divergent evolution. A common immunoglobulin fold is preserved by the maintenance of the hydrophobic character of key internal sites in cylindrical domains differing substantially in amino acid sequences. Modifications of the functions of the domains have occurred as a result of the rotational allomerism. Substitutions among important residues in both the three- and four-chain layers are required for a stable transition of one allomer to the other. The collective findings are incorporated into a possible evolutionary pathway, as illustrated in Fig. 7.

ACKNOWLEDGMENT

This work was supported by the U. S. Energy Research and Development Administration.

REFERENCES

1. Schiffer, M., Hardman, K. D., Wood, M. K., Edmundson, A. B., Hook, M. E., Ely, K. R., and Deutsch, H. F. (1970): *J. Biol. Chem.,* 245:728.
2. Edmundson, A. B., Wood, M. K., Schiffer, M., Hardman, K. D., Ainsworth, C. F., Ely, K. R., and Deutsch, H. F. (1970): *J. Biol. Chem.,* 245:2763.
3. Edmundson, A. B., Schiffer, M., Ely, K. R., and Wood, M. K. (1972): *Biochemistry,* 11:1822.
4. Ely, K. R., Girling, R. L., Schiffer, M., Cunningham, D. E., and Edmundson, A. B. (1973): *Biochemistry,* 12:4233.
5. Schiffer, M., Girling, R. L., Ely, K. R., and Edmundson, A. B. (1973): *Biochemistry,* 12:4620.
6. Edmundson, A. B., Ely, K. R., Girling, R. L., Abola, E. E., Schiffer, M., and Westholm, F. A. (1974): *Prog. Immunol. II,* 1:103.
7. Edmundson, A. B., Ely, K. R., Girling, R. L., Abola, E. E., Schiffer, M., Westholm, F. A., Fausch, M. D., and Deutsch, H. F. (1974): *Biochemistry,* 13:3816.
8. Edmundson, A. B., Ely, K. R., Abola, E. E., Schiffer, M., and Panagiotopoulos, N. (1975): *Biochemistry,* 14:3953.
9. Deutsch, H. F. (1971): *Seibutsu Butsuri Kagaku,* 16:73.
10. Deutsch, H. F., and Suzuki, T. (1971): *Ann. NY Acad. Sci.,* 190:472.
11. Fett, J. W., Deutsch, H. F., and Smithies, O. (1973): *Immunochemistry,* 10:115.
12. Fett, J. W., and Deutsch, H. F. (1974): *Biochemistry,* 13:4102.
13. Fett, J. W., and Deutsch, H. F. (1975): *Immunochemistry,* 12:643.
14. Ein, D., and Fahey, J. L. (1967): *Science* 156:947.
15. Hess, M., Hilschmann, N., Rivat, L., Rivat, C., and Ropartz, C. (1971): *Nature* [*New Biol.*], 234:58.
16. Poljak, R. J., Amzel, L. M., Avey, H. P., Becka, L. N., and Nisonoff, A. (1972): *Nature* [*New Biol.*], 235:137.
17. Poljak, R. J., Amzel, L. M., Avey, H. P., Chen, B. L., Phizackerley, R. P., and Saul, F. (1973): *Proc. Natl. Acad. Sci. USA,* 70:3305.
18. Amzel, L. M., Poljak, R. J., Saul, F., Varga, J. M., and Richards, F. F. (1974): *Proc. Natl. Acad. Sci. USA,* 71:1427.
19. Poljak, R. J., Amzel, L. M., Chen, B. L., Phizackerley, R. P., and Saul, F. (1974): *Proc. Natl. Acad. Sci. USA,* 71:3440.
20. Poljak, R. J. (1975): *Nature,* 256:373.
21. Padlan, E. A., Segal, D. M., Spande, T. F., Davies, D. R., Rudikoff, S., and Potter, M. (1973): *Nature* [*New Biol.*], 245:165.
22. Segal, D. M., Padlan, E. A., Cohen, G. H., Rudikoff, S., Potter, M., and Davies, D. R. (1974): *Proc. Natl. Acad. Sci. USA,* 71:4298.
23. Padlan, E. A., and Davies, D. R. (1975): *Proc. Natl. Acad. Sci. USA,* 72:819.
24. Davies, D. R., Padlan, E. A., and Segal, D. M. (1975): *Annu. Rev. Biochem.,* 44:639.
25. Strosberg, A. D., Fraser, K. J., Margolies, M. N., and Haber, E. (1972): *Biochemistry,* 11:4978.
26. Epp, O., Colman, P., Fehlhammer, H., Bode, W., Schiffer, M., Huber, R., and Palm, W. (1974): *Eur. J. Biochem.,* 45:513.

27. Fehlhammer, H., Schiffer, M., Epp, O., Colman, P. M., Lattman, E. E., Schwager, P., Steigemann, W., and Schramm, H. J. (1975): *Biophys. Struct. Mech.* 1:139.
28. Wang, B.-C., and Sax, M. (1974): *J. Mol. Biol.,* 87:505.
29. Eisen, H. N., Little, J. R., Osterland, C. K., and Simms, E. S. (1967): *Cold Spring Harbor Symp. Quant. Biol.,* 32:75.
30. Underdown, B. J., and Eisen, H. N. (1971): *J. Immunol.,* 106:1431.
31. Schubert, D., Jobe, A., and Cohn, M. (1968): *Nature,* 220:882.
32. Schubert, D., Roman, A., and Cohn, M. (1970): *Nature,* 225:154.
33. Potter, M. (1972): *Physiol. Rev.,* 52:631.
34. Riesen, W., and Morell, A. (1972): *Immunochemistry,* 9:979.
35. Singer, S. J., and Doolittle, R. F. (1966): *Science,* 153:13.
36. Hill, R. L., Delaney, R., Fellows, R. E., Jr., and Lebovitz, H. E. (1966): *Proc. Natl. Acad. Sci. USA,* 56:1762.
37. Blake, C. C. F. (1975): *Nature,* 257:447.
38. Seon, B.-K., Roholt, O. A., and Pressman, D. (1972): *J. Biol. Chem.,* 247:2151.
39. Glenner, G. G. (1973): *Br. J. Haematol.,* 24:533.
40. Glenner, G. G., Terry, W., Harada, M., Isersky, C., and Page, D. (1971): *Science,* 172:1150.
41. Glenner, G. G., Ein, D., Eanes, E. D., Bladen, H. A., Terry, W., and Page, D. L. (1971): *Science,* 174:712.
42. Linke, R. P., Zucker-Franklin, D., and Franklin, E. C. (1973): *J. Immunol.,* 111:10.
43. Linke, R. P., Tischendorf, F. W., Zucker-Franklin, D., and Franklin, E. C. (1973): *J. Immunol.,* 111:24.
44. Eanes, E. D., and Glenner, G. G. (1968): *J. Histochem. Cytochem.,* 16:673.
45. Bonar, L. C., Cohen, A. S., and Skinner, M. M. (1969): *Proc. Soc. Exp. Biol. Med.,* 131:1373.
46. Termine, J. D., Eanes, E. D., Ein, D., and Glenner, G. G. (1972): *Biopolymers,* 11:1103.
47. Pick, A. I., and Osserman, E. F. (1968): In: *Amyloidosis,* p. 100, edited by E. Mandema, L. Ruinen, J. H. Scholten, and A. S. Cohen. Excerpta Medica, Amsterdam.
48. Solomon, A., and McLaughlin, C. L. (1969): *J. Biol. Chem.,* 244:3393.

Antibodies in Human Diagnosis and Therapy,
edited by Edgar Haber and Richard M. Krause.
Raven Press, New York © 1977.

Polymorphism of Antibodies Against a Single Determinant in Inbred Mice

Frank F. Richards

I should like to describe some work that several students, Jerome Zeldis and Ancil Jones, have been doing in our group, under Dr. Varga's supervision. It relates to the polyfunctionality of individual antibodies. They have been following the inheritance of a single V-region in an inbred strain of mice. This bears on the old question of whether there are in the ovum all the genes necessary for antibody diversity, or whether in between the ovum and the immunocompetent organism there is expansion in gene number. If you make N (the gene number expansion factor) very small, for example, 1.003, then you are essentially saying that there is the occasional mutation, but that the number of genes is not changed to any great extent between the ovum and the adult. But if you make the factor N 10 or larger, then in addition to the original gene present in the germ line you must have an additional 9 genes present in the adult. I recognize that each one of the mutations may not effect specificity; nevertheless, if one believes that a somatic process is responsible for antibody diversity, this should be reflected in changes in V-region specificity. We are looking at the inheritance of one individual V-region in an inbred mouse line to see if there is any change in the antigen specificity.

You cannot follow the inheritance of a single V-region if you use the specificity of the V-region for a single small determinant as a marker. At any one time in the mouse's immune response, there are perhaps 100 to 200 different anti-Dnp antibodies expressed and you cannot distinguish between them by isoelectric focusing. It is much simpler to follow the inheritance of double-binding antibodies because for any pair of antigens there are many fewer antibodies.

In fact, in BALB/c mice there are only two or perhaps three cell clones that produce antibodies that bind both Dnp plus menadione. Once the two or three antibodies are identified, it is easy to characterize them further by looking at other binding characteristics they possess. The K_0 for Dnp and for menadione can be measured, and their position on isoelectric focusing gels can be determined. In addition, you can look at their reaction with anti-idiotype antibodies produced against myeloma protein having similar binding activities. Finally, the inhibition by hapten of the idiotype-anti-idiotype reaction can be examined, until there are about five or six different binding functions to measure. The statistical chance that any one V-region shares all these characteristics with another V-region becomes small. Double-binding antibodies may be induced by methods we have described previously.

Using Dnp and menadione as our antigen pair, we find that in BALB/c mice there are only two clones of cells that produce antibodies binding both Dnp and menadione. One antibody forms a triplet of lines on IEF at pH 7.5 and we call this the DM-1 pattern. The other forms a triplet at pH 6.3 and we call this the DM-2 pattern. The hapten-binding constants are different for DM-1 and DM-2 although they are the same for each band within each DM-1 or DM-2 pattern. The interesting point is that individual BALB/c mice, when they express double-binding antibodies, show either the DM-1 or the DM-2 pattern; we have not yet found a mouse that shows both patterns. When we bred DM-1 mice, those members of the F_1 generation that expressed double-binding antibodies showed only the DM-1 pattern. The DM-2 mice behaved in the same way. Heterologous crosses produced both DM-1 and DM-2 pattern mice. This seemed to suggest that we had different structural genes for antibodies within an inbred mouse strain. We found this very hard to believe. Jerry Zeldis then did the crucial experiment. He overloaded the IEF gels and overexposed them. Under these conditions he found that DM-1 pattern mice also have very low quantities of DM-2 pattern antibodies and vice versa. It now appears that there are differences of gene expression responsible for these patterns, and that there are different high-responder and low-responder clones present in individual BALB/c mice. We are also now beginning to spot-check individual DM-1 and DM-2 antibodies in F_1 and F_2 generations to see if we can detect changes in binding patterns.

The thing that has really diverted us from this is the finding of this curious polymorphism by which antibodies against a single determinant in a single inbred mouse strain may come from a high-responder or a comparatively low-responder clone. Now what is very interesting is the parallel of this phenomenon to the IR gene system in which there are no individual clones, but rather entire groups of high-responder or low-responder clones.

One could speculate that the IR gene phenomenon developed from the inbreeding of this type of polymorphism when the mouse strains developed, but there is no evidence for this.

Prospects for the Chemical Synthesis of Antibodies

Antibodies in Human Diagnosis and Therapy,
edited by Edgar Haber and Richard M. Krause.
Raven Press, New York © 1977.

Prospects for the Chemical Synthesis of Antibodies

Bruce Merrifield

Two remarkable stories have been presented concerning the primary structure and the three-dimensional structure of antibodies and the ways in which antigens are bound to them. The next logical stage in the chemistry of this important class of compounds is their synthesis. In principle this would provide the ultimate proof of structure, but, more important, the synthesis of all or part of an immunoglobulin would provide a way to test and extend the information on antigen-binding sites that has been obtained by other methods. A detailed structure-function study should be possible through the design and synthesis of suitable analogs.

The first important question to consider is whether or not it is reasonable to think in terms of the synthesis of a molecule with 1,200 amino acid residues, or 400, or even 200. It is clear that these are very large molecules compared with the ordinary organic compounds one is used to dealing with in the laboratory. I think it is fair to conclude that 200 residues would be an upper limit to consider at this time by any method now available. Fortunately, it has become possible to obtain the variable regions of both heavy and light chains of certain immunoglobulins that can be reassociated to form functional binding sites. Therefore, we are now in a position to carry out meaningful synthetic experiments on protein chains of only 100 residues. I think this really is a feasible goal.

Two general approaches to the synthesis of proteins are possible: by classic or by solid-phase methods. In the former the reactions are carried out in solution and each intermediate is isolated, purified, and characterized before proceeding to the next step. It is, of course, a very big task to prepare over 100 separate pure compounds, and such an undertaking will require a major commitment. This approach has been extremely successful in the synthesis of peptides up to about 40 residues (for example, the 21- and 30-residue chains of insulin and the 39 residue chain of ACTH). Furthermore, intensive work has been under way for some time on proteins such as RNase T_1, lysozyme, proinsulin, and myoglobin, and the completion of some of these syntheses can be expected soon.

The solid-phase method is appreciably easier and faster to perform, but in its simplest form it does not allow for purification of intermediates. It therefore imposes an added burden of purification at the end of a synthesis. Solid-phase synthesis has also been applied to insulin and ACTH, to the proteins RNase A, RNase T_1, staphylococcal nuclease, acyl carrier protein, trypsin inhibitor and myoglobin, and to a number of other interesting biologically active peptides.

The three laboratories represented in this section of the volume that have started synthetic work on the immunoglobulins have all selected the solid-phase method for their initial studies. The reader should therefore bear in mind that this section will give a rather biased view and will not deal adequately with all synthetic methods.

The first presentation discusses work from our laboratory and deals primarily with the various strategies of solid-phase synthesis, some of the special chemical problems associated with them, and some of the ways in which antibodies can be studied by these techniques. The next two chapters describe in greater detail some actual synthetic studies on the variable regions of certain immunoglobulins. These are all preliminary results and they should be looked at in the context of the title of this section, "Prospects for the Chemical Synthesis of Antibodies."

Antibodies in Human Diagnosis and Therapy,
edited by Edgar Haber and Richard M. Krause.
Raven Press, New York © 1977.

Strategies for Solid-Phase Synthesis of Immunoglobulin Domains

Bruce W. Erickson and David E. Krieger

The Rockefeller University, New York, New York 10021

The chemical synthesis of antibodies is an ambitious goal that has only recently received serious attention. Current efforts to explore the combining site of antibodies by the chemical synthesis of immunoglobulin domains are a logical extension of recent progress in three separate areas of research. First, the covalent structure of immunoglobulins has been precisely described through chemical sequencing studies. Second, the three-dimensional arrangement and interactions of immunoglobulin domains have been clearly portrayed through X-ray crystallographic studies. Third, the chemical synthesis of enzymes and other proteins has been achieved through the solid-phase method of peptide synthesis. Knowledge of the correct sequence of amino acids present in an immunoglobulin domain is necessary before synthesis of the domain or its analogues can begin. Similarly, knowledge of the spatial relations between sequentially distant residues in the antibody combining site is important for understanding the influence of structural changes on the efficiency of antigen binding.

The covalent and three-dimensional structures of the combining sites of several immunoglobulins are described in the preceding chapters of this volume. This chapter outlines several strategies that should be useful for exploring the structure of immunoglobulin domains through solid-phase synthesis. The solid-phase method is described briefly and some chemical tactics for the synthesis of proteins are presented. Then the relevant structural features of the variable domain of the heavy chain of the phosphorylcholine-binding immunoglobulin M603 are reviewed and specific strategies for the assembly of this domain are proposed. Finally, a new and general strategy for affinity purification of polypeptides assembled by the solid-phase method is explained and demonstrated, and its possible application to the synthesis of an immunoglobulin domain is outlined.

SOLID-PHASE PROTEIN SYNTHESIS

The Solid-Phase Method

The assembly of a peptide from its constituent amino acids using solution techniques is a laborious process that often suffers from low yields due to mechanical losses and limited solubility of intermediates. Twelve years ago Merrifield (27) demonstrated a rapid procedure that avoids these problems by keeping the growing peptide chain covalently attached to an insoluble solid support. The basic idea is illustrated by the dipeptide synthesis outlined in Fig. 1. The carboxyl-terminal amino acid of the desired peptide is attached to small beads of cross-linked polystyrene by a benzyl ester bond. After the α-amino group is deprotected and neutralized, it is coupled with the carboxyl group of the next amino acid by formation of a peptide bond. The three steps of N^{α}-deprotection, neutralization, and coupling comprise one synthetic cycle, which can be repeated until the desired peptide has been assembled. Finally, the peptide is cleaved from the solid support and the remaining protecting groups are removed. After removal from the support, the peptide can be purified by conventional physical methods.

The solid-phase method offers several advantages over conventional solution techniques. The insoluble peptide intermediates can be separated from soluble reagents and by-products by repeated filtering and washing of the solid support. This operation avoids the mechanical losses normally encountered during isolation and purification of synthetic intermediates in solution. Since the soluble reagents can be used in excess, each reaction can be driven to completion and relatively high overall yields can be achieved. Stepwise assembly of the peptide chain is further simplified and accelerated by conducting all of the reactions in the same reaction vessel. As a result, each of the operations involved in extending the peptide chain can be carried out automatically. Solid-phase synthesis has been applied to more than 70 classes of peptides and more than a dozen proteins (17), including the 124-residue enzyme ribonuclease (21).

Proteins can be assembled on a solid support by two basic strategies. Stepwise assembly involves the addition of many protected amino acids, whereas fragment condensation involves addition of a few protected peptides. Specific chemical tactics required to achieve each of these synthetic strategies are discussed below.

The C-to-N Stepwise Strategy

Proteins can be assembled on a solid support by the stepwise addition of protected amino acids, as shown schematically in Fig. 2. The common C-to-N stepwise strategy involves attaching the carboxyl-terminal residue to the support and sequentially extending the polypeptide toward the amino terminus by addition of suitably protected and activated amino acids. When

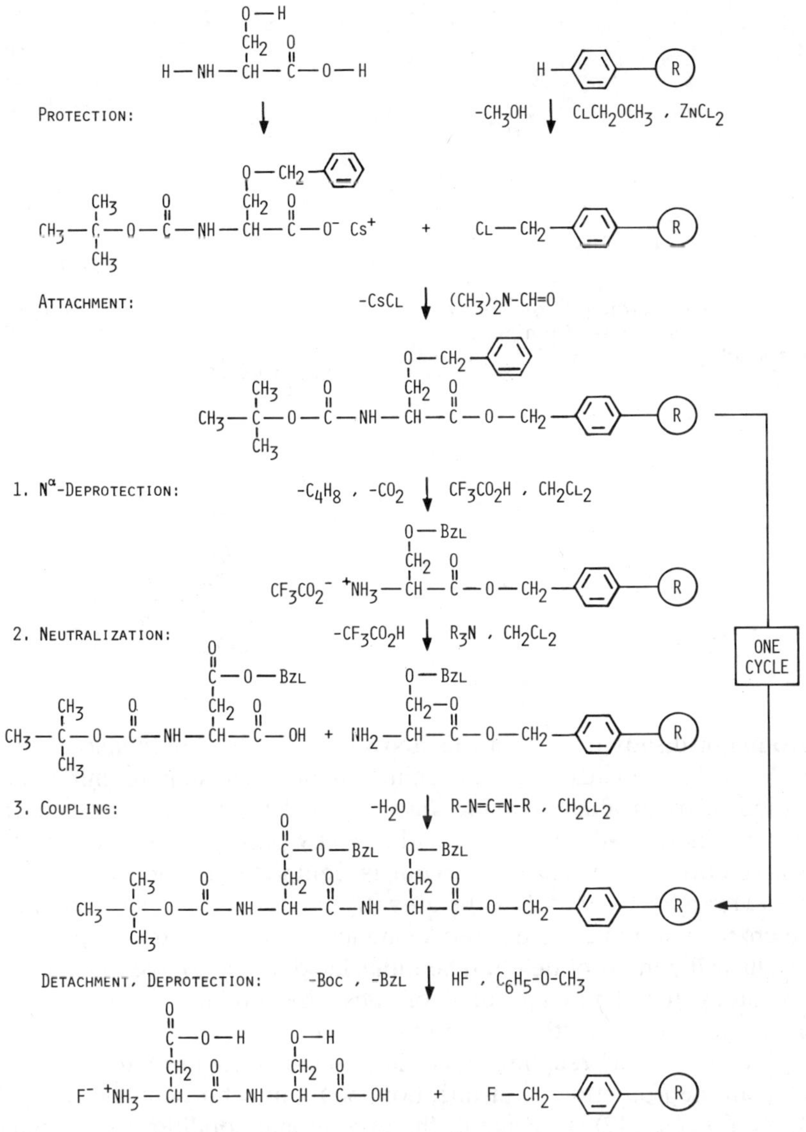

FIG. 1. A synthesis of the dipeptide aspartylserine by the solid-phase method.

the desired protein has been assembled, it is cleaved from the support and purified. This strategy has three major advantages. Only 20 readily available protected amino acids are needed to assemble any protein. These relatively inexpensive reagents can be used in significant excess to drive the coupling reactions to completion. Finally, racemization does not occur when the proper protecting groups and coupling methods are used.

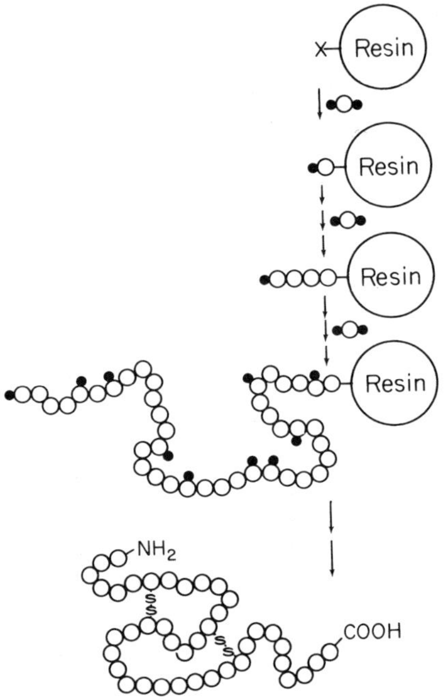

FIG. 2. General strategy for stepwise assembly of polypeptides from protected amino acids.

Two major disadvantages of the stepwise strategy should also be considered. Very high yields must be achieved at each step to minimize the formation of by-products lacking only a few of the residues present in the desired product. In addition, support-bound by-products cannot be removed until assembly of the desired protein is complete. If each amino acid is coupled at random to 99.5% of the growing chains, after 120 cycles only half of the chains will have the desired sequence of amino acid residues and the other half will consist of deletion peptides lacking one to five residues (1). A new strategy for removing some of these by-products is described and demonstrated in the section on affinity purification.

A wide variety of reaction conditions have been used to perform the attachment, deprotection, neutralization, coupling, and cleavage steps of the assembly process (17). In general, the appropriate conditions are limited by the nature of the protecting groups used to mask the α-amino, α-carboxyl, and side-chain functional groups. Stepwise assembly of the desired protein requires two levels of protecting group stability. During each cycle the temporary N^{α}-protecting group should be selectively removed under conditions that leave the C^{α}-link to the solid support and all side-chain protecting groups intact.

The essential elements of the Boc—benzyl tactic for stepwise synthesis are shown in Fig. 3. This common stepwise tactic uses the *tert*-

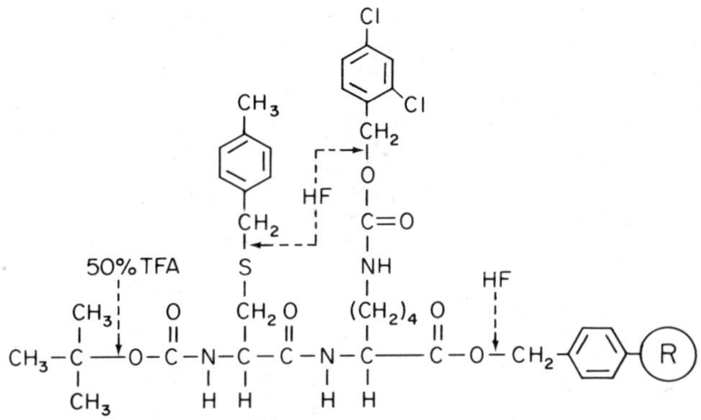

FIG. 3. The Boc—benzyl tactic for stepwise synthesis of unprotected peptides.

butyloxycarbonyl (Boc) group for N^{α}-protection and benzylic groups for both side-chain and C^{α}-protection. The Boc group is selectively removed at each cycle by 20 to 50% trifluoracetic acid (TFA) in dichloromethane. Suitably substituted benzyl (Bzl) protecting groups are stable in this medium but are removed by liquid hydrogen fluoride. This Boc—benzyl tactic has been used in most of the protein syntheses performed to date, including synthesis of the 124-residue enzyme ribonuclease A (21).

The use of suitably substituted benzylic groups is an important element in the success of this tactic. If a very small proportion of a side-chain protecting group is lost each time the N^{α}-Boc group is removed, the cumulative results can be disastrous. For example, the results calculated (16) for loss of protecting groups during assembly of the 214-residue light chain of the human IgG1 Eu (12) are listed in Table 1. The central entry of this table is illustrative. The light chain contains 18 threonine residues spaced throughout the chain. After each threonine residue is added to the growing peptide, the benzyl group protecting its β-hydroxyl group is exposed to 50% TFA for 20 min during each subsequent cycle until assembly is complete. The total acid exposure of all 18 threonine residues is equivalent to exposing one threonine residue to 50% TFA for 602 hr. Based on the rate of deprotection of O^{β}-benzylthreonine in solution (14), this treatment should remove 13.9% of the benzyl group. Correspondingly, 13.9% of the fully assembled light chains would have lost one of the 18 benzyl groups protecting its threonine residues. Once deprotected, the threonine hydroxyl group is available to couple with an activated amino acid and thus initiate the growth of a peptide branch at the deprotected threonine. The remaining 86.1% of the synthetic light chains would have maintained complete protection of the 18 threonine hydroxyl groups throughout assembly. This level of stability for a side-chain group is marginally acceptable.

The top of Table 1 shows four protected amino acids commonly used for solid-phase synthesis of peptides and even proteins. Similar calculations (14)

TABLE 1. *Synthesis of the 214-residue light chain of human immunoglobulin Eu: Calculated stability of the side-chain protecting groups to CF_3CO_2H*

Type of protected residue	Residues of this type	Exposure to CF_3CO_2H (hr)[a]	Protein remaining fully protected (mole %)	
			50% CF_3CO_2H	2% CF_3CO_2H
Tyr(Bzl)	9	344	0.0	99.2
Lys(Z)	15	608	0.0	99.2
Cys(4-CH_3OBzl)	5	216	53.0	99.9
Ser(Bzl)	32	1,130	64.7	>99.9
Thr(Bzl)	18	602	86.1	>99.9
Lys(3-ClZ)	15	608	98.9	>99.9
Ser(2-ClBzl)	32	1,130	>99.2	>99.9
Tyr(2,6-Cl_2Bzl)	9	344	>99.8	>99.9
Cys(4-CH_3Bzl)	5	216	>99.9	>99.9

[a] Based on N^{α}-deprotection for 20 min per cycle.

indicate that use of Boc for N^{α}-protection and 50% TFA for N^{α}-deprotection during each cycle would also remove at least one O-benzyl group of tyrosine and one N^{ϵ}-benzyloxycarbonyl (Z) group of lysine from every light chain at some time during its assembly. This result would be especially disastrous for lysine because the growth of a peptide branch from the ϵ-amino group of a prematurely deprotected lysine residue is an efficient side reaction (15). In addition, about half of the light chains would lose one S-4-methoxybenzyl group from cysteine and one-third would lose an O^{β}-benzyl group from serine during assembly. The combined losses of these and other benzylic protecting groups would doom such a synthesis of the light chain to failure.

Two options are available that avoid this low but repetitive loss of side-chain protecting groups. First, the Boc—benzyl tactic becomes feasible if the benzyl groups are modified to make them more stable in acid. This is readily achieved by incorporating one or two electron-withdrawing chloro substituents to the benzyl group or, in the case of cysteine, by substituting a weaker electron-donating substituent. Thus minimal loss of side-chain deprotection would occur if the ϵ-amino group of lysine were protected by 3-chlorobenzyloxycarbonyl (15), the β-hydroxyl of serine by 2-chlorobenzyl (16), the phenolic hydroxyl of tyrosine by 2,6-dichlorobenzyl (14), and the thiol of cysteine by 4-methylbenzyl (14,16), as shown at the bottom of Table 1.

The second option is the use of an N^{α}-protecting group that is removed by a much weaker acid. For example, the N^{α}-2(4-biphenylyl)-2-propyloxycarbonyl (Bpoc) group (38) is readily removed in 20 min by 2% TFA, a reagent that is about 1,000 times milder than 50% TFA (16). If Bpoc were used instead of Boc for N^{α}-protection, Table 1 indicates that use of 2% TFA for N^{α}-deprotection would result in negligible loss of benzyl from tyrosine, serine, or threonine, of benzyloxycarbonyl from lysine, and of 4-methoxybenzyl from cysteine during assembly of the 214-residue light chain. This Bpoc—Bzl tactic has recently been used for the synthesis of several small proteins, including the solid-phase synthesis of a 50-residue fragment of ovine lipotropin (49).

The Fragment Condensation Strategy

Proteins can also be assembled by the condensation of peptide fragments. One variation of this approach that makes extensive use of the solid-phase method is outlined in Fig. 4. A short fragment of the desired protein is assembled on a solid support and removed as a protected peptide, which is coupled to another support-bound peptide. Further protected peptide fragments are added until the desired protein is fully assembled.

Fragment condensation offers two major advantages. Once a protected peptide intermediate is removed from the support, it can be purified before being added to the growing polypeptide. In addition, purification of the desired protein is relatively easy because the by-products will lack not just a few residues but one or more entire peptide fragments. Synthesis of any one protein requires the preliminary synthesis of several specific protected frag-

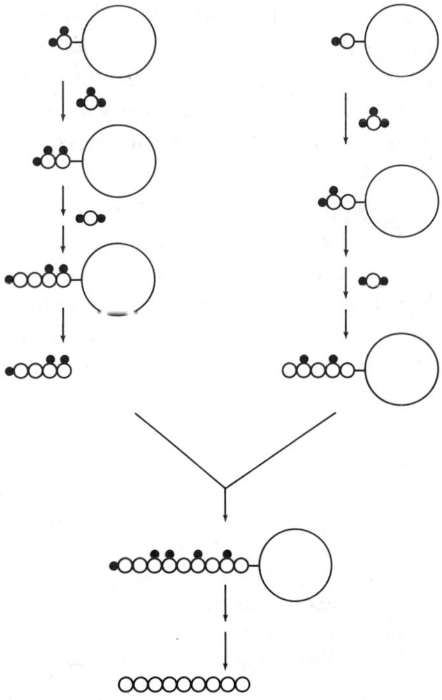

FIG. 4. The general strategy for fragment condensation of protected peptides.

ments. Since the preparation of each fragment requires a significant investment of time and material, these valuable intermediates are used in only slight excess during coupling with the growing polypeptide, which makes quantitative coupling very difficult. Finally, the carboxyl-terminal residue of the protected peptide usually undergoes significant racemization during coupling unless it is glycine or proline. Although the strategy of fragment condensation on solid supports has been less explored than stepwise assembly, certain problems have been overcome. For example, the 58-residue basic trypsin inhibitor has been successfully assembled by condensing four protected peptides containing 9 to 19 residues with a support-bound dipeptide (48).

Synthesis of the protected peptide fragments requires not two but three levels of protecting group stability because the side-chain protecting groups should be stable to the reagent that cleaves the peptide from the resin. The major components of the Bpoc—*tert*-butyl—phenacyl tactic for stepwise synthesis of protected peptides are shown in Fig. 5. This fragment condensation strategy uses the Bpoc group for N^{α}-protection, *tert*-butyl and Boc groups for side-chain protection, and a support-bound phenacyl ester (28,46) for C^{α}-protection. The Bpoc group is selectively removed by 0.5 to 2% TFA in dichloromethane. The fully protected peptide is selectively cleaved from the support by a nucleophile, such as potassium hydroxide, hydrazine, or sodium thiophenoxide. After all of the protected peptide fragments have been assembled into the desired protein, the side-chain *tert*-butyl and Boc protecting groups are removed by 50% TFA. An alternative to use of the phenacyl ester link to the solid support is use of a 2-nitrobenzyl ester link, which is cleaved by photolysis (35). Milder tactics of this type are only now being applied to the solid-phase synthesis of proteins.

SOLID-PHASE STRATEGIES FOR IMMUNOGLOBULIN DOMAINS

The chemical synthesis of an antibody is a formidable goal. Fortunately, the 1,300 residues of a typical immunoglobulin are present not in a single polypeptide chain but in two copies each of a light chain containing about

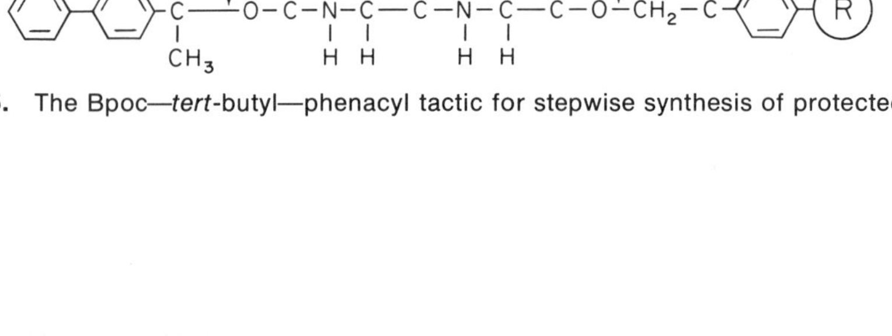

FIG. 5. The Bpoc—*tert*-butyl—phenacyl tactic for stepwise synthesis of protected peptides.

210 residues and a heavy chain containing about 440 residues. As shown in Fig. 6, the light and heavy chains are composed of two and four globular domains, respectively, containing 110 to 120 residues (13). The amino-terminal variable domains of the light and heavy chains (V_L and V_H) are responsible for the ability of the antibody to bind its antigen (24). The variable domains each contain three hypervariable regions (47) that together form the antigen-binding site (6).

The effects of structural changes on antigen binding can thus be explored by chemical synthesis of antibody variable domains. A completely synthetic approach requires the synthesis of both the V_L and V_H domains of a single antibody and their association to form an active F_v complex (22). Synthesis of only one variable domain will be useful in those cases where the complementary variable domain is available through limited proteolysis (22) or hydrolysis (18). Alternatively, the synthetic variable domain of one chain may bind selectively to the variable domain of the complementary chain and thus form an active L-chain—V_H or H-chain—V_L complex.

Structure of a Heavy Chain Variable Domain

Chemical sequencing and X-ray diffraction studies provide essential structural information for the strategic design of an antibody domain synthesis. A case in point is the mouse myeloma protein M603, one of a series of IgA kappa proteins that bind phosphorylcholine (33). The three-dimensional structure of the F_{ab} fragment of M603 has been determined to 3.1 Å (10,11,37), and the amino acid sequence has been determined for the 120-residue V_H domain (36) and for the first 55 residues of the V_L domain (37).

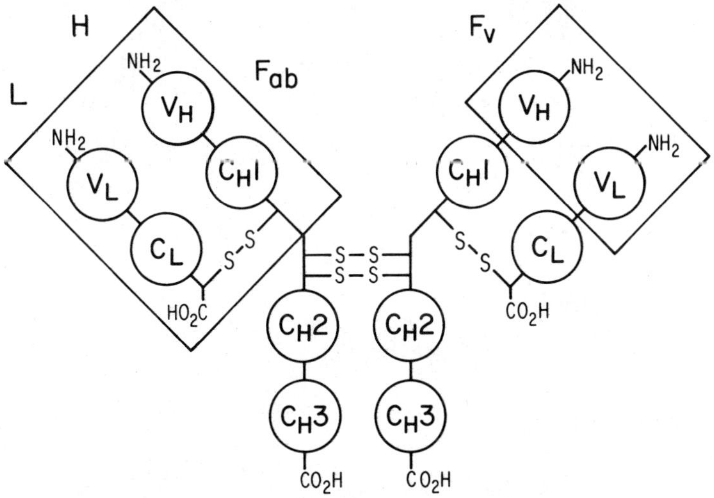

FIG. 6. The domain structure of a typical immunoglobulin.

Chemical synthesis of a domain must begin with a specific (and presumably correct) sequence of amino acid residues. A diagram of the M603 V_H domain that combines the identity of each residue with some three-dimensional information is shown in Fig. 7. The problem of distinguishing between aspartic acid and asparagine residues or between glutamic acid and glutamine residues often requires special efforts. [We are grateful to Doctors Rudikoff and Potter for informing us of the amide assignments at previously unresolved positions (36).]

The basic structure of the M603 V_H domain from X-ray diffraction studies (10,11,37) is a sandwich-like arrangement of two layers composed of straight segments of the polypeptide chain. Adjacent segments within each layer run antiparallel to one another and often adopt a β-pleated sheet conformation. The inner layer, which faces the V_L domain, consists of the five segments indicated by horizontal stripes in Fig. 7. The outer layer is comprised of the four segments shown by vertical stripes. The two-dimensional arrangement of Fig. 7 can be mentally converted into a rough three-dimensional structure in two steps. First, the 1–26 outer loop is lifted up and the 93–120 inner loop is positioned below it so that Cys[98] lies directly below Cys[22]. Second, the

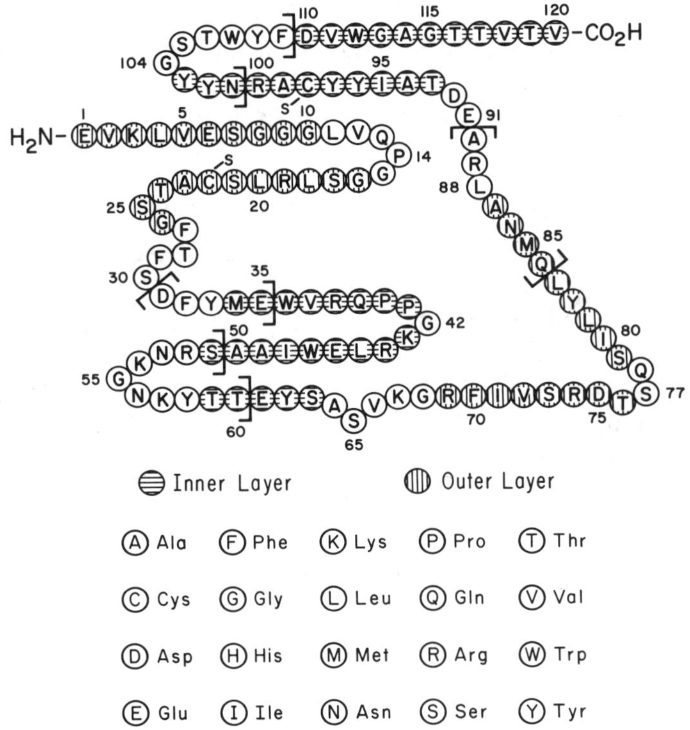

FIG. 7. Primary structure and stretches of β-pleated sheet for the V_H domain of immunoglobulin M603.

64–87 outer loop is bent up and over residues 27–63 by pivoting on Ala^{64} and Ala^{87} so that Ser^{77} lies over Ser^{30}.

Phosphorylcholine [$HOP\bar{O}_2$-O-CH_2-CH_2-$N^+(CH_3)_3$] and antigens bearing this hapten bind to M603 in a large wedge-shaped cavity between the strongly interacting V_L and V_H domains (10,11,37). The walls of the antigen-binding site are lined exclusively with hypervariable residues located in the L1 and L3 regions of V_L and the H1, H2, and H3 regions of V_H. The phosphorylcholine moiety is bound more closely to the heavy chain with the quaternary choline group deeply buried and the phosphate group near the exterior. Six residues from V_H involved in antigen binding are Tyr^{33} and Glu^{35} from H1, Arg^{52}, Lys^{54}, and Glu^{61} from H2, and Trp^{107} from H^3, which are shown in Fig. 8. The phosphate group forms strong hydrogen bonds with the hydroxyl of Tyr^{33} and an amino group of Arg^{52}. The negative charge of the phosphate group interacts ionically with the positively charged side chain of Arg^{52} and possibly of Lys^{54}. The methyl and methylene groups of the choline moiety make van der Waals contacts with the aromatic rings of Tyr^{33} and Trp^{107}, while the positive charge of its quaternary ammonium group interacts ionically with the carboxylate side chains of Glu^{35} and Glu^{61}.

Another indication of the importance of the antigen-binding residues is provided by comparing the primary sequences of similar antibodies. At present, V_H sequences are available for five phosphorylcholine-binding immunoglobulins (M603, M167, T15, S107, H8) isolated from the highly inbred BALB/c strain of mice (32). The V_H domains of these proteins are quite similar, varying at only 16 of 120 residues. The variable positions are shown in Fig. 8 by slanted lines. The variability outside the bracketed hypervariable

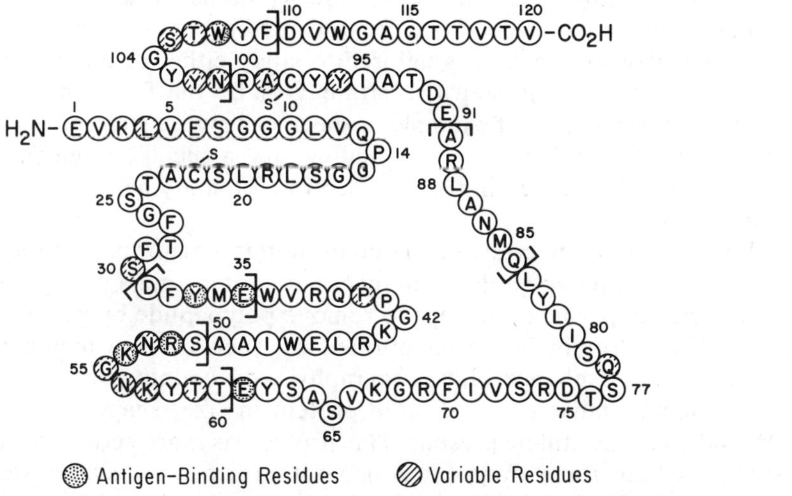

FIG. 8. Antigen-binding and variable residues of the V_H domains of M603 and four other phosphorylcholine-binding antibodies.

regions, such as positions 4, 39, 78, and 96, is probably due to genetic drift within the BALB/c mice. Ten of the variable positions lie within the hypervariable regions H2 (residues 51–60) and H3 (residues 101–109), which contribute to the antigen-binding site. Both H1 (residues 31–35) and He (residues 94–90), a region that is hypervariable in man (5) but is outside the antigen-binding site, were constant in all five immunoglobulins. The importance of Tyr^{33}, Glu^{35}, Arg^{52}, Lys^{54}, and Glu^{61} in the binding of phosphorylcholine is suggested by the fact that these residues are conserved in all five V_H domains.

Parallel Stepwise Synthesis of the Domain

The impetus behind a project to synthesize an immunoglobulin domain varies with the investigator. Some protein chemists are mainly interested in learning if present synthetic procedures permit the successful synthesis of a polypeptide with the chemical and biological properties expected or already known for the natural domain. Others are more concerned with synthesizing a series of domains that differ from the natural prototype by one or more residues of interest and then correlating the activity of each analogue with its structure. These reasons for synthesis of an immunoglobulin domain are complementary, since adequate synthetic procedures are necessary before a series of domain analogues can be prepared and, in turn, the major justification of synthesizing a naturally occurring domain is to provide access to analogues.

Two major immunological questions that can be studied by synthesis of variable domain analogues are (a) the structural features important for the strong interaction between the V_H and V_L domains and (b) the structural features that confer specificity for the binding of antigen to the V_H—V_L complex. Domain analogues tailored to study either of these questions can be synthesized by the stepwise strategy or by the fragment condensation strategy. For the sake of example, a specific stepwise strategy is described in this section for studying antigen binding, and a specific fragment condensation strategy is outlined in the next section for studying the V_H—V_L interaction.

Purification of the crude synthetic protein is a major part of the synthesis. Assembly of a polypeptide chain by either strategy must usually be followed by a series of purification steps to remove polypeptide by-products. Even if a synthetic protein has been purified by such diverse techniques as gel filtration, ion-exchange chromatography, and countercurrent distribution, by-products similar to the desired protein in size, shape, charge, and hydrophilicity may still be present. This problem is more serious for a stepwise synthesis than for a fragment condensation scheme because stepwise by-products may lack only one residue but the condensation byproducts usually lack an entire peptide. Thus it is desirable to include in a stepwise scheme features that decrease the number of by-products, allow a diagnosis of protein purity, and minimize the importance of impurities during interpretation of bioassay results. Two strategies to accomplish these latter goals are described below, and an affinity purification strategy for reducing the number of by-products is described in the section on affinity purification.

Double Resin Strategy

Biological comparison of a natural product with a synthetic analogue can be deceptive if the synthetic material contains a small amount of a by-product that is a potentiator or inhibitor of the biological activity. A more reliable comparison involves examination of two synthetic analogues synthesized and purified under essentially identical conditions, since the minor impurities would generally have quite similar structures and their biological effects would cancel out.

Two domain analogues can potentially be assembled under essentially identical conditions by the tactic of using two solid supports. The support commonly used for solid-phase synthesis of proteins (17,21,27) is small beads of polystyrene cross-linked with 1 to 2% divinylbenzene, which is designated Ⓡ. Another support that has comparable chemical behavior (41) is small beads bearing noncross-linked chains of polystyrene grafted onto poly(trifluorochloroethylene), or Kel-F, which is denoted by Ⓚ. These two resins have significantly different densities and thus can be separated cleanly from one another at any step of a parallel synthesis by differential flotation. For example, when suspended in dichloromethane, the Ⓡ resin floats and the Ⓚ resin sinks. Since the two types of beads are readily mixed by agitation during solid-phase synthesis, they experience essentially identical chemical environments when together in the same reaction vessel.

We have recently used this double resin strategy for preparing two peptides derived from the anaphylatoxin C3a, which is the amino-terminal 77-residue fragment of the third component of human serum complement (23). As is shown in Fig. 9, a ^{14}C-labeled tripeptide was assembled on the cross-linked resin Ⓡ and a ^{3}H-labeled tetrapeptide was separately prepared on the graft copolymer Ⓚ. Then the peptide–resins were combined and simultaneously extended by five residues in parallel stepwise fashion. After the resins were separated by differential flotation, the peptides were cleaved from the resins and purified by gel filtration and ion-exchange chromatography on carboxymethyl cellulose. Both the amino-terminal [^{14}C]C3a-(70–77)-octapeptide and the [^{3}H]nonapeptide analogue bearing carboxyl-terminal glycine were isolated in pure form. The double resin strategy has previously been used to prepare (a) two samples of the amino-

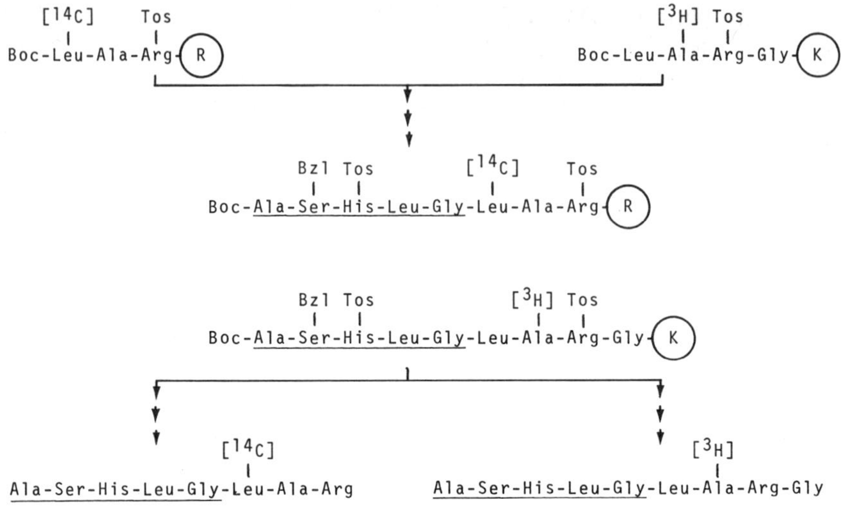

FIG. 9. Parallel stepwise synthesis of C3a-(70-77)-octapeptide and an analogue by the double resin strategy.

terminal dodecapeptide of bovine parathyroid hormone (41) and (b) the undecapeptide substance P plus an analogue with norleucine in place of the carboxyl-terminal methionine amide (44).

In principle, this strategy can be extended to allow the parallel stepwise synthesis of two domain analogues, as illustrated in Fig. 10. One resin could be used for assembly of the V_H domain of M603 and the other for assembly of the V_H domain of T15, another phosphorylcholine-binding myeloma protein (31). The V_H domain of T15 is specifically interesting because the T15 immunoglobulin has idiotypic determinants very similar or identical to antibodies from BALB/c mice raised against phosphorylcholine-containing antigens (7,8). The primary structures of M603 and T15 differ only at positions 55 (Gly vs. Ala) and 57 (Lys vs. Asp) of the second hypervariable region H2 and at positions 101 (Asn vs. Asp) and 105 (Ser vs. Ser-Tyr) of H3. Figure 10 shows a scheme for adding 107 of the 120–121 residues simultaneously in parallel stepwise fashion. Thus the tripeptide resins are combined and 11 synthetic cycles are performed. After the pentadecapeptide–resins are separated, five residues are added to the M603 intermediate while six residues are added to the T15 intermediate. The peptide–resins are recombined, extended by 43 residues under essentially identical conditions, and separated to allow residues 57–55 to be added. After the resins are combined for the third time, 52 residues are added in parallel stepwise manner. Finally, the protein–resins are separated to allow the last two residues to be added.

This scheme allows the assembly of two analogous domains for about one-fourth more effort than is required to assemble one domain alone. More important, about 90% of all residues and about 65% of the 25 residues in the

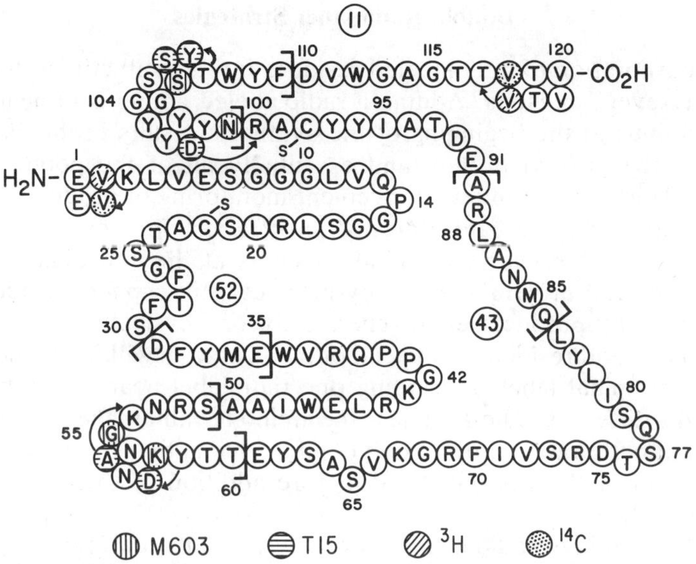

FIG. 10. Parallel stepwise synthesis of the V_H domains of M603 and T15 using the double resin and four radiolabel strategies.

hypervariable regions H1, H2, and H3 would be added under essentially identical conditions. Thus any by-products resulting from side reactions during these synthetic cycles would probably be formed in comparable amounts on each resin. Since the V_H domains of M603 and T15 have quite similar primary structures, the synthetic domains could be purified using the same series of separation techniques. For any minor polypeptide by-product contaminating one synthetic domain, an analogous by-product would probably also contaminate the other. Analogous by-products should influence the binding of phosphorylcholine to the V_H—V_L complex (or binding of V_H to V_L) similarly. Thus the effect of structural differences on binding activity is measured more reliably by comparing two domain analogues synthesized through such a parallel stepwise strategy than by comparing two synthetic domains prepared independently or by comparing a synthetic domain with a natural domain. The importance of comparing domain analogues synthesized in parallel increases as the difficulty of purifying the synthetic domains increases.

The cross-linked resin Ⓡ has been used as the solid support for synthesis of proteins as long as 188 residues (26) and the grafted resin Ⓚ was employed for several 34-residue peptides (34,42,43). These resins are readily separated by their difference in density when they each bear a 12-residue peptide. If the number of chains attached to each support is kept relatively small (50 to 150 μmoles/g of support), they should still be separable when each bears a 120-residue domain.

Double Radiolabel Strategies

The incorporation of radiolabeled residues into a synthetic protein can be useful for several reasons. Adding a radiolabeled amino acid near the carboxyl terminus at the beginning of the synthesis results in this label being present in the desired protein and essentially all of the polypeptide by-products. This strategy allows convenient monitoring of products with little or no ultraviolet absorbance during purification. In contrast, adding a radiolabeled residue near the amino terminus at the end of the synthesis labels the desired protein and any by-products still capable of adding residues. But all by-products either chemically or physically terminated and thus unable to add residues at the end of the synthesis will fail to incorporate the amino-terminal label. By adding one radiolabel near the carboxyl terminus and a different radiolabel near the amino terminus, the desired protein will contain both labels. But all polypeptide chains that became terminated after addition of the first label and before addition of the second will be singly labeled.

This two-terminus radiolabeling strategy provides an internal probe for homogeneity of the synthetic domain, namely, the ratio of the specific activity of the amino-terminal label to that of the carboxyl-terminal label. During the series of purification steps following removal of the assembled protein from the resin, the ratio of amino-terminal label to carboxyl-terminal label will increase as the terminated or truncated polypeptide by-products are separated from the desired protein. Thus one criterion of homogeneity would be purification of the synthetic product to constant relative radioactivity. For example, the scheme for assembly of the M603 domain shown in Fig. 10 shows addition of [^{14}C]valine at position 118 during synthetic cycle 2 and addition of [^{3}H]valine at position 2 during synthetic cycle 118. As the ^{14}C-labeled by-products are separated from the doubly labeled V_H domain, the ^{3}H/^{14}C ratio will increase. When several purification techniques fail to increase the ^{3}H/^{14}C ratio further, the synthetic domain has been purified to an operational state of purity.

The two-terminus labeling strategy is applicable to both the stepwise synthesis of one domain or parallel stepwise synthesis of two analogous domains. Another double radiolabel strategy is specifically applicable to the double resin strategy. A major assumption involved in using two resins for parallel synthesis is that complete separation of the resin beads actually occurs. This assumption can be tested for any two peptide–resins if they are radiolabeled differently. For example, the C3a peptide-resins shown in Fig. 9 were separately labeled as Boc-[^{14}C]Leu-Ala-Arg (Tos)-Ⓡ and Boc-Leu-[^{3}H]Ala-Arg(Tos)-Gly-Ⓚ. Our ability to separate these peptide–resins without cross-contamination was shown by double-label liquid scintillation counting. Comparison of the ^{3}H and ^{14}C disintegrations per minute observed for each peptide–resin before mixing and after mixing in dichloromethane and separation by flotation showed that contamination of Ⓡ by Ⓚ and vice versa was lower than the 1% limit of detection. A more precise experimental measure of cross-contamination would involve direct combustion of the resins to [^{3}H]water and [^{14}C]carbon dioxide followed by liquid scintillation counting of the separated products. Alternatively, after hydrolysis the [^{14}C]leucine and [^{3}H]alanine could be separated by ion-exchange chromatography and individually counted.

Low levels of cross-contamination can also be detected by amino acid analysis if each peptide–resin contains one amino acid not present in the other (44). The scheme for parallel stepwise synthesis shown in Fig. 10 requires that the peptide–resins be separated cleanly at three points: after addition of residues of Thr^{106}, of Tyr^{58}, and of Lys^{3}. At each of these points it is important to show that cross-contamination does not occur. But the criterion of amino acid composition is not useful at any of these points. After addition of Thr^{106}, the pentapeptide–resins have identical amino acid composition. After addition of Tyr^{58}, the M603 intermediate contains 8 threonine, 6 serine, and 8 tyrosine residues per chain while the T15 intermediate contains 7 threonine, 7 serine, and 9 tyrosine residues. The asparagine–aspartic acid difference at position 101 is not useful, since both residues appear as aspartic acid on acid hydrolysis. After addition of Lys^{3} the situation is even worse.

The resins bearing M603 or T15 intermediates can be readily distinguished, however, by incorporating different radiolabels near their carboxyl termini. The scheme in Fig. 10 employs [^{14}C]valine at position 118 of the M603 intermediate and [^{3}H]valine for the T15 precursor. Thus the combined peptide–resins can be separated at residues 106, 58, and 3 or at any intermediate step when samples of the separated peptide–resins are needed for amino acid analysis. In each case complete separation can be shown by double-label counting or by combustion and counting.

Both the two-terminal and carboxyl-terminal strategies of double labeling can be applied to the parallel stepwise synthesis of two domain analogues. The peptide–resins are labeled differently before being combined for the first time and are again labeled differently but with the labels reversed after being separated for the last time. The simplest example of this four-label strategy is shown in Fig. 10. At position 118 Boc-[^{14}C]Val is added to the M603 precursor and Boc-[^{3}H]Val to the T15 intermediate. After 115–116 synthetic cycles and the addition of Lys^{3}, the peptide–resins are separated for the last time. Once separated, the individual polypeptide–resins need no longer be distinguished radiochemically. So the same reagents used at position 118 are also again used at position 2 but in reverse. Boc-[^{3}H]Val is added to the nearly completed M603 V_H-resin and Boc-[^{14}C]Val to the T15 V_H-resin. Thus during purification of the cleaved proteins, the $^{3}H/^{14}C$ ratio will increase to constant activity for M603 V_H and the $^{14}C/^{3}H$ ratio will increase for T15 V_H.

The opportunity to label both termini of both domains using two radiolabeled forms of the same protected amino acid is convenient, since the small-scale preparation of only one protected amino acid need be mastered. In general, however, the four-label strategy for parallel synthesis could be accomplished using as many as four different radiolabels in four different protected amino acids added at four different positions. But since greater effort is required to suitably protect labeled amino acids with side-chain functional groups requiring protection, labeled derivatives of bifunctional amino acids (glycine, alanine, valine, proline, leucine, isoleucine, phenylalanine) are generally used. Furthermore, the use of ^{3}H and ^{14}C is indicated because these radiolabels are long-lived, readily available, and relatively inexpensive.

Parallel stepwise synthesis of two domain analogues using the double-resin strategy, the four-label strategy, and the affinity purification strategy is currently being examined. Variations of the scheme shown in Fig. 10 would allow substitution of any or all of the hypervariable residues by other residues of interest. Correlation of the antigen-binding activity of various domain analogues with their synthetically produced structural differences should furnish precise information about the contribution of individual residues to the specificity of the antigen-binding site.

Fragment Condensation of Both Half Domains

The second binding phenomenon that can be studied with synthetic domains is the interaction between light and heavy chains. Studies with intact H and L chains from two mouse IgA proteins that both bind dinitrophenyl haptens have shown that homologous recombinants are preferentially formed over heterologous recombinants (29) and that the latter exhibit very little anti-DNP activity (4). Thus antigen-binding activity depends upon effective L—H interactions even when the heterologous chains are derived from antibodies with similar antigen-binding sites. The specific contribution that F_V makes to the L—H interaction could be explored by examining the binding of a synthetic V_H domain with its homologous natural L chain and with several closely related L chains. Alternatively, a series of synthetic analogues could be bound individually or competitively with the natural L chains.

The specific residues involved in the L—H interaction are clearly revealed by X-ray diffraction studies (10,11). For example, the 13 residues of the M603 V_H domain that interact with the V_L domain are shown in Fig. 11. Interestingly, these interdomain residues, which mainly bear hydrophobic and aromatic side chains, are clustered in and around hypervariable region H3 and to a lesser extent between H1 and H2. Thus the contribution of individual residues to the L—H interaction could be conveniently explored by a synthetic strategy that introduced amino acid substitutions in these regions, such as fragment condensation.

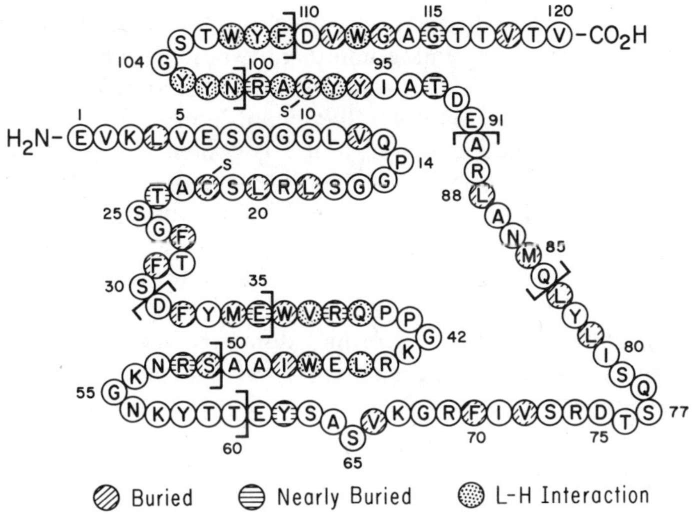

FIG. 11. The predominantly hydrophobic intradomain and interdomain residues of the M603 V_H domain.

The 13 interdomain hydrophobic residues are effectively hidden from contact with the aqueous medium by van der Waals interactions with the first hypervariable region of the V_L domain. The bilayer structure of the V_H domain is similarly maintained by a group of 24 residues that are completely buried in the domain interior and 8 other residues that are nearly buried, as shown in Fig. 11. Unlike the interdomain residues these 32 buried intradomain residues are not clustered but scattered throughout the entire domain. With respect to the three-dimensional structure, the outer layer (1–27 plus 65–90) contributes 14 intradomain residues and the inner layer (28–64 plus 91–120) provides 18. From the alternate perspective of primary structure, the amino-terminal half domain (1–64) furnishes 17 intradomain hydrophobes and the carboxyl-terminal half domain (65–120) affords 15.

This relatively even distribution of the bilayer-inducing internal hydrophobes suggests that the bilayer structure would remain essentially intact if the polypeptide chain were cleaved into two half domains by scission of the Ala^{64}-Ser^{65} peptide bond. Such a modification would leave the β-pleated sheets that comprise the inner and outer layers intact and still covalently linked through the disulfide bond between Cys^{22} and Cys^{28}. The nicked V_H domain (V_H') would consist of the amino-terminal half domain (H_N) and the carboxyl-terminal half domain (H_C). The inner layer of the bilayer would be composed of three polypeptide stretches from H_N and two from H_C and the outer layer of two stretches from each half domain (*cf.* Fig. 7).

The structural analysis that leads to the consideration of half domains extends the classic dissection of immunoglobulins into monomers, chains, and domains. The functional role of the half domains is effectively masked by the role of the V_H domain itself, which in turn is only half of the globular F_v complex, the primary crystallographic entity. Independent of the present structural role of the V_H half domains, they may have been important in the genetic history of immunoglobulins. Significant internal homology has been observed between the sequences near the two highly conserved cysteine residues of V_H domains and this has been explained by the hypothesis that domains arose by duplication of an ancestral gene coding for a half domain (3).

Conditions that permit the cleavage of a variable domain into two half domains are presently unknown. But construction of a functional nicked domain by disulfide bridging of two half domains can be studied by total chemical synthesis. A scheme for assembly of both half domains of M603 V_H' by fragment condensation is shown in Fig. 12. Assembly of the 64-residue H_N would involve the sequential coupling of four protected peptides containing 11 to 16 residues to a solid support bearing the (56–64) -nonapeptide. The usual problem of racemization of the carboxyl-terminal residue during activation and coupling of protected peptides is avoided for H_N by selecting peptide intermediates with carboxyl-terminal glycine. This tactic also permits the synthesis of all four protected peptides from a single Gly-resin. Assembly of the 56-residue H_C might involve coupling of two glycine-terminal nonapeptides to a heptapeptide–resin followed by two isoleucine-terminal peptides containing 15 and 16 residues. The lack of glycine or proline residues in the region 70–100 precludes the use of a racemization-free tactic. But the amount of racemization incurred at the carboxyl-terminal isoleucine residues could be readily assessed by measur-

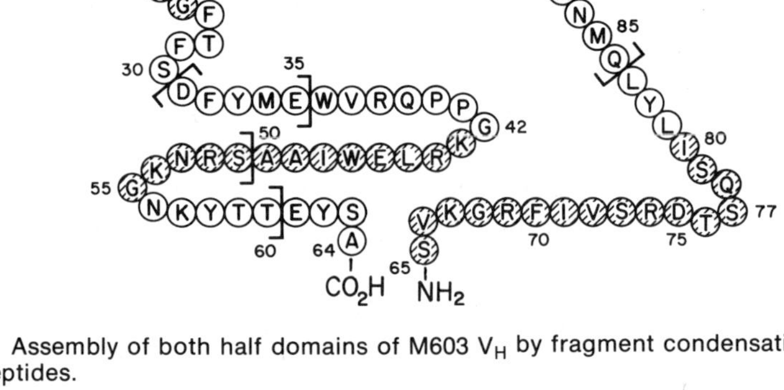

FIG. 12. Assembly of both half domains of M603 V_H by fragment condensation of protected peptides.

ing the relative amount of the enantiomer allo-isoleucine produced on hydrolysis of the synthetic products.

After assembly, detachment and purification, the half domains would be combined and mildly oxidized to form disulfide-linked dimers. The heterologous dimer H_N—H_C (V_H') could be isolated by ion-exchange chromatography and compared with intact V_H. This fragment condensation scheme could also provide the intact V_H domain by a variation requiring the preparation of only one additional protected peptide. Thus the H_C-resin would be coupled with the protected (56–64)-nonapeptide followed by the four protected peptides used in the assembly of H_N. In addition, analogues of the intact V_H or either half domain could be prepared by replacing one or more of the protected peptides used for M603 by peptide analogues.

Selection of specific protected peptides for a solid-phase synthesis by fragment condensation should avoid situations where synthetic problems can be expected. For example, the decision to end the first protected peptide for H_C at Ser^{105} and begin the second protected peptide at Gly^{104} was partially based on a specific problem encountered during assembly of the hydrophobic (98–111)-tetradecapeptide containing the H3 region of M603 (40). The fully protected peptide–resin is shown in Fig. 13. During most synthetic cycles we observed that only two coupling steps were required to react essentially all of the free amino groups. But the coupling reactions for adding Asn^{101} and then Arg^{100} to the decapeptide–resin were quite slow and inefficient, the addition of Boc-Asn(Mbh) requiring six coupling steps and coupling of Boc-Arg(Tos) needing five. Subsequent additions of Ala^{99} and Cys^{98}, however, were each complete after two coupling steps.

This temporary reluctance of the peptide–resin to react with amino acid reagents can be ascribed to decreased accessibility of the α-amino group of the peptide–resin. Two structural features may contribute to burying the amino groups of Tyr^{102} and Asn^{101}. First, residues 103–107 participate in a 180° bend in the crystalline state (37). If a similar bend exists in the protected peptide–resin, the peptide chain would bend back on itself so that the amino groups of Asn^{101} and Arg^{100} are close to Asp^{110} and the resin backbone. Second, the protected peptide–resin contains two stretches of residues

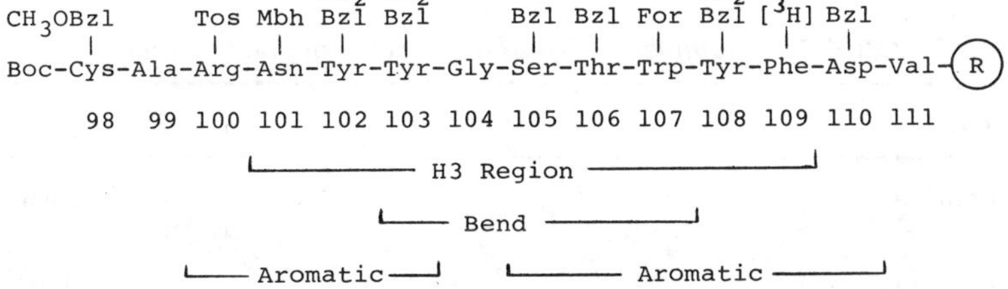

FIG. 13. Two aromatic stretches and a potential bend in the H3 region of fully protected M603 V_H-(98-111)-tetradecapeptide.

bearing aromatic side chains (100–103 and 105–110). Interaction of these aromatic stretches through van der Waals forces would reinforce any tendency of the chain to bend back on itself. The addition of Boc-Asn(Mbh) only increases the opportunity for aromatic interactions, since the amide-protecting group Mbh contains two benzene rings. Addition of Boc-Arg(Tos), however, evidently extends the chain sufficiently to prevent adoption of a bent configuration, which allows Boc-Ala to couple normally. Steric hindrance during coupling could be avoided in this case by isolating the aromatic stretch 105–110 in one protected peptide and stretch 100–103 in another by beginning one fragment at Gly^{104}, as shown in Fig. 12. Similar reasoning supports the selection of Gly^{15}, Gly^{26}, Gly^{42}, and Gly^{55} to begin peptide fragments for assembly of half domain H_N.

AFFINITY PURIFICATION AFTER SOLID-PHASE ASSEMBLY

A major problem of solid-phase peptide synthesis is the formation of undesired peptide by-products. The desired peptide can often be separated from the major by-products by such time-honored techniques as gel filtration, ion-exchange chromatography, electrophoresis, and countercurrent distribution. Purification is more difficult for larger peptides because small differences in size, charge, or hydrophobicity are harder to detect analytically or to exploit preparatively.

Synthetic peptide by-products are conveniently divided into three classes based on the nature of the structural defect: modified peptides, addition peptides, and deletion peptides. These by-products are illustrated in Table 2 with peptides that might be formed during synthesis of the M603-(114-120)-hexapeptide. Modified peptides differ from the desired peptide by containing a chemically modified residue. In the example, the terminal dipeptide is not covalently bound to the amino group of Thr^{116} but to its side-chain hydroxyl group. Addition peptides contain a residue not present in the desired peptide. The extra glycine residue shown could arise by formation and coupling of a dipeptide during attempted coupling of an amino acid. Deletion peptides lack a residue present in the desired peptide. In the example, the terminal glycine residue failed to couple.

TABLE 2. *Examples of by-products from peptide synthesis*

Product class	Example
Desired peptide	H-Gly-Thr-Thr-Val-Thr-Val-OH
Modified peptide	H-Gly-Thr-O H-Thr-Val-Thr-Val-OH
Addition peptide	H-Gly-Gly-Thr-Thr-Val-Thr-Val-OH
Deletion peptide	H-Thr-Thr-Val-Thr-Val-OH
Terminated deletion peptide	Ac-Thr-Thr-Val-Thr-Val-OH

Deletion peptides can be further segregated according to the position of the deletion: amino-terminal, internal, carboxyl-terminal. Amino-terminal deletion peptides, which are missing a stretch of one or more residues at the amino terminus, include chains that have permanently stopped growing in the middle of the synthesis. Internal deletion peptides, which lack a nonterminal stretch of one or more residues, arise from chains that temporarily stopped growing only to restart after one or more synthetic cycles. If the peptide chains that fail to couple with the activated amino acid are blocked by an effective terminating agent during each synthetic cycle, these potential internal deletion peptides are converted into terminated amino-terminal deletion peptides. In the example, a chain that failed to incorporate the glycine residue was terminated by acetylation.

This section outlines a strategy and two tactics for affinity purification of solid-phase peptides that allow separation of the desired peptide from all terminated deletion peptides. This general method is independent of the length or sequence of the desired peptide. Affinity purification is illustrated by syntheses of a 14-residue peptide from ribonuclease and a 37-residue fragment of histone H4.

The Affinity Purification Strategy

After solid-phase assembly of the desired peptide is complete and before the peptide is cleaved from the solid support, the desired peptide and other growing peptides can be chemically distinguished from terminated deletion peptides and other nongrowing chains. This chemical distinction can be utilized to separate growing peptides from nongrowing peptides after cleavage from the support. The general strategy is shown in Fig. 14. Before

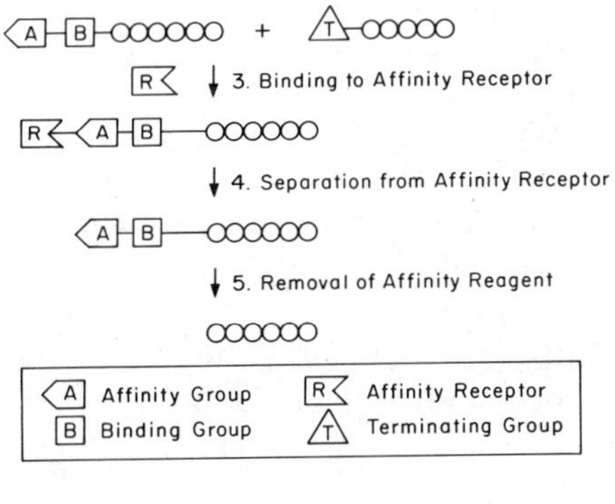

FIG. 14. A general strategy for affinity purification of solid-phase peptides.

cleavage of the synthetic peptide from the support, an affinity reagent A-B consisting of an affinity group A and a binding group B is selectively attached to the growing chains. After cleavage from the solid support, the affinity-labeled peptides are separated from terminated peptides and other peptides lacking the affinity reagent by selective interaction with an affinity receptor R. Once the affinity-labeled peptides are separated from the affinity receptor, the affinity reagent is removed to obtain the desired peptide.

The components of the affinity system must meet certain criteria for successful application of the method. The affinity reagent must be selectively and efficiently attached to the growing chains through a bond that later can be selectively and efficiently cleaved to restore the desired peptide. The affinity receptor must selectively and reversibly bind the affinity group of affinity-labeled peptides. The affinity group and the binding group must withstand the chemical reactions and physical manipulations involved, especially cleavage from the synthetic support.

This strategy permits the separation of the desired peptide from terminated deletion peptides and other chains lacking the affinity reagent. Terminated peptides include chains inadvertently blocked during synthesis, such as chains bearing an amino-terminal glutamine residue that cyclized to pyroglutamic acid. In addition, other chains are intentionally blocked by a terminating agent to prevent the formation of internal deletion peptides, which can be very difficult to separate from the desired peptide. Finally, some nongrowing chains might be chemically reactive but physically inaccessible due to a particular configuration of the protected peptide or to their location in a particularly congested region of the solid support. The affinity purification method should efficiently remove all such nongrowing chains. This strategy does not, however, remove modified or addition peptides that are still growing at the end of solid-phase assembly. The formation of these by-products must be minimized or avoided by independent procedures.

The Organomercurial Cys-Met Tactic

One example of an affinity reagent that meets the requirements of the affinity purification strategy is the dipeptide cysteinyl-methionine. The carboxyl group of methionine functions as the binding group and the thiol group of cysteine as the affinity group. Cys-Met can be selectively and efficiently attached to the terminal amino group of the desired peptide through an amide bond, which can be selectively and efficiently cleaved by cyanogen bromide (39) to restore the free terminal amino group. Organomercurial agarose (9) can serve as the affinity receptor to bind the thiol group selectively and reversibly through a covalent mercury—sulfur bond.

A specific protocol that employs these components is shown in Fig. 15. Although Cys-Met could be coupled as a protected dipeptide, it is more convenient to add methionine and then cysteine separately using the con-

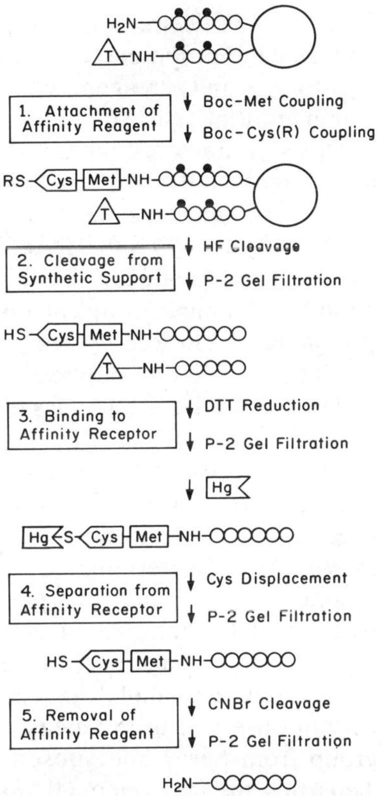

FIG. 15. The organomercurial Cys-Met tactic for affinity purification of peptides lacking cysteine and methionine.

ventional reagents and procedures of solid-phase synthesis. After the peptide chains are cleaved from the support with liquid hydrogen fluoride, which liberates the protected thiol group of cysteine, they are gel filtered to remove nonpeptide by-products of low molecular weight. Then the peptide mixture is treated with dithiothreitol to ensure that the thiol group is reduced. After removal of the reducing agent by gel filtration, the peptide solution is immediately applied to a column of organomercurial agarose, a beaded carbohydrate matrix to which *p*-chloromercuribenzoic acid is covalently linked. Each Cys-Met–peptide is selectively bound through its thiol affinity group to an organomercurial affinity receptor. The column is eluted with aqueous buffer until all peptides lacking a thiol group have been removed. Then the Cys-Met–peptides are displaced from the column by elution with an organic thiol, such as cysteine, which is then removed from the Cys-Met–peptide by gel filtration. Finally, the Cys-Met affinity reagent is selectively and efficiently removed from the desired peptide by treatment with cyanogen bromide, which converts the methionine residue into homoserine and liberates the amino group of the desired peptide. A final gel filtration to remove excess reagent and the cysteinyl-homoserine dipeptide furnishes the desired peptide plus any other peptides derived from chains still growing at the end of solid-phase assembly. Peptides that retained the Cys-Met affinity reagent can be removed by recycling through the organomercurial agarose column.

This organomercurial Cys-Met tactic for affinity purification has been used successfully to purify two biologically active peptides, bovine pancreatic ribonuclease–(111-124)-tetradecapeptide (20) and calf thymus [Lys([^{14}C]Ac)^{12}, Lys([^{3}H]Ac)^{16}]histone H4–(1–37)-heptatriacontapeptide (25). At the end of each synthetic cycle during assembly of the ribonuclease peptide, the amino groups of any peptide chains that had failed to couple were permanently blocked by acetylation with [^{3}H]acetic anhydride. Thus potential internal deletion peptides were converted into radiolabeled terminated deletion peptides. After assembly of the desired 14-residue peptide was complete, Boc-Met and then Boc-Cys(4-CH_3OBzl) were added to the growing chains and the protocol of Fig. 15 was carried out.

The separation of the Cys-Met–tetradecapeptide from the tritiated terminated deletion peptides was monitored by ultraviolet absorbance, as shown in Fig. 16. The terminated deletion peptides were eluted near the void volume of the column as a peak containing most of the radiolabel. A small fraction of the terminated peptides that had bound nonspecifically to the organomercurial agarose was eluted by changing the buffer. Finally, elution with mercaptoethanol displaced the Cys-Met–peptides. The small amount of tritium observed in this peak can be explained by slow loss of the benzyl group from Ser^{123} and subsequent acetylation of its free hydroxyl group. Thus this method removed essentially all of the terminated deletion peptides, which amounted to about 5 mol% of the peptide chains. After removal of Cys-Met, the tetradecapeptide was combined with natural ribonuclease-(1–118) to reconstitute 85 to 90% of the ribonuclease activity observed using the pure tetradecapeptide.

The amino-terminal 37-residue fragment of histone H4 was also synthe-

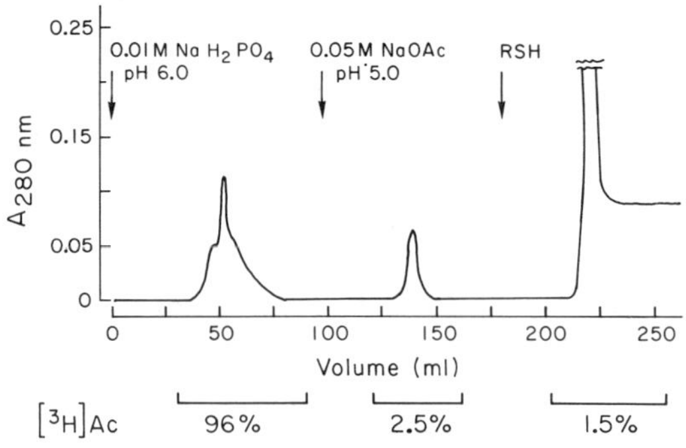

FIG. 16. Affinity purification of Cys-Met–ribonuclease-(111-124) on a column of organomercurial-agarose.

sized by the stepwise solid-phase method and purified by the organomercurial Cys-Met tactic. This peptide is a substrate for calf thymus histone deacetylase, which cleaves the acetyl groups attached to the epsilon amino groups of Lys^{12} and Lys^{16} (25). To distinguish the acetyl groups at these positions, the peptide was synthesized with $[^{14}C]$acetyl at Lys^{12} and $[^{3}H]$acetyl at Lys^{16}. Unreacted chains were blocked at the end of most synthetic cycles by acetylation with nonradiolabeled *N*-acetylimidazole. Since synthetic difficulties were encountered at several points during assembly of the peptide, the Cys-Met protocol was examined not only as a test of the affinity purification strategy but also as an important initial step in purification of the synthetic peptide.

The separation of terminated deletion peptides from Cys-Met–histone H4-(1–37) was monitored by total radioactivity, as shown in Fig. 17. About 80% of the radiolabeled peptides lacked the amino terminal Cys-Met moiety and passed directly through the column. The desired Cys-Met–peptide was removed from the organomercurial agarose column by elution with aqueous cysteine. The purification accomplished by the affinity purification method is graphically shown in Fig. 18 by the gel filtration profiles of the synthetic peptide before and after the affinity column. Since the first radiolabeled residue was the $[^{3}H]$acetyl lysine at position 16, only peptides longer than 21

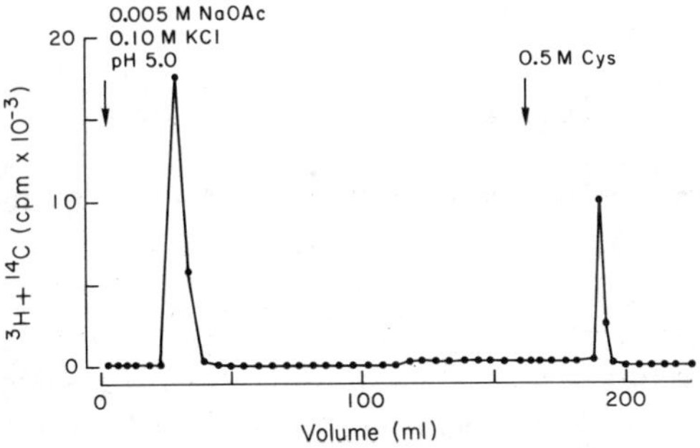

FIG. 17. Affinity purification of Cys-Met-[Lys($[^{14}C]$Ac)^{12}, Lys($[^{3}H]$Ac)^{16}] histone H4 (1-37) using organomercurial agarose.

FIG. 18. Gel filtration of Cys-Met-histone H4-(1-37) on Bio-Gel P-2 before **(a)** and after **(b)** affinity purification on organomercurial agarose.

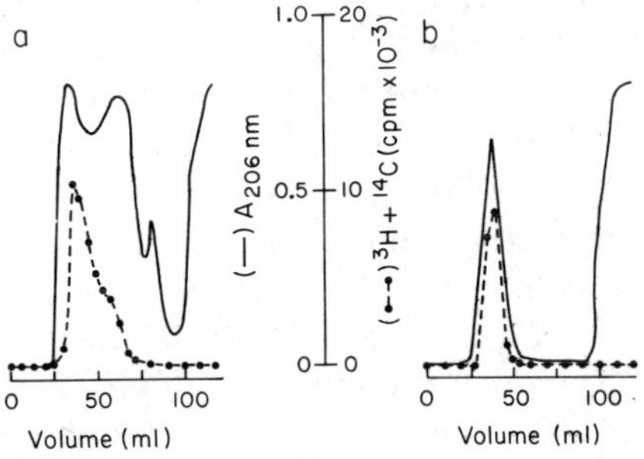

residues are radiolabeled. Panel *a* reveals three UV-active peaks in front of the salt peak. Most of the radiolabeled peptides are found in the first peak, which occurred at the void volume of the P-2 column and thus should only contain peptides larger than about 25 residues. The second peak contained the remaining radiolabeled peptides and most of the synthetic material as judged by ultraviolet absorbance of the peptide bond at 206 nm. In contrast, panel *b* shows that the Cys-Met–peptides retained by the affinity column eluted as a single peak at the void volume. All of the peptides smaller than 25 residues and even some larger peptides had been removed.

The histone peptide contains a unique serine residue at the amino terminus and a single threonine residue at position 30 near the carboxyl terminus. Thus the ratio of these diagnostic residues after acid hydrolysis provides an independent criterion of homogeneity. All peptide chains longer than seven residues will contain one threonine residue, but only chains that are still growing at the end of the synthesis will contain serine. The Thr/Ser ratio was 5.7 to 1 before the affinity column, which indicates that only about 20% of the chains were still growing at the end of the assembly process. But the ratio was 1.1 to 1 after the affinity column, which demonstrates that practically every Cys-Met–peptide contained both Ser^1 and Thr^{30}. Thus essentially all of the terminated deletion peptides and other nongrowing chains were removed by the organomercurial Cys-Met method of affinity purification.

The Dimeric Cys-Met Tactic

Application of the organomercurial Cys-Met tactic to an antibody domain or any other polypeptide containing cysteine or methionine residues introduces major complications. If the thiol groups of the cysteine residues are not masked during the affinity purification procedure, peptide by-products containing cysteine but lacking the amino-terminal Cys-Met moiety will also bind to the organomercurial agarose column. Furthermore, if the sulfide groups of the methionine residues are not blocked, the cyanogen bromide treatment to remove Cys-Met will also fragment the desired peptide after each methionine residue. Thus the cysteine and methionine residues should be protected by groups that survive not only assembly of the desired peptide but also each step of the affinity purification procedure.

The presently available protecting groups most compatible with this goal are the S-acetamidomethyl (Acm) group (45) for cysteine, which was used during solid-phase synthesis of human calcitonin (30), and the S-oxide group for methionine, which was employed for synthesis of bovine pancreatic ribonuclease (21). The Cys(Acm) residues are stable during solid-phase synthesis, hydrogen fluoride cleavage, reduction with a thiol, and cyanogen bromide treatment; the Met(O) residues are stable during synthesis, hydrogen fluoride cleavage, binding to organomercurial-agarose, and cyanogen bromide cleavage. But the organomercurial Cys-Met tactic will generally fail for polypeptides containing Cys(Acm) or Met(O) residues because the Acm group is removed by mercuric reagents and the sulfoxide is reduced to the sulfide by thiols, which are needed to reduce the Cys-Met–peptides and to elute them from organomercurial agarose.

This dilemma is resolved by replacing the organomercurial column by another affinity receptor. One Cys-Met–peptide molecule can effectively act as an affinity receptor for another such molecule by formation of a disulfide bond between their amino-terminal cysteine residues. This dimeric Cys-Met tactic is illustrated for the V_H domain of M603 in Fig. 19. The Cys^{22} and Cys^{98} residues are protected by acetamidomethyl groups and Met^{34} and Met^{38} are masked as the sulfoxide during assembly and purification. After cleavage

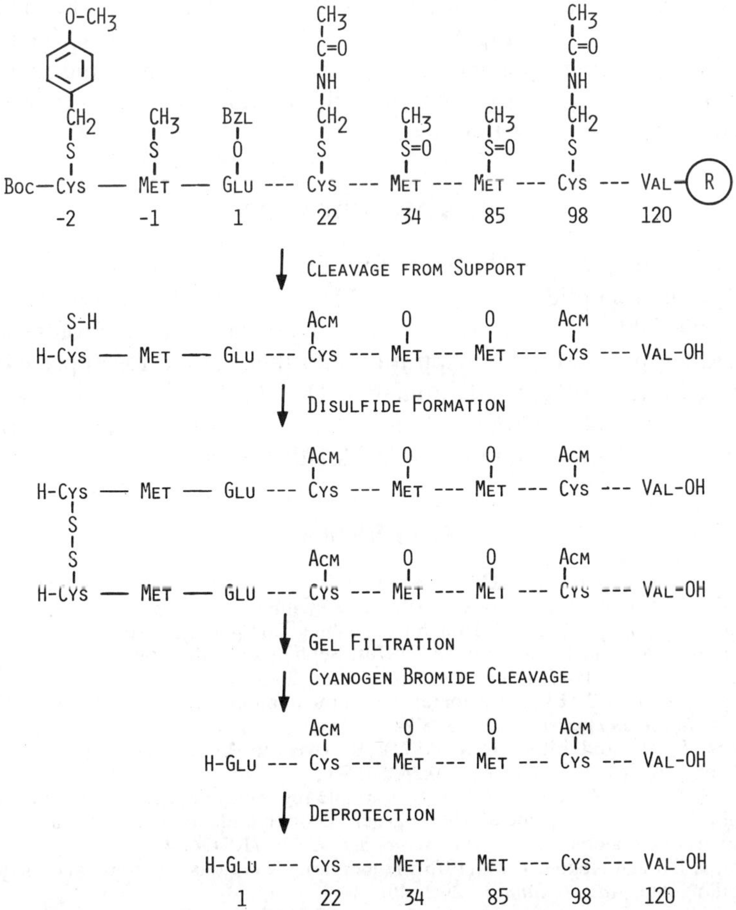

FIG. 19. The dimeric Cys-Met tactic for affinity purification of the M603 V_H domain and other proteins containing cysteine or methionine.

from the solid support, the free thiol group of Cys^{-2} is air oxidized to the disulfide. Then the 244-residue dimers are separated by gel filtration from the 122-residue monomer and any smaller by-products lacking Cys-Met. After cyanogen bromide cleavage of the dimer, the Met(O) residues are reduced with a thiol and the Cys(Acm) residues are deprotected with mercuric acetate to provide the desired immunoglobulin domain.

Several other promising groups can be envisaged in the role of affinity receptor, affinity group, or binding group. For example, the strong but reversible noncovalent binding between avidin and biotin could be exploited by using avidin-agarose (2,19) as the affinity receptor and a biotin derivative, such as N^{ϵ}-biotinyllysine (biocytin), as the affinity group. Moreover, a photolabile group could be used as the binding group. The two Cys-Met tactics described above, however, have the distinct advantage of using commercially available components and well-established chemical reactions. Thus the organomercurial Cys-Met tactic is recommended for polypeptides lacking cysteine and methionine and the dimeric Cys-Met tactic should be useful for immunoglobulin domains and other proteins containing cysteine or methionine.

ACKNOWLEDGMENTS

We thank Mrs. Dianne Dampney for her superb technical assistance with the complement peptides and Peter S. Tippett for his lively collaboration on the hypervariable peptide. We are especially grateful to Professor R. B. Merrifield for providing the inspiration and the method for approaching the chemical synthesis of antibody domains. This work was supported in part by Grant-in-Aid 74-844 from the American Heart Association and by Grant AM 01260 from the United States Public Health Service.

REFERENCES

1. Baas, J. M. A., Beyerman, H. C., van de Graaf, B., and de Leer, E. W. B. (1971): The relative statistical distribution of peptides in stepwise synthesis. In: *Peptides 1969,* edited by E. Scoffone, pp. 173–176. North-Holland Publishing Company, Amsterdam.
2. Bodanszky, A., and Bodanszky, M. (1970): Sepharose-avidin column for the binding of biotin or biotin-containing peptides. *Experientia,* 26:327.
3. Bourgois, A. (1975): Evidence for an ancestral immunoglobulin gene coding for half a domain. *Immunochemistry,* 12:873–876.
4. Bridges, S. H., and Little, J. R. (1971): Recovery of binding activity in reconstituted myeloma proteins. *Biochemistry,* 10:2525–2530.
5. Capra, J. D., and Kehoe, J. M. (1974): Variable region sequences of five human immunoglobulin heavy chains of the V_HIII subgroup: definitive identification of four heavy chain hypervariable regions. *Proc. Natl. Acad. Sci. USA,* 71:845–848.
6. Capra, J. D., and Kehoe, J. M. (1975): Hypervariable regions, idiotypy, and the antibody-combining site. *Adv. Immunol.,* 20:1–40.
7. Claflin, J. L., Lieberman, R., and Davie, J. M. (1974): Clonal nature of the immune response to phosphorylcholine. II. Idiotypic specificity and binding characteristics of anti-phosphorylcholine antibodies. *J. Immunol.,* 112:1747–1756.

8. Cosenza, H., and Köhler, H. (1972): Specific inhibition of plaque formation to phosphorylcholine by antibody against antibody. *Science,* 176:1027–1029.
9. Cuatrecasas, P. (1970): Protein purification by affinity chromatography. Derivatizations of agarose and polyacrylamide beads. *J. Biol. Chem.,* 245:3059–3065.
10. Davies, D. R., Padlan, E. A., and Segal, D. M. (1975): Three-dimensional structure of immunoglobulins. *Annu. Rev. Biochem.,* 44:639–667.
11. Davies, D. R., Padlan, E. A., and Segal, D. M. (1975): Immunoglobulin structures at high resolution. *Contemp. Top. Mol. Immunol.,* 4:127–155.
12. Edelman, G. M., Cunningham, B. A., Gall, W. E., Gottlieb, P. D., Rutishauser, U., and Waxdal, M. J. (1969): The covalent structure of an entire γG immunoglobulin molecule. *Proc. Natl. Acad. Sci. USA,* 63:78–85.
13. Edelman, G. M., and Gall, W. E. (1969): The antibody problem. *Annu. Rev. Biochem.,* 38:415–466.
14. Erickson, B. W., and Merrifield, R. B. (1973): Acid stability of several benzylic protecting groups used in solid-phase peptide synthesis. Rearrangement of *O*-benzyltyrosine to 3-benzyltyrosine. *J. Am. Chem. Soc.,* 95:3750–3756.
15. Erickson, B. W., and Merrifield, R. B. (1973): Use of chlorinated benzyloxycarbonyl protecting groups to eliminate N^{ϵ}-branching at lysine during solid-phase peptide synthesis. *J. Am. Chem. Soc.,* 95:3757–3763.
16. Erickson, B. W., and Merrifield, R. B. (1974): Stability of side-chain protecting groups in solid-phase peptide synthesis. *Isr. J. Chem.,* 12:79–85.
17. Erickson, B. W., and Merrifield, R. B. (1976): Solid-phase peptide synthesis. In: *The Proteins, Third Edition, Vol. 2,* edited by H. Neurath and R. H. Hill, pp. 255–527. Academic Press, New York.
18. Fraser, K. J., Poulsen, K., and Haber, E. (1972): Specific cleavage between variable and constant domains of rabbit antibody light chains by dilute acid hydrolysis. *Biochemistry,* 11:4974–4977.
19. Green, N. M., and Toms, E. J. (1973): The properties of subunits of avidin coupled to sepharose. *Biochem. J.,* 133:687–700.
20. Gutte, B., Lin, M., Caldi, D. G., and Merrifield, R. B. (1972): Reactivation of des(119-, 120-, or 121–124) ribonuclease A by mixture with synthetic COOH-terminal peptides of varying lengths. *J. Biol. Chem.,* 247:4763–4767.
21. Gutte, B., and Merrifield, R. B. (1971): The synthesis of ribonuclease A. *J. Biol. Chem.,* 246:1922–1941.
22. Hochman, J., Inbar, D., and Givol, D. (1973): An active antibody fragment (Fv) composed of the variable portions of heavy and light chains. *Biochemistry,* 12:1130–1135.
23. Hugli, T. E. (1975): Human anaphylatoxin (C3a) from the third component of complement. Primary structure. *J. Biol. Chem.,* 250:8293–8301.
24. Inbar, D., Hochman, J., and Givol, D. (1972): Localization of antibody-combining sites within the variable portions of heavy and light chains. *Proc. Natl. Acad. Sci. USA,* 69:2659–2662.
25. Krieger, D. E., Levine, R., Merrifield, R. B., Vidali, G., and Allfrey, V. G. (1974): Chemical studies of histone acetylation. Substrate specificity of a histone deacetylase from calf thymus nuclei. *J. Biol. Chem.,* 249:332–334.
26. Li, C. H., and Yamashiro, D. (1970): The synthesis of a protein possessing growth-promoting and lactogenic activities. *J. Am. Chem. Soc.,* 92:7608–7609.
27. Merrifield, R. B. (1963): Solid phase peptide synthesis. I. The synthesis of a tetrapeptide. *J. Am. Chem. Soc.,* 85:2149–2154.
28. Mizoguchi, T., Shigezane, K., and Takamura, N. (1969): Bromoacylpolystyrene, a new type of polymer support for solid phase peptide synthesis. *Chem. Pharm. Bull. (Tokyo),* 17:411–412.
29. Olander, J., and Little, J. R. (1975). Preferential homologous recombination of H and L chains from mouse myeloma proteins which bind Dnp ligands. *Immunochemistry,* 12:383–387.
30. Ontjes, D. A., Roberts, J. C., Hennessy, J. F., Burford, H. J., and Cooper, C. W. (1973): Biological and immunological activity of a thyrocalcitonin produced by the solid-phase method of peptide synthesis. *Endocrinology,* 92:1780–1785.
31. Potter, M. (1972): Immunoglobulin-producing tumors and myeloma proteins of mice. *Physiol. Rev.,* 52:631–719.

32. Potter, M. (1976): *This volume.*
33. Potter, M., and Leon, M. A. (1968): Three IgA myeloma immunoglobulins from the BALB/c mouse: precipitation with penumococcal C polysaccharide. *Science,* 162:369–371.
34. Potts, J. T., Jr., Tregear, G. W., Keutmann, H. T., Niall, H. D., Sauer, R., Deftos, L. J., Dawson, B. F., Hogan, M. L., and Aurbach, G. D. (1971): Synthesis of a biologically active N-terminal tetratriacontapeptide of parathyroid hormone. *Proc. Natl. Acad. Sci. USA,* 68:63–67.
35. Rich, D. H., and Gurwara, S. K. (1975): Preparation of a new *o*-nitrobenzyl resin for solid-phase synthesis of *tert*-butyloxycarbonyl-protected peptide acids. *J. Am. Chem. Soc.*, 97:1575–1579.
36. Rudikoff, S., and Potter, M. (1974): Variable region sequence of the heavy chain from a phosphorylcholine binding myeloma protein. *Biochemistry*, 13:4033–4038.
37. Segal, D. M., Padlan, E. A., Cohen, G. H., Rudikoff, S., Potter, M., and Davies, D. R. (1974): The three-dimensional structure of a phosphorylcholine-binding mouse immunoglobulin Fab and the nature of the antigen binding site. *Proc. Natl. Acad. Sci. USA*, 71:4298–4302.
38. Sieber, P., and Iselin, B. (1968): Selective acidolytic cleavage of aralkoxycarbonyl-amino protecting groups. *Helv. Chim. Acta*, 51:614–622.
39. Spande, T. F., Witkop, B., Degani, Y., and Patchornik, A. (1970): Selective cleavage and modification of peptides and proteins. *Adv. Protein Chem.*, 24:97–260.
40. Tippett, P. S. (1975): Structural-specificity relationships of the immunoglobulin molecule and the solid phase peptide synthesis of two antigen-binding peptides. B. A. Thesis, Kalamazoo College, Kalamazoo, Michigan.
41. Tregear, G. W. (1972): Graft copolymers as insoluble supports in peptide synthesis. In: *Chemistry and Biology of Peptides,* edited by J. Meienhofer, pp. 175–178. Ann Arbor Science Publishers, Ann Arbor, Michigan.
42. Tregear, G. W., van Rietschoten, J., Greene, E., Keutmann, H. T., Niall, H. D., Reit, B., Parsons, J. A., and Potts, J. T., Jr. (1973): Bovine parathyroid hormone: Minimal chain length of synthetic peptide required for biological activity. *Endocrinology*, 93:1349–1353.
43. Tregear, G. W., van Rietschoten, J., Greene, E., Niall, H. D., Keutmann, H. T., Parsons, J. A., O'Riordan, J. L. H., and Potts, J. T., Jr. (1974): Solid-phase synthesis of the biologically active *N*-terminal 1–34 peptide of human parathyroid hormone. *Hoppe Seylers Z. Physiol. Chem.*, 355:415–421.
44. van Rietschoten, J., Tregear, G. W., Leeman, S., Powell, D., Niall, H., and Potts, J. T., Jr. (1975): Simultaneous synthesis of two peptide analogs on different insoluble supports. In: *Peptides 1974*, edited by Y. Wolman, pp. 113–115. Halsted Press, New York.
45. Veber, D. F., Milkowski, J. D., Denkewalter, R. G., and Hirschmann, R. (1968): The synthesis of peptides in aqueous medium. IV. A novel protecting group for cysteine. *Tetrahedron Lett.*, pp. 3057–3058.
46. Weygand, F. (1968): Bromoacetylated polystyrene as a starting material for peptide synthesis on a solid support. In: *Peptides 1968,* edited by E. Bricas, pp. 183–184. North-Holland Publishing Company, Amsterdam.
47. Wu, T. T., and Kabat, E. A. (1970): An analysis of the sequences of the variable regions of Bence Jones proteins and myeloma light chains and their implications for antibody complementarity. *J. Med. Chem.*, 132:211–250.
48. Yajima, H., Kiso, Y., Okada, Y., and Watanabe, H. (1974): Synthesis of the basic trypsin inhibitor from bovine pancreas (Kunitz and Northrop) by fragment condensation on a polymer support. *J. Chem. Soc. Chem. Commun.*, pp. 106–107.
49. Yamashiro, D., and Li, C. H. (1974): Synthesis of a pentekontapeptide with high lipolytic activity corresponding to the carboxyl-terminal fifty amino acids of ovine β-lipotropin. *Proc. Natl. Acad. Sci. USA*, 71:4945–4949.

Antibodies in Human Diagnosis and Therapy,
edited by Edgar Haber and Richard M. Krause.
Raven Press, New York © 1977.

Approach to the Chemical Synthesis of Antibodies

David Givol, Moshe Gavish, Rina Zakut, Jacqueline Sharon, and Meir Wilchek

Departments of Chemical Immunology and Biophysics, The Weizmann Institute of Science, Rehovot, Israel

Antibodies form a family of closely related proteins whose variability, as far as binding specificity is concerned, is confined to the N-terminal 110 residues (V_L or V_H domains) of both L and H chains (1, 2). The principal variability is localized in three hypervariable regions in each V domain (3), which are clustered together to form the antibody combining site (4, 5). The rest of the sequence in the V domains may be regarded as constructing a rigid framework, whose three-dimensional structure is very similar in different antibodies, and to which are attached the hypervariable loops forming the specific site (6). Hence, antibodies are a random collection of homologous polypeptide chains with diverse specificity. Due to these random variations it is not yet possible to correlate precisely sequence and specificity, or to predict which combining site will be generated by a particular sequence. It is also not clear today what the relationship is between replacements in hypervariable regions and in framework residues. It is conceivable that synthetic homologues of antibodies will enable a systematic step-by-step replacement of amino acids and preparations of well-planned homologues which will help to elucidate the contribution of each segment in the V domain to the overall structure. It is hoped that, if the synthetic methods will be found suitable for the synthesis of V domains, they can be utilized to prepare synthetic antibodies with known specificity.

It should be realized that the fact that the antibody site is formed by two peptide chains puts an additional constraint on the synthetic product, namely, the recognition of another peptide chain. Thus a synthetic V region should (i) fold correctly, (ii) interact with the other V domain to form an Fv fragment, and (iii) contain the correct sequences to contribute contact residues in the site. These features, on the other hand, provide an efficient means for the isolation of the correct sequences from the synthetic mixture. The main scheme of such an isolation procedure will be the association of native V_H (or V_L) with the correct polypeptides from a synthetic V_L (or V_H) mixture, and the affinity chromatography of this mixture on an insoluble ligand column. Hence, our effort was concentrated on the Fv fragment from protein 315 and the association of its peptide chains V_L and V_H. The conditions found for the correct association of V_L and V_H and of their renaturation after the reduction of disulfide bonds in urea, are particularly efficient in our system. Preliminary results of testing these conditions for the association of a synthetic V_L preparation with native V_H suggest that the synthetic product may contain approximately 0.1% of active material.

Fv FRAGMENTS

The experience with the present methods of solid phase synthesis of proteins was predominantly with polypeptides of sizes no longer than 120–150 residues, such as RNase (7), lysozyme (8), or growth hormone (9). Previously we reported the preparation of Fv fragment from protein 315, a mouse myeloma IgA, with high affinity to nitrophenyl ligands (10,11). This fragment has a molecular weight of 25,000 and retains full binding activity. It is composed of V_L and V_H, each 110–120 residues long, and therefore provides the suitable material to attempt chemical synthesis of V domains. The only other Fv fragment reported was prepared from human IgM (12) without known binding activity. We have tried extensively to prepare Fv fragment by enzymic digestion of other mouse myeloma proteins, without success so far. Presumably, the unique type of L chain (λ_2) in protein 315 is responsible for rendering this protein susceptible to pepsin digestion, which will split only the constant domains in its Fab, leaving an intact Fv. However, it is possible to prepare half light chains, V_L domains (13–15) as well as V_H domains (16), and it is conceivable that such fragments will associate with the other chain, and after a suitable enzymic digestion may yield an active Fv.

Indeed, we were able to prepare, by this indirect way, Fv from the mouse myeloma IgA XRPC-25 which binds nitrophenyl ligands. The L chain was digested with trypsin at pH 8.1 to yield V_L and the digest recombined with the H chain to yield molecules with binding activity. These molecules were digested with papain and fractionated on a Sephadex G-75 column to yield an Fv fragment. The molecular weight determined by sedimentation and diffusion was 24,000 and the K_A as determined by equilibrium dialysis with DNP lysine was 2×10^5 similar to that of the intact IgA. On SDS gel two peptide chains with molecular weights around 12,000 were obtained. This Fv was obtained in 35% of the theoretical yield expected with respect to the original light chain. This study demonstrates that the major difficulty in obtaining Fv lies in the interaction between V and C domains in intact Fab. If this is circumvented by using isolated L chains, an Fv composed of V_L and V_H can be obtained.

ASSOCIATION OF V_L AND V_H TO FORM AN ACTIVE Fv

We reported previously (11) that in 8 M urea Fv dissociates into V_L and V_H which can be separated on DEAE-cellulose in urea. When a 1:1 mixture of the isolated chains in urea was diluted into PBS there was a full regain of binding activity (11).

The kinetics of the association process between V_L and V_H to form an active Fv was followed by monitoring their fluorescence. The emission spectra of V_L and V_H in 8 M urea and in PBS is given in Fig. 1. A red shift of 5 nm is observed for both chains. In addition, there is a 2.3-fold increase in the fluorescence intensity of V_L in 8 M urea, whereas no such a change is found in the fluorescence of V_H. Similarly, fluorescence enhancement of 55% is also observed for Fv in 8 M urea, along with a red shift of 11 nm (Fig. 2). It is also shown that Fv reconstituted from urea has the same emission spectrum as native Fv. Hence the association and renaturation of Fv (or a mixture of V_L and V_H) from 8 M urea can be followed by the decrease in protein fluorescence. The results of such studies, at different protein concentrations are presented in Fig. 3. It is shown that at a relatively high protein concentration (2×10^{-6}) the renaturation of Fv is composed of two distinct steps: (i) a rapid step which cannot be resolved by the regular manual mixing and comprises about 30% of the regain of native fluorescence, and (ii) a slow process which takes place in about 15 min. It is also shown that, as protein concentration in the renaturation process decreases, the rapid step becomes more prominent and at 1.6×10^{-8} M the entire association process is due to

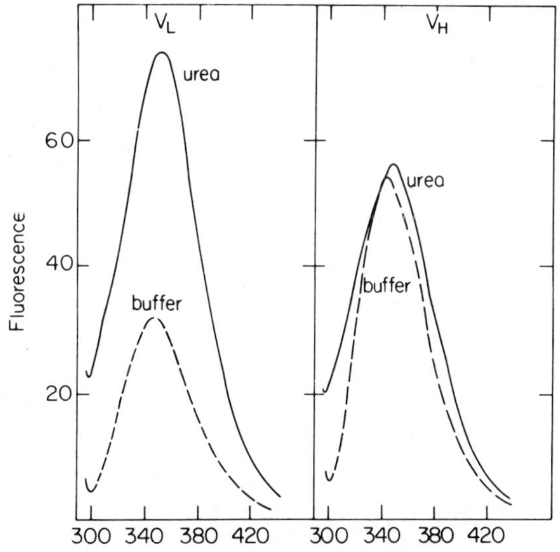

FIG. 1. Effect of 8 M urea on the fluorescence of V_L and V_H. Protein concentration was 33 μg/ml. Excitation wavelength was 280 nm.

FIG. 2. Fluorescence emission spectra of Fv in buffer (native) and in 8 M urea, and of Fv reconstituted (recons) from 8 M urea. Protein concentration was 66 μg/ml. The spectrum of reconstituted Fv was taken 45 min after its 20-fold dilution from 8 M urea.

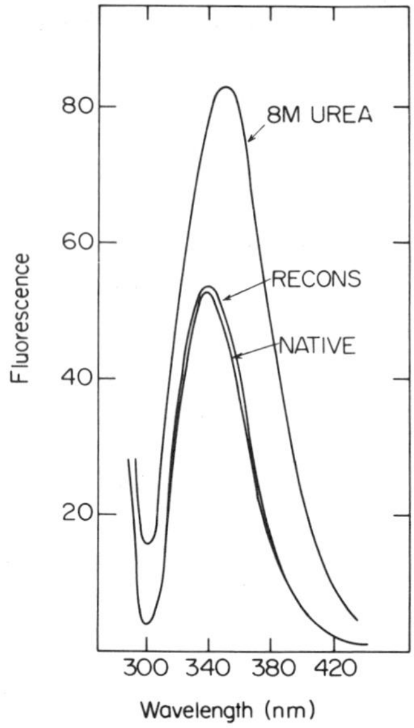

the rapid process only (Fig. 3). The kinetics of regain of binding activity were analyzed in these reactions by a rapid titration with equivalent amounts of DNP-lysine, measuring the fluorescence quenching due to the binding of DNP lysine to Fv. At all protein concentrations the kinetics of regain of native fluorescence were identical with that of regain of binding activity.

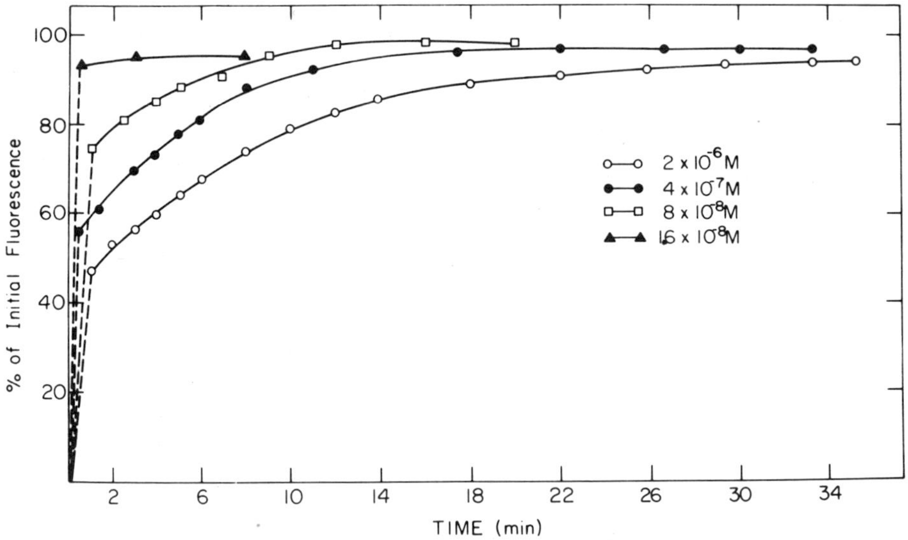

FIG. 3. Kinetics of renaturation of Fv from 8 M urea. Fv was diluted 20-fold from 8 M urea, to yield different protein concentrations. The decrease of protein fluorescence at 375 nm was recorded and the data are expressed as regain of native fluorescence.

This study demonstrates that the association constant between V_L and V_H is greater than 10^8 and that the association process is more efficient at low protein concentrations. We assume that the slow renaturation at higher protein concentration is due to aggregates and "wrong" associates, which slowly reshuffle to the "correct" more stable Fv conformation. The results also show that the association process is rapid and can be analyzed immediately (~ 30 sec) after mixing V_L and V_H. The amount that can be detected in solution is at the range of 0.5–1 μg/ml of active Fv. If, however, the mixture contains even 99.9% of an inactive or unrelated protein, affinity chromatography on DNP-lysine-Sepharose column can be used to bind the active Fv. In this purification procedure Fv in concentrations of 0.1 μg/ml can be isolated.

OXIDATION AND RENATURATION OF REDUCED Fv

Since the expected synthetic V_L will be obtained in the reduced form, it was necessary to find conditions for renaturation of Fv from a completely unfolded state. Various attempts to oxidize and renature Fv, after reduction in 8 M urea, by removing the reagents on Sephadex G-25, or by dilution and dialysis, did not yield active Fv. Since either V_L or V_H each contain only one disulfide bond, we anticipated that upon oxidation in 8 M urea no "wrong" disulfide bonds could be generated, since only the single intrachain disulfide bonds would be formed. It is recalled that, when such urea oxidized ("urox") derivative of RNase was prepared, a completely inactive protein with wrong disulfide bonds was obtained (17). This urox-RNase was converted to an active RNase by disulfide interchange induced with a low level of mercaptoethanol (17, 18). In the case of Fv we found that "urox-Fv," prepared by dialysis of reduced Fv against 8 M urea to remove the mercaptoethanol, can be subsequently renatured by dilution into PBS exactly as described in the previous section. Similarly "urox-V_L" can associate efficiently with V_H to yield an active Fv (Fig. 4). The advantage of this procedure is that reduction and oxidation can be performed at high protein concentrations (5 mg/ml), whereas in other cases of oxidation and renaturation of proteins, high dilution was essential (19).

SEQUENCE OF V_L

The sequence of the N-terminal 115 residue of L_{315} was used as published (20). It was, however, necessary to establish the positions of amides in six positions (residues 39, 40, 81, 83, 96), which were not given in this sequence. Fortunately, all the unknown Glx and Asx were localized in thermolytic peptides containing either Tyr or His. Hence the Pauli-positive peptides of a thermolytic digest of V_L were isolated and sequenced. This established amides at positions 40 (Gln), 81 (Gln), and 96 (Asn) and acids at positions 39 (Glu), 43 (Asp), and 83 (Glu). The sequence used in the synthesis is presented in Fig. 5.

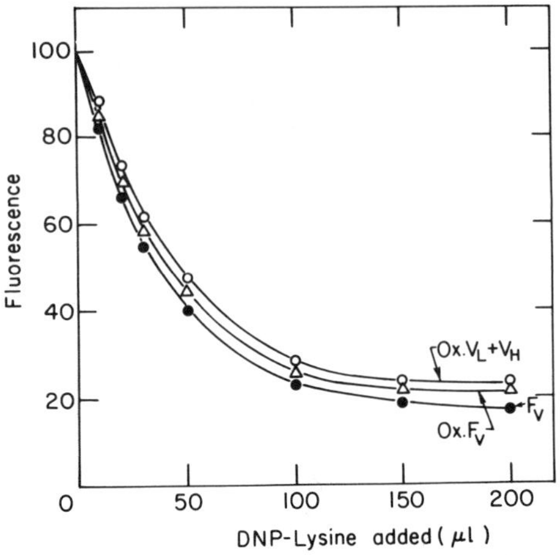

FIG. 4. Fluorescence quenching analysis of the binding of DNP-lysine by urea-oxidized Fv or V_L after renaturation. Fv or V_L were reduced with β-mercaptoethanol in 8 M urea and dialyzed against 8 M urea, to oxidize disulfide bonds. These "urox-V_L," after combination with V_H, or "urox-Fv" were diluted 20-fold into PBS. Final protein concentrations were 50 μg/ml. DNP-lysine was 5×10^{-5} M.

SOLID PHASE SYNTHESIS OF V_L

The general procedures of solid phase methods were followed (21,22). The synthesis was begun with 3 g of 1% crosslinked resin (Schwartz Mann) to which the C-terminal Thr (0.28 mm/g) was attached and continued with stepwise addition of protected amino acids until the 115 amino acid chain

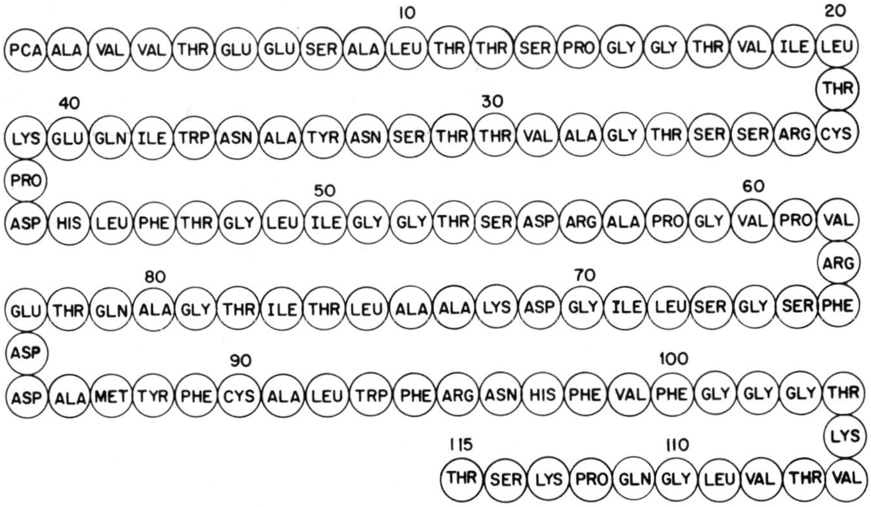

FIG. 5. The sequence of V_L of protein 315 [Dugan et al. (20)] with location of amides (see text).

was assembled. All amino acids were protected in the α-amino position with the Boc group and the side chain blocking groups were: Asp (OBzl), Glu (OBzl), Cys (MeOBzl), Ser (Bzl), Tyr (Bzl), Lys (Z), Arg (NO_2), His (Boc). Coupling was affected with dicyclohexylcarbodiimide, except for Asn and Gln where nitrophenylesters were used. Coupling was catalyzed with 1-hydroxybenzotriazole. Boc groups were removed with TFA:CH_2Cl_2:anisole:ethanedithiol (45:50:2:5). The coupling and deblocking of Boc groups were monitored by ninhydrin reaction (23). The course of synthesis was followed by hydrolysis and analysis of a sample (approx. 10 mg) at intervals of 5 residues during the synthesis. The final product was removed from the resin with HF in the presence of anisole (24). After removal of the HF at 0° under vacuum the product was precipitated with ether, washed several times, dried and suspended in 10% acetic acid. The solution was lyophilized to yield 6 gr of peptide material (60% yield).

CHARACTERIZATION OF SYNTHETIC V_L

The isolated product was analyzed by several physical and chemical criteria. The amino acid analysis of synthetic V_L is shown in Table 1. Notable features were low recovery of Cys (only 1 mm) and of Trp (not shown in the table). The reduced content of Trp was also verified by the low intensity of emission spectrum of the product which was only 25% that of native V_L. Also the major emission peak was at 360 nm and only a shoulder appeared at 335 nm which is the emission maximum of native V_L (Fig. 6). The solubility of the synthetic product at neutral pH was only 15%, whereas V_L is completely soluble in the same conditions. Upon chromatography on Sephadex

TABLE 1. *Amino acid composition of V_L and synthetic V_L ($V_{L(s)}$)*

Residue	V_L	$V_{L(s)}$	Δ
Lys	4	4.8	+0.8
His	2	1.8	
Arg	4	2.8	−1.2
Asp	8	8.3	
Thr	16	15.1	−0.9
Ser	9	8.2	−0.8
Glu	8	8.9	+0.9
Pro	5	6.5	+1.5
Gly	14	14.6	
Ala	10	10.6	
½ Cys	2	1.07	−0.93
Val	9	7.1	−1.9
Met	1	0.96	
ILe	5	4.3	−0.7
Leu	8	8.4	
Tyr	2	1.97	
Phe	6	7.4	+1.4
	113	112.8	

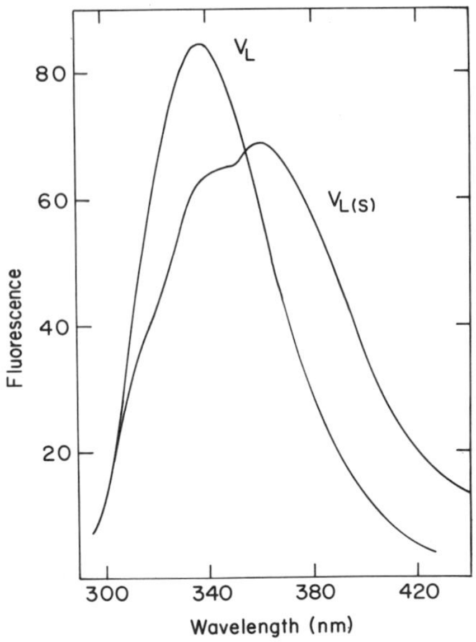

FIG. 6. Fluorescence emission spectrum of synthetic V_L ($V_{L(S)}$) and V_L. Excitation was at 280 nm. Proteins were dissolved in 0.1 M acetic acid to give 0.1 OD at 280 nm. The sensitivity of the fluorimeter in recording $V_{L(S)}$ spectrum was threefold higher than that of V_L.

G-50, most of the synthetic V_L emerged at the front of the column, and only 10% of the material appeared at the position of V_L. On SDS gel electrophoresis the synthetic product migrated as a continuous smear within the molecular weight range of 8000–20,000, whereas V_L appears as a discrete band of a 12,000 MW (Fig. 7).

Peptide map analysis (Fig. 8) showed some correlation between peptides derived from native and from synthetic V_L. Staining of the peptides with specific stains for tyrosine, histidine, and arginine demonstrated that the

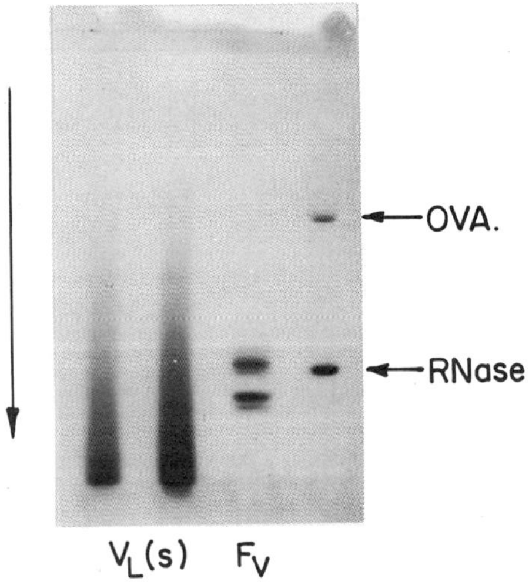

FIG. 7. SDS-polyacrylamide (12.5%) gel of Fv and synthetic V_L.

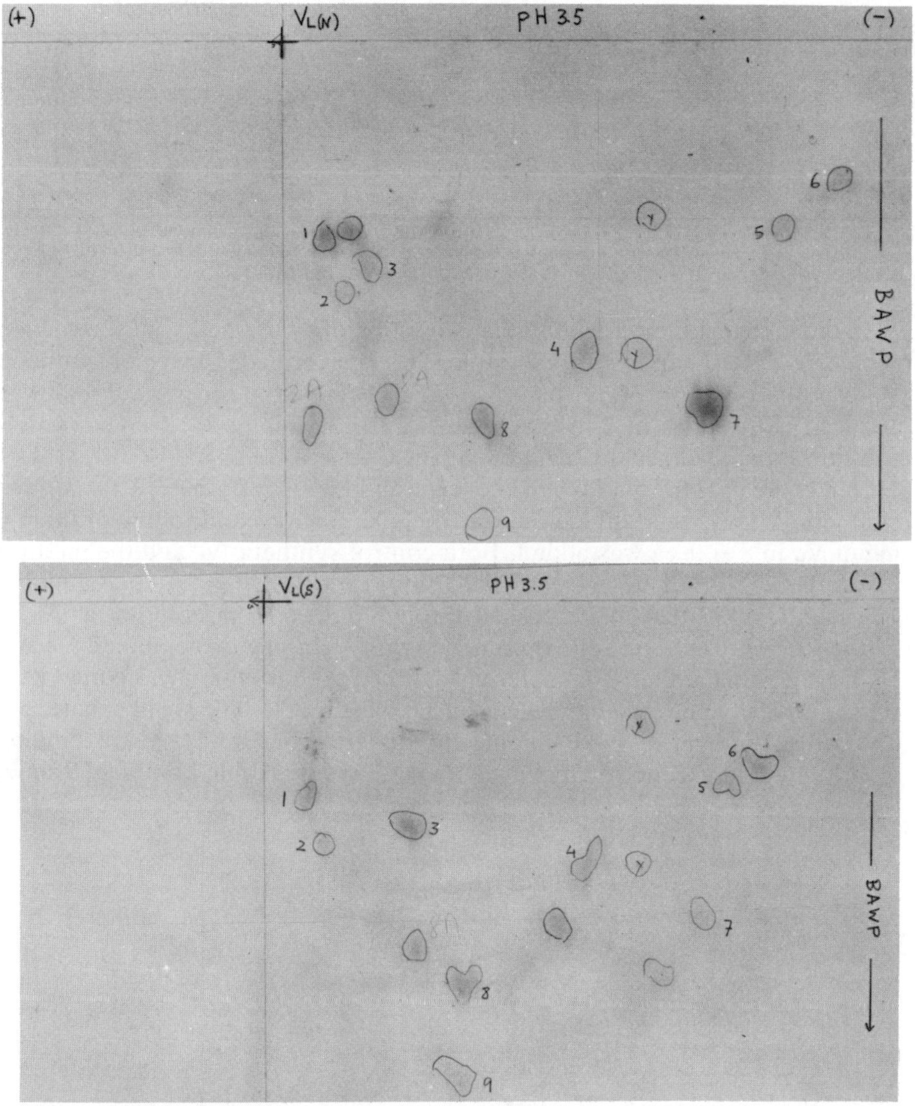

FIG. 8. Peptide map analysis of V_L and synthetic V_L ($V_{L(S)}$). Reduced and alkylated samples of V_L and $V_{L(S)}$ were digested with trypsin and thermolysin. The digest was fractionated by high-voltage electrophoresis at pH 3.5 in one dimension and by chromatography in Butanol:acetic acid:water:pyridine (15:3:12:10) in the second dimension. Numbers indicate peptides with similar mobilities. **Top:** V_L. **Bottom:** Synthetic V_L.

peptide containing Tyr_{88} is present in very low amount in the synthetic V_L. Otherwise, the histidine peptides were identical, and good correlation was also observed for arginine containing peptides. The $\epsilon^{1\%}_{280}$ of synthetic V_L was 1.8 as compared with 1.0 of native V_L.

TABLE 2. *Specific absorption of $V_{L(s)}$ – V_H recombinant on DNP-Sepharose*

Mixture	% Specifically Absorbed
V_LV_H	80
V_H	5
$V_{L(S)}$	0.7
$V_{L(S)}V_H$	1.2

The various fractions were passed through DNP-Sepharose and eluted with DNP-glycine. $V_{L(S)}$ is synthetic V_L. See text for experimental conditions.

In order to test the possibility that a fraction of synthetic V_L may associate with native V_H the following experiment was performed: 100 mg of synthetic V_L was dissolved in 30 ml of 0.1 M Tris HCl–8 M urea, pH 8.0 and was reduced with 0·1 M of 2-mercaptoethanol for 30 min at 37° C. The protein solution was oxidized by dialysis against several changes of 0.1 M Tris HCl–8 M urea pH 8.0 until the SH concentration was lower than 10^{-4} = M. The synthetic V_L was then divided into two equal portions and to one of them 4 mg of V_H in 8 M urea was added. Both control synthetic V_L and the mixture of synthetic V_L and V_H were diluted 20-fold into PBS and after remaining overnight at 4°C they were passed through a 1 × 3 cm column of DNP–lysine–Sepharose. Control experiments with V_H alone or combined V_L and V_H were run in parallel. The material adsorbed to the column was eluted with 0.05 M DNP glycine and dialyzed against 0.02 M NaCl–0.01 M phosphate, pH 7.4. After removal of DNP-glycine the material was analyzed for binding activity by fluorescence quenching with DNP-lysine. Table 2 and Fig. 9 show

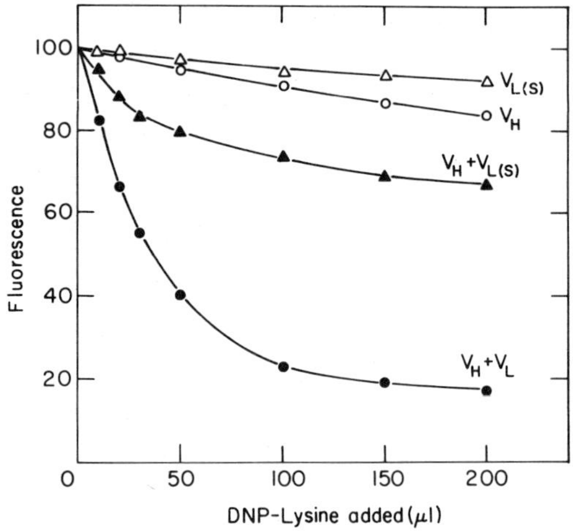

FIG. 9. Fluorescence quenching analysis of the binding of DNP lysine to a recombinant of V_H and synthetic V_L. Control of V_H, synthetic V_L, or recombinant of V_L and V_H are given. All samples were obtained after affinity chromatography on DNP-lysyl-Sepharose and elution with DNP-glycine. For recoveries see Table 2. $V_{L(S)}$, synthetic V_L. Other conditions as in Fig. 4.

the results of this experiment. It is shown that about 1% of material is absorbed from the mixture of synthetic V_L and V_H or synthetic V_L alone. On this level of recovery it is difficult to discriminate between specific and nonspecific binding to the DNP-lysine-Sepharose. However, the material eluted from either V_H or synthetic V_L alone did not show any fluorescence quenching with DNP-lysine, whereas the material eluted from the mixture of V_H and synthetic V_L showed some fluorescence quenching, corresponding to about 15% of that shown by reconstituted V_L and V_H (Fig. 9). It is therefore possible to suggest that if the synthetic V_L contains some correct sequences which can associate with V_H to yield an active Fv, they do not represent more than 0.1% of the synthetic product.

CONCLUDING REMARKS

The studies presented demonstrated that methodology to prepare Fv fragments of antibodies with intact binding site is available. Procedures were developed for separation of their component chains (V_L and V_H) and for their efficient renaturation in high yields, after complete reduction of disulfide bonds. By these procedures, it is not difficult to detect 0.1% of active V_L in a mixture of unrelated proteins. Experiments with a synthetic product which was synthesized according to the sequence of V_L of protein 315 suggest that this material may contain 0.1% of correct sequences which can associate with native V_H to yield an active Fv fragment. We believe that a program to study the structural basis of antibody specificity by synthetic homologues cannot be fruitful unless at least 10% of the synthetic material were able to recombine with native V_H to yield an active Fv. Only then could the effect of different replacements on activity be analyzed with confidence. This constitutes a great challenge to the synthetic methods, since we believe that the improvement of these methods is the key step for further progress toward a synthetic antibody.

SUMMARY

Our approach toward a chemical synthesis of antibodies is based on the availability of Fv fragment of protein 315 which contains two peptide chains (V_L and V_H) each of approximately 110 residues. The isolation of another Fv fragment from mouse myeloma IgA XRPC-25 is reported. This is based on the preparation of V_L from light chain, its association with the heavy chain and digestion of the recombined molecules with papain. The Fv obtained from XRPC-25 protein has a molecular weight of 25,000 and K_A of 2×10^5 toward DNP-lysine, similar to that of the intact IgA. This method may open a general way to prepare Fv fragments.

The kinetics of V_L–V_H association to form an active Fv was studied and found to be rapid and more efficient at low (10^{-8} M) protein concentration. Conditions for reoxidation and renaturation of Fv or V_L after reduction in 8

M urea are described. This involves removal of the reducing agent and oxidation of the single intrachain disulfide bond in urea. The urea-reoxidized V_L recombine with V_H to yield a completely active Fv.

Solid phase synthesis according to Merrifield was employed to prepare V_L of protein 315. The product was analyzed by various physical and chemical methods. Some features of its peptide maps corresponded to native V_L. On the other hand, by several criteria this product was different from native V_L, the most notable ones were its aggregation and lack of discrete bands on SDS acrylamide gels. The association of the synthetic V_L with native V_H was studied, according to the above mentioned results, and suggest that synthetic V_L may contain 0.1% of material capable of binding to V_H to form an active Fv.

ACKNOWLEDGMENTS

This study was supported by NIH grant No. 4 ROI AI 11453 and by a grant from the Israel Commission for Basic Research.

REFERENCES

1. Gally, J. A., and Edelman, G. M. (1972): The genetic control of immunoglobulin synthesis. *Annu. Rev. Genet.,* 6:1–46.
2. Givol, D. (1975): Affinity labeling and topology of the antibody combining site. *Essays Biochem.,* 10:73–103.
3. Wu, T. T., and Kabat, E. A. (1970): An analysis of the sequences of the variable regions of Bence Jones protein and myeloma light chains and their implications for antibody complementarity. *J. Exp. Med.,* 132:211–250.
4. Amzel, L. M., Poljak, R. J., Saul, F., Varga, J. M., and Richards, F. F. (1974): The three dimensional structure of a combining region-ligand complex of immunoglobulin. New at 3.5 Å resolution. *Proc. Natl. Acad. Sci. USA,* 71:1427–1430.
5. Segal, D. M., Padlan, E. A., Cohen, G. H., Rudikoff, S., Potter, M., and Davies, D. R. (1974): The three-dimensional structure of a phosphorylcholine-binding mouse immunoglobulin Fab and the nature of the antigen binding site. *Proc. Natl. Acad. Sci. USA,* 71:4298–4302.
6. Padlan, E. A., Davies, D. R., Pecht, I., Givol, D., and Wright, C. (1975): The antigen binding site of MOPC 315 immunoglobulin: A model building study. *Biochemistry (submitted for publication).*
7. Gutte, B., and Merrifield, R. B. (1971): The synthesis of ribonuclease A. *J. Biol. Chem.,* 246:1922–1941.
8. Sharp, J. J., Robinson, A. B., and Kamen, M. D. (1973): Synthesis of a polypeptide with lysozyme activity. *J. Am. Chem. Soc.,* 95:6097–6108.
9. Li, C. H., and Yamashiro, D. (1970): The synthesis of a protein possessing growth-promoting and lactogenic activities. *J. Am. Chem. Soc.,* 92:7608–7609.
10. Inbar, D., Hochman, J., and Givol, D. (1972): Localization of the antibody combining site within the variable portions of heavy and light chains. *Proc. Natl. Acad. Sci. USA,* 69:2659–2662.
11. Hochman, J., Inbar, D., and Givol, D. (1973): An active antibody fragment (Fv) composed of the variable portions of heavy and light chains. *Biochemistry,* 12:1130–1135.
12. Kakimoto, K., and Onou, K. (1974): Characterization of the Fv fragment isolated from a human immunoglobulin M. *J. Immunol.,* 112:1373–1382.
13. Solomon, A., and McLaughlin, C. L. (1969): Bence-Jones proteins and light chains of immunoglobulins. *J. Biol. Chem.,* 244:3393–3404.

14. Karlsson, F. H., Peterson, P. A., and Berggard, I. (1969): Properties of halves of immunoglobulin light chains. *Proc. Natl. Acad. Sci. USA,* 64:1257–1263.
15. Seon, B.-K., Roholt, D. A., and Pressman, D. (1972): Differences in the enzymatic digestibility of the variable and constant halves of Bence-Jones protein with temperature. *J. Biol. Chem.,* 247:2151–2155.
16. Mole, L. E., Geiger, M. D., and Koshland, M. E. (1975): The isolation and characterization of the V_H domain from rabbit heavy chains of different a locus allotype. *J. Immunol.,* 114:1442–1448.
17. Givol, D., Goldberger, R. F., and Anfinsen, C. B. (1964): Oxidation and disulfide interchange in the reactivation of reduced ribonuclease. *J. Biol. Chem.,* 239:PC 3114–PC 3116.
18. Givol, D., De Lorenzo, F., Goldberger, R. F., and Anfinsen, C. B. (1965): Disulfide interchange and the three-dimensional structure of proteins. *Proc. Natl. Acad. Sci. USA,* 53:676–684.
19. Epstein, C. J., Goldberger, R. F., Young, D. M., and Anfinsen, C. B. (1962): A study of the factors influencing the rate and extent of enzymic reactivation during reoxidation of reduced ribonuclease. *Arch. Biochem. Biophys.* [*Suppl.*], 1:223–231.
20. Dugan, E. S., Bradshaw, R. A., Sims, E. S., and Eisen, H. N. (1973): Amino acid sequence of the light chain of a mouse myeloma protein (MOPC–315). *Biochemistry,* 12:5400–5416.
21. Stewart, J. M., and Young, J. D. (1969): *Solid Phase Synthesis.* Freeman, San Francisco, Calif.
22. Meienhoffer, J. (1973): Peptide synthesis: A review of the solid phase methods. In: *Hormonal Proteins and Peptides,* Vol. II, pp. 45–267, edited by C. H. Li. Academic Press, N.Y.
23. Kaiser, E., Colescott, R. L., Bossinger, C. D., and Cook, P. I. (1970): Color test for detection of free terminal amino groups in the solid phase synthesis of peptides. *Anal. Biochem.,* 34:595–598.
24. Sakakibara, S., Shimonishi, Y., Kishida, T., Ikada, M., and Sugihara, H. (1967): Use of anhydrous hydrogen fluoride in peptide synthesis. *Bull. Chem. Soc. Jap.,* 40:2164–2167.

Antibodies in Human Diagnosis and Therapy,
edited by Edgar Haber and Richard M. Krause.
Raven Press, New York © 1977.

Approaches to the Synthesis of an Antibody-Combining Site

James Burton, Mario S. Rosemblatt, and Edgar Haber

Cardiac Unit, Massachusetts General Hospital, Boston, Massachusetts 02114

Heterologous antibodies are now used clinically in man where the risk of adverse immunologic reaction is acceptable because of the gravity of the condition under treatment. The scope of this form of immunotherapy would be greatly extended, even to routine clinical use, if the antibody fragments could be rendered nonimmunogenic. Advances in understanding the three-dimensional structure of the antigen-binding site coupled with improvements in synthetic technique make preparation of nonimmunogenic antibodies a real possibility. Such a synthesis has not yet been completed in our laboratory, but both the tools and the techniques are at hand. The purpose of this chapter is to discuss both our approach to the chemical synthesis of a digoxin-specific antibody and specific points unique to the renaturation and purification of this protein.

CLASSIC CHEMICAL SYNTHESES OF PROTEINS

From the first preparation of the nonapeptide oxytocin (1) to the assembly of proteins containing in excess of 100 residues (2,3), the ability of peptide chemists to produce ever longer sequences has expanded steadily. Of the 16 syntheses of proteins now reported or in progress, five involve condensation of fragments produced by classic chemical methods of synthesis. These are ribonuclease A S-protein (3), cytochrome C (4), ribonuclease T_1 (5), trypsin inhibitor II [Kazal] (6), trypsin inhibitor [Kunitz and Northrop] (7), and proinsulin (8). There can be little question that synthesis of a protein by classic methods requires far more effort than preparation by solid-phase techniques. The repeated purification of intermediate peptides, however, yields a product that may be rendered homogeneous with little additional work. With solid-phase techniques, a major expenditure of effort is often required to effect the purification of the crude product. As will be shown, immunoglobulin fragments possess properties that enable special, very powerful techniques to be used for their purification. These properties make solid-phase synthesis a reasonable approach for the preparation of an antibody.

SOLID-PHASE SYNTHESES OF PROTEINS

Ten different syntheses yielding more than 20 proteins have been performed by solid-phase methods. Beginning with the initial preparation of ferredoxin (9), syntheses have been reported for: cytochrome C (10); acyl carrier protein (11); ribonuclease A (2); trypsin inhibitor (12); lysozyme (13); cobra toxin (14); ribonuclease T_1 (15); growth hormone analog (16); and β-lipotropin (17).

Two solid-phase syntheses will be discussed briefly because of their relevance to the proposed synthesis of an antibody. Ribonuclease A (2) was prepared by solid-phase synthetic techniques, and, because of the detailed biochemical characterization of the product, it must be considered the standard against which other syntheses are compared. Reagents, protected amino acid derivatives, apparatus, and reaction schedules used in this synthesis have all been improved. A problem was uncovered, however, that still influences the proposed synthesis of an antibody. Eighty-three percent of the growing peptide chain was lost from the support in the course of the ribonuclease A synthesis (an average of 1.4% of the growing chains at each cycle). This was presumably caused by cleavage of the anchoring ester bond during the various deprotection stages. If a protein is synthesized to prove a proposed structure or to permit the activity of an analog to be tested, yield is not of major importance. A small amount of material will normally suffice for biochemical assay. Clinical application, however, requires a high yield of product. Gram amounts of highly purified material are needed for even the most preliminary evaluation. For this reason an optimal yield must be sought from the earliest stages of synthesis.

An important step in the purification of synthetic ribonuclease A was proteolytic digestion. This effected a 10-fold purification of the product and was important in obtaining highly active material. This functional approach to purification resolved a mixture which may well have been intractable to purification by conventional chromatographic techniques.

The synthesis of acyl carrier protein (ACP) (11) is also of particular interest. Again, an enzymatic method for functional purification was employed to yield an active product. The enzyme *holo* ACP-synthetase was used to attach a pantetheine moiety to recognizable synthetic ACP sequences. The correct sequences were thus selected from a rather complex mixture of synthetic peptides.

STRATEGY FOR THE SYNTHESIS OF AN ANTIGEN-COMBINING SITE

The synthesis of the entire immunoglobulin G molecule is a formidable task. The light chain consists of more than 200 amino acid residues and the heavy chain contains more than twice that number. Fortunately, only the variable region of each of these chains is needed to form the domain that binds antigen. Hochman et al. (18) have shown that a fragment produced by enzymatic digestion of mouse myeloma protein MOPC-315 is capable of binding dinitrophenol with the same affinity as the parent molecule. This fragment (Fv) consists of the variable region of the light and the heavy chain, bound to each other by noncovalent interactions. The polypeptide chains, each approximately 100 residues long, may be separated after denaturation. Recombination without significant loss of antigen-combining activity may then be effected. Extrapolation of this idea to other antibodies would allow the synthesis of an antibody to be reduced to preparation of two polypeptide chains, each containing approximately 100 amino acid residues.

Fragmentation of other antibodies by pepsin digestion does not yield an Fv fragment as was reported for MOPC-315. Our strategy for the preparation of such fragments from rabbit antibodies involves the selective cleavage of heavy and light chains followed by recombination of isolated variable regions. A previous demonstration that rabbit light chain can be cleaved by limited acid treatment between residues 109 (aspartic acid) and 110 (proline) in the switch region (19) indicates that an intact variable region may be obtained from light chains of the b4 allotype. The acid-cleaved light chain recombines with homologous heavy chain to yield a fully active product. More recently, we have shown that the variable region of the heavy chain may also be isolated (M.S. Rosemblatt and E. Haber, *in preparation*). A rabbit antibody specific for type III pneumococcal polysaccharide was subjected to limited reduction and aminoethylation. After brief digestion with trypsin and gel filtration under denaturing conditions, a fragment was obtained, which was shown by amino acid sequence analysis to extend between residues 11 (leucine) and 123 (lysine). This fragment retained the allotypic determinants present on the parent heavy chain and, by sequence analysis, did not show evidence of internal tryptic cleavage. On recombination with rabbit light chain, a major fraction of the initial antigen-combining sites were recovered. It is reasonable to assume that V_H will recombine equally well with V_L isolated from the homologous light chain.

Synthesis of one of these variable segments followed by combination with the other of either of natural or synthetic origin should produce an antigen-binding molecule.

CHEMICAL SYNTHESIS OF IMMUNOGLOBULIN 3368-V_L

The solid-phase synthesis of V_L from 3368, a rabbit antipneumococcal type III specific antibody (20), has now been completed by standard solid-phase techniques. Amino acid side-chain protecting groups are: tyrosine, *O*-2,6-dichlorobenzyl; lysine, N^{ϵ}-2-chlorobenzyloxycarbonyl; serine and threonine, *O*-benzyl ether; glutamic and aspartic acid, benzyl ester; and cysteine, S-benzyl. The N^{α}-amino function was protected with the *t*-butoxycarbonyl group. Glutamine and aspargine were added as the *o*-nitrophenyl active esters (21). One percent cross-linked resin (0.75 mmole Cl^-/g) was lightly esterified (8.6%) with *t*-butoxycarbonylglycine. The complete synthesis was done without interruption over a period of 21 days, employing manually operated apparatus. Rate of addition of residues averaged 5.33 per 24-hr period. The accelerated procedure for dichlorohexylcarbodiimide couplings introduced by Corley et al. (22) was employed. Completeness of coupling was monitored by the modified ninhydrin (23) and fluorescamine tests (24).

Evaluation of this synthesis revealed two major problems. First, as discussed in connection with the synthesis of ribonuclease A, a significant loss of peptide from the polymer was obvious. Solvolysis of the anchoring ester bond has been reported frequently in the synthesis of long peptides (12). Amino acid analysis showed that over the course of the synthesis 82% of the growing peptide chains were lost, averaging 1.48% of the peptide chains at each cycle.

The second problem was the sudden inaccessibility of 40% of the peptide chains between addition of the ninth and tenth residues. Quantitative addition of the phenylalanine residue at position 98 to the preceding glycyl residue could not be obtained in two separate syntheses. The resin gave a negative test with ninhydrin and fluorescamine after coupling. The incomplete addition may have been caused by incomplete deprotection (25). After addition of asparagine-92, the full complement of growing peptide chains once again became available. Presumably these data may be accounted for by the matrix solvation hypothesis (26).

Cleavage of the antibody fragment from the supporting matrix by treatment with HF in the presence of indole, mercaptoethanol, and anisole yielded the crude immunoglobulin in 42% yield. Scavengers and truncated sequences were removed by dialysis. Amino acid analysis of this material is in good agreement with theoretical values except for low recoveries of methionine and cysteine.

TECHNICAL INNOVATIONS EVALUATED FOR THE SYNTHESIS OF V_L

Attachment of the Growing Peptide to the Polymer by More Stable Bonds

Two approaches have been developed to yield more stable bonds between the growing protein and the polymer. A method of attachment, pioneered by Meienhofer and Trzeciak (27), has now been expanded by Glass et al. (28). This involves attachment of the growing peptide to the resin through a side-chain functional group. Linkage groups other than the standard α-carboxyl ester are employed. The penultimate residue in 3368-V_L is lysine-107. Attachment of the growing peptide chain to the polymer through the ϵ-amino group of this residue is an attractive possibility. Erickson and Merrifield (29) demonstrated that the N^{ϵ}-benzyloxycarbonyl protecting group does not possess adequate stability to serve as the anchoring bond. Bromination of the benzyl moiety, however, greatly stabilizes this protecting group (30). Synthesis of a polymer analogous to that reported by Meienhofer and Trzeciak (27) in which the benzyl group is brominated has now been completed. The synthetic dipeptide Boc-Lys-Gly-NH_2 has been attached to this matrix and elongated by standard solid-phase techniques to yield the C-terminal tetrapeptide of 3368-V_L. Early tests indicate that the peptide is attached very strongly to the polymer. The resin is now being evaluated in a more extended synthesis.

The second approach involves attachment of glycine-108 to a benzhydrylamine resin (31). The final product can be removed from the polymer by treatment with HF to yield the protein amide. Because glycine-108 is normally attached to an aspartyl residue, synthesis of the V_L fragments with a C-terminal amide is not expected to affect biological activity. Synthesis of the model tetrapeptide Boc-Val-Val-Lys(2Cl-Z)-Gly on both the benzhydrylamide and standard benzyl ester resins followed by treatment with trifluoroacetic acid –dichloromethane (1:1) shows the former peptide polymer bond to be about 10 times as stable as the benzyl ester bond.

Quantitation of Deprotection

Measurement of amino groups at each stage of a synthesis usually shows a steady decrease in the number available for reaction (11,32). Precipitous drops may occur at a given step (26). A marked decrease occurred in the synthesis of 3368-V_L that was reproduced in subsequent syntheses. The yield of incorporation of amino acids between residues 108 and 100 was uniform and nearly quantitative. Between residues 99 and 91, 40% of the peptide chains were unavailable for reaction. At residue 90, full incorporation was again observed. At least part of the final crude product must lack residues 91–99. Possibly the growing, relatively polar protein affects the properties of the supporting matrix. At certain critical stages in the synthesis, parts of the peptidyl polymer may not swell sufficiently to allow full reaction by added reagents (26). Changes in the nature of the support have partially mitigated this problem. Thus, use of a 1% cross-linked resin allows synthesis of the octadecapeptide angiotensinyl–bradykinin (33). Limitations in the use of even less highly cross-linked, more friable resin have been discussed by Stewart (34).

Because of the problem of an unanticipated major decrease in incorporation of residues, monitoring the extent of deprotection is essential. Prolongation of deprotection time or a change of reagents may increase reaction yield (25). The ninhydrin method (35) used to measure deprotection is insufficiently quantitative for accurate determination of available amino groups.

More quantitative techniques reported by Dorman and Love (36), Brunfeld and co-workers (32), or Niall et al. (37) may be adapted, although a modification of the first method can be used with little change in the present procedure (26). This method employs the ionic binding of the γ-emitting $^{36}Cl^-$ to protonated amino groups, followed by washing and then release of bound isotope on neutralization. The quantity of ionically bound ^{36}Cl can be related directly to the number of amino groups available after deprotection. A complex 24-step cycle is suggested if the procedure is applied to determine the amino groups remaining after the coupling reaction (26), but it may be simplified considerably when this method is used to measure available amino groups after deblocking. The greater number of amino groups to be measured under these circumstances minimizes the problem of nonspecifically absorbed radionuclide. It is a method that may be incorporated into a conventional synthetic procedure.

Purification of Synthetic V_L

Crude synthetic V_L obtained after HF cleavage contains a mixture of correct sequences, deletion peptides, and truncated sequences. Purification may be effected by conventional chromatographic techniques as well as by methods peculiarly suited to immunoglobulin fragments. The variety and power of these specific techniques make solid-phase synthesis an attractive method for preparation of antibody fragments.

Purification by Enzymatic Digestion

This purification technique (2) depends on the differential susceptibility of correct and incorrect synthetic products to enzymatic digestion, perhaps based on the ability of the correct sequence to form the compact structure of the native molecule. We have now demonstrated that native antibody light chain is stable to 1% (w/w) trypsin at 37°C. On the other hand, light chain in which the amino groups have been modified by citraconylation is hydrolyzed to a mixture of peptides under these conditions. This observation suggests that trypsin will preferentially degrade some synthetic products where native structure has been disturbed. Gel filtration should remove low molecular weight digestion products from the desired synthetic product.

Purification by Association with Heavy Chain

The association of antibody light and heavy chains is dependent on stringent requirements. X-Ray crystallographic studies (38,39) show that residues 36, 38, 43, 44, 87, and 98 in the human light chain are in contact with residues in the heavy chain. These are invariant or nearly invariant residues. Deletions at most places in a synthetic sequence should displace these contact residues in such a manner that they would be unlikely to interact with the homologous position in the heavy chain. If the chains were out of register, they should not interact strongly and individual chains could be separated from V_L–V_H complex by gel filtration. This approach may provide a powerful tool for isolation of correct sequences.

Fractionation by Affinity Chromatography Utilizing Idiotypic Antibody

Idiotypic antibodies are specific for determinants in the complementarity (hypervariable) regions of an immunoglobulin. Such antibodies may be readily raised in either a homologous (40) or a heterologous species. These antibodies may be immobilized on Sepharose and then utilized to fractionate the synthetic V_H–V_L recombinant by affinity chromatography.

Purification by Affinity Chromatography with Antigen

We have previously described methods for antibody fractionation utilizing antigen-affinity chromatography (41). These methods can now be applied to purify that fraction of semisynthetic V_L–V_H recombinant which binds antigen. Gradient elution utilizing an increasing concentration of hapten may separate synthetic products with lower affinity for antigen from those possessing optimal binding characteristics.

SYNTHESIS OF NONIMMUNOGENIC ANTIBODIES OF KNOWN SPECIFICITY

The synthesis of a functional antibody fragment is a procedure that may now be pursued with reasonable assurance of success. The ability to chemically synthesize an antibody in addition makes possible preparation of multiple analogs of the basic sequence. One objective of this research is to prepare antibody analogs of defined specificity that are nonimmunogenic in man.

Two important principles of antibody structure have emerged within the past several years. First, analysis of the amino acid sequences of many light- and heavy-chain variable regions shows that the principle sites of sequence variation are confined to three short segments in homologous positions on both polypeptide chains. Both affinity-labeling and X-ray crystallographic studies confirm that these hypervariable regions are directly involved in antigen binding. Second, a comparison of X-ray models shows that the only significant variation in three-dimensional structure of an antibody occurs within the hypervariable segments. The structure of the supporting framework is constant. Since the framework appears to be relatively invariant between species (38), it is intriguing to speculate on the exchange of hypervariable regions among antibodies originating from different species. The attractiveness of this concept lies in the facility with which rabbit antibodies of varying specificity may be obtained for sequencing coupled with the need for antibodies of human structure for therapeutic use. Recognition of the variable region of an immunoglobulin as foreign between species depends mainly on variations in the framework residues. Antibodies to the hypervariable regions (idiotypic antibodies) are relatively difficult to raise.

Thus, a hybrid antibody consisting of rabbit complementarity regions integrated into a human framework may have low immunogenic potential in man. If these speculations bear fruit, many different combining-site specificities revealed by studies of homogeneous rabbit antibodies may be combined with human variable-region framework sequences to produce antibody fragments of therapeutic utility.

ACKNOWLEDGMENT

This work was supported by USPHS grant HL-14150.

REFERENCES

1. DuVigneaud, V., Ressler, C., Swan, J. M., Katsoyannis, P. G., and Gordon, S. (1953): The synthesis of an octapeptide amide with the hormonal activity of oxytocin. *J. Am. Chem. Soc.*, 75:4879–4880.
2. Gutte, B., and Merrifield, R. B. (1971): The synthesis of ribonuclease A. *J. Biol. Chem.*, 246:1922–1941.
3. Hirschmann, R., Nutt, R. F., Veber, D. F., Vitali, R. A., Varga, S. L., Jacob, T. A., Holly, F. W., and Denkewalter, R. G. (1969): Studies on the total synthesis of an enzyme. V. The preparation of enzymatically active material. *J. Am. Chem. Soc.*, 91:507–508.
4. Romovacek, H., Drabarek, S., Kawasaki, K., Dowd, S. R., Obermyer, R., and Hofmann, K. (1974): Studies on polypeptides. Synthesis of a cysteine protected nonatriconta peptide corresponding to position 66-104 of the peptide chain of ribonuclease T1. *Int. J. Pept. Protein Res.*, 6:435–445.
5. Yanaihara, N., Yanaihara, C., Dupuis, G., Beacham, J., Camble, R., and Hofmann, K. (1969): Studies on polypeptides. XLII. Synthesis and characterization of seven fragments spanning the entire sequence of ribonuclease T_1. *J. Am. Chem. Soc.*, 91:2184–2185.
6. Rocchi, R., Benassi, M., Guarneri, M., Guggi, A., and Tomatis, R. (1974): Synthetic peptides spanning the entire sequence of porcine pancreatic secretory trypsin inhibitor II (Kazal). *Int. J. Pept. Protein Res.*, 6:391–396.
7. Yajima, G., and Kiso, Y. (1974): Studies on peptides. XLIV. Synthesis of the basic trypsin inhibitor from bovine pancreas (Kunitz and Northrop) by the fragment condensation procedure on polymer support. *Chem. Pharm. Bull. (Tokyo)*, 22:1087–1094.
8. Zahn, H., Berndt, H., Föhles, J., Branderburg, D., Schumutzki, W., and Naithani, Z. K. (1974): Progress in the synthesis of human proinsulin and of a proinsulin model. *Angew. Chem. Int. Ed.*, 13:419.
9. Bayer, E., Jung, G., and Hagenmaier, H. (1968): Total synthesis of ferredoxin. I. Synthesis of the amino acid sequence of clostridium pasteurianum ferredoxin. *Tetrahedron*, 24:4853–4860.
10. Sano, S., and Kurihara, M. (1969): Synthesis of an analog of horse heart cytochrome C by the solid phase method. *Hoppe Seylers Z. Physiol. Chem.*, 350:1183–1187.
11. Hancock, W. S., Prescott, D. J., Marshall, G. R., and Vagelos, P. R. (1972): Acyl carrier protein. XVIII. Chemical synthesis and characterization of a protein with acyl carrier protein activity. *J. Biol. Chem.*, 247:6224–6233.

12. Noda, K., Terada, S., Mitsuyasu, N., Waki, M., Kato, T., and Izumiya, N. (1971): Synthesis of a peptide with basic pancreatic trypsin inhibitory activity. *Naturwissenschaften,* 58:147–148.
13. Sharp, J. J., Robinson, A. B., and Kamen, M. D. (1973): Synthesis of a polypeptide with lysozyme activity. *J. Am. Chem. Soc.*, 95:6097–6108.
14. Aoyagi, H., Yonezawa, H., Takahasi, N., Kato, T., Izumiya, N., and Yang, C.-C. (1972): Synthesis of a peptide with cobrotoxin activity. *Biochim. Biophys. Acta*, 263:823–826.
15. Waki, M., Mitsuyasu, N., Terada, S., Matsuura, S., Kato, T., and Izumuji, N. (1974): Synthesis of polypeptides corresponding to ribonuclease T_1 and its analog [59-tyrosine]-ribonuclease T_1. *Biochem. Biophys. Res. Commun.*, 61:576–582.
16. Li, C. H., and Yamashiro, D. (1970): The synthesis of a protein possessing growth-promoting and lactogenic activity. *J. Am. Chem. Soc.*, 92:7608–7609.
17. Yamashiro, D., and Li, C. H. (1974): Synthesis of a pentekontapeptide with high lipolytic activity corresponding to the carboxyl-terminal fifty amino acids of ovine β-lipotropin. *Proc. Natl. Acad. Sci. USA*, 71:4945–4949.
18. Hochman, J., Inbar, D., and Givol, D. (1973): An active antibody fragment (Fv) composed of the variable portions of heavy and light chains. *Biochemistry*, 12:1130–1135.
19. Poulsen, K., Frazer, K. J., and Haber, E. (1972): An active derivative of rabbit antibody light chain composed of the constant and the variable domains held together only by a native disulfide bond. *Proc. Natl. Acad. Sci. USA*, 69:2495–2499.
20. Margolies, M. N., Cannon, L. E., Strosberg, A. D., and Haber, E. (1975): Diversity of light chain variable region sequences among rabbit antibodies elicited by the same antigens. *Proc. Natl. Acad. Sci. USA*, 72:2180–2184.
21. Bodansky, M., Kondo, M., and Lin, C. Y. (1974): *o*-Nitrophenyl esters of benzyloxycarbonylamino acids and their application to the synthesis of peptide chains by the in situ technique. *J. Org. Chem.*, 39:444–447.
22. Corley, L., Sachs, D. H., and Anfinsen, C. B. (1972): Rapid solid-phase synthesis of bradykinin. *Biochem. Biophys. Res. Commun.*, 47:1353–1359.
23. Burton, J., Poulsen, K., and Haber, E. (1975): Competitive inhibitors of renin. Inhibitors effective at physiological pH. *Biochemistry*, 14:3892–3898.
24. Felix, A., and Jimenez, M. H. (1974): Usage of fluorescamine as a spray reagent for thin-layer chromatography. *J. Chromatogr.*, 89:361–364.
25. Chou, F. C.-H., Chawla, R. K., Kibler, R. F., and Shapira, R. (1971): Incomplete deblocking as a cause of failure sequence in solid phase peptide synthesis. *J. Am. Chem. Soc.*, 93:267–268.
26. Hancock, W. S., Prescott, D. J., Vagelos, P. R., and Marshall, G. R. (1973): Solvation of the polymer matrix. Source of truncated and deletion sequences in solid phase synthesis. *J. Org. Chem.*, 38:774–781.
27. Meienhofer, J., and Trzeciak, A. (1971): Solid-phase synthesis with attachment of peptide to resin through an amino acid side chain: [8-lysine]-vasopressin. *Proc. Natl. Acad. Sci. USA*, 68:1006–1009.
28. Glass, J. D., Talansky, A., Grzonka, Z., Schwartz, I. L., and Walter, R. (1974): Bidirectional solid-phase peptide synthesis. Extention of the dinitrophenylene bridging method to cysteine-containing peptides. *J. Am. Chem. Soc.*, 96:6476–6480.
29. Erickson, B. W., and Merrifield, R. B. (1974): Stability of side-chain protecting groups in solid phase peptide synthesis. *Isr. J. Chem.*, 12:79–85.
30. Yamishiro, D., and Li, C. H. (1973): Protection of tyrosine in solid-phase peptide synthesis. *J. Org. Chem.*, 38:591–592.
31. Pietta, P. G., and Marshall, G. R. (1970): Amide protection and amide supports in solid-phase peptide synthesis. *J. Chem. Soc.* [*Sect. D*], 650–651.
32. Brunfeldt, K., Roepstorff, P., and Thomsen, J. (1969): Process control in the solid phase peptide synthesis by titration of free amino groups. *Acta Chem. Scand.*, 23:2906–2907.
33. Stewart, J. M., and Matsueda, G. R. (1972): Some problems in solid-phase peptide synthesis. In: *Chemistry and Biology of Peptides*, edited by J. Meienhofer, pp. 221–224. Ann Arbor Press, Ann Arbor, Michigan.
34. Stewart, J. M. (1973): The challenge to polymer chemists in solid phase peptide synthesis. *NATO Adv. Study Inst. Ser C.* 412–433.
35. Kaiser, E., Colescott, R. L., Bossinger, C. D., and Cook, P. I. (1970): Color test for detection

of free terminal amino groups in the solid-phase synthesis of peptides. *Anal. Biochem.*, 34:595–598.

36. Dorman, L. C. (1969): Non-destructive method for the determination of completeness of coupling reactions in solid phase peptide synthesis. *Tet. Lett.* 28:2319–2321.
37. Niall, H. D., Tregear, G. W., and Jacobs, J. (1972): Automated Edman degradation monitoring of solid-phase peptide synthesis. In: *Chemistry and Biology of Peptides*, edited by J. Meienhofer, pp. 695–699. Ann Arbor Press, Ann Arbor, Michigan.
38. Padlan, E. A., and Davies, D. R. (1975): Variability of three-dimensional structure in immunoglobulins. *Proc. Natl. Acad. Sci. USA*, 72:819–823.
39. Poljak, R. L., Amzel, L. M., Avey, H. P., Chen, B. L., Phizackerley, R. P., and Saul, F. (1973): Three-dimensional structure of the Fab′ fragment of a human immunoglobulin at 2.8 Å resolution. *Proc. Natl. Acad. Sci. USA*, 70:3305–3310.
40. Huser, H., Haimovich, J., and Jaton, J.-C. (1975): Antigen binding and idiotypic properties of reconstituted immunoglobulins G derived from homogeneous rabbit anti-pneumococcal antibodies. *Eur. J. Immunol.*, 5:206–210.
41. Cheng, W., Fraser, K., and Haber, E. (1973): Antibody fractionation by affinity chromatography. *J. Immunol.*, 111:1677–1689.

Antibodies in Human Diagnosis and Therapy,
edited by Edgar Haber and Richard M. Krause.
Raven Press, New York © 1977.

Future Prospects for Conventional Peptide Synthesis in Solution

Johannes Meienhofer

Opportunities and constraints in antibody synthesis by the solid-phase method (8,9) have been discussed in detail by Drs. Erickson, Burton, and Givol, and exemplified by preparations of variable light-chain domains. It should be borne in mind that any peptide synthesis of such an order of magnitude cannot be expected to yield fully, or even highly, active products at first trial. In fact, a total chemical synthesis of fully active insulin (10) was achieved 10 years after the announcements of the first synthetic insulin preparations, which contained 0.1 to 1% of activity (3,5,7).

This comment briefly outlines the present state of the art and future prospects for conventional peptide synthesis in solution, the alternative to the solid-phase procedure. Advantages of conventional synthesis lie in its potential for obtaining products of higher purity, and for preparing larger quantities—aspects that become particularly critical and important in discussing synthesis of antibodies for future therapeutic application. On the other hand, conventional procedures do not yet allow routine synthesis of protein-sized peptides (containing more than 100 amino acid residues),[1] and are impeded by being exceedingly laborious and therefore very slow. An advantageous combination of methods lies in rapid solid-phase synthesis of small protected peptides to serve as intermediates for conventional synthesis of well-characterized larger peptides (9,14); in this manner the best features of each method are exploited. The speed and convenience of solid-phase synthesis is utilized for preparation of tri- to decapeptide fragments, which can still be highly purified by existing fractionation techniques; the relatively few fragment condensations are then conducted by conventional procedures, which allow for product purification and analysis after each coupling and for large-scale preparation.

This combination approach has been adapted by S. S. Wang in our department for synthesis of a part of pituitary growth hormone (sequence region 96–135).[2] A *p*-alkoxybenzyl alcohol resin or the known hydroxymethyl resin was used for solid-phase synthesis of a series of small protected peptides, which were rigorously purified to homogeneity as judged by several independent analytical criteria (13). Most of these intermediates were crystalline, which allowed for repeated recrystallization from organic solvents—a procedure that is still the most powerful purification procedure for protected peptides. Those compounds that do not crystallize are rapidly purified by preparative high-performance liquid chromatography systems developed in our laboratory (1). In this manner the fragments shown in Fig. 1 were

[1] The one notable exception, synthesis of the S-protein of ribonuclease A by the Merck group (2), has thus far rather confirmed the restraints than provided a general breakthrough to protein synthesis.

[2] Ovine pituitary growth hormone (96–135): H - Val - Phe - Thr - Asp - Ser - Leu - Val - Phe - Gly - Thr - Ser - Asp - Arg - Val - Tyr - Glu - Lys - Leu - Lys - Asp - Leu - Glu - Glu - Gly - Ile - Leu - Ala - Leu - Met - Arg - Glu - Leu - Glu - Asp - Val - Thr - Pro - Arg - Ala - Gly - OH (6).

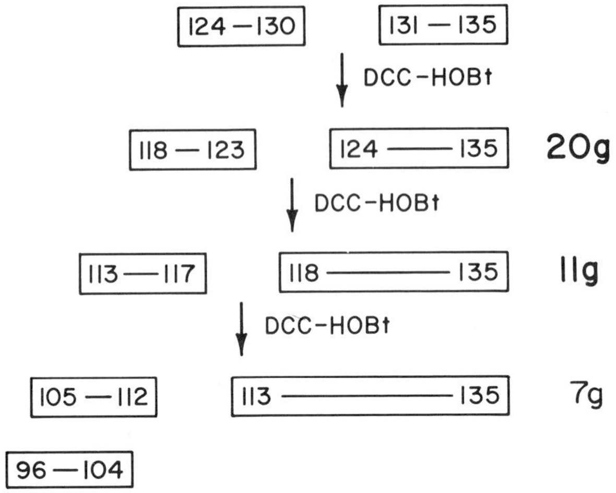

FIG. 1. Scheme of synthesis (13) of sequence region 96–135 of ovine pituitary growth hormone by combined solid-phase and conventional peptide synthesis. Fragments prepared by solid-phase procedures are condensed by conventional methods in solution. DCC, dicyclohexylcarbodiimide; HOBT, hydroxybenzotriazole.

prepared in large quantity in a relatively short time by repeated solid-phase syntheses.

The intermediates are presently being condensed (Fig. 1) by conventional synthesis in solution with careful purification and characterization after each coupling. The dodecapeptide (sequence 124–135) was obtained in crystalline form in a 20-g batch. Even the octadecapeptide (sequence 118–135) crystallized after purification; 11 g of product was obtained. The tricosapeptide has been prepared, and condensation of the remaining intermediates is in progress for an expected yield of 1 to 2 g of the desired growth hormone fragment 96–135.

Thus "combined solid-phase and conventional peptide synthesis" in this manner allows preparation of moderate parts of proteins within reasonable time. With the present state of the art this approach might be extended to the synthesis of peptides containing 60 to 70 amino acid residues.[3]

The problem for conventional synthesis remains how to proceed from here, i.e., how to condense peptides containing 50 or more amino acid residues. Efficient coupling of such large peptides has not yet been mastered, which constitutes one of the major obstacles in the field. Difficulties arise from the second order kinetics of peptide coupling, which require higher concentrations for rapid condensation of larger molecules. However, in practice the concentrations of larger peptides are dramatically reduced because of their sharply decreased solubility in suitable solvents. At the low concentrations obtained because of limited solubility, peptide bond formation becomes very slow, and irreversible intramolecular side reactions or solvent interactions of first order kinetics take over. It would not help to develop new reagents for more powerful carboxyl group activation, since the available coupling methods are actually very powerful. Indeed, stronger carboxyl activation would be counterproductive by greatly increasing the danger of racemization.

[3] Peptides of this size have been synthesized using conventional procedures entirely (2,11).

What one would like to have is a sort of molecular clamp for pulling the reactive groups of two fragments close together and convert the formation of the peptide bond into a concentration-independent intramolecular rearrangement (Fig. 2). In fact, this type of approach is being actively pursued in the laboratories of I. Ugi (12) in Munich and D. S. Kemp (4) at MIT, using quite different methods. In Ugi's "four-component condensation," an isonitrile and an aldehyde undergo a series of fast addition reactions with the carboxyl and amine components producing an intermediate which forms the desired peptide bond by an intramolecular rearrangement. Removal of the auxiliary N-substituent provides the peptide. In Kemp's "amine capture," Schiff base formation with salicylaldehyde derivatives brings the amine component into close proximity with the carboxyl component. After activation and Schiff base reduction, the peptide bond is formed by intramolecular rearrangement. Again, the auxiliary bridging moiety must be cleaved off. Both approaches still face certain difficulties with solvent limitations and reagent preparation, but there are good indications that these will soon be overcome. Most importantly, it remains to be demonstrated that these procedures will be capable of joining large fragments efficiently and with acceptable yields. Both methods have the potential of revo-

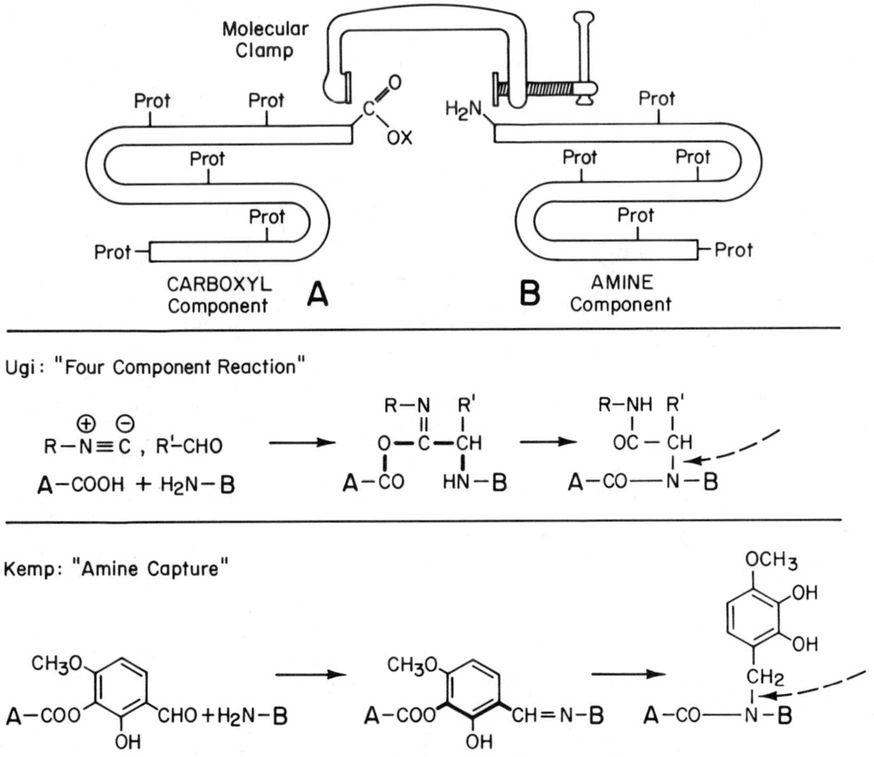

FIG. 2. Novel approaches to peptide fragment condensation. The Ugi reaction (12) and amine capture method (4) produce clamp-type intermediates that bring the activated carboxyl group and the amine group of protected (Prot) peptide fragments A and B into close proximity. The desired peptide bonds are then formed by intramolecular rearrangement.

lutionizing peptide synthesis. At the very least they indicate directions for future research efforts.

In conclusion, future chemical synthesis of proteins of acceptable homogeneity will still require a great deal of imaginative and intensive methodological work in developing higher levels of sophistication and efficiency in both solid-phase synthesis and conventional fragment condensation, as well as in fractionation and purification methods. However, if peptide chemists are given this opportunity,[4] and if they are granted sufficient time (perhaps 10 or more years) to put as much effort into synthesis as has been spent on isolation, sequence analysis, and conformational studies, then antibodies might well be synthesized.

REFERENCES

1. Gabriel, T. F., Michalewsky, J., and Meienhofer, J. (1976): *J. Chromatogr.* (*In press.*)
2. Hirschmann, R., Nutt, R. F., Veber, D. F., Vitali, R. A., Varga, S. L., Jacob, T. A., Holly, F. B., and Denkewalter, R. G. (1969): *J. Am. Chem. Soc.*, 91:507–508.
3. Katsoyannis, P. G., Fukuda, K., Tometsko, A., Suzuki, K., and Tilak, M. (1964): *J. Am. Chem. Soc.*, 86:930–932.
4. Kemp, D. S., Roberts, D., Hoyng, C., Grattan, J., Vellaccio, F., and Reczek, J. (1975): In: *Peptides: Chemistry, Structure, Biology,* edited by R. Walter and J. Meienhofer, pp. 295–305. Ann Arbor Sci. Publ., Ann Arbor, Michigan.
5. Kung, Y. T., Du, Y. C., Huang, W. T., Chen, C. C., Ke, L. T., Hu, S. C., Jiang, R. Q., Chu, S. Q., Niu, C. I., Hsu, J. Z., Chang, W. C., Cheng, L. L., Li, H. S., Wang, Y., Loh, T. P., Chi, A. H., Li, C. H., Shi, P. T., Yieh, Y. H., Tang, K. L., and Hsing, C. Y. (1965): *Sci. Sinica*, 14:1710–1716.
6. Li, C. H., Gordon, D., and Knorr, J. (1973): *Arch. Biochem. Biophys.*, 156:493–508.
7. Meienhofer, J., Schnabel, E., Bremer, H., Brinkhoff, O., Zabel, R., Sroka, W., Klostermeyer, H., Brandenburg, D., Okuda, T., and Zahn, H. (1963): *Z. Naturforsch.*, 18B:1120–1121.
8. Merrifield, R. B. (1963): *J. Am. Chem. Soc.,* 85:2149–2154.
9. Merrifield, R. B. (1973): In: *The Chemistry of Polypeptides,* edited by P. G. Katsoyannis, pp. 335–361. Plenum Press, New York.
10. Sieber, P., Kamber, B., Hartmann, A., Jöhl, A., Riniker, B., and Rittel, W. (1974). *Helv. Chim. Acta*, 57:2617–2621.
11. Storey, H. T., Beacham, J., Cernosek, S. F., Finn, F. M., Yanaihara, C., and Hofmann, K. (1972): *J. Am. Chem. Soc.*, 94:6170–6178.
12. Ugi, I., Arora, A., Burghard, H., Eberle, G., Eckert, H., George, G., Gokel, G., Herlinger, H., v. Hinrichs, E., Hoffmann, P., Kleimann, H., Klusacek, H., Hsiao-Ling, L., Marquarding, D., Nah, H-S, Offermann, K., Rehn, D., Stüber, S., Tamasi, M., Urban, R., Wackerle, L., Zahr, S., v. Zychlinsky, H. (1975). In: *Peptides 1974*, edited by Y. Wolman, pp. 71–91. Wiley, New York.
13. Wang, S. S., and Kulesha, I. D. (1975): *J. Org. Chem.*, 40:1227–1234.
14. Wang, S. S., and Merrifield, R. B. (1971): In: *Peptides 1969,* edited by E. Scoffone, pp. 74–83. Elsevier, New York.

[4] Opportunities for such fundamental work will probably have to be provided by the larger immunological research groups or institutions. This author does not know of any funding agency that would at present support this type of work.

Prospects for Cell Cultures in the Production of Antibodies

Antibodies in Human Diagnosis and Therapy,
edited by Edgar Haber and Richard M. Krause.
Raven Press, New York © 1977.

Prospects for Cell Cultures in the Production of Antibodies

N. A. Mitchison

Zoology Department, University College, Gower Street, London W.C.1

That tailor-made antibody would be of immense value in medicine and biochemistry has long been accepted. It has been clear also that the production of such antibodies would almost certainly have to be carried out *in vitro* for reasons of economy and purity. But only within the last year has the appropriate technology advanced to the point where detailed discussion is worthwhile. This, then, is a most appropriate time for our symposium.

Production has not yet been carried out on an industrial scale. Nevertheless, laboratory scale production *in vitro* is proving so successful that it is perfectly reasonable to contemplate scaling up to the point where clinical trials could be made. That we should be able to think in this way is due in large part to the efforts of the authors in this volume. My own efforts will be confined first to considering the advantages and limitations of *in vitro* antibody production, and second, to a discussion of our long-distance objectives.

At first sight the main advantage of *in vitro* production might seem to be the potential for making available large amounts of antibody at low cost. Here I am skeptical, because we have to compare *in vitro* systems with what is already available within the body. We are already endowed with a remarkably flexible and powerful system which operates, in economic terms, at zero marginal cost. We have no hope of beating what this system does best and no reason to wish to do so.

If we restrict our aims then, there are two kinds of advantages to be expected of *in vitro* production. One concerns its "tailor-made" quality. *In vivo* production, as it occurs during the normal immune response, is notable for its heterogeneity. Antibodies produced *in vivo* vary over a wide range in their capacity to combine with the antigen which elicits them. They vary also in those parts of the molecule outside the combining site which are important in determining the functions that antibodies can perform within the body. While it is true that the immunochemistry of 5 years ago may have somewhat overestimated this heterogeneity of affinity and immunoglobulin class, these rules certainly apply to our major responses to pathogens. Here they serve the important function of providing us with a response that is sufficiently flexible to cope with the extraordinary variability which pathogens display. Their variability is matched by ours, and evolution has kept the balance.

This very flexibility, however, proves a disadvantage in some instances. One is immunological disease, where we may wish to introduce antibody of restricted heterogeneity in order to control an aberration of the immune system. We may need to avoid the blunderbuss effect of introducing antigens. Thus anti-Rh antibody can be used to prevent hemolytic disease of the newborn. Another example is organ transplantation where anti-I^A antibody can be used in a specific manner to prevent or delay graft rejection. Here the principal rival method, introduction of antigen to induce immunological tolerance, may not only produce the unwanted opposite effect of immunization but is also relatively slow. A third example is in the possible use of antibodies to autogenous hormones or enzymes. These molecules may occur at concentrations well below the average affinity of antibodies produced in a normal immune response but may nevertheless be accessible to antibodies deliberately selected for high affinity. A fourth example is prophylactic immunization with impure bacterial or viral vaccines.

All vaccines in current use probably perturb the immune system more than necessary, and in some cases, such as rabies, this is a serious hazard. One approach is to use more and more highly purified microbial antigens. Another is to use antibodies produced *in vitro* (although the speed of action in rabies, for example, provides a more important reason for their use).

The second kind of advantage of *in vitro* production lies in our ability to control the tempo of passive immunization. Here again the tempo of active immunization has been set by evolution to cope with microbial infection. Often this is for the best; but in two of the examples that have been cited—enhancing graft survival and rabies treatment—antibodies are required before they are normally produced.

The other side of tempo control is the ease with which passive immunity can be reversed, simply by ceasing to administer antibody. This advantage applies with special force in what I believe to be the most important potential use of *in vitro* production, the control of human fertility. The results which have so far been obtained in abolishing fertility in female monkeys by active immunization with human chorionic gonadotrophin are extremely encouraging (1). The most promising antigen is the C-terminal 30-peptide of the β subunit of HCG, coupled to a protein carrier; this part of the molecule is unique, unlike the first 115 amino acids which share much of their sequence with luteinizing hormone. In the long run, as I have argued elsewhere (2), control of the sex ratio will prove the most acceptable form of fertility control, and an immunological approach to this is now reasonable via the H-Y antigen on spermatozoa. The drawback of all present strategies based on active immunization is their irreversibility. This can be circumvented by passive immunization. The requirement for prolonged administration need not itself constitute a serious drawback, now that depot systems are becoming available with zero-order release kinetics (3).

Two further points concerning the possible use of anti-HCG made *in vitro* are worth noting. The capacity of *in vitro* methods to provide high-affinity antibody will be needed, because the hormone occurs at low concentrations. In the long run, passive immunization is preferable to active because it avoids genetical selection of immunological cripples (2).

Assuming that our strategy is to administer restricted antibodies of human immunoglobulin type, the major limitation that I can foresee is the antiidiotype response. This possibility will require further investigation. My hunch is that we shall not find out how important the antiidiotype network is until we have repeated immunology in Ly-2,3-deficient mice.

What of the more distant future? The last year or two has seen a proliferation of factors which purport to control the immune response, which are produced *in vitro* by T cell subsets and by macrophages, and which have molecular weights of 50,000 or less. These factors are probably at least as important as antibodies in control of the immune response, and in some ways may be more important even than antigen. Their low molecular weight makes them better candidates than immunoglobulins for genetic engineering. They have not yet been completely isolated or well characterized, mainly because of the low concentrations at which they are encountered and at which they are active. Yet their very combination of low natural concentrations and high activity adds to their appeal as possible therapeutic agents. One can foresee a rich future for large scale *in vitro* production of these factors, first to investigate their role *in vivo,* and eventually for use in man.

REFERENCES

1. Stevens, V. C. (1974): Fertility control through active immunization using placenta proteins. *Acta Endocrinol.* [*Suppl.*] (*Kbh.*), 194.
2. Mitchison, N. A. (1974): Long-term hazards in immunological methods of fertility control. *Acta Endocrinol.* [*Suppl.*] (*Kbh.*), 194.
3. Diczfalusy, E. (editor) (1974): Steps in the human reproductive process susceptible to immunological interference. *Acta Endocrinol.* [*Suppl.*] (*Kbh.*), 194.

Antibodies in Human Diagnosis and Therapy,
edited by Edgar Haber and Richard M. Krause.
Raven Press, New York © 1977.

Obtaining Homogeneous Antibody of Desired Specificity from Fragment Cultures

Norman R. Klinman, Gary P. Segal,
Walter Gerhard, Thomas Braciale, and Ronald Levy

Department of Pathology,
University of Pennsylvania, Medical School,
Philadelphia, Pennsylvania 19174

Although only slightly more than a decade has passed since the first confirmed reports of antibody synthesis by cells in culture, many of the aspirations for this approach to the analysis of the immune system have already been fulfilled. The progression of these studies has seen the advent of highly sensitive assays for quantitating and characterizing antibody and antibody-producing cells, as well as a variety of microculture techniques for maximizing cell responses (1–8). The goal of these studies, however, has generally been to characterize the immune response itself rather than the stimulating antigens. Thus there has been little consideration of the ultimate usefulness of such specific approaches to the diagnosis and therapy of human disease, other than what may be gained by an increased understanding of the immune mechanism. This is not surprising, since the quantity of antibody obtained in culture is miniscule and stimulation has been achieved with a relatively restricted range of antigens. Thus the rich and varied antibody harvest from *in vivo* systems, given available purification and adsorption procedures, have seemed a far more practical source for appropriate antibodies. However, serum antibodies often represent highly cross-reactive populations of antibodies reflecting a vast composite of individual antibody specificities directed against a large array of antigenic determinants. Rendering such populations uniformly specific by adsorption procedures requires an a priori knowledge of the antigenic composition of the array of substances used for adsorption.

Recent *in vitro* studies in our laboratory of the immune response to bacterial, viral, and tumor antigens have indicated that *in vitro* production of antibodies may circumvent such difficulties and may prove to be clinically useful: (i) as sensitive diagnostic probes for the presence of pathogens or antibodies reactive to them; (ii) as tools for defining antigenic relationships between related pathogens, a process potentially crucial for prophylactic immunizations; and (iii), in the long run, as weapons for the therapeutic arsenal either as direct agents against pathogens or tumors, or as carriers of substances highly toxic for such cells. As mentioned above, the relatively small quantities of antibody produced *in vitro* constitutes an obvious disadvantage. However, for the aforementioned proposals (i) and (ii), small quantities of antibody should prove adequate, and for (iii) either small quantities of antibody may prove adequate or methods for increasing the amount of antibody obtained *in vitro* may eventually be realized. The advantages which may accrue from obtaining antibodies for these purposes *in vitro* are so varied and extensive that indeed they may eventually obviate quantitative considerations.

METHODOLOGY

In order to discuss the potential advantages of obtaining antibodies *in vitro* for therapy and diagnosis, I have chosen to emphasize the *in vitro* fragment culture system currently used in our laboratory. Fragment culture techniques have proven particularly useful in providing measurable and analyzable quantities of antibody. Such production of antibody *in vitro* permits a variety of experimental manipulations not possible *in vivo*. For instance, it is possible to remove most of the antigen from the system after the period of stimulation. In addition, during the period of maximal antibody synthesis in organ fragment cultures (days 10–15), it appears that antibody can be obtained in relatively pure form from culture. Once antibody synthesis is maximal, serum (and protein)-free medium appears sufficient to support cultures and the majority of immunoglobulin released is, in some cases, the desired monoclonal antibody (G. Segal and N. Klinman, *unpublished observation*). Since it is possible to label antibody as it is synthesized by adding labeled amino acids to the culture, it is possible to obtain culture fluids containing a relatively purified antibody synthesized in the relative absence of antigen and, if desired, in a highly labeled form. This would eliminate the necessity of externally labeling antibody by potentially denaturing procedures, and, furthermore, would obviate the necessity for the purification of specific antibody, a process which for antibodies to most cellular antigens is extremely difficult, unproductive, and often highly deleterious.

Perhaps the single most important advantage of this approach is the potential for generating an enormous array of specific antibodies *in vitro*. This is particularly true for the culturing of the progeny of isolated precursor cells; each of which yields monoclonal antibodies. Within this system, not only is it possible to obtain antibodies of a single immunoglobulin class, but theoretically the specificity of the desired antibody can be chosen from among any of the many clonotypes (individual specificities) responsive to any given antigenic determinant. Thus, operationally, it is possible to select, among the antibodies reactive to a critical determinant, those clonotypes of highest affinity or least cross-reactivity for other specified related determinants.

To date three general forms of organ fragment culture have been utilized in a variety of laboratories. These are depicted graphically in Fig. 1. In the first method (Fig. 1A), the spleen or lymph nodes of nonimmune or immune animals are diced into 1-mm cubes, put into culture, and stimulated *in vitro*. Such fragments, upon stimulation, may give rise to antibodies by the seventh to tenth day of culture. When primary fragment cultures (i.e., derived from nonimmune donor animals) are used, antibody production may occur at low levels—often not significantly higher than in nonstimulated cultures, i.e., natural antibody (2,5,6,9). Fragment cultures from spleens of previously immunized animals have been much more useful for the production of antibody *in vitro*. This form of stimulation has been successfully used in several laboratories; and significant quantities of antibody can be obtained (5,6,9,10–12). Using this method, one can obtain more than 0.1 mg of antibody per day specific for a variety of antigens including bacteria and viruses as well as hapten-protein conjugates.

A more difficult, but potentially more rewarding, technique has been the adoptive transfer of cells into irradiated recipients prior to *in vitro* organ fragment culturing of the recipient's spleen. This technique has been utilized extensively in our laboratory. The advantage of this method is that the donor

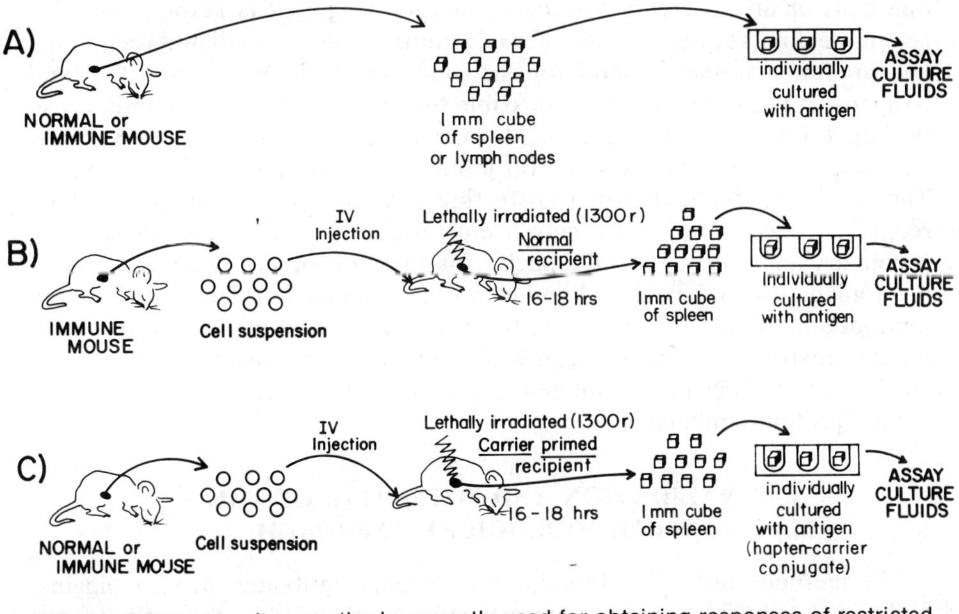

FIG. 1. Fragment culture methods currently used for obtaining responses of restricted clonality *in vitro*.

cells can be transferred at a limiting dilution, and so the responding fragments can be predicted to contain at most one precursor cell. Under such conditions, the antibody produced is monoclonal. These experiments have taken two forms. In the first case (Fig. 1B), cells from an immune donor are injected into nonimmune irradiated recipients. Fragment cultures stimulated *in vitro* give rise to secondary clones, and antibody can readily be obtained to a variety of antigens including tumor antigens, viruses, and hapten-protein conjugates. In the second method (Fig. 1C), nonimmune or immune cells are injected into carrier-primed irradiated recipients. This modification (i) increases the yield of antibody *produced in culture,* (ii) increases the number of responding secondary cells, and (iii) permits the stimulation of nonprimed B cells to subsequent monoclonal antibody production. The limitation in this latter technique is that the antigen has to be presented in the form of a hapten-carrier conjugate. That is, it has not yet been possible to immunize to the antigen itself and obtain vigorous primary responses to that antigen by donor cells when cultured *in vitro*.

All three of these techniques may be used effectively for obtaining significant quantities of antibody under different situations. The first technique, the use of spleen or lymph fragments derived directly from immune or nonimmune animals, has been most useful for responses in which a limited number of cells may be responsive to a given antigenic determinant (5,9,12). Since few cells respond, the response of spleen fragments even from primed mice, appears to have an "all-or-none" quality, presumably indicating that one cell, or at most very few cells, in each fragment is responding. The second technique, i.e., the transfer of immune cells to irradiated recipients, has proven extremely useful for the analysis of clones specific for viral antigens. It has not as yet been possible to carrier-prime to viral antigens so that the transfer of secondary cells to normal recipients has proven the most effective way of obtaining antiviral monoclonal antibody responses (9,11). The third technique, that of transferring cells to carrier-primed irradiated recipients, has proven extremely useful in the analysis of the response of secondary and primary B cells specific for hapten-protein conjugates, including conjugates using bacterial antigens such as phosphorylcholine as a haptenic determinant (13–15). Thus, from a combination of these three techniques, restricted responses can be obtained *in vitro* against a variety of antigens, including tumor antigens, viral antigens, bacterial antigens, and hapten-protein conjugates.

VALIDATION AND POTENTIAL OF THE METHODOLOGICAL APPROACH

The methods used for obtaining monoclonal antibodies *in vitro* against biologically significant antigens may best be exemplified by the antiinfluenza virus response obtained in mouse splenic fragment cultures (9,11). As described above, a limiting dilution of cells from a previously immunized donor mouse was transferred intravenously to a lethally irradiated syngeneic recipient. The donor cells were obtained from mice which had been previously immunized intraperitoneally or intranasally. The results obtained after these two methods of immunization were similar. The fragments were incubated during the initial culture period of 3 days with purified virus incorporated into the culture medium. Thereafter, the fragments were kept *in vitro* in virus-free medium for 30 days while changing the culture medium at 3 to 4 day intervals. Antiviral antibodies produced at any stage of this culture period were detected by a radioimmunoassay, which utilized purified virus coupled to bromoacetylcellulose as an immunoadsorbent and ^{125}I-labeled antimouse immunoglobulin as detecting reagent. Standard curves for quantitating the antibody were established using antiviral antibody. Figure 2 demonstrates that the response to influenza virus had an all or none character and that the number of responding fragments is linearly related to the number of transferred cells. Thus it was anticipated that the antibody from the majority of positive spleen fragments was produced by the clonal progeny of a single stimulated influenza-specific B cell. Three findings indicate the monoclonal nature of the antibody response. Figure 3 demonstrates that the isoelectric focusing pattern of antibodies obtained from such cultures was quite restricted as compared to the antibody obtained *in vivo* against the influenza virus. Second, these antibodies reacted, in general, only with one of the three major antigenic components of egg-grown influenza virus. That is, cross-reactivity patterns indicated that, while serum antibodies specific for these viruses were directed against the hemagglutinin as well as the neuraminidase and host components, monoclonal antibodies were invariably reactive with only one of these three components. Third, as shown in Table 1, the cross-reactivity patterns of the anti-PR8 hemagglutinin antibodies, when tested for their binding to related influenza virus strains, were highly restricted. Thus, compared to the broad cross-reaction exhibited by "monospecific" antihemagglutinin antisera from immune mice, these

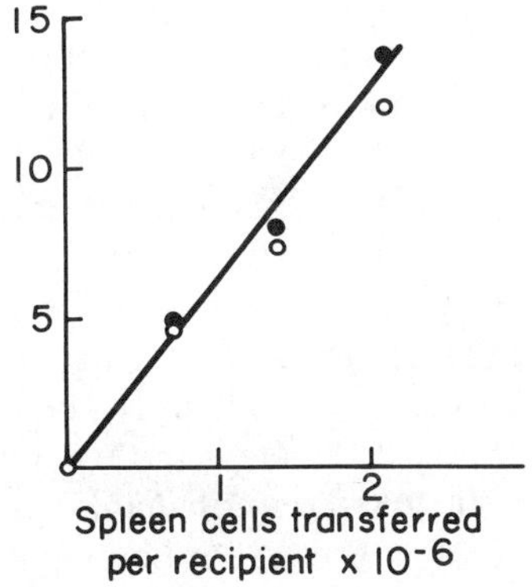

FIG. 2. Frequency of positive cultures at varying doses of transferred cells. Ordinate: Average number of detected antiviral antibody-producing spleen fragments per recipient spleen (*open dots*) and estimated number of virus-specific precursor B cells per recipient spleen (*filled dots*). This number was calculated by using the Poisson formula

$$P\,(x) = \frac{e^{\mu}\mu^{x}}{x!}$$

to estimate the fractions of fragments containing more than one precursor cell (x). The estimate of the average number of stimulatable precursor cells per fragment (μ) was based on the fraction of nonresponding fragments: $P(0) = e^{\mu}$. (From ref. 11.)

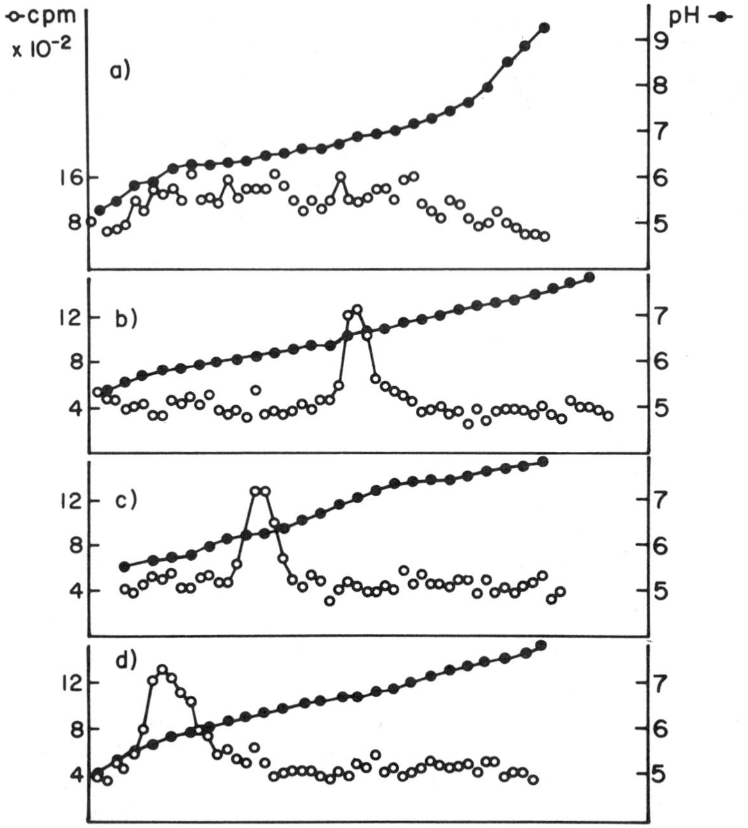

FIG. 3. Isoelelectric focusing spectra of anti-PR_8 antibodies produced *in vivo* and *in vitro*. (**a**):35–70 ng of serum antibody and (**b, c, d**) antibodies produced *in vitro* were focused in Ampholine-sucrose columns. The pH (*filled dots*) and antibody content, given as counts per minute (cpm) as detected in RIA (*open dots*), in the 50 μl successive eluate fractions. (From ref. 11.)

monoclonal antibodies exquisitely discriminate many distinct similarities and dissimilarities among the viral hemagglutinins. It is this latter quality that makes the availability of such highly restricted antibodies of enormous value for the analysis of the influenza virus antigenic structure and potentially for predicting the effectiveness of immunization procedures.

As mentioned above, perhaps the most valuable aspect of the *in vitro* monoclonal response for the production of relevant antibodies is the availability of a vast array of specific antibody preparations for any given antigen. Indeed, only the inherent restriction in the repertoire of cells responsive to such determinants limits the array of specific antibodies available. As shown in Table 1, a great variety of distinct clonotypes are responsive to the PR-8 hemagglutinin determinants. Further analyses have re-

TABLE 1. *Heterogeneity of the antihemagglutinin response*

Test sample	Anti-PR8 antibody (ng) binding to immunoadsorbent				
	PR8(H0)	WSE(H0)	MEL(H0)	BEL(H0)	CAM(H1)[a]
Serum	14.5	3.8	4.7	2.5	2.8
Clone no.					
1	5.4	5.0	7.2	8.6	—[b]
2	14.0	—	—	—	—
3	16.9	17.7	—	13.4	18.3
4	14.5	—	13.9	—	—
5	10.3	10.6	13.1	—	13.5
6	4.6	5.1	—	—	3.7
7	2.4	—	3.7	2.8	—

[a] Influenza strains and assay method as in ref. 11.
[b] Less than 0.2 ng.

vealed at least 14 distinct specificities of monoclonal antibodies reactive to these determinants. This number is clearly a minimum estimate since it is limited by the cross-reacting strains of viruses currently available for testing. Where finer discrimination has been possible more than a thousand different clonotypes have been found responsive to certain haptens (15–17) and even the clonotypes responsive to the phosphorylcholine determinants of *Pneumococcus* may represent more than 50 clonotypes (13,14; P. Gearhart and N. Klinman, *unpublished observation*).

With this description of the basic techniques of lymphoid tissue fragment culture and the potential array of available specific antibodies in mind, it may be useful to examine some preliminary studies in another area of potential usefulness of *in vitro* techniques for providing antibodies of restricted specificity, the *in vitro* humoral response to murine sarcoma-induced mouse tumors (MSC) (5). In this particular system conditions for the transfer of responsive cells to irradiated recipients have not yet been worked out. Thus, the available data relate to responses of splenic fragments derived directly from spleens of nonimmune and immune mice (mice previously inoculated with viable syngeneic tumor cells) after appropriate stimulation *in vitro*. Culture fluids from these fragments were assayed for antibody specific for MSC related antigenic determinants using a modified radioimmunoassay which utilized transformed cell monolayers as immunoadsorbents (5,18). Both nonimmune mice and mice whose tumors regressed responded and the antibodies derived from such cultures were compared to serum antibodies in their capacity to bind to a number of cell line immunoadsorbents. Figure 4 demonstrates the cross-reactivity of antibodies obtained against the tumor antigens. The top panel demonstrates the degree of cross-reactivity of serum antibodies against each of the cell lines which are cross-reactive with MSC. Serum contains a highly complex

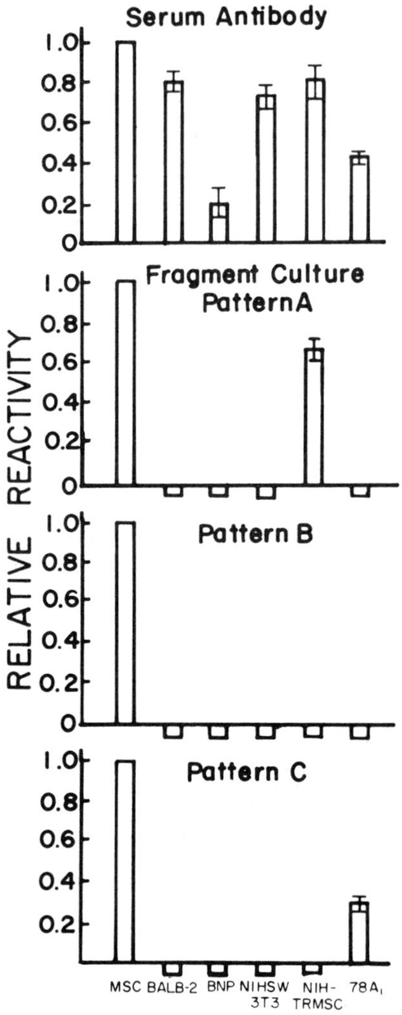

FIG. 4. Top panel: Cross-reactivity patterns of serum anti-MSC antibody from BALB/c regressor mice after injection with viable MSC cells. Mice were bled on day 57 after injection and reactivity of the serum against each of six cell line immunoadsorbents was determined using labeled [I^{125}]anti-Fab. Activity against each immunoadsorbent was expressed as the fraction of the reactivity against MSC. (Brackets include standard error.) **Patterns A, B, C:** Cross-reactivity patterns of *in vitro* synthesized antibodies. Immunoadsorbents included monolayers of MSC as well as: (i) BALB-2: a primary culture of BALB/c embryonic fibroblasts prepared from eviscerated 15-day-old BALB/c embryos. Cultures were not oncogenic when 10^6 cells were injected into adult BALB/c mice. Each cell preparation was maintained in culture for five passages in order to generate an immunoadsorbent and then discarded. (ii) $78A_1$: an MSV producer line derived from Wistar rat cells transformed *in vitro* by MSV-M. The line was kindly provided by Dr. Jorge Ferrar, University of Pennsylvania. (iii) BNP: a nonproducer cell line derived by infecting BALB, 3T3 cells with MSV. The line was kindly provided by Dr. Anthony Girardi, Wistar-Institute. (iv) NIHSW3T3: Normal 3T3 cells prepared from Swiss mouse embryo cultures were also provided by Dr. Girardi. (v) NIHTRMSC: an MSV producer cell line prepared by infecting NIHSW3t# with MSV freshly harvested from MSC culture fluids. Infection was carried out in the presence of DEAE-Dextran (Pharmacia) 25 μg/ml). (See ref. 5.)

mixture of antibodies which showed a great deal of cross-reactivity with all of the cell lines tested. Patterns A, B, and C of Fig. 4 show the cross-reactivity patterns obtained from cultures. The cross-reactivity patterns of these *in vitro* produced antibodies are highly restricted relative to the antibodies obtained *in vivo.* Some cultures clearly contained antibodies of more than a single specificity. Others contained antibodies which showed only a single type of reactivity. Thus it is possible to obtain a large variety of antibodies of highly selected specificity even against tumor antigens by appropriate *in vitro* techniques.

The extension of these studies to provide antibodies of restricted specificity for human tumor antigens is currently under investigation by Drs.

Levy and Segal at Stanford University. It is conceivable that mouse fragment cultures will produce antibodies specific for human tumors. While the vast majority of antibody derived from xenogeneic immunizations may be directed against species-specific determinants or other nonrelevant determinants, there is no a priori reason to exclude the possibility that clones will be identifiable which produce antibody reactive only with tumor-specific antigens.

APPLICATION TO DIAGNOSIS AND THERAPY

Assuming that antibodies against pathogenic agents or tumor cells are available from *in vitro* techniques, to what use might these antibodies now be put? Perhaps the most obvious use of antibodies of a known restricted specificity would be diagnostic. It could be conceived that, if a highly labeled antibody of a known relevant specificity and an immunoadsorbent containing that antigenic specificity were available, the homologous binding could be inhibited by minute amounts of appropriate antigen or antibody. By setting up such antibody antigen systems, it should be possible to detect minute quantities of an antigenic material either in tissues or in serum, and by the same token to detect minute quantities of antibody reactive to such a determinant. This might serve as a general technique for diagnosis, not only of the presence of viral, tumor, and bacterial antigens, but also the presence of early antibody responses against such agents.

Several years ago Fazekas de St. Groth postulated that antigenic variations in the influenza system might occur in a predictable fashion with new epidemic variants arising directly from preexisting strains (19). He proposed that the selective immunization of individuals with a presumed future epidemic strain (selected in the laboratory) might provide a means of preventing future influenza epidemics and pandemics. As potentially exciting as such prognostic investigations may prove, they rely on the data used to predict the changes in viral determinants and presumed antigenic relationships. Yet the preliminary analyses using monoclonal antiinfluenza antibodies indicate that many of the previously assumed relationships may be more complex than previously realized. Thus it is conceivable that further analysis with these highly specific antibodies and a variety of derivative strains of the influenza virus will eventually enable correct antigenic relationships to be established between the various derivative strains of influenza virus. Given this information, the process of antigenic variation in such pathogens as influenza virus and perhaps other viruses could be forecast and thus proper prophylaxis initiated.

Finally, the question arises as to the potential application of *in vitro*-derived antibodies for the therapy of human disease. The demonstrated capacity to obtain restricted antibody populations which are highly purified and highly specific for tumor antigens may be an excellent starting point for such an approach. We have already mentioned the possibility of obtaining antibodies to human tumor-specific antigenic determinants in mouse fragment cultures. It should be possible to test putative tumor antigen-specific monoclonal antibodies, not only against the relevant tumor antigens, but simultaneously against normal cells or serum components derived from the same patient. Only those antibodies which react specifically with the tumor alone could then be isolated. While the occurrence of such clones may be infrequent, once detected these clones could provide the antibody needed. The availability of such antibodies, particularly mouse antibodies against human tumors, may provide, not only excellent probes for the analysis of human tumor antigens and their interrelationships, but highly effective therapeutic reagents as well. A number of investigators have recently succeeded in attaching drugs or other toxic agents to antitumor antibodies in the hope that the antibody will impart a greater specificity to these toxic agents, directing their activity against the tumor cell (20–25). For example, Dr. Levy in collaboration with co-workers at the Weizmann Institute, has found a method of binding daunomycin and adriamycin, two of the most active cancer chemotherapeutic agents, to antitumor immunoglobulins, preserving most of the activity of both the antibody and the drug (24). These drug–antibody conjugates were shown to mediate a specific cytotoxic effect *in vitro*. The most difficult problem with such "magic bullets" remains the lack of antitumor antibodies with sufficient specificity and purity for effective use as carriers of drugs *in vivo*. Highly selected monoclonal antibodies produced relatively pure *in vitro* and screened for their inability to be adsorbed by normal cells or circulating inhibitors may be such reagents.

The availability of antibody from *in vitro* stimulation opens up a wide array of potentially useful methods for obtaining antibodies against human pathogens which can be used in diagnosis or therapy. It is, however, debatable whether the quantities of antibodies that may be obtained from these methods will render them useful for the stated purposes. If such antibodies can be highly labeled, then tiny amounts may be sufficient for diagnostic purposes. If, however, larger quantities of antibody are necessary there may be means available for expanding a selected clone. When clones are permitted to produce antibody for a month or more, they will often produce tens of micrograms of antibody. This may still be too little antibody. Other methods, therefore, may be essential for expanding such clones. The anticipation that such clones could be expanded has been dampened by experimental findings. In general it has not been possible, in our hands, to transfer clones either *in vitro* or *in vivo,* to provide larger quantities of antibody. It may be that such procedures are doomed to failure, since there is no assurance that an antibody-producing clone also contains precursors to antibody-forming cells.

Other methods of expanding clonal production *in vitro* may be attempted. These include the hybridization of antibody forming cells with long term culture lines, or the transformation of antibody producing clones by agents such as the Abelson virus. Experiments in both of these directions are currently being carried out in our laboratories, and with Drs. Kaprowski and Sklar. It is hoped that such procedures will provide large amounts of monoclonal antibody. It should be pointed out that if clones can be expanded, then clones derived by cell suspension culture techniques may prove at least as useful as fragment cultures (7,8,26). To date such cultures have proven as efficient as fragment cultures in both the number of clonal precursor cells stimulated (8) and the amount of antibody produced per antibody-forming cell (26). Fragment cultures, however, have permitted greater clonal expansion and longevity, thus providing sufficient antibody for many of the stated purposes. Obviously, successful procedures for expanding monoclonal antibody production *in vitro* would increase enormously the amount of antibody available from either method, thus enhancing their potential usefulness.

ACKNOWLEDGMENTS

This work was supported by U.S. Public Health Service Grants A1-08778-06 and CA-15822-02.

REFERENCES

1. Tao, T. W., and Uhr, J. W. (1966): Primary-type antibody response *in vitro*. *Science*, 151:1096.
2. Globerson, A., and Auerbach, R. (1965): Primary immune reactions in organ cultures. *Science*, 149:991.
3. Dutton, R. S. (1967): *In vitro* studies of immunological responses of lymphoid cells. *Adv. Immunol.*, 6:253.
4. Klinman, N. R. (1969): Antibody with homogeneous antigen binding produced by splenic foci in organ culture. *Immunochemistry*, 6:757.
5. Segal, Gary P., and Klinman, N. R. (1976): Defining the heterogeneity of anti-tumor antibody responses. *J. Immunol.*, 116:1539.
6. Nakamura, I., Ray, A., and Makela, O. (1973): Oligomeric IgA: The major component of the *in vitro* primary response of mouse spleen fragments. *J. Exp. Med.*, 138:973.
7. Maarbrook, J., and Haskill, J. S. (1974): The *in vitro* response to sheep erythrocytes by mouse spleen cells: Segregation of distinct events leading to antibody formation. *Cell. Immunol.*, 13:12.
8. Cosenza, H., Quintans, J., and Lefkovits, I. (1975): Antibody response to phosphorylcholine *in vitro*. I. Studies on the frequency of precursor cells, average clone size and cellular cooperation. *Eur. J. Immunol.*, 5:343.
9. Braciale, T. J., Gerhard, W., and Klinman, N. R. (1976): The analysis of the *in vitro* humoral immune response to influenza virus. *J. Immunol.*, 116:827.
10. Segal, S., Globerson, A., Feldman, M., Haimovich, S., and Sela, M. (1970): *In vitro* induction of a primary response to the dinitrophenyl determinant. *J. Exp. Med.*, 131:93.
11. Gerhard, W., Braciale, T., and Klinman, N. R. (1975): The analysis of the monoclonal immune response to influenza virus I. Production of monoclonal anti-viral antibodies *in vitro*. *Eur. J. Immunol.*, 5:720.
12. Maccario, A. J. L., Conway de Macario, E., Francheschi, C., and Celada, F. (1971): Attempt to separate antibody-forming cell clones of different affinity for one antigenic determinant on a macromolecule. *Adv. Exp. Med. Biol*, 12:365.

13. Gearhart, P. J., Sigal, N. H., and Klinman, N. R. (1975): Heterogeneity of the BALB/c anti-phosphorylcholine antibody response at the precursor cell level. *J. Exp. Med.*, 141:56.
14. Sigal, N. H., Gearhart, P. J., and Klinman, N. R. (1975): The frequency of phosphorylcholine-specific B cells in conventional and germfree BALB/c mice. *J. Immunol.*, 114:1354.
15. Klinman, N. R., and Press, J. L. (1975): The B cell specificity repertoire: Its relationship to definable subpopulations. *Transplant. Rev.*, 24:49.
16. Kreth, A. W., and Williamson, A. R. (1973): The extent of diversity of anti-hapten antibodies in inbred mice: Anti-NIP (4-hydroxy-5-iodo-3-nitrophenacetyl) antibodies in CBA/H mice. *Eur. J. Immunol.*, 3:141.
17. Pink, J. R. K., and Askonas, B. A. (1974): Diversity of antibodies to cross-reacting nitrophenyl haptens in inbred mice. *Eur. J. Immunol.*, 4:426.
18. Goldstein, L. T., Klinman, N. R., and Manson, L. A. (1973): A Microtest radioimmunoassay for non-cytotoxic tumor-specific antibody to cell surface antigens. *J. Nat. Cancer Inst.*, 51:1713.
19. Fazekas de St. Groth, S. (1967) Cross recognition and cross reactivity. *Cold Spring Harbor Symp. Quant. Biol.*, 32, 525.
20. Moolten, F. L., Capparell, N. J., and Cooperband, S. R. (1972): Antitumor effects of antibody–diphtheria toxin conjugates: Use of hapten-coated tumor cells as an antigenic target. *J. Natl. Cancer Inst.*, 49:1057.
21. Philpott, G. W., Shearer, W. T., Bower, R. J., and Parker, C. W. (1973): Selective cytotoxicity of hapten-substituted cells with an antibody-enzyme conjugate. *J. Immunol.*, 111:921–929.
22. Ghose, T., and Nigam, S. P. (1972): Antibody as carrier of chloramucil. *Cancer*, 29:1398–1400.
23. Davies, D. A. L., and O'Neill, G. J. (1973): *In vivo* and *in vitro* effects of tumor specific antibodies with chlorambucil. *Br. J. Cancer,* 28:Suppl. I.
24. Hurwitz, E., Levy, R., Maron, R., Wilchek, M., Arnon, R., and Sela, M. (1975): The convalent binding of daunomycin and adriamycin to antibodies, with retention of both drug and antibody activities. *Cancer Res.*, 35:1178.
25. Levy, R., Hurwitz, E., Maron, R., Arnon, R., and Sela, M. (1975): The specific cytotoxic effects of daunomycin conjugated to anti-tumor antibodies. *Cancer Res.*, 35:1182.
26. Braun, D. G., Quintans, J., Luzzati, A. L., Lefkovits, I., and Read, S. E. (1976): Antibody response of rabbit blood lymphocytes in vitro. Kinetics, clone size, and clonotype analysis in response to streptococcal group polysaccharide antigens. *J. Exp. Med.,* 143:360.

Antibodies in Human Diagnosis and Therapy,
edited by Edgar Haber and Richard M. Krause.
Raven Press, New York © 1977.

Potential Usefulness of Monoclonal Culture Techniques

Charles W. Parker

I would like to make a few general comments about the potential usefulness of a monoclonal culture technique that has been described in the initial analysis of an antigen and in immunoassay applications. I think this system has considerable potential in the elucidation of antigenic structure because of the ability to obtain homogenous antibodies with specificity for individual antigenic domains on the immunogen. In addition, in dealing with a complex mixture of antigens and no easy means of purification, the ability to obtain an antibody with specificity only for the antigen might be very useful. Once the antigen has been partially characterized and purified, the usefulness of antibodies produced by this method is more questionable, considering the very limited quantities of antibody available. In almost all of the immunoassays in everyday use, the antibodies involved are heterogenous in their combining properties. By and large, when an effort is made to maximize sensitivity by use of high dilutions of antibody, only the highest affinity antibodies are participating appreciably in binding. Thus, in effect you are selecting for those antibodies with the most favorable binding characteristics. In addition, there is the option of absorbing the antisera or of partially degrading the antigen if immunologic cross-reactivity is a problem. For routine immunoassay purposes, therefore, the advantage of having a homogenous antibody preparation is more theoretical than real since even heterogenous antibodies may give a sensitive linear inhibition curve on a semi-log plot of bound labeled antigen versus the log of the unlabeled antigen concentration. Even though antibodies obtained by conventional methods of immunization are likely to be heterogenous, highly satisfactory immunoassay results can usually be obtained.

Antibodies in Human Diagnosis and Therapy,
edited by Edgar Haber and Richard M. Krause.
Raven Press, New York © 1977.

In Vitro Production of Antibody by Viral Transformed Cell Lines

Paul H. Black and Vincent R. Zurawski, Jr.

Department of Medicine, Massachusetts General Hospital and Department of Medicine, Harvard Medical School, Boston, Massachusetts 02114

A major advance in the diagnosis and therapy of many human disorders would be achieved if one could establish continuous human cell lines in culture which produced homogeneous antibody to a specific antigen. Immunoglobulin-secreting cell lines derived from mouse myelomas and lymphomas have been established; in some cases the immunoglobulin produced by such lines is of monoclonal origin (1). In addition, human plasmacyte cell lines (2) and lymphoma (3–5) cell lines which produce immunoglobulin have been established. Although certain myeloma proteins, particularly from the mouse, have been found to have antihapten or antipolysaccharide binding activity by screening many antigens, it is not known whether the antigen bound represents the original immunogen or a cross-reacting moiety. Thus, alternative measures must be sought for the continuous production of antibodies of predetermined specificity. Some success in producing structurally homogeneous antibody preparations has been achieved by hyperimmunization with bacterial vaccines (6,7). However, production of such antibody is dependent upon continuous antigen stimulation and long-term culture of such cells has not been achieved.

In an attempt to establish a cell line programmed to produce antibody against a specific antigen and capable of indefinite growth *in vitro,* viral transformation studies of cells obtained from animals immunized with a specific antigen(s) were carried out. In this chapter, we shall review our past and present efforts to "fix" the antibody-producing potential of an immunocompetent cell by transformation with an oncogenic virus. In addition, we will speculate on the medical implications of the production *in vitro* of specific antibody.

TRANSFORMATION OF RAT LYMPH NODE CELLS

Initial attempts to produce a continuous cell line capable of producing monospecific antibody involved rats hyperimmunized with dinitrophenyl bovine gamma-globulin (8). Lymph node cells obtained from such rats were infected with simian virus 40 (SV40) immediately or 3 days after stimulation with the mitogen phytohemagglutinin (PHA) or specific antigen (dinitrophenyl bovine gamma-globulin). Transformed cells were observed growing on the glass surface of 4 cultures (3 unstimulated, 1 stimulated with PHA) 34–46 days after infection and cell lines were established. Neither specific antibodies nor immunoglobulin in the supernatant fluid of transformed cell cultures or extracts of transformed cells was detected by radioimmunoelectrophoresis or hemagglutinination-inhibition assays. Furthermore, transformed cells labeled with [^{14}C]leucine failed to incorporate labeled amino acid into immunoglobulin molecules as assayed by immunoelectrophoresis and agarose gel electrophoresis combined with autoradiography; these tests were performed on concentrated supernatant fluids and disrupted cells. Thus, the transformed cells did not appear able to synthesize immunoglobulin. Further attempts to establish a relationship between the transformed cells and lymphoid cells were inconclusive. Thus, the identity of the target cells for transformation in these experiments remained uncertain.

TRANSFORMATION OF SPLEEN CELLS FROM A HYPERIMMUNIZED RABBIT

We reasoned that successful transformation of an antibody-producing cell would be more likely to occur with cells from the spleen of a rabbit hyperimmunized with a bacterial vaccine. Rabbits repeatedly injected intravenously with pneumococcal vaccines produce large amounts of antibody of restricted heterogeneity (7). Marked plasma cell infiltration occurs in the spleens of such animals (9). Such an organ would provide a rich source of antibody-producing cells for viral transformation. We therefore established cultures of rabbit spleen as well as bone marrow cells and infected them with purified SV40 virus (10). There was no evidence of proliferation of the suspended cells in any of the cultures as determined by periodic cell counts. On the 114th day after infection of a spleen cell culture, one focus of transformed cells was observed growing on the glass surface of the culture vessel. This focus of cells increased in size; it was then dispersed with trypsin and a cell line, designated TRSC-1, was established. A clone, designated TRSC-1-1, was subsequently isolated from the mass culture TRSC-1.

CHARACTERISTICS OF THE SV40 TRANSFORMED RABBIT SPLEEN CELLS (TRSC-1)

Morphologic and Virologic Features

The TRSC-1 cells are polygonal in shape and contain many multinucleated syncytical giant cells (see Fig. 1); such features are characteristic of SV40 transformed cells (11). They can be adapted to grow in suspension, but adhere to glass or plastic surfaces quite well where they may pile up and form multilayered colonies. A unique feature of the transformed spleen cells is the frequent occurrence of curved or indented nuclei. The modal number of chromosomes in TRSC-1 cells is in the subtetraploid range. The cells contained the SV40-specific T antigen in the nuclei of nearly all the cells by immunofluorescence staining, and infectious SV40 virus could be rescued from such cells by fusion with permissive monkey kidney cells utilizing Sendai fusion factor (12). These features, together with the fact that the cells can be propagated in continuous culture,* attest to the transformed nature of the cells.

Collagen or reticulum was not detected in the TRSC-1 cells when stained with the Masson or Wilder technique, respectively, nor did these cells have the characteristics of mast cells upon staining with toluidine blue. Electron-microscopic examination of the cells revealed high densities of polyribosomes in the cytoplasm. In addition, dilated cisternae of rough endoplasmic reticulum filled with a proteinaceous material were observed in many cells. Such structures were not observed in SV40 transformed fibroblasts. The cells did not have the typical appearance of plasma cells, however, and were quite large relative to normal lymphocytes or lymphoblasts.

Immunologic Features

Evidence for immunoglobulin production was sought from cultures of transformed cells. No IgG could be detected on the cell surface by immunofluorescence staining. Culture fluids and extracts of labeled or unlabeled transformed cells, prepared by five cycles of freezing and thawing, were clarified, dialyzed, lyophilized, and reconstituted in phosphate-buffered saline. Immunoelectrophoresis was performed with supernatant fluids from subconfluent, as well as confluent, cultures of [^{14}C]leucine labeled TRSC-1 cells and a cell extract; all samples were combined with normal rabbit serum as "carrier." Precipitin lines were formed with goat antisera against rabbit γ, α, and μ and the plates submitted to autoradiography. After a 3–6 month exposure, labeling of the IgG, but not the IgA or

* TRSC-1 cells are currently at passage 65 with no observable differences in the cell line.

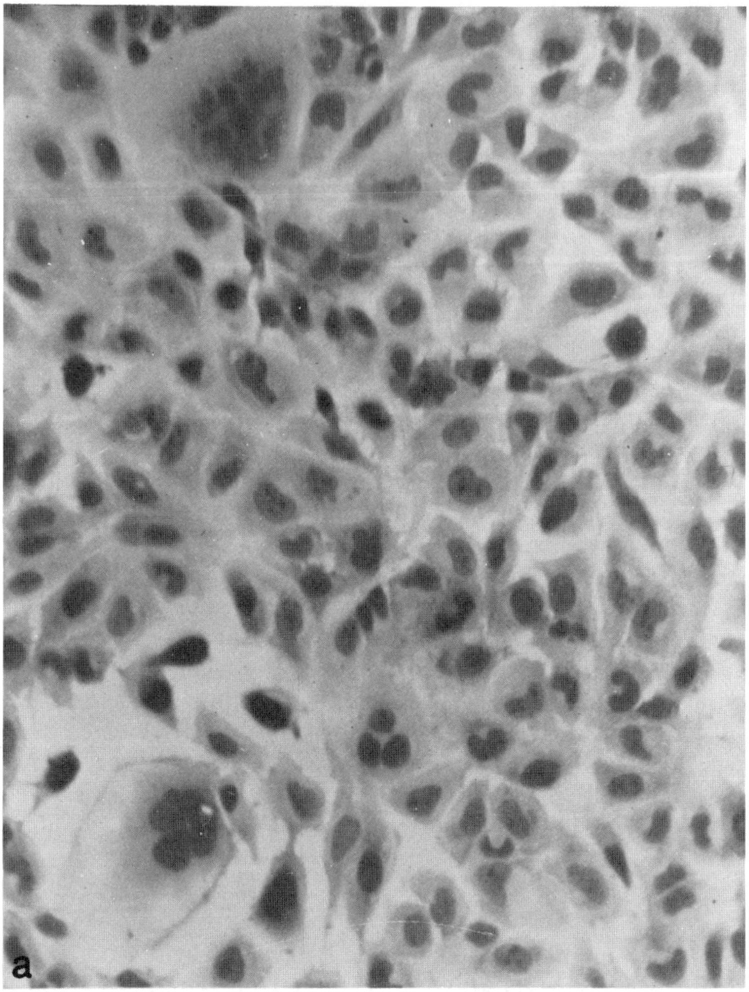

FIG. 1a. Monolayer culture of TRSC-1 cells at the 5th passage level. Stained with hematoxylin and eosin. Note large epithelial cells, bandlike nuclei, and multinucleated syncytial giant cells. ×85.

IgM precipitin arc, of both the culture media and cell extract of TRSC-1 cells was observed (see Fig. 2). Normal rabbit kidney cells or SV40 transformed rabbit kidney cells did not label proteins with the antigenic characteristics of rabbit immunoglobulin as determined by radioimmunoelectrophoresis. In addition, agarose gel electrophoresis in the presence of an [^{125}I]S3-polysaccharide-protein conjugate of extracts of TRSC-1 cells indicated binding in the gamma globulin region. In order to quantitate the total IgG present in both the labeled culture fluids and cell extracts, precipitation with sheep antisera against rabbit IgG was carried out. This procedure precipitated 1.2

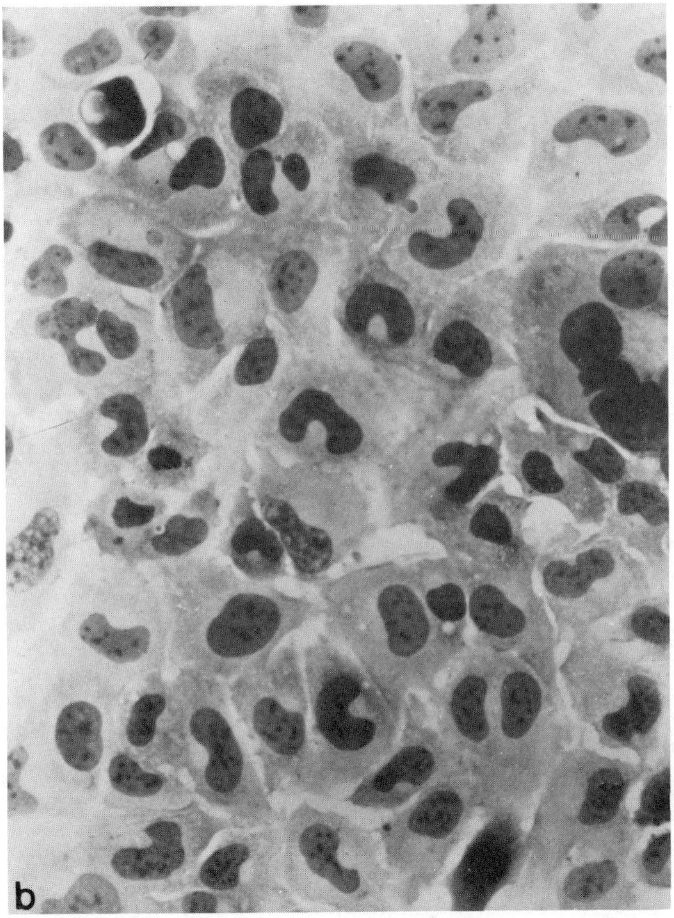

FIG. 1b. Higher power magnification (×365.5) of TRSC-1 cells (Fig. 1a).

and 1.4% of the labeled protein in extracts of TRSC-1 and TRSC-1-1, respectively; 0.7% of labeled protein was precipitated from the culture fluids of confluent TRSC-1 cells. These procedures indicated that TRSC-1 cells produce rabbit immunoglobulin of the IgG class (10).

In order to determine whether the rabbit IgG produced by TRSC-1 cells had specific antibody specificity, immunoadsorbtion studies were carried out (13). S3 and S8 polysaccharides were purified from types III and VIII pneumococcal vaccines as described (7,14). Either S3 or S8 polysaccharide was coupled to bovine gamma globulin; the protein-polysaccharide conjugates were then coupled to cyanogen bromide-activated Sepharose 4B. A portion of the S3 adsorbent was inactivated by exposure to 8 M urea—1 N HCl for 24 hr at 50°C. In Table 1, the results of the immunoadsorbtion studies performed with extracts of cells labeled with [^{3}H]leucine are given.

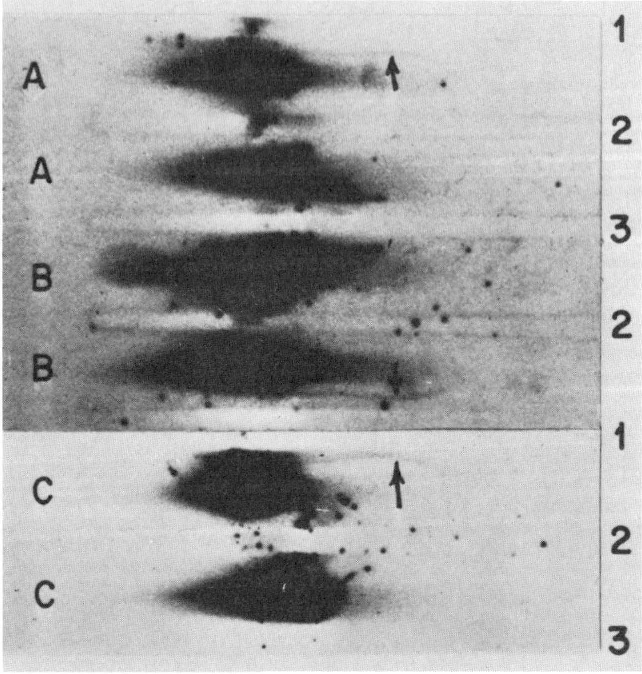

FIG. 2. Autoradiography of immunoelectrophoresis of labeled transformed cell cultures. Growth to 90% confluence (*wells A*); growth to complete confluence (*wells B*); cell extract (*wells C*). Precipitin lines developed with anti-rabbit γ (*troughs 1*), α (*troughs 2*), and μ (*troughs 3*) heavy chain antisera. Duration of exposure, 6 months. Labeled amino acid was incorporated into rabbit IgG only (*arrows*) (anode to left).

TABLE 1. *Binding of extracts of labeled cells to various immunoadsorbents*

		Eluate (cpm)	
Extract applied	Immuno-adsorbent	After 1st adsorption (1.0 M HOAC)[a]	After 2nd adsorption (1.0 M HOAC)[a]
TRSC-1	S3	49,700	43,500
	S8	13,400	1,700
	Inactive S3	11,800	800
TRK-73	S3	8,800	1,800

Culture fluids and cell extracts were dialyzed extensively and applied to 5 ml columns of S3, S8, or inactivated S3 immunoadsorbents. The columns were then washed with PBS. Thereafter, elutions were carried out with 1 N acetic acid, the eluates were neutralized, concentrated, and combined with 5 mg of purified anti-S3 antibodies as carrier. Adsorption to and elution from the immunoadsorbents were then repeated, and the radioactivity of the eluates was determined.

[a] HOAC = acetic acid.

TRSC-1 extracts bound significantly to an S3, but not to an S8 or inactivated S3 immunoadsorbent. An extract of a transformed rabbit kidney cell line did not bind significantly to an S3 immunoadsorbent. The amount of TRSC-1 extract bound represented approximately 1% of the total labeled proteins in the cell extract. The recovery of labeled protein, therefore, corresponded to that obtained by precipitation of rabbit IgG with sheep antirabbit IgG and suggests that all or nearly all of the IgG produced represents antibody to S3.

Isoelectric focusing in polyacrylamide gels was carried out with both unfractionated cell extracts or culture fluids and the immunospecifically purified fractions (eluates from immunoadsorbtion columns) (see Fig. 3 and 4). Whereas the former preparations showed many labeled peaks, the latter fractions showed a single major peak. The isoelectric points of the peaks from the extract and culture fluid were similar. Also shown in Fig. 4 is the isoelectric focusing pattern of the purified serum antibodies from the restricted responder rabbit whose spleen cells were subsequently transformed.

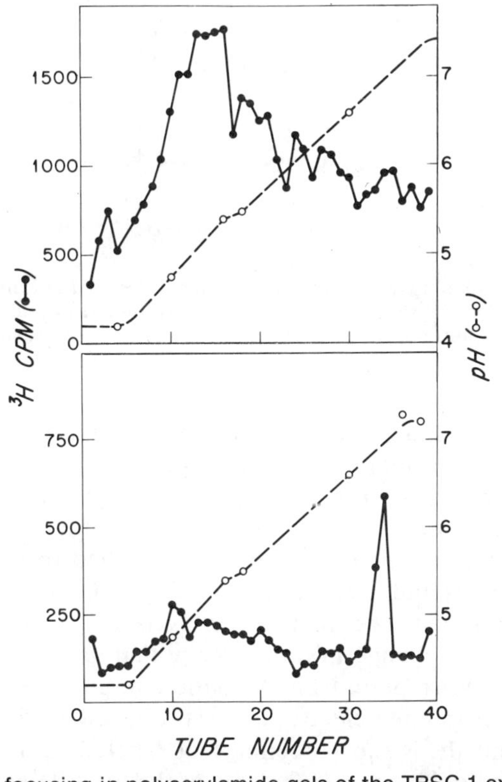

FIG. 3. Isoelectric focusing in polyacrylamide gels of the TRSC-1 extract proteins (**top**) and of S3-immunoadsorbent purified cell extract proteins (**bottom**). Cell extracts or supernatant fluids before and after elution from S3 immunoadsorbents were lyophilized and redissolved in an acrylamide-urea ampholyte solution before polymerization. (See ref. 13 for details.)

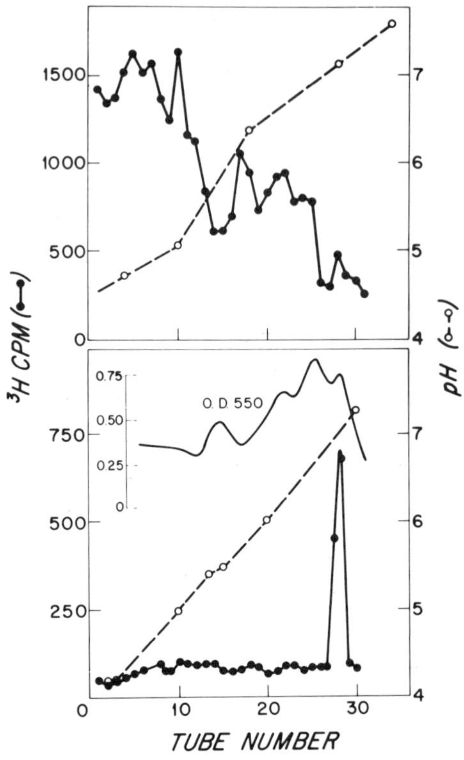

FIG. 4. Isoelectric focusing in polyacrylamide gels of TRSC-1 culture fluid proteins (**top**) and of S3 immunoadsorbent purified culture fluid proteins (**bottom**). The optical scan of the stained gel of purified serum antibodies from the restricted responder rabbit is also shown (**bottom**).

One of the four bands identified by scanning the stained gel corresponds in isoelectric point to the major radioactive peak obtained from TRSC-1 extracts and culture fluids. The isoelectric homogeneity suggests that the product of a single clone was examined. The fact that similar results were obtained with the clonal line TRSC-1-1, established from TRSC-1, further indicates that the initial transformation event likely involved a single antibody-producing cell, presumably a plasma cell or lymphocyte, which resulted in a marked change in morphology and growth characteristics.

Because of the long period of incubation required to develop the autoradiographs after immunoelectrophoresis of labeled TRSC-1 extracts and culture fluids, and the same percentage of labeled protein coprecipitated with rabbit IgG in liquid media (1.2–1.4% of the total ^{3}H-labeled proteins), it is evident that the transformed cells produce only small amounts of antibody. It is of interest that less immunoglobulin was detected in the culture fluid than in the corresponding cell extracts. This differs from most myeloma

TABLE 2. *Treatment of TRSC-1 cells with various mitogens*

Mitogens	Dose range (μg/ml)	Exposure range (hr)	Recovery range (hr)
PWM	0–200[a]	2–48[b]	0–24 (2,6,12,24)
LPS (*Salmonella*)	0–200	24	0–24 (2,6,12,24)
LPS (*E. coli*)	0–150	24	0–24 (2,6,12,24)
PHA-M	0–100	2–48[b]	0–24 (2,6,12,24)
Con A (purified)[c]	0–100	2–48[b]	0–24 (2,6,12,24)

Uptake of [^{3}H]thymidine amounted to approximately 10% of the total counts added in both test and control cultures.

Abbreviations: PWM, pokeweed mitogen (Gibco); LPS, lipopolysaccharide (Gibco, Difco); PHA, phytohemagglutinin (Gibco); Con A, concanvalin A (Sigma).

[a] Doses utilized were: 0, 2, 5, 10, 20, 50, 100, 150, 200 μg/ml.

[b] Exposure times were: 2, 6, 12, 24, 48 hr.

[c] Purified according to Cunningham et al. (38).

cultures where the majority of immunoglobulin produced is secreted and immunoglobulin production may comprise approximately 10–20% of the cells' protein. The specific reason(s) for such limited production of immunoglobulin in TRSC-1 cells is not known. It is unlikely that a heterogeneous population of cells is present, since the data presented indicate that the population consists of the progeny of a single cell.

ATTEMPTS TO INCREASE ANTIBODY PRODUCTION

The absolute amount of rabbit immunoglobulin produced in the culture medium (15 ml–100 mm plastic petri dish) is approximately 20–30 ng/ml after 7–14 days in culture as determined by radioimmunoassay (RIA),* or approximately 2–5 ng/ml/10^6 cells during this time period. Various attempts were made to increase the amount of antibody synthesized and/or secreted. The results of experiments in which various mitogens were utilized are given in Table 2. TRSC-1 cells were exposed to various concentrations of the lectin for varying lengths of time. Since higher doses of some of the lectins were toxic to the cells, recovery periods of 2, 6, 12, or 24 hr were allowed prior to assays for mitogenesis and measurement of immunoglobulin secretion. There was no indication of cell proliferation as determined by cell counts relative to control, nor of mitogenesis as determined by examination

* Radioimmunoassay was performed as follows: TRSC-1 culture fluid or control medium was added to a tube. Then ^{125}I-labeled rabbit gamma-globulin (RGG) mixed with goat anti-RGG was added. After equilibrium was attained, the soluble complex was precipitated with donkey antigoat antibody in the presence of goat gamma-globulin as carrier. The precipitate and supernatant fluid were then counted. Known amounts of RGG were added to one group of tubes with 10% fetal calf serum in phosphate buffered saline to obtain a standard curve for each experiment. The amount of cold RGG present in the culture fluid was calculated from the amount of labeled RGG displaced.

of cells and uptake of [^{3}H]thymidine after a 1-hr exposure at the end of the recovery period. There was also no increase in the production of rabbit immunoglobulin as determined by coprecipitation with sheep antirabbit gamma globulin or RIA.

Other agents which might stimulate antibody synthesis and/or secretion were tried as follows: cyclic GMP or dibutryl cyclic GMP (5×10^{-4} M), cyclic AMP or dibutryl cyclic AMP (10^{-3} M), S3 and S8 polysaccharides (0.1, 1, 10, 100 μg/ml), S3 vaccine (at pneumococci:TRSC-1 ratios of: 0.25:1; 1:1; 2:1; 2.5:1; 10:1; 20:1; 25:1; 100:1; 200:1), mitomycin C (0.05–2.0 μg/ml) or conditioned medium from thymocyte suspension cultures or thymocyte monolayer cultures, established from thymus cells obtained from an animal hyperimmunized with S3 polysaccharide. Direct cocultivation experiments were also carried out with thymocyte suspension and monolayer cultures and TRSC-1 cells. The latter experiments were done to test cooperativity between thymic factors and antibody-producing cells. Again, as determined by coprecipitation or RIA, neither mitogenesis nor increase in the production of rabbit immunoglobulin occurred. These results suggest that some cellular lesion exists such that immunoglobulin production cannot be increased by the agents utilized. It is possible that TRSC-1 cells lack receptors for the mitogens or antigens tested. Other possible reasons include the selection of the cell line from a precursor cell rather than a plasma cell, the absence of antigen in the culture, or an inability to secrete antibody, which may inhibit production. Viral transformation may have resulted in such dedifferentiation that antibody production and secretion may have diminished significantly. It is of interest that the lymphomas, leukemias, reticulum cell sarcomas, and osteogenic sarcomas induced in hamsters after intravenous inoculation of SV40 differed markedly in their degree of differentiation (15). Thus, different transformation events might produce transformed cells with varying degrees of preservation of differentiated function. It is possible that other clones of transformed cells might produce more antibody than TRSC-1.

CURRENT EXPERIMENTS AND POTENTIAL UTILIZATION

In collaboration with Dr. Edgar Haber, we have initiated a series of experiments in which we are attempting to create a multiple myeloma or a functional equivalent *in vivo* in animals producing antibody of restricted heterogeneity. Extirpation and explantation of the malignant tissue could result in the establishment of a continuous culture producing monospecific antibody. If the culture is heterogeneous and composed of several antibody-producing cells, indicating that more than one transformation event has occurred, cloning procedures will separate out the various clones of antibody-producing cells. We will utilize the rabbit, immunized with pneumococcal vaccine, and producing antibody of restricted heterogeneity.

The viruses that will be utilized include *Herpes saimiri,* which produces a malignant lymphoma in the rabbit (16), SV40 which is able to transform rabbit cells in culture and for which B cells have receptors (10,17–19), *Herpes sylvilagus* which causes a lymphoproliferative disease in cottontail rabbits (20,21), and Epstein Barr virus, a human B cell-tropic virus (22,23), which can transform lymphocytes into continuous lines which produce immunoglobulin for long periods of time.

Another line of investigation we are pursuing is that of transformation of an immunocompetent, human cell which has been sensitized to a specific antigen *in vitro*. Methods have been described whereby lymphocytes have been sensitized to whole cells (24) or soluble antigens (25); assays for antibody production can also be easily carried out (24). If the incidence of cell sensitization is low, or one desires a more concentrated population of programmed cells, techniques such as affinity chromatography (27) or use of the Herzenberg laser cell sorter (28) are currently available for concentration of sensitized cells.

With a sensitized population of cells transformation with an oncogenic virus could be carried out. A human B-topic virus such as EBV is especially well suited, since receptors for EBV are present on human B cells (19). In this way one might produce an autonomous cell line capable of producing homogeneous antibody of desired specificity. Should the transformed line produce small amounts of virus, separation of virus from antibody could be effected by chromatographic techniques as well as by differential inactivation procedures such as heat, pH treatment, formalin, etc. Viruses might also be eliminated by viral antiserum.

Many uses for human antibody can be envisioned. Use of human antibody would, of course, be safer than use of heterologous antibody with its attendant risks of anaphylaxis and other allergic reactions. A vehicle for the production of large amounts of human antibody would provide a source of antibody which is not antigenic and which could be utilized repeatedly in the same patient for diagnosis and therapy. Together with Dr. E. Haber we are currently sensitizing cells to a digoxin-carrier conjugate. Transformation of such a cell could result in the continuous production of antibody to digoxin. Use of the Fab fragment of sheep antidigoxin antibody has already been shown to reverse the electrophysiological and inotropic effects of digoxin (29), and has been efficacious in reversing digitalis intoxication in human patients (30).

Continuous production of human antibody would be of enormous benefit in the prophylaxis and treatment of certain human infectious diseases, particularly viral, where the effectiveness of present-day chemotherapy has not been conclusively demonstrated. Much progress has been made in the fractionation of various virus proteins and the identification of those involved in neutralization. Antiserum prepared against such proteins results in neutralization of the infectivity of the virus. Since these proteins or glycoproteins can be solubilized, it might be possible to sensitize a human lymphocyte population against them and transform the programmed cell with a virus. Establishment of a continuous cell line producing neutralizing antibody to these purified viral constituents would aid immeasurably in the treatment or prophylaxis of several virus diseases. With human hepatitis B infection, administration of gamma-globulin has proved effective in the prophylaxis of the disease. Likewise, treatment of vaccinia gangrenosa and disseminated herpes zoster with hyperimmune globulin has proved effective in containing the spread of these diseases. Human interferon has proved effective in both prophylaxis and treatment of some viral diseases (31,32). Most interferon preparations are not pure, and, although considerable progress has been made in its purification, current procedures are time-consuming and expensive (33). Sensitization of an immunocompetent cell to interferon with subsequent transformation might result in the production of a cell line synthesizing specific antibody to interferon which could aid immeasurably in interferon purification by the use of affinity chromatography.

The present studies may have relevance for cancer diagnosis and immunotherapy. There is currently evidence that tumor antigens or antigen–antibody complexes block lymphocytes thereby preventing effective cell-mediated tumoricidal activity (35). Deblocking sera may act by competing for such antigens or antigen–antibody complexes (36). If such a mechanism proves to be operative in humans, production of antibody by cells sensitized to tumor antigens, which can be isolated and purified and are soluble, may have a place in the treatment of human malignancy. Deblocking sera has been used therapeutically in rats with good results (37). Antibodies directed at tumor antigens may also prove useful in identifying metastatic tumor foci.

Production of human antibody against antigens, the antisera to which have been raised in heterologous species, would also be of importance. Human antilymphocyte serum, for example, is currently raised in horses, and its repeated use frequently produces untoward allergic sequellae.

In addition to the above, an antibody-producing cell line capable of indefinite growth *in vitro* may prove capable of providing large amounts of homogeneous antibody and perhaps cell surface receptors for detailed structural studies, especially in determining the structural basis of antibody specificity. Further studies of the effect of antigen on antibody production as well as delineation of the mechanisms involved in synthesis, processing, and secretion of immunoglobulin might also be made. The influence of mutational events and genctic drift on the nature of the antibody product might also be determined.

SUMMARY

We sought to produce a continuous cell line which is capable of producing specific homogeneous antibody. Spleen cells derived from a rabbit hyperimmunized with a type III pneumococcal vaccine and producing antibody of restricted heterogeneity were transformed with the oncogenic virus SV40. A cell line, capable of indefinite growth *in vitro,* was found to synthesize an immunoglobulin, which bound specifically to an S3 immunoadsorbent column, and which, after elution, showed a single band on isoelectric focusing which corresponded in isoelectric point to one of the serum anti-S3 antibodies of the donor rabbit. Thus, a committed cell can be brought into continuous culture subsequent to transformation with an oncogenic virus. Morphologic, virologic, and immunologic aspects of the transformed cells are described. The implications of these findings with respect to the large-scale production of human antibodies for diagnosis and treatment of various human diseases are discussed.

ACKNOWLEDGMENTS

The work presented was done in collaboration with Drs. J. J. Collins, K. J. Bloch, A. D. Strosberg, and E. Haber. We thank Dr. John Kimball for providing rabbit L-27 and Dr. Ann Dvorak for the electron-microscopic analysis.

REFERENCES

1. Potter, M. (1972): Immunoglobulin-producing tumors and myeloma proteins of mice. *Physiol. Rev.*, 52:631–719.
2. Jobin, M. E., Fahey, J. L., and Price, Z. (1974): Long term establishment of a human plasmacyte cell line derived from a patient with IgD multiple myeloma. I. Requirement of a plasmacyte-stimulating factor for the proliferation of myeloma cells in tissue culture. *J. Exp. Med.*, 140:494–507.
3. Wakefield, J. D., Thorbecke, G. J., Old, L. J., and Boyse, E. A. (1967): Production of immunoglobulins and their subunits by human tissue culture cell lines. *J. Immunol.*, 99:308–319.
4. Fahey, J. L., Finegold, I., Rabson, A. S., and Manaker, R. H. (1966): Immunoglobulin synthesis *in vitro* by established human cell lines. *Science*, 152:1259–1261.
5. Finegold, I., Fahey, J. L., and Dutcher, T. F. (1968): Immunofluorescent studies of immunoglobulins in human lymphoid cells in continuous culture. *J. Immunol.*, 101:366–373.
6. Osterland, C. K., Miller, E. J., Karakawa, W. W., and Krause, R. M. (1966): Characteristics of streptococcal group-specific antibody isolated from hyperimmune rabbits. *J. Exp. Med.*, 123:599–614.
7. Pincus, J. H., Jaton, J. C., Bloch, K. J., and Haber, E. (1970): Antibodies to Type III and Type VIII pneumococcal polysaccharide: Evidence for restricted structural heterogeneity in hyperimmunized rabbits. *J. Immunol.,* 104:1143–1148.
8. Collins, J. J., Bloch, K. J., and Black, P. H. (1970): Transformation of rat lymph node cells by simian virus 40: Attempt to "fix" specialized function. *J. Natl. Cancer Inst.*, 45:1097–1104.
9. Pincus, J. H., Jaton, J. C., Bloch, K. J., and Haber, E. (1970): Properties of structurally restricted antibody to type VIII pneumococcal polysaccharide. *J. Immunol.*, 104:1149–1154.
10. Collins, J. J., Black, P. H., Strosberg, A. D., Haber, E., and Bloch, K. J. (1974): Transformation by simian virus 40 of spleen cells from a hyperimmune rabbit: Evidence for synthesis of immunoglobulin by the transformed cells. *Proc. Natl. Acad. Sci. USA,* 71:260–262.

11. Black, P. H., Rowe, W. P., and Cooper, H. L. (1963): An analysis of SV40-induced transformation of hamster kidney tissue *in vitro*. II. Studies of 3 clones derived from a continuous line of transformed cells. *Proc. Natl. Acad. Sci. USA*, 50:847–854.
12. Burns, W., and Black, P. H. (1968): Analysis of simian virus 40-induced transformation of hamster kidney tissue *in vitro*. V. Variability of virus recovery from cell clones inducible with Mitomycin C and cell fusion. *J. Virol.*, 2:606–609.
13. Strosberg, A. D., Collins, J. J., Black, P. H., Malamud, D., Wilbert, S., Bloch, K. J., and Haber, E. (1974): Transformation by simian virus 40 of spleen cells from a hyperimmune rabbit: Demonstration of production of specific antibody to the immunizing antigen. *Proc. Natl. Acad. Sci. USA*, 71:263–264.
14. Campbell, J. H., and Papperheimer, A. M., Jr. (1966): Quantitative studies of the specificity of anti-pneumococcal polysaccharide antibodies types III and VIII-1. Isolation of oligosaccharide antibodies from acid and from enzymatic hydrolysates of S3 and S8. *Immunochemistry*, 3:195–212.
15. Diamandopoulos, G. Th. (1973): Induction of lymphocytic leukemia, lymphosarcoma, reticulum cell sarcoma, and osteogenic sarcoma in the Syrian Golden hamster by oncogenic DNA simian virus 40. *J. Natl. Cancer Inst.*, 50:1347–1365.
16. Daniel, M. D., Meléndez, L. V., Hunt, R. D., King, N. W., Anver. M., Fraser, C. E. D., Barahona, H., and Baggs, R. B. (1974): Herpes saimiri. VII. Induction of malignant lymphoma in New Zealand white rabbits. *J. Natl. Cancer Inst.*, 53:1803–1807.
17. Black, P. H., and Rowe, W. P. (1963): SV40 induced proliferation of tissue culture cells of rabbit, mouse and porcine origin. *Proc. Soc. Exp. Biol. Med.*, 114:721–727.
18. Coe, J. E., and Green, I. (1975): B-cell origin of hamster lymphoid tumors induced by simian virus 40. *J. Natl. Cancer Inst.*, 54:269–270.
19. Jondal, M., and Klein, G. (1973): Surface markers on human B and T lymphocytes. II. Presence of Epstein-Barr virus receptors on B lymphocytes. *J. Exp. Med.*, 138:364–372.
20. Hinze, H. C. (1971): Induction of lymphoid hyperplasia and lymphoma-like disease in rabbits by *Herpesvirus sylvilagus*. *Int. J. Cancer*, 8:514–522.
21. Wegner, D. L., and Hinze, H. C. (1974): Virus-host cell relationship of *Herpesvirus sylvilagus* with cottontail rabbit leukocytes. *Int. J. Cancer*, 14:567–575.
22. Miller, G., and Lipman, M. (1973): Release of infectious Epstein-Barr virus by transformed marmoset leukocytes. *Proc. Natl. Acad. Sci. USA*, 70:190–194.
23. Leibold, W., Flanagan, T. D., Menezes, J., and Klein, G. (1975): Induction of Epstein-Barr virus-associated nuclear antigen during *in vitro* transformation of human lymphoid cells. *J. Natl. Cancer Inst.*, 54:65–68.
24. Mischell, R. I., and Dutton, R. W. (1967): Immunization of dissociated spleen cell cultures from normal mice. *J. Exp. Med.*, 126:423–442.
25. Kapp, J. A., Pierce, C. W., and Benacerraf, B. (1973): Genetic control of immune responses *in vitro*. *J. Exp. Med.*, 138:1107–1120.
26. Jerne, N. K., and Nordin, A. A. (1963): Plaque formation in agar by single antibody-producing cells. *Science*, 140:405.
27. Schlossman, S. F., and Hudson, L. (1973): Specific purification of lymphocyte populations on a digestible immunoadsorbent. *J. Immunol.*, 110:313–315.
28. Bonner, W. A., Hulett, H. R., Sweet, R. G., and Herzenberg, L. A. (1972): Fluorescence activated cell sorting. *Rev. Sci. Instrum.*, 43:404–409.
29. Curd, J., Smith, T. W., Jaton, J. C., and Haber, E. (1971): The isolation of digoxin specific-antibody and its use in reversing the effects of digoxin. *Proc. Natl. Acad. Sci. USA*, 68:2401–2406.
30. Smith, T. W., Butler, V. P., Jr., and Haber, E. (1976): *This volume*.
31. Merigan, T. C., Hall, T. X., Reed, S. E., and Tyrrell, D. A. J. (1973): Inhibition of respiratory virus infection by locally applied interferon. *Lancet*, 1:563–567.
32. Panusark, C., Stanley, E. D., Dirda, V., Rubenis, M., and Jackson, G. G. (1974): Prevention of illness from rhinovirus infection by a topical interferon inducer. *N. Engl. J. Med.*, 291:57–61.
33. Anfinsen, C. B., Bose, S., Corley, L., and Gurari-Rotenan, D. (1974): Partial purification of human interferon by affinity chromatography. *Proc. Natl. Acad. Sci. USA*, 71:3139–3142.
34. Huang, J. W., Hejna, C. I., Sulkowski, E., Carter, W. A., Silver, G. H., Munayyer, H., and Came, P. (1975): The human interferon–albumin interaction: The influence of albumin conformation. *Virology*, 65:268–271.

35. Hellstrom, K. E., and Hellstrom, I. (1970): Immunologic enhancement as studied by cell culture techniques. *Annu. Rev. Microbiol.*, 24:373–398.
36. Hakala, T. R., and Lange, P. H. (1974): Serum-induced lymphoid cell-mediated cytotoxicity to human transitional cell carcinomas of the genitourinary tract. *Science*, 184:795–797.
37. Bansal, S. C., and Sjögren, H. O. (1971): Unblocking serum activity *in vitro* in the polyoma system may correlate with antitumor effects of antiserum *in vivo*. *Nature* [*New Biol.*], 233:76–78.
38. Cunningham, B. A., Wang, J. L., Pflumm, M. N., and Edelman, G. M. (1972): Isolation and proteolytic cleavage of the intact subunit of concanavalin A. *Biochemistry,* 11:3233–3239.

Antibodies in Human Diagnosis and Therapy,
edited by Edgar Haber and Richard M. Krause.

Synthesis of Streptococcal Group-Specific Antibody by Peripheral Blood Lymphocytes

Dietmar Braun

It is possible to stimulate peripheral blood lymphocytes (PBL) obtained from previously hyperimmunized rabbits *in vitro* in the presence of the corresponding bacterial vaccines (1). Under such conditions, *de novo* synthesis of streptococcal group-specific antibody is achieved if the PBL came from rabbits that had been hyperimmunized up to 15 months previously. The response depends on the presence of antigen in the form of whole heat-killed bacteria and on an optimal cell concentration of 2×10^6/ml per culture (2). Under these conditions, a day-42 culture may contain 1 specific plaque-forming cell among 20 viable cells (Fig. 1). The secretory rate of these plaque-forming cells was estimated to reach 2.4×10^4 molecules sec^{-1} (2). Clonotype analysis with culture supernatants revealed complete overlaps of the sum of these antibodies with the corresponding spectrotypes of the hyperimmune sera (Fig. 2) (1,2). This result is evidence for the presence of a complete set of precursor cells in the peripheral blood that had been stimulated previously to produce group-specific antibody.

A frequency analysis of these precursor cells—determined as precursor units (2)—in the blood revealed that a complete unit, containing a precursor B cell, was

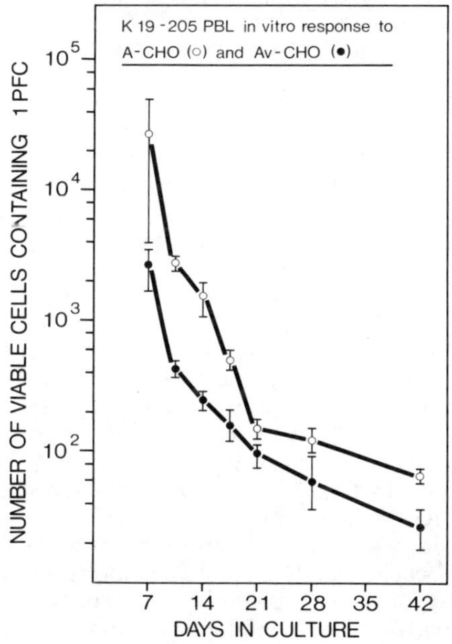

FIG. 1. Enrichment of specific plaque-forming cells (PFC) to the streptococcal groups A (A-CHO) and A-variant (Av-CHO) polysaccharides in cultures of PBL from rabbit K19-205 previously hyperimmunized against streptococcal groups A-variant and A vaccines. Data were obtained from three independent experiments using the same pool of PBL for stimulation with groups A and A-variant vaccines. At day 42, 1 PFC specific for the Av-CHO was present among 20 to 30 viable cells, and 1 PFC specific for the A-CHO was found among 60 to 80 viable cells.

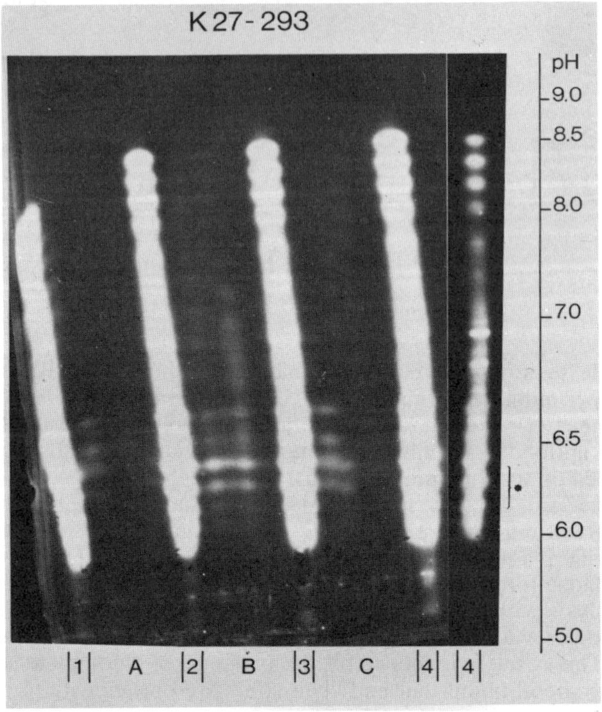

FIG. 2. Analytical IEF patterns of 2° antiserum (1–4) and of pools from microculture supernatants of K27-293 PBL. Patterns were developed by autoradiography after binding of the Av-CHO-tyr-^{131}I. *A,* 6 pools of supernatants from 10 cultures, each containing 10^4 PBL that were cultured; *B,* 6 pools of 20 cultures each with 2×10^4 PBL; *C,* 6 pools of 10 cultures each with 2×10^4 PBL. Asterisk marks the position of the dominant clonotype (14.5 mg Ab/ml) in the antiserum.

present in a frequency of 1 to 3×10^{-5} PBL. From this a recirculating repertoire of 3×10^4 precursor cells specific for the streptococcal group A-variant polysaccharide was estimated for peripheral blood, each of which gave rise to a clone size of 128 to 256 under conditions of microculture (2). This clone size and a secretory rate of 2×10^4 antibody molecules sec^{-1}/cell together with a recirculating repertoire of 3×10^4 precursor cells is not sufficient to explain antibody levels on the order of 30 mg/ml (Fig. 3) in rabbit hyperimmune sera induced against streptococcal vaccines (3,4). Based on the fact that a rabbit may have 10^{10} lymphocytes of which 10^9 are present in the blood, and given the assumption that the distribution of lymphocytes in blood and lymphoid organs is analogous, an increase of specific antibody levels up to 30 mg/ml/week by active immunization at an assumed active secretion time of 7 days per plasma cell would require that each precursor cell approaches a clone size of 3 to 6×10^5, or a total of 9.9×10^9 to 1.9×10^{10} plasma cells. The alternative explanation would take into account an uneven distribution of precursor cells in blood and lymphatic tissue. For example, it cannot be excluded that a residential fraction of lymphocytes is responsible for a major part of antibody production *in vivo,* which occurs mainly in lungs and spleen (5).

Estimation of precursor frequencies and the analysis of clonotypes triggered *in vitro* under conditions of microculture provided information with regard to the principle that might determine clonal dominance. This is illustrated for one experi-

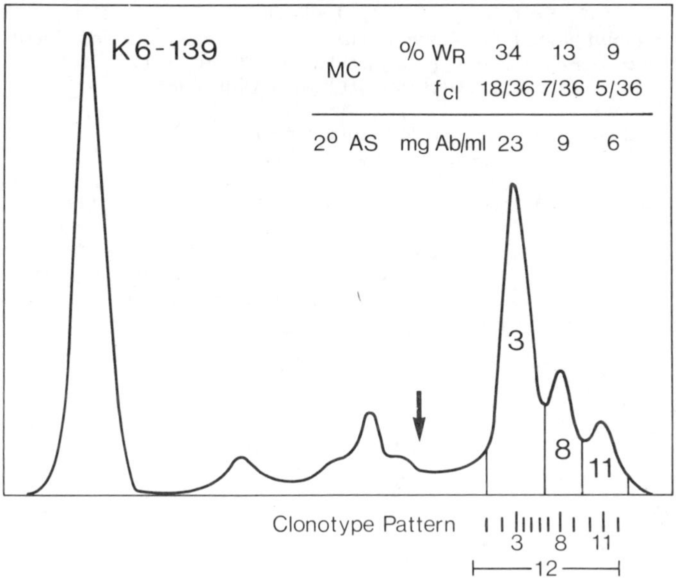

FIG. 3. Densitometric tracing of the microzone electrophoretic pattern of the 2° antiserum (2° AS) of K6-139. The arrow indicates the position of application. A schematic diagram of the number (12) and the position of clonotypes by IEF in the 2° immune serum are shown at the bottom. Clonal antibodies 3, 8, and 11 dominated the *in vivo* response (concentration of clonotype 3 > 8 > 11). MC, microculture; f_{cl}, frequency of clonotypes 3, 8, and 11 identified in microcultures; % W_R, percent of reactive culture-well supernatants by the spot test on Av-CHO-O-St-SRBC that contained clonotypes 3, 8, and 11. The frequency of shared clonotype patterns in the hyperimmune serum and in microculture supernatants suggests a hierarchy in the number of clonal precursor cells.

ment with PBL obtained from rabbit K6-139 (Fig. 3). Here, one precursor cell specific for the streptococcal group A-variant polysaccharide was present among 3.5×10^4 PBL. Isoelectric focusing analysis of culture-well supernatants identified the three major clonotypes 3, 8, and 11 of the hyperimmune sera in 34%, 13%, and 9%, respectively, of all wells that contained antistreptococcal group A-variant antibody. From this the minimal frequencies of the precursor cells for clonotypes 3, 8, and 11 were computed among PBL; according to this computation, cells of clonotype 3 had a frequency of 10^{-5}, those of clonotype 8 a frequency of 4×10^{-6}, and those of clonotype 11 a frequency of 2.6×10^{-6} (2). These data infer that the other nine clonotypes that had been identified in the hyperimmune serum must have had lower precursor frequencies. Therefore, it was concluded that clonal hierarchy established as memory *in vivo* is persistent and penetrant *in vivo* (6) and *in vitro* leading to locked-in clonal antibody patterns (2). Such apparent clonal hierarchy infers a parallel expansion of specific memory cells and cells maturing into plasma cells under conditions of antigen contact (7).

REFERENCES

1. Read, S. E., and Braun, D. G. (1974): *In vitro* antibody response of primed peripheral blood lymphocytes to group A-variant streptococcal polysaccharides. *Eur. J. Immunol.*, 4:422–426.

2. Braun, D. G., Quintans, J., Luzzati, A. L., Lefkovits, I., and Read, S. E. (1976): Antibody response of rabbit blood lymphocytes *in vitro*. Kinetics, clone size and clonotype analysis in response to streptococcal group polysaccharide antigens. *J. Exp. Med.,* 143:360–371.
3. Krause, R. M. (1970): The search for antibodies with molecular uniformity. *Adv. Immunol.,* 12:1–56.
4. Braun, D. G., and Jaton, J.-C. (1974): Homogeneous antibodies: Induction and value as probe for the antibody problem. *Curr. Top. Microbiol. Immunol.,* 66:29–76.
5. Braun, D. G., and Krause, R. M. (1969): Die Morphologie von Kaninchenorganen nach Immunisierung mit Streptokokken-Vaccine. *Z. Immunitaetsforsch.,* 139:104–118.
6. Cramer, M., and Braun, D. G. (1975): Immunological memory: Stable IgG patterns determine *in vivo* responsiveness at the clonal level. *Scand. J. Immunol.,* 4:63–70.
7. Braun, D. G., Huser, H., and Riesen, W. F. (1976): Variability patterns of anti-polysaccharide antibodies. In: *The Generation of Antibody Diversity*, edited by A. J. Cunningham, Academic Press, London.

Antibodies in Human Diagnosis and Therapy,
edited by Edgar Haber and Richard M. Krause.
Raven Press, New York © 1977.

Comments on Immunoglobulin Production by Cell Cultures

Matthew Scharff

I think something should be said about the technical approach. Almost all the cell lines that I know of that stick to glass and produce immunoglobulin produce only very small amounts. One example is that of Paraf. Another is the fusion that Periman reported between MOPC-315 and L-cells. Still another is described by Black in the preceding chapter. I believe these were all selected or obtained as sticking cells.

In these instances, there was the opportunity to look for cells that would not stick. For example, this is true in the fusion situation. Milstein's chapter deals with this point. It may well be that the stuck cells are the wrong ones to consider, that the ability of the cells to flatten out may preordain them to being poor secretory or producing cells. It is much easier to deal with the stuck cells, which is why it is being done. But the proper way to do the experiment may be to use cells that will not stick.

The other thing I want to mention is the problem of using a specific antigen and what you might get. Several years ago, Reuben Baumal, Barry Bloom, and I reported starting human lymphoid cell lines with a variety of antigens. The experiment was to set up a situation where you did not automatically get the cells into culture when you added EB virus to them. Then, to take cells from patients that were sensitive to various antigens, we used patients who showed cellular immunity to SKSD, to tuberculin, or in one case to diphtheria toxoid. We then set up the initiation of cultures using a specific antigen. For someone who was sensitive to SKSD but not to tuberculin, we used SKSD with tuberculin as a control. In all cases we had controls with or without EB virus.

We obtained cell lines only when the antigen to which the donor was sensitive and EB virus were mixed with the cells. We had 100% take of cell lines with the specific antigen, and all of these cell lines produced large amounts of immunoglobulin, more than the usual lymphoid cell lines. In fact, they looked almost like myelomas in the percent of immunoglobulin they were producing. However, in no case were we able to detect any antibody against the antigens that were used. Admittedly, the assays for anti-tuberculin antibody or anti-SKSD antibody are not that sensitive, but this was the result we obtained.

What I assume happened was that we stimulated many T-cells to replicate and produce factors by adding antigen. These factors plus antigen induced the differentiation of B-cells which went into culture and produced large amounts of immunoglobulin. There may have been a few cells producing specific antibody but we were unable to detect them or to clone them out.

Antibodies in Human Diagnosis and Therapy,
edited by Edgar Haber and Richard M. Krause.
Raven Press, New York © 1977.

Antibodies à La Carte

A. R. Williamson

Department of Biochemistry, University of Glasgow, Glasgow G12 8QQ, Scotland

The first attempt to artificially produce antibodies to order may have been Pauling and Campbell's experiment (1) in which mixed γ-globulins were denatured and renatured in the presence of antigen. The authors claimed to have produced antihapten (e.g., *p*-azophenylarsonate) antibody which would precipitate hapten–protein complexes and antipolysaccharides. That experiment was based on the prediction from the instructional hypothesis that the γ-globulin molecule folds to fit the antigen to which the response is being mounted. We are now aware from our knowledge of protein structure that this is not possible.

The question to be posed in this chapter is whether, in the light of current hypotheses concerning antibody production, we can predict ways of producing *in vitro* antibody of given specificity and thereby present a range of antibodies for clinical use. Production of antibodies *in vitro* requires the use of specific components, either protein, RNA, or DNA and so I will also consider the production of these components *in vivo*.

First, let us briefly examine our current understanding or misunderstanding of immune responses. We currently accept a selectionist view of the immune response. Each antibody is produced by a clone of cells and each clone produces a single antibody. Clonal selection is effected initially by antibody, but clonal population dynamics involve other regulating factors. The specificity of the clonal precursor is determined by selection of V genes from a multiple array and joining of one V_L gene to the appropriate C_L gene and one V_H gene to any of the C_H genes. It is as yet unclear whether the entire set of V regions which an individual could potentially express are encoded in germ-line V genes or whether a more limited set undergoes somatic mutations to give rise to more extensive V region diversity upon which antigenic selection acts (2). Somatic variation of the type needed to generate diversity has not yet been observed in selected clones of antibody-forming cells (3,4). Each clone produces a single antibody characterized by the V_L and V_H regions, though possibly differing in class during clonal development. All cells of the clone appear to carry V_L and V_H genes, of normal stability to mutation, coding for the V regions of the antibody produced by the clone. Switches of antibody class during clonal expansion imply either relocation of a V_H gene or a transcriptional control event.

A major contribution to the extent of antibody diversity comes from the association of V_L and V_H region in most possible combinations; thus p V_L regions and q V_H regions could (and probably do) make $p \times q$ different antibody combining sites (5,6). In addition each V_LV_H combination appears to exhibit multiple specificities. An estimate of 100 shared specificities per V_LV_H pair has been offered (7). This reduces the number of antibodies needed to cover the range of diversity of activity. Specificity of induced antibody is accounted for by the population of different molecules present in any given antiserum; binding to the eliciting antigen will be dominant, since this activity is shared by all molecules in the population and cross-reactivity will only be seen if a significant proportion of the molecules present can bind another antigen. Homogeneous antibodies will show cross-reactivities unexpected on the basis of conventional cross reactions seen with heterogeneous antibody populations. This must be borne in mind when we are able to select individual antibodies *à la carte* rather than accepting the *soupe du jour* obtainable upon immunization.

A host of ways of manipulating antibody production both *in vivo* and *in vitro* could be devised and a few will be suggested here. Some are immediately feasible, others appear more futuristic, but if need is sufficient, then technical difficulties will no doubt be surmounted. The most limiting feature in producing antibodies *à la carte* would be finding ways to increase the relatively low level of production of antibody which most methods afford.

PRODUCTION OF ANTIBODIES *IN VIVO*

It is assumed that for human therapy it is desirable to use antibodies of human origin. Production of antibodies in one person for treatment of another has ethical problems, but is not unreasonable in certain instances, e.g., Rh-negative male volunteers might be used to produce anti-Rh antibodies to be used in treatment of Rh-negative mothers immediately postpartum.

In many instances, antibodies raised in heterologous species may be satisfactory. Immunization of outbred animals, and often of different individuals of the same inbred strain, elicits a different population of antibodies each time. Efforts could be made to control the antibody population either by exploiting natural mechanisms or by a more extensive manipulation of the immune system.

Natural Population Control

1. Preferential primary selection of a particular clone is seen in certain inbred strains of mice (8). The initial response to some antigens is dominated by a particular antibody. Good examples are the responses to carbohydrates such as α1:3 dextran (9) and streptococcal A carbohydrate (10), or to the haptens phosphorylcholine (11), *p*-azophenyl arsonate (12), and 3-nitro-5-idio-4-hydroxyphenylacetyl (13). The finding of a preferential clonal response is often associated with a particular inbred strain of mouse and a useful antibody selected in this way would be serendipitous.

Responses to the carbohydrates and to phosphorylcholine apparently show preferential primary clonal selection because the repertoire of antibodies specific for these antigens is small in the appropriate inbred strains. Measurements of the repertoire size can be made (2) and could be of predictive value in selecting a suitable source for antibody of a given specificity.

2. Clonal dominance describes a situation in an ongoing response in which a single clone (or a small number of clones) of antibody-forming cells is preferentially expanded despite the presence of a heterogeneous population of cells capable of responding to the antigen (4,15). The response then consists of antibody which is homogeneous or of very restricted heterogeneity. Clonal dominance is of practical value, since (i) it would appear possible for any clone to be made dominant, (ii) dominance can be established during a multiclonal response, and (iii) dominance can persist for lengthy periods. The factors governing dominance have been discussed elsewhere (8). The antibody produced by a dominant clone is thought to play an important part in maintaining dominance. The experimental establishment of dominance by selected clones would appear to be worth pursuing.

Experimental Selection and Propagation of Clones

1. Monoclonal responses may be selected by cross-immunization with different antigens. This makes use of the multiple shared specificities of antibody combining sites by selection for the characteristic fine specificity of a given site.

2. Suppression of unwanted clones might be effected by the opposite of cross-immunization. Antigen could be coupled to a molecule such as a bacterial polysaccharide to give a tolerogenic form of the antigen (8,16).

3. Cloning can be effected by limiting cell dilution, using cell transfer with subsequent propagation *in vivo* or *in vitro*.

4. Fractionation of clonal precursor cells can be effected by affinity

chromatography (17,18) or on a cell-sorting machine (19). Antibody secreting cells might also be fractionated.

5. Propagation of selected clones by antigen driven clonal expansion could be superseded by transformation of either the clonal memory cells or of the antibody-secreting cells to yield stable cell lines. For human cells Epstein-Barr virus is used to produce lymphoid cell lines (20); for murine cells Abelson virus has been effective both *in vivo* (21) and *in vitro* (22) yielding lymphomas and plasmacytomas, while SV-40 has in one instance been used to transform rabbit antibody-producing cells (23). Cell lines secreting antibody at a high rate would seem the obvious aim. However, of possibly greater value would be a cell line of memory cell origin which maintained the property of responding, either to antigen or a less specific stimulus, and differentiating into a high-rate secreting line. The best analogies here are erythroleukemic cell lines derived by transformation with Friend virus (24). These cells can be induced by a number of agents such as dimethylsulfide (25) or butyric acids (25) to differentiate along the erythroid pathway and produce large amounts of hemoglobin.

CELL-FREE PRODUCTION OF ANTIBODY

Cell-free systems for production of protein molecules start with information contained in macromolecules of specific biological origin. Any of the systems supplied above might be used to supply specific biological information. The many nonspecific components required for the processes of DNA, RNA, or protein synthesis can be obtained from the most convenient source: viral, bacterial, plant, or animal. I will consider systems according to the level at which specific biological information is obtained from the cell.

Polypeptide

The multispecificity of a single combining site suggests that individual antibodies might be selected by one antigen but have useful specificity for an apparently unrelated antigen.

Multiple shared specificities of myeloma proteins have been revealed by an elegant screening method (27). The observed shared specificities of myeloma proteins have been shown to reflect the existence of antibodies which can be elicited by suitable cross-immunization (28). Screening of myeloma proteins and elicited antibodies against antigens of clinical importance and against simple haptens might predict synthetic immunogens to be used to elicit clinically useful antibodies.

Another possibility is to mimic the fact that antibody diversity makes use of most V_L and V_H combinations and to produce new antibody specificities *in vitro* by assembling different combinations of L and H chains (a modern version of the Pauling and Campbell experiment).

The technology here is quite simple. A large number (say 100) human (or murine) myeloma proteins could be purified and the L and H chains prepared. Reassociation of L and H chains in all possible combinations would yield 10^4 antibodies which could be screened for useful activities. Once useful combinations have been discovered other techniques could be sought to perpetuate production of the required L and H chains. Large-scale growth of the myeloma tumor cells in culture would be desirable, but may not be possible for human tumors.

RNA

Isolation of mRNA coding for L and H chains and production of those polypeptides by cell-free translation of the mRNA have been demonstrated, starting from murine myeloma cells as the mRNA source (29–34) and using a variety of translation systems; e.g., rabbit reticulocyte lysate, Krebs ascites extract, a reconstituted system using partially purified components (35) of wheat germ extract or oocytes of *Xenopus laevis*. Generally in these experiments microgram quantities of mRNA are being translated, and, in the most efficient cell-free systems, only a few rounds of translation per mRNA molecule are achieved during approximately 1 hr. One round of synthesis yields a weight of polypeptide of about 10% the weight of mRNA. The exception to this generalization is the oocyte which accepts milligram to microgram amounts of RNA by microinjection (36). The rate of translation for globin mRNA in oocytes is 15–20 molecules/hr per mRNA molecule (37). Translation can be maintained for 2 weeks (38), so conceivably up to 50 μg quantities of polypeptide might be produced per oocyte. It should be noted for comparison that 1 μg of total mRNA might be obtained from 10^7 antibody-secreting cells which would produce antibody at the rate of about 2,000 molecules/cell-sec adding up to 0.5mg/day.

The quantity of specific protein which might be produced in the oocyte is an extrapolation of various estimates. The system has not to my knowledge been used to produce useful amounts of heterologous proteins, and measurements have usually followed incorporation of radioactive amino acids.

The purity of mRNA preparations used in cell-free translation systems should not be critical. While it might be desirable to partially purify immunoglobulin mRNAs, and in some instances, to separate mRNA-L and mRNA-H, it would be easier, and generally satisfactory, to purify the translation products.

With most cell-free systems individual H and L chains are recovered and no assembly to immunoglobulin molecules is observed (the wheat germ system yields large amounts of incomplete polypeptide chains, formed apparently by premature termination) (39). Again the oocyte is exceptional. When total myeloma cell mRNA is injected the translation products assemble into four chain molecules almost as efficiently as in the plasma cell (40).

In none of these systems is mRNA usually recoverable. For extensive increase in the yield of polypeptide from a given amount of isolated mRNA one might attempt to replicate the RNA. In this case it would be more efficient to start with moderately pure mRNA-L or mRNA-H. One method would be to make a complete cDNA copy using viral reverse transcriptase (41), and then to use the cDNA to generate more mRNA. The cDNA could be amplified biologically by making a complementary DNA strand with bacterial DNA polymerase 1 and inserting the double-stranded DNA molecule into a plasmid or λ phage which would then replicate in bacteria (42).

DNA

A single VC gene constitutes a very small fraction (1 part in 10^7) of genome DNA so transcription of naked DNA would yield negligible specific mRNA from that gene. Transcription of chromatin from an antibody-producing cell should continue to show specific transcription of the antibody genes, but such a system *in vitro* is unlikely to be efficient enough for mRNA production on a useful scale.

Purification of specific genes is potentially the most rewarding exercise and is technically feasible. Considerable purification (10^4-fold) of the SV-40 DNA from an artificial mixture with *E. coli* DNA has been achieved by conventional hybridization procedures (43). Ultimate purification and concomitant amplification of a gene can be achieved by molecular cloning techniques. The isolated DNA molecules are linked into bacterial plasmids or into specifically tailored deletion mutants of λ phage, which are then grown in bacteria (42). Clones are selected and screened for the presence of the specific DNA sequence under purification. It should then be possible to grow the phage or plasmid containing the gene in bulk. Such procedures have been successfully used to purify individual genes from *Drosophila melanogaster* (44) and the genes coding for rRNA form *Xenopus laevis* (45).

Transcription and translation of isolated genes *in vitro* would yield the corresponding polypeptide. It would, however, be more efficient to express the gene *in vivo*. Currently it seems unlikely that complete expression of mammalian gene could be effected in a bacterial host so the prospect of being able to have antibody-producing *E. coli* in one's gut is remote.

A more promising vector from the viewpoint of gene expression *in vivo* would be a virus such as SV-40. Suitable deletion mutants of SV-40 have been made (45). Presumably the vector could be used either with or without its transforming capacity. Virus grown in permissive or semipermissive cells would be used for gene amplification and hopefully complete expression. Alternatively, a transforming vector might be used to insert an antibody gene into the genome of the host cell. The hope would then be that clones of antibody-producing cells could be selected.

An intriguing possibility is that V genes may be isolated in a form in which they can be inserted next to chromosomal C genes by the processes used in normal VC joining. Cells might thus be programmed to make a desired antibody specificity. This notion springs from current ideas on the joining of V and C genes which view the V gene as a transposable element analogous to such elements in plant and insect systems (46). Models for V gene excision to form a transposable episome have either assumed a generalized recombination mechanism (47) which would lead to increased diversity, or a specific excision analogous to λ prophage formation. No evidence has been found in amino acid sequences of V regions for a general recombinational model. A specific excision mechanism seems probable. Most models require a specific insertion involving a specific nucleotide sequence in both the episome and the C gene and probably a specific protein (48). Further study of the VC joining process, either at the initial programming of the stem cell or during switches of class as a clone develops, may reveal discrete possibilities for manipulation of the episomal V gene.

Working directly at the DNA level raises the possibility of coupling V genes from animal genomes to C genes of the human genome. This might be effected either by joining the two genes in a viral vector or by inserting the V gene adjacent to a C gene in a human cell.

CONCLUSION

The methods suggested here for production of specific antibodies *in vitro* have ranged from the immediately practical to the futuristic. Most of the methods will only supply small quantities of specific antibody and must remain *hors d'oeuvre* on the menu. The main course may be supplied by fusion of antibody-producing cells with myeloma tumors to yield cell lines producing any desired antibody specificity (49).

REFERENCES

1. Pauling, L., and Campbell, D. H. (1942): *J. Exp. Med.,* 76:211–220.
2. Williamson, A. R. (1976): *Annu. Rev. Biochem.,* 45:467.
3. Milstein, C. et al. (1974). In: *Cellular Selection and Regulation in the Immune Response,* edited by G. M. Edelman. Raven Press, New York.
4. Scharff, M. D. et al. (1975): In: *Molecular Approaches to Immunology,* edited by E. E. Smith and D. S. Ribbons. Academic Press, New York.
5. Gally, J. A., and Edelman, G. M. (1972): *Annu. Rev. Genet.,* 6:1–46.
6. Poljak, R. J. (1975): *Nature,* 256:373–376.
7. Richards, F. F. et al. (1975): *Science,* 187:130–137.
8. Williamson, A. R., Zitron, I. M., and McMichael, A. J. (1976): *Fed. Proc.,* 35 (*in press*).
9. Blomberg, B., Geckler, W. R., and Weigert, M. (1972): *Science,* 177:178–180.
10. Eichmann, K. (1972): *Eur. J. Immunol.,* 2:301–307.
11. Claflin, J. L., and Davie, J. M. (1974): *J. Immunol.,* 113:1678–1684.
12. Pawlak, L., and Nisonoff, A. (1973): *J. Exp. Med.,* 137:855–869.
13. McMichael, A. J. et al. (1976): *Immunogenetics* (*in press*).
14. Haber, E. (1968): *Annu. Rev. Biochem.,* 37:497–520.
15. Askonas, B. A., and Williamson, A. R. (1972): *Nature,* 238:339–341.
16. Mitchell, G. F., Humphrey, J. H., and Williamson, A. R. (1972): *Eur. J. Immunol.,* 2:460–467.

17. Wigzell, H., and Andersson, B. (1969): *J. Exp. Med.,* 129:23–36.
18. Truffa-Bachi, P., and Wofsy, L. (1970): *Proc. Natl. Acad. Sci. USA,* 66:685–692.
19. Bonner, W. A. et al. (1972): *Rev. Sci. Instrum.,* 43:404–409.
20. Steel, C. M. (1972): *J. Natl. Cancer Inst.,* 48:623–628.
21. Sklar, M. D. et al. (1975): *Nature,* 253:550–552.
22. Rosenberg, N., Baltimore, D., and Scher, C. D. (1975): *Proc. Natl. Acad. Sci. USA,* 72:1932–1936.
23. Collins, J. J. et al. (1974): *Proc. Natl. Acad. Sci. USA,* 71:260–262.
24. Friend, C. et al. (1971): *Proc. Natl. Acad. Sci. USA,* 68:378–382.
25. Tanaka, M. et al. (1975): *Proc. Natl. Acad. Sci. USA,* 72:1003–1006.
26. Leder, A., and Leder, P. (1975): *Cell,* 5:319–322.
27. Varga, J. M., Lande, S., and Richards, F. F. (1974): *J. Immunol.,* 112:1565–1570.
28. Varga, J. M., Konigsberg, W. H., and Richards, F. F. (1973): *Proc. Natl. Acad. Sci. USA,* 70:3269–3274.
29. Stavnezer, J. et al. (1974): *J. Mol. Biol.,* 88:43–63.
30. Brownlee, G. G. et al. (1973): *Nature* [*New Biol.*], 244:236–240.
31. Swan, D., Aviv, H., and Leder, P. (1972): *Proc. Natl. Acad. Sci. USA,* 69:1967–1971.
32. Tonegawa, S., and Baldi, I. (1973): *Biochem. Biophys. Res. Commun.,* 51:81–87.
33. Schecter, I. (1974): *Biochemistry,* 13:1875–1885.
34. Mach, B., Faust, C. H., and Vassalli, P. (1973): *Proc. Natl. Acad. Sci. USA,* 70:451–455.
35. Schreier, M., and Staehelin, T. (1973): *J. Mol. Biol.,* 73:329–349.
36. Gurdon, J. B. (1974): *The Control of Gene Expression in Animal Development.* Clarendon Press, Oxford.
37. Gurdon, J. B. et al. (1971): *Nature,* 233:177–182.
38. Gurdon, J. B., Lingrel, J. B., and Marbaix, G. (1973): *J. Mol. Biol.,* 80:539–551.
39. Schechter, I. et al. (1975): *Science,* 188:160–162.
40. Stevens, R. H., and Williamson, A. R. (1972): *Nature,* 239:143–146.
41. Efstratiadis, A. et al. (1975): *Cell,* 4:367–378.
42. Nathans, D., and Smith, H. O. (1975): *Annu. Rev. Biochem.* 44:273–293.
43. Shih, T. Y., and Martin, M. A. (1973): *Proc. Natl. Acad. Sci. USA,* 70:1697–1700.
44. Wensink, P. C. et al. (1974): *Cell,* 4:315–325.
45. Morrow, J. F. et al. (1974): *Proc. Natl. Acad. Sci. USA,* 71:1743–1747.
46. Fincham, J. R. S., and Sastry, G. R. K. (1974): *Annu. Rev. Genet.,* 8:15–50.
47. Gally, J. A., and Edelman, G. M. (1970): *Nature,* 227:341–348.
48. Williamson, A. R., and Fitzmaurice, L. C. (1975): In: *The Generation of Antibody Diversity: A New Look,* edited by A. J. Cunningham. Academic Press, New York and London.
49. Köhler, G., and Milstein, C. (1975): *Nature,* 256:495–497.

Antibodies in Human Diagnosis and Therapy,
edited by Edgar Haber and Richard M. Krause.
Raven Press, New York

A Synthetic Approach for Constructing an Artificial Gene

Johannes Meienhofer

In connection with the suggestions made by Williamson (*preceding chapter*) for incorporating DNA coding for immunoglobulin light chains into SV 40 mutants by transduction, I should like to mention a synthetic approach that is being developed by Dr. A. L. Nussbaum and his colleagues in our department. To explore the feasibility of constructing an artificial gene for introduction into a target cell, a double-stranded DNA fragment coding for the S-peptide-(2-14)-tridecapeptide of bovine pancreatic ribonuclease A was selected as the target (3). This required synthesis of two chains, each 45 nucleotides in length, including minimal start-and-stop control triplets. Chemical oligonucleotide synthesis provided four fragments for each 45-mer (1,2,6,7,10,11), which were subsequently joined enzymatically (5), using polynucleotide ligase from bacteriophage T_4 with the aid of suitable overlap fragments (splints). Both synthetic strands (plus and minus strand, Fig. 1) were characterized by several independent analytical procedures.

We are currently engaged in providing this 45-mer duplex with single-stranded extensions of the 5′-termini. These extensions contain the appropriate recognition sequences (base-pairing complementarity) for Eco-R_1-restricted SV40 DNA (9). After annealing the modified 45-mer with Eco-R_1-restricted DNA, the aim is to convert the resultant circular aggregate into the closed covalent double-stranded form using polynucleotide ligase.

Techniques for the propagation of foreign DNA in prokaryotic (8) and eukaryotic (4) systems are now well advanced. These procedures, it is hoped, will permit the replication of this synthetic "minigene," the study of its transcription controls, if any, and its possible translation into protein.

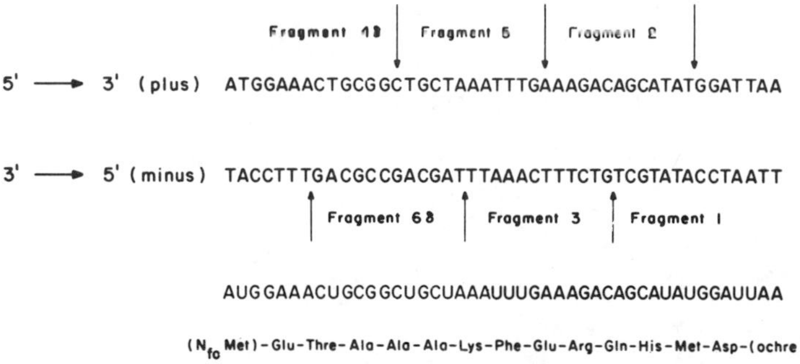

FIG. 1. Relationship of S-peptide-(2-14), corresponding "messenger" (RNA), and synthetic DNA duplex sequences. Chemically synthesized fragments are delineated by vertical arrows.

REFERENCES

1. Cook, A. F., De Czekala, A. M., Gabriel, T. F., Harvey, C. L., Holman, M. J., Michalewsky, J. E., and Nussbaum, A. L. (1973): *Biochim. Biophys. Acta*, 324:433–439.
2. Cook, A. F., Heimer, E. P., Holman, M. J., Maichuk, D. T., and Nussbaum, A. L. (1972): *J. Am. Chem. Soc.*, 94:1334–1340.
3. Finn, F. M., and Hofmann, K. (1973): *Acc. Chem. Res.*, 6:169–176.
4. Ganem, D., Nussbaum, A. L., Davoli, D., and Fareed, G. C. (1976): Submitted to *Cell.*
5. Harvey, C. L., Olson, K., De Czekala, A. M., and Nussbaum, A. L. (1975): *Nucleic Acids Res.*, 2:2007–2020.
6. Harvey, C. L., Wright, R., Cook, A. F., Maichuk, D. T., and Nussbaum, A. L. (1973): *Biochemistry*, 12:208–214.
7. Heimer, E., Ahmad, M., Roy, S., Ramel, A., and Nussbaum, A. L. (1972): *J. Am. Chem. Soc.*, 94:1707–1713.
8. Kedes, L. H., Chang, A. C. Y., Houseman, D., and Cohen, S. N. (1975): *Nature*, 255:533–538.
9. Morrow, J. F., Cohen, S. N., Chang, A. C. Y., Boyer, H. W., Goodman, H. M., and Helling, R. B. (1974): *Proc. Natl. Acad. Sci. USA*, 71:1743–1747.
10. Poonian, M. S., Nowoswiat, E. F., and Nussbaum, A. L. (1972): *J. Am. Chem. Soc.*, 94:3992–3997.
11. Poonian, M. S., Nowoswiat, E. F., Tobias, L., and Nussbaum, A. L. (1973): *Bioorg. Chem.*, 2:322–336.

Antibodies in Human Diagnosis and Therapy,
edited by Edgar Haber and Richard M. Krause.
Raven Press, New York © 1977.

Cell Fusion and the Derivation of Cell Lines Producing Specific Antibody

C. Milstein and G. Köhler

Medical Research Council, Laboratory of Molecular Biology, Hills Road, Cambridge CB2 2QH, England

Immunoglobulins consist of V regions (variable) concerned with antibody specificity and C regions (constant) which define effector functions common to all members of a given class, such as activation of complement, fixation to the skin, etc. It is generally believed that there are separate pools of V and C region genes, and that each chain is the result of the expression of two genes, one from each pool (1). The mechanism whereby this integration is achieved has important implications in the possible *in vitro* manufacture of predefined specific antibodies. At Cambridge we have devoted considerable efforts trying to define the level at which V–C integration is achieved (2–4), and one of the approaches has been the use of cell fusion between immunoglobulin-producing lines.

Since Ig-producing cells normally express only one set of heavy and light chains, our first question was whether it was possible to develop cell lines capable of simultaneous expression of multiple Ig chains. The obvious way to test such a possibility was the use of cell fusion techniques. We soon discovered that fusion between Ig-producing cells was a very low-frequency event, and that it was essential to use powerful selection procedures to eliminate unfused parent cells. We adopted an approach developed by Littlefield (5) and summarized in Fig. 1. Appropriate stable mutants with low reversion rates are required. One (5′-bromodeoxyuridine-resistant line) is probably defective in the enzyme thymine kinase, and the other (8-azaguanine-resistant) in hypoxanthine-guanine phosphoribosyl transferase. They are, respectively, unable to use thymidine and hypoxanthine from the culture medium. Neither of the two cell lines are, therefore, able to grow in selective HAT medium, where HAT symbolizes the simultaneous presence of hypoxanthine, aminopterin, and thymidine. Both cells, however, complement each other when fused, and the hybrid lines are the only ones capable of sustained growth. Other selective procedures have also been developed. In particular, one of them (6) uses a double mutant which can be

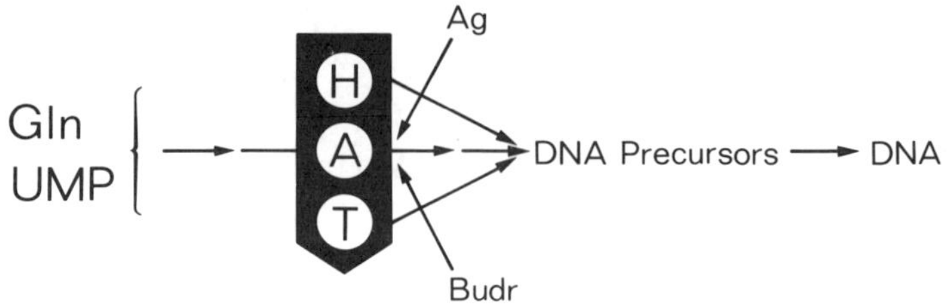

FIG. 1. Selection of myeloma hybrids. The procedure requires suitable parents that are unable to utilize hypoxanthine (H) or thymidine (T) from the medium when the *de novo* synthesis of these DNA precursors is blocked by aminopterin (A). Mutant cell lines are selected by their ability to survive in the presence of the killer analogues 8-azaguanine (Ag) and 5′-bromodeoxyuridine (Budr). Resistant lines, which cannot grow in HAT medium, are selected so that complementary hybrids survive (5).

fused with wild-type cell lines. The double mutant which we have prepared is Budr-resistant (dies in HAT), but is also resistant to high concentrations of ouabain. The hybrid (but not the double mutant) will survive in HAT and in the presence of ouabain, which would kill any wild-type parental line. We have used this procedure in some recent hybridization experiments.

The derivation of cell lines resulting from fusion of two Ig-producing cells (7) demonstrated that the hybrid lines simultaneously expressed all parental Ig-chains. It was therefore possible to investigate putative cytoplasmic events leading to V–C integration.

HYBRID MYELOMA CELLS SCRAMBLE CHAINS BUT DO NOT SCRAMBLE V–C REGIONS

The first successful experiment was done a few years ago with a mouse–rat cell hybrid (Fig. 2). The parental rat cell only produced a kappa chain. The mouse parental line produced a complete immunoglobulin of the kappa type, but synthesis was not balanced and free light chains were also secreted. This system was designed to test V–C integration of the light chains, since both V_L and C_L regions differed in the parental lines. It suffered the disadvantage that any species-specific effects would be missed. In more recent experiments (Fig. 3) we have studied hybrids of two cell lines both from Balb/C mouse origin, which differed in the Ig subclass produced. They also differed in the fact that the kappa chain synthesis was balanced in one line but not in the other.

The results obtained in all cases were similar in the sense that all the parental components were detected in the hybrids, and that, in addition, there were new products. We were able to show that these new products were the result of random association of light and heavy chains from both parental origins. (We use the term random in a rather loose way, only in a qualitative sense.) This was demonstrated by isoelectrofocusing analysis of the separated chains (8) which revealed the parental chains but no extra

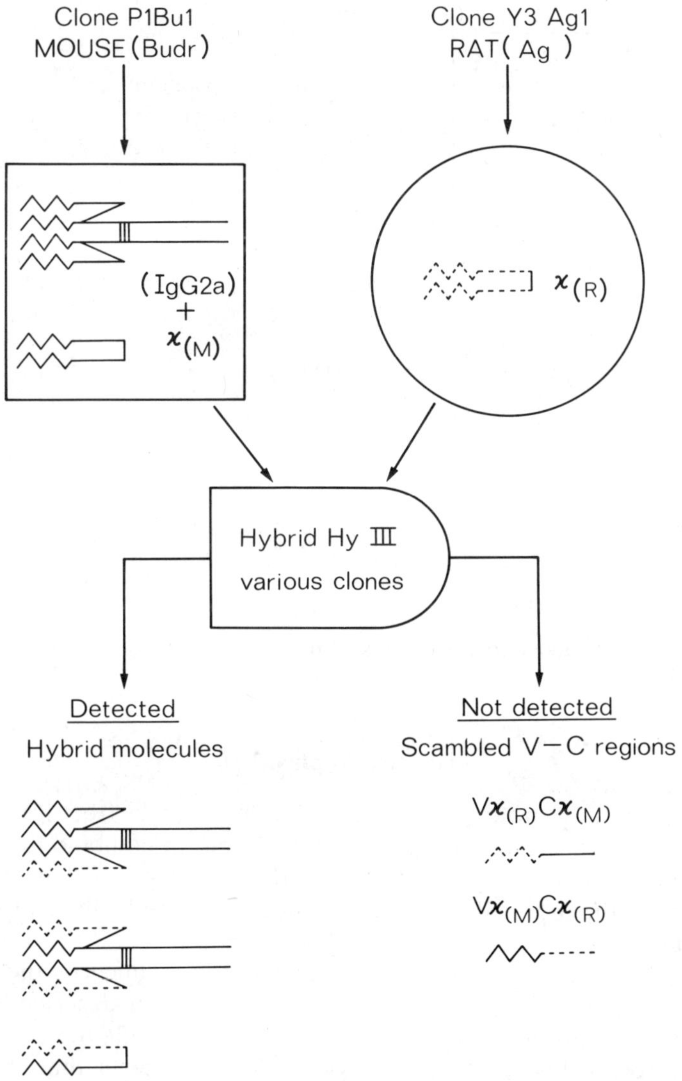

FIG. 2. Immunoglobulin secreted by hybrid cells derived by fusion of mouse and rat myeloma lines (7).

bands. Within the limitations of the experimental analysis, we concluded that no new V–C combinations occurred and that the new products were the result of new combinations of light and heavy chains. Such new components could not be detected by mixing parental cells; therefore, they were characteristic products of hybrid cells. Such hybrid immunoglobulins must be the result of intracellular assembly, and point to the fact that *each cell* is capable of multiple chain synthesis.

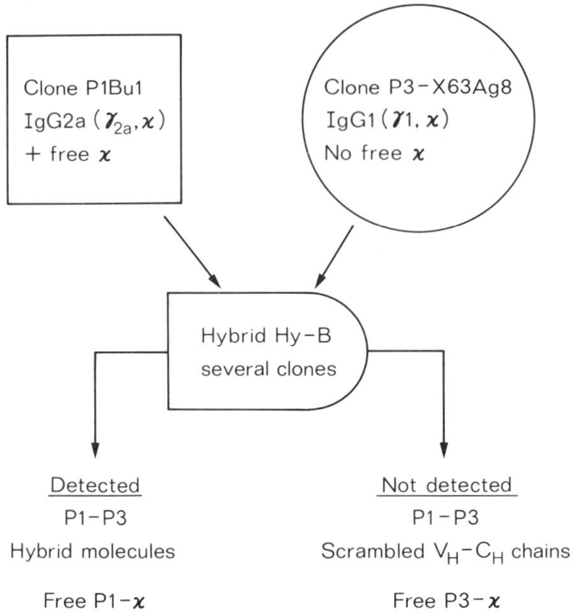

FIG. 3. Immunoglobulins secreted by hybrid cells derived by fusion of two Balb/C myelomas (8).

Genetic Implication

It is interesting to note that, even in the presence of free light chains of mouse origin (Fig. 2), the rat light chain was able to combine to the mouse heavy chain, which gives an indication that there may be no significant restrictions to the intracellular assembly of any theoretically available H–L chain pair.

Another point brought out by these experiments concerns the balanced synthesis of light and heavy chains. In the experiment described in Fig. 3 one parental line produced an excess of light chains, while the other did not. Neither property was lost in the hybrid. This suggests that the rate of synthesis and/or processing of light chains is not under the control of structures which can equally recognize both kappa chains. It is possible that the V region itself plays a critical role in this control.

As far as the mechanism of V–C integration is concerned we were unable to detect "scrambled" chains involving interchange between a V region of one parent and a C region from the other. This indicates that the integration does not involve diffusible transcription products. We have presented structural evidence for the existence of integrated mRNAs for both light (3) and heavy (2) chains. The results on hybrid cells indicate to us that the RNA precursors of the cytoplasmic mRNA are also already integrated. Integration therefore appears to be at the DNA level, either by a translocation event or during transcription. The final solution to this problem must come from an analysis of the linear arrangement of the DNA. The question is whether in antibody-committed cells the expressed V gene is covalently contiguous to the C gene. If so, this may be an example, which, even if special, is nevertheless very important, in which differentiation from stem cell to plasma cell is the result of changes in the primary structure of the DNA.

THE DERIVATION OF HYBRID LINES SECRETING ANTIBODY OF A PRESELECTED SPECIFICITY

From a more practical point of view, the experiments described above suggested to us that it ought to be possible to fuse myeloma cells and normal antibody-producing cells, so that the hybrid cell could carry the antibody-producing capacity of one parent and the ability to grow *in vitro*, or as tumors, from the other parent. For a number of reasons we chose sheep red blood cells (SRBC) as the antigen. Spleen cells from immunized Balb/C mice were fused with an azaguanine-resistant cell line derived from MOPC 21 plasmacytoma. The cultures were kept for several weeks in tissue culture selective medium in which only hybrid cells multiply. The presence of anti-SRBC antibody-producing cells among the hybrids was monitored by Jerne plaque assays (9). Clones were grown in soft agar and tested for specific antibody production by a suitable modification (8) of the assay procedure. Purified clones (Fig. 4) were used for further experiments. When purified clones were recloned, over 90% of the subclones could be shown to be active against SRBC. However, when individual cells were tested, over 50% of the cells failed to give plaques. Considerable differences in size and morphology of the individual plaques were apparent. These observations were somewhat surprising. We do not know whether they are due to technical problems or whether they reflect real differences in antibody production at different stages of cell growth.

Three SRBC-specific hybrid lines, derived from different pools of spleen cells, have been studied in some detail and code named Sp1, Sp2, and Sp3 (Fig. 5). (Further numbers added to these code names indicate the steps in the purification of derived clones.) The nature of the secreted products has been studied by isoelectric focusing the supernatant fluid of cells cultured in medium containing radioactive amino acids. The pattern of radioactive bands revealed by autoradiography is very complex, probably the result of hybrid molecules containing chains from both parental origin (8). Bands which correspond to the MOPC 21 parental pattern are detectable, although they represent a small proportion of the total radioactivity. Upon reduction and separation of chains, however, the patterns are largely simplified. The MOPC 21 light and heavy chain components become very prominent. Also prominent is the presence of other components, presumably representing

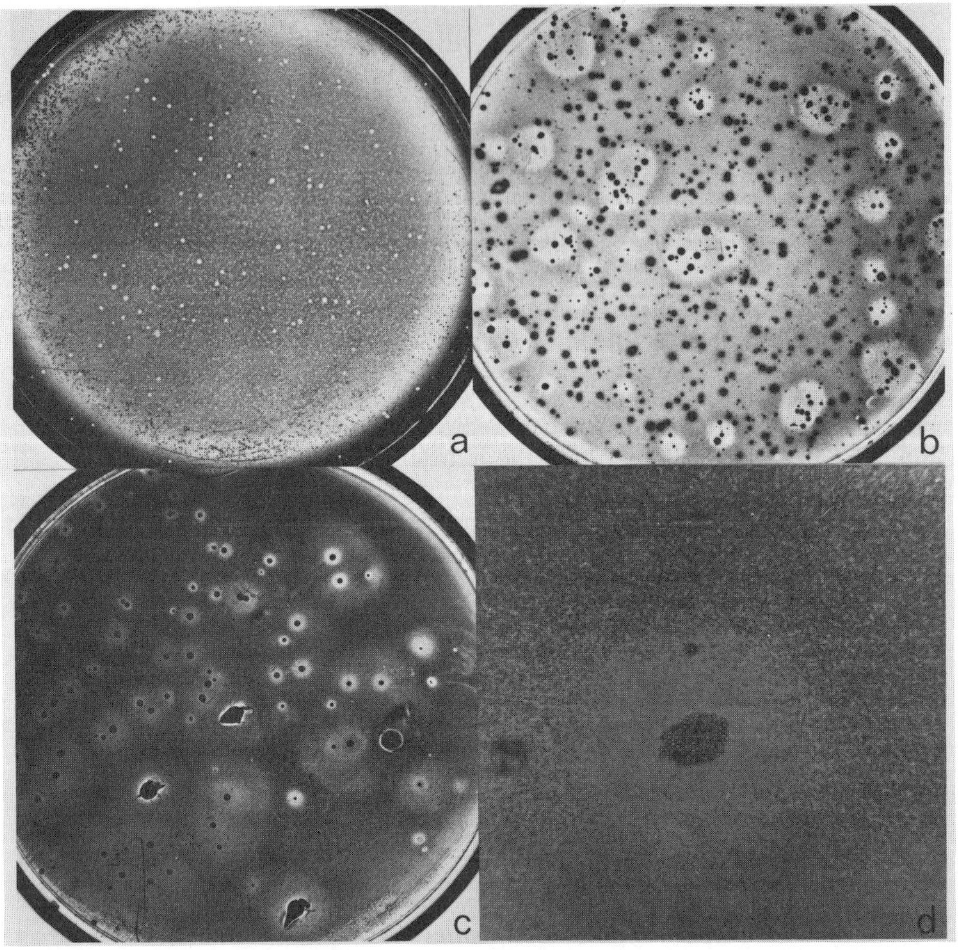

FIG. 4. Detection and isolation of hybrid clones secreting anti-SRBC antibodies. **a:** Single cell-plaques produced by hybrid cells. **b:** Plaques produced by unfractionated (Sp1) hybrid clones grown in soft agar. Clone-plaques are revealed by a modification (8) of the classical Jerne plaque assay (9). The black spots are clones and the halos show areas of lysis around specific clones secreting anti-SRBC antibodies. **c:** Clone-plaques produced by clone Sp1/7. **d:** Clone-plaque seen under the microscope. (Taken from ref. 8.)

the Ig chains which are responsible for the anti-SRBC activity (Figs. 6 and 7).

Variant clones which have lost the capacity to lyse SRBC can be easily detected and isolated. We have studied some such variants and, as expected, found that the loss of antibody activity was associated with the loss of some bands. In Fig. 6 it can be seen that the loss of activity was associated with the loss of a heavy chain from clone Sp2/3-3.

Studies of the class of antibody produced by the different clones are in

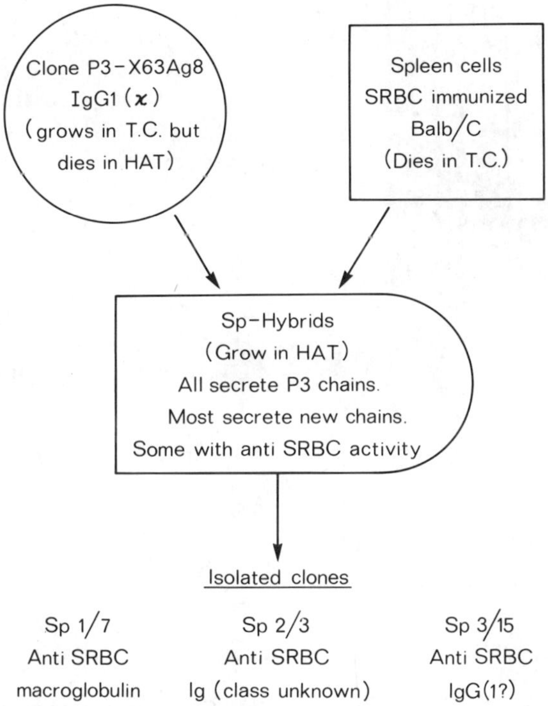

FIG. 5. Immunoglobulins secreted by hybrid cells derived by fusion of a Balb/C myeloma line and spleen cells from a Balb/C mouse immunized with sheep red blood cells (SRBC). TC, tissue culture (8).

progress. Clone Sp1/7 gave direct plaques, and this suggested that it produced an IgM antibody. Lysis is inhibited by anti-IgM and the secreted products include a component whose mobility in SDS gels in the presence of mercaptoethanol indicates a μ chain. Clone Sp3/15 gave indirect plaques which could be developed by an anti-MOPC 21 antiserum. This was also true when the antiserum was absorbed with an IgG2a(κ) myeloma protein (AdjPC5), while an anti-AdjPC5 antiserum absorbed with MOPC 21 protein gave a negative result. Sp3/15 may represent, therefore, an IgG1 antibody, and we are eager to subject the protein to chemical typing to confirm it.

Clone Sp2/3 also gave indirect plaques, and variant clones with no antibody activity have been isolated. The IEF pattern of the secreted products shows the disappearance of a heavy background as their salient feature (Fig. 6, left). In SDS-reducing gels the corresponding heavy chains appear as a doublet of somewhat higher molecular weight than the γ1 chain (Fig. 6, right). Indirect plaques were developed with unabsorbed antiserum, but when the antisera were absorbed with nonhomologous class proteins the lysing activity was abolished. The antibody produced by this clone did not seem to be an IgG1, IgG2a, IgG3, IgA, or IgM. Preliminary results showed

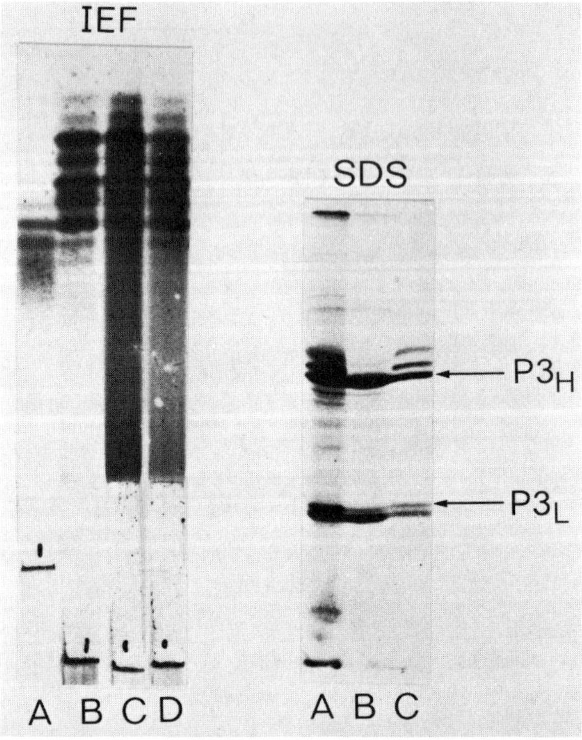

FIG. 6. Autoradiograph of labeled components secreted by purified subclones of hybrid line Sp2/3-3. **Left: A,** P3 paternal clone; *B*, *C*, and *D*, three subclones of Sp2/3-3. One of these subclones (Sp2/3-3-4, gel *B*) was selected as a variant clone which had lost the anti-SRBC activity. **Right:** A comparison of the secreted products of clone Sp2/3-3 (*A*) and of its partially purified anti-SRBC antibody (*C*) with the inactive subclone Sp2/3-3-4 (*B*). The arrows show the approximate position of nonradioactive chains from purified MOPC 21 (*P3*) myeloma protein run as control in the same gel. Fractionation was by isoelectric focusing (pH 5–9) **(left)** and acrylamide gel electrophoresis in SDS under reducing conditions **(right)**.

that subclass specific IgG2b antiserum (kindly provided by Dr. Parkhouse) revealed indirect plaques, suggesting that, in this case, the antibody may be of the IgG2b subclass. The fact that what seems to be the component responsible for the activity appears as a streak in isoelectric focusing suggests that it is a fairly insoluble molecule. Whatever the final identification of classes and subclasses is, it is clear that we have been able to isolate clones producing antibodies directed against a predetermined antigen and carrying different effector functions.

In these early studies we have used a myeloma parental line which itself produced an immunoglobulin. For practical purposes the presence of the MOPC 21 parental chains is an inconvenience. From our model studies with the two myeloma parents (HyB hybrids, Fig. 3), it was clear that variants

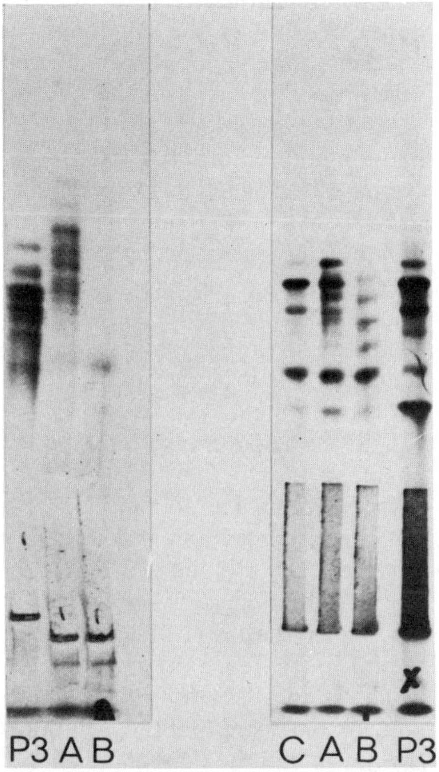

FIG. 7. Isolation of subclones which have lost the paternal MOPC 21 H chain. Subclones of Sp1/7-2-6 were screened by isoelectric focusing for changes in the pattern of secreted radioactive components as described previously (Fig. 6) **(left)** and by isoelectric focusing under reducing conditions which separate heavy and light chains **(right)**. *P3*, parental line; *A* and *B*, subclones of the hybrid Sp1/7-2-6 giving direct plaques; *C*, a subclone which has lost this activity. IEF does not reveal the 19S IgM in the unreduced samples **(left)**. The components in the reduced samples **(right)** present in pattern *A* and *C*, but absent in *B*, represent the MOPC 21 heavy chain.

which have lost one or another parental chain were reasonably frequent (8). Although no precise measurement of such frequency has been made, we estimate this to be in the order of 10^{-2}–10^{-3}. By random screening of clones by the isoelectric focusing procedure we have already been able to isolate variant clones (Fig. 7) from the three anti-SRBC hybrid lines which have lost the expression of the parental MOPC 21 heavy chain. Therefore, selection of lines in which only the specific antibody is expressed does not seem to be a problem.

HYBRIDS BETWEEN Ig SECRETING CELLS AND NONSECRETING VARIANTS

To avoid the presence of inactive immunoglobulin, the parental myeloma tissue culture cell line could be replaced by cell lines of other origins. However, not all lines may be suitable, and relevant model experiments have been reported using fibroblasts (10,11), L cells (12), and lymphomas (13).

An alternative way to avoid the presence of nonspecific antibody would be the use of nonsecreting variants of myeloma lines (14,15) as parental cells for fusion. But first it has to be shown that the "mutation" which led to nonexpression of the myeloma protein does not hinder the expression of another immunoglobulin in a hybrid cell. Further, it must be shown that fusion does not restore the expression of the lost chains.

Model experiments with known parental lines were done in two sets of experiments. Clone P3-NSI-1/Ag4-1 was derived from a MOPC 21 nonsecreting clone which had been studied previously (16) and shown to synthesize P3(κ) light chain. This chain is destroyed intracellularly and no secreted immunoglobulin could be detected. The clone was fused with P1Bu1 which secretes IgG2a and excess P1(κ) light chains (Figs. 8 and 9). Conclusions drawn from the results were: (i) all the chains expressed in the parental lines remained so in the hybrids; (ii) the silent heavy chain of the parental P3 line remained silent in the hybrid; (iii) the ability to secrete P3 light chain was restored in the hybrid, but only as part of the IgG2a molecule. In the hybrid cell, as in the parental clone, the P1 light chain could be secreted as free light chains.

Another nonsecretor clone, P3-NSIII-2, belongs to the type which does not contain detectable amounts of intracellular light or heavy chains. It was derived from a clone (P3-NSII-1, ref. 16) which was shown to contain an inactive light-chain mRNA. This clone was hybridized with clone P1Bu1-Ou1 which dies in HAT and is resistant to high ouabain concentrations. The hybrid line is resistant to ouabain and grows in HAT (Fig. 8). The secreted products of these hybrids were identical to those of the P1 parent. The conclusions were that the "mutations" in the NSIII variant leading to loss of the heavy and the light chains could not be complemented in the fused cells. Nonsecreting lines of this type therefore appear to be suitable to fuse with cells from an immunized donor to produce hybrid lines capable of producing pure antibody-specific molecules.

FUTURE PROSPECTS

The experiments which we have described suggest that hybridization techniques may be a method of choice to derive antibody-producing cells capable of permanent growth in tissue culture or as tumors in animals. All of the anti-SRBC-producing clones described above are easy to grow in Balb/C mice as tumors, because we have used Balb/C parental cells. The serum from these tumor-bearing mice have high titers of the specific antibody. If cells from different genetic backgrounds are used, it seems likely that they will grow as tumors in nude mice.

In the fusion experiments described, the fraction of hybrid cells which had anti-SRBC activity was quite high. In one experiment, from 157 independent hybrids, 15 had anti-SRBC activity. We are at present extending these studies to other antigens to establish that the procedure has general applicability. Preliminary studies with the hapten TNP have been encouraging and

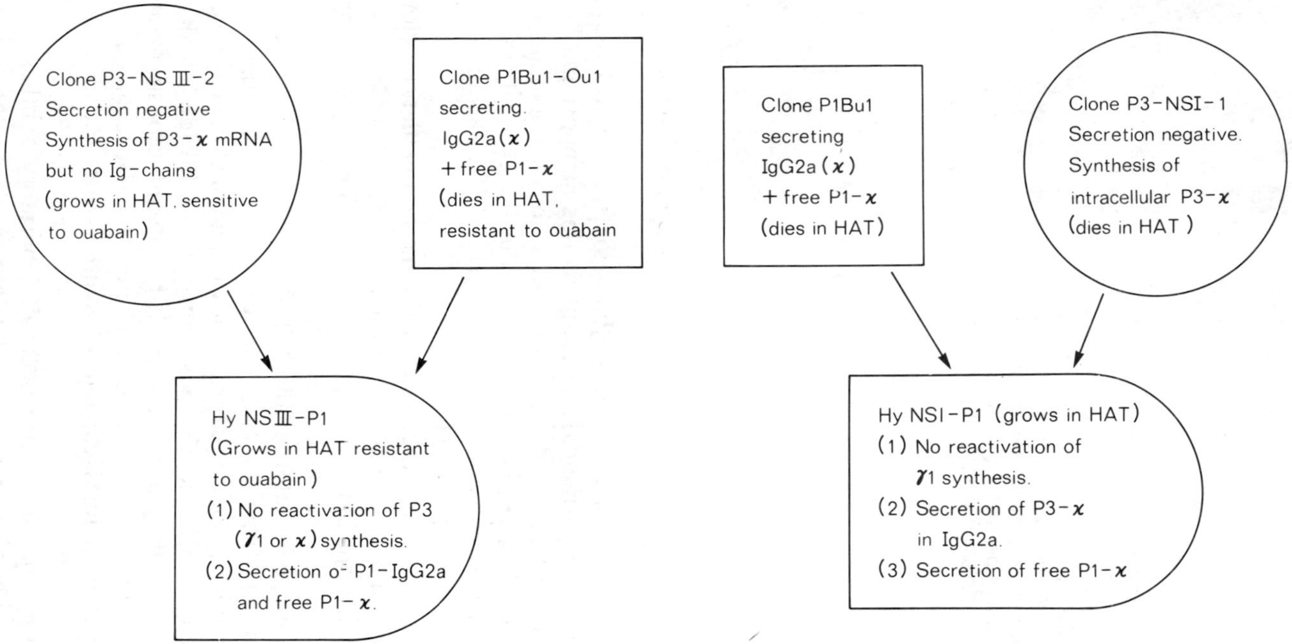

FIG. 8. Immunoglobulins secreted by hybrid cells derived by fusion of secreting and nonsecreting variants of myeloma cells.

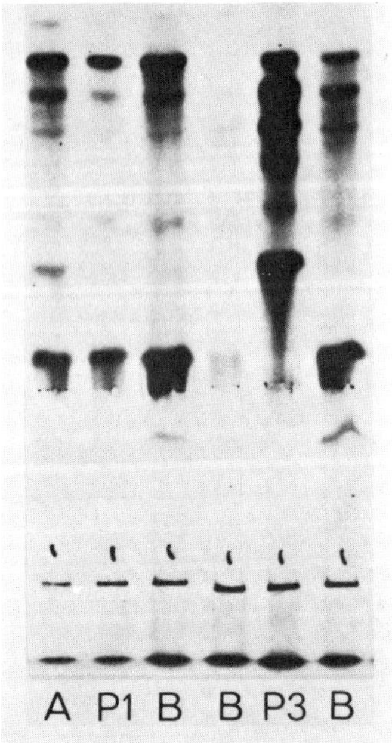

FIG. 9. Autoradiograph of labeled components secreted by the hybrid clones described in Fig. 8. Fractionation was by IEF under reducing conditions which separate heavy and light chains. *P1* and *P3* are the secreted chains from the parental wild-type lines. Products from different hybrid clones: HyNSI-P1 (*A*) and HyNSIII-P1 (*B*).

TNP hybrid clones have been detected, but not yet isolated and established as stable pure lines.

The possibility of using similar procedures to derive hybrid lines secreting human antibodies is particularly relevant to this symposium. There are a number of practical problems to be considered; for instance, the availability of suitable cells from immune donors and the preferential loss of human chromosomes in cases of mouse–human hybrids (24). Relevant model experiments are very encouraging (17).

One of the most powerful features of the method is that, by cloning, it is easy to derive cell lines synthesizing pure antibody molecules, starting with a heterogeneous population of hybrids. This becomes an ideal approach to study complex antigenic structures without the requirement of elaborate separation of antigenic determinants. This approach could be very valuable to the preparation of specific reagents; for instance, for antigens related to the histocompatibility locus. A further advantage of established cell lines is their unlimited supply and the possibility of worldwide standardization. Other products of medical interest might be advantageously manufactured starting from permanent tissue culture lines. Fusion techniques could be valuable in preparing lines capable of synthesizing, for example, human hormones.

In considering the future prospects of laboratory synthesis of specific antibodies, we would like to say a few words about the use of genetic engineering technology to develop suitably modified bacterial systems. This has obvious economic potential. There are a number of problems which will have to be solved in this context. Some of these problems apply not only to antibody synthesis, but more generally to the synthesis of any eukaryotic protein in bacterial systems. Problems, for instance, related to the differences between eukaryotes and prokaryotes in terms of transcriptional and translational machinery and to the specific selection of the correct DNA fragments to be inserted in the appropriate bacterial host. But, in addition, there are specific difficulties concerning the antibody system which are worth mentioning. Ig chains are synthesized on membrane bound polysomes (18–20). Free polysomes are also active, but the synthesized products in cell-free systems contain an extra fragment in the light (21) [and probably heavy (20,22)] chains. A membrane enzyme may be required for the conversion of these precursor molecules (23). Furthermore, the importance of subcellular structures for correct assembly of Ig chains is not fully understood. More important is that V and C genes are unlikely to be contiguous in germ-line DNA. Hopefully, differentiation from stem cells to plasma cells proceeds through changes in the primary sequence of the DNA. An integrated DNA fragment could then be isolated and inserted in the bacterial host. The heterogeneity of the plasma cell population makes the isolation of such a fragment practically impossible. If the factor required for the purification of a germ-line gene is 10^6, the purification factor for the isolation of a specific somatically integrated V–C gene, using spleens from immune donors as starting material, may be 10^9–10^{12}. This assumes that in this tissue the frequency of a specific plasma cell expressing the required antibody gene is between 10^{-3} and 10^{-6}. We hope that the isolation of hybrid lines which carry these specific genes will go a good way toward simplifying this formidable task.

REFERENCES

1. Millstein, C., and Munro, A. J. (1973): Genetics of immunoglobulins and of the immune response. In: *Defence and Recognition,* edited by R. R. Porter, pp. 199–228. Butterworths, London.
2. Cowan, N. J., Secher, D. S., and Milstein, C. (1976): Purification and sequence analysis of the mRNA coding for an immunoglobulin heavy chain. *Eur. J. Biochem.,* 61:355–368.
3. Milstein, C., Brownlee, G. G., Cartwright, E. M., Jarvis, J. M., and Proudfoot, N. J. (1974): Sequence analysis of immunoglobulin light chain messenger RNA. *Nature*, 252:354–359.
4. Milstein, C., Secher, D. S., Cowan, N. J., Harrison, T. M., Cotton, R. G. H., and Brownlee, G. G. (1974): Immunoglobulin mRNA, immunoglobulin mutants or the integration of two genes into one polypeptide. In: *Cellular Selection and Regulation in the Immune Response*, pp. 245–264, edited by G. M. Edelman. Raven Press, New York.
5. Littlefield, J. W. (1964): Selection of hybrids from matings of fibroblasts *in vitro* and their presumed recombinants. *Science*, 145:709.
6. Baker, R. M., Brunette, D. M., Mankovitz, R., Thompson, L. H., Whitmore, G. F., Siminovitch, L., and Till, J. E. (1974): Ouabain-resistant mutants of mouse and hamster cells in culture. *Cell*, 1:9–21.
7. Cotton, R. G. H., and Milstein, C. (1973): Fusion of two immunoglobulin-producing myeloma cells. *Nature*, 244:42–43.

8. Köhler, G., and Milstein, C. (1975): Continuous cultures of fused cells secreting antibody of predefined specificity. *Nature*, 256:495–497.
9. Jerne, N. K., and Nordin, A. A. (1963): Plaque formation in agar by single antibody-producing cells. *Science*, 140:405.
10. Bevan, M. J., Parkhouse, R. M. E., Williamson, A. R., and Askonas, B. A. (1972): Biosynthesis of immunoglobulins. *Prog. Biophys. Mol. Biol.*, 25:131–162.
11. Coffino, P., Knowles, B., Nathenson, S. G., and Scharff, M. D. (1971): Suppression of immunoglobulin synthesis by cellular hybridization. *Nature* [*New Biol.*], 231:87–90.
12. Periman, P. (1970): IgG synthesis in hybrid cells from an antibody-producing mouse myeloma and an L cell substrain. *Nature*, 228:1086–1087.
13. Mohit, B. (1971): Immunoglobulin G and free kappa-chain synthesis in different clones of a hybrid cell line. *Proc. Natl. Acad. Sci. USA*, 68:3045–3048.
14. Coffino, P., and Scharff, M. D. (1971): Rate of somatic mutation in Ig production by mouse myeloma cells. *Proc. Natl. Acad. Sci. USA*, 68:219–223.
15. Cotton, R. G. H., Secher, D. S., and Milstein, C. (1973): Somatic mutation and the origin of antibody diversity. Clonal variability of the immunoglobulin produced by MOPC 21 cells in culture. *Eur. J. Immunol.*, 3:135–140.
16. Cowan, N. J., Secher, D. S., and Milstein, C. (1974): Intracellular immunoglobulin chain synthesis in non-secreting variants of a mouse myeloma: Detection of inactive light chain mRNA. *J. Mol. Biol.*, 90:691–701.
17. Schwaber, J., and Cohen, E. P. (1974): Pattern of immunoglobulin synthesis and assembly in a human-mouse somatic cell hybrid clone. *Proc. Natl. Acad. Sci. USA*, 71:2203–2207.
18. Harrison, T. M., Brownlee, G. G., and Milstein, C. (1974): Studies on polysome–membrane interactions in mouse myeloma cells. *Eur. J. Biochem.*, 47:613–620.
19. Vassalli, P., Lisowska-Bernstein, B., Lamm, M. E., and Benacerraf, B. (1967): Studies on cell-free synthesis of rat immunoglobulins, II. Synthesis of immunoglobulin and of antibody to the dinitrophenyl hapten. *Proc. Natl. Acad. Sci. USA*, 58:2422–2429.
20. Schmeckpeper, B. J., Cory, S., and Adams, J. M. (1974): Translation of immunoglobulin mRNAs in a wheat germ cell-free system. *Mol. Biol. Rep.*, 1:335–363.
21. Milstein, C., Brownlee, G. G., Harrison, T. M., and Mathews, M. B. (1972): A possible precursor of immunoglobulin light chains. *Nature* [*New Biol.*], 239:117–120.
22. Cowan, N. J., and Milstein, C. (1973): The translation *in vitro* of mRNA for immunoglobulin heavy chains. *Eur. J. Biochem.,* 36:1–7.
23. Cowan, N. J., Harrison, T. M., Brownlee, G. G., and Milstein, C. (1973): The cell-free synthesis of immunoglobulin chains. *Biochem. Soc. Trans.,* 1:1247–1250.
24. Ruddle, F. H. (1973): Linkage analysis in man by somatic cell genetics. *Nature,* 242:165–169.

Present and Future Applications of Humoral Antibodies in Human Therapy

Antibodies in Human Diagnosis and Therapy,
edited by Edgar Haber and Richard M. Krause.
Raven Press, New York © 1977.

Antibodies and the Cancer Problem

Herman N. Eisen

Center for Cancer Research and Department of Biology, Massachusetts Institute of Technology and the Harvard–MIT Joint Program for Health Sciences and Technology, Cambridge, Massachusetts 02139

Tumor immunology rests on three propositions: (i) Tumor cells have distinctive antigens ("tumor-associated antigens," TAAs); those on the cell surfaces are particularly significant because of their accessibility to antibodies, lymphocytes, and macrophages. (ii) TAAs can induce diverse immune responses in the autochthonous host or, in the case of transplantable tumors, in syngeneic hosts; the responsiveness of these hosts indicates that TAAs are either totally lacking in normal cells, or so rarely expressed by normal cells that they do not establish solid tolerance. (iii) Host immune responses to TAAs exert a major influence on the growth and metastases of tumors. By analogy with immune responses to invasive microbes or tissue allografts, it is assumed that the main effect of immune responses to TAAs is to suppress tumor growth.

These propositions lead to two expectations: (i) Individuals with pronounced immunological deficiencies—due either to congenital or genetic abnormalities or to immunosuppressive therapy—should be especially prone to develop cancer. (ii) Appropriate immune responses against tumor antigens, if induced at the proper time, should have a protective or even a therapeutic effect in controlling cancer.

In this chapter I should like to examine some of these notions, particularly with regard to the role of antibodies in the detection of tumor antigens and in the modulation of tumor cell growth. We begin with TAAs, which are conveniently discussed from the point of view of chemically induced tumors, virus-induced tumors, and spontaneous tumors.

TUMORS INDUCED BY CHEMICAL CARCINOGENS

TAAs on these tumors were first demonstrated by transplantation tests (1–4). Chemically induced transplantable tumors were used to establish immunity in syngeneic hosts, either by injecting irradiated tumor cells (intact but unable to divide) repeatedly or by allowing a small transplant of viable cells to grow into a small (~1 cm) tumor mass, which was then excised or destroyed by ligation at the base (obliterating nutrient blood vessels). Both procedures induced transplantation resistance: treated animals rejected approximately 10- to 1,000-fold more tumor cells than sufficed to establish progressively growing (eventually fatal) tumors in most untreated controls.

The striking feature of the induced resistance was its specificity. The rejection reaction was manifest only with the particular tumor used as immunogen. Other tumors, induced by the same carcinogen in the same inbred strain, or even in the same autochthonous host, were generally not rejected, or only infrequently (1–3). The uniqueness of TAAs on these tumors has been even more strikingly emphasized in the recent report that treatment of a clone of cultured fibroblasts with 3-methylcholanthrene yielded 17 transformed subclones, each able to induce individually specific transplantation immunity (5).

The chemical and genetic basis for the great diversity of TAAs on these tumors is totally mystifying. These antigens are not essential for the malignant phenotype: this is evident both from their diversity and from the retention of the malignant phenotype after these TAAs are lost, as sometimes happens during prolonged serial passage through syngeneic hosts (probably as a result of immune selection). Attempts to isolate and characterize these antigens have been hindered by difficulties in preparing potent monospecific antisera.

In addition to their individually unique TAAs, tumors induced by chemical carcinogens often have in common other cell surface TAAs. Some of these shared antigens are of viral or fetal origin (see below).

TUMORS INDUCED BY VIRUSES

With respect to TAAs, these tumors are of two kinds: those that continue to produce oncogenic virus (RNA tumor viruses of mice, chickens, etc.) and those that do not (tumors induced by polyoma, SV40, and other transforming DNA viruses). RNA virus particles budding from the cell surface are coated by virion envelope glycoproteins, which can act as TAAs even when the virus is endogenous (i.e., part of the strain's normal genome) and is expressed in the normal host (6). Some of these tumors also produce an additional cell surface glycoprotein which seems not to be on or in mature virions (glycoprotein of 100,000 daltons, gp 100, of Rous sarcoma infected cells; ref. 7). Thus tumors induced by a particular oncogenic RNA virus usually share common glycoprotein TAAs; these are relatively strong immunogens, eliciting formation of cytotoxic antibodies in syngeneic hosts (see below).

Nonproductive tumors due to DNA oncogenic viruses also have surface TAAs that are the same in the tumors induced by a particular virus. Even the tumors induced in different species (e.g., in hamsters and in mice by SV40) can share a surface TAA, suggesting specification by the viral genome (8).

Viral TAAs on tumor cells may, however, reflect a coincidental rather than a causal infection. For instance, cells transformed by a murine sarcoma virus may lack demonstrable TAAs and acquire those of the murine leukemia (MLV) type only after a subsequent productive infection by MLV (e.g., ref. 9). In another example, a rat sarcoma induced by a chemical carcinogen acquired strongly immunogenic TAAs of the Friend virus, due to a subsequent infection by that virus (10). In our laboratory, we are currently studying 3-methylcholanthrene-induced murine leukemias with highly immunogenic TAAs as a result of MLV production, which is probably due to activation of an endogenous leukemia virus (11).

Immunization with irradiated fetal cells can generate transplantation resistance to some SV40-induced tumors and can diminish the incidence of these tumors in hamsters treated at birth with SV40 virus (12). This suggests that some of the TAAs on the virus-induced tumor cells are *fetal antigens*: i.e., cell surface antigens that are expressed normally on certain fetal cells during particular phases of embryonic development and little if at all in adults unless they bear certain tumors (12). Fetal antigens also seem to be present on chemically induced tumors. The best known tumor-related fetal antigens—carcinoembryonic antigen (CEA) and α-fetoprotein (AFP)—are secreted profusely: their detection in human sera, rather than on cell surfaces, provides the basis for widely adopted diagnostic procedures (13,14).

SPONTANEOUS TUMORS

Spontaneously arising carcinomas and sarcomas of mice and rats generally seem to lack demonstrable TAAs, either by serological or transplantation tests (15–17). The reasons may derive from the rule, based on observations of chemically induced tumors, that the longer the latent period between application of carcinogen and appearance of a tumor mass, the less immunogenic the resulting tumor (4). Long latent period offers increased opportunity for host immune selection against the more immunogenic TAAs. This would favor the emergence of tumor variants whose TAAs have little or no immunogenicity in the autochthonous host (or in syngeneic animals). Spontaneous tumors could well have exceptionally long latent periods.

Reduced immunogenicity could arise from the paucity of TAAs per cell, lack of suitable carrier determinants, or lack of an effective immune response (Ir) gene. The last possibility is illustrated by the so-called X.1 antigens of certain mouse leukemias: these TAAs are immunogenic in BALB/c × C57BL/6 hybrids but not in purebred BALB/c, the strain of origin (18). It is also possible that tumors appearing after a long period of latency are selected for reduced susceptibility to immune destruction, rather than for reduced immunogenicity of their TAAs.

Whatever the reason, spontaneously arising tumors of inbred animals represent particularly cogent models for human tumors, which also generally elicit feeble immune responses (if any) in autochthonous hosts. The key to immunological manipulation of all these tumors may well rest on developing the ability to cope with exceedingly weak TAAs.

MOUSE MYELOMA TUMORS

TAAs on myeloma tumor cells seem to be exceedingly weak (or even nonexistent), and in the following sections we wish to consider how some of the foregoing generalities apply to these tumors (plasmacytomas). Many of these tumors have been produced in BALB/c mice by M. Potter (19); smaller numbers have recently been raised in NZB mice by Warner (20). The tumors usually arise in the peritoneal cavity after a long latent period characterized by chronic granulomatous peritonitis induced by injecting diverse agents (mineral oil, leucite shavings, etc.). The latent period has recently been shortened by infection with Abelson (MLV) virus (21).

Rollinhoff, Rous, and Warner (22) reported that myeloma tumors arising in NZB mice readily established transplantation immunity in NZB or in (NZB × BALB/c)F_1 hybrids, whereas BALB/c myeloma tumors were generally incapable of inducing immunity in these hybrids or in purebred BALB/c. Furthermore, the immunity induced by an NZB myeloma extended to most other NZB myelomas and to only a rare BALB/c myeloma. The difference seems to be related to viral infection: the NZB tumors contain C-type particles and produce murine leukemia virus (MLV); BALB/c myelomas on the other hand do not produce this virus though they usually contain defective A-type particles. The strain difference is in keeping with the frequent spontaneous expression of endogenous MLV genes in NZB mice and the infrequent expression of these genes in BALB/c mice (23).

However, the myeloma proteins on BALB/c myelomas can serve as TAAs, and for several reasons they are attractive model tumor antigens. (i) Like fetal and endogenous virus TAAs, myeloma proteins (or, more likely, other immunoglobulins with their idiotypes) are probably produced not only by tumor cells but sporadically and at low levels by rare, normal plasma cells; the concentrations attained by these proteins are probably not sufficient to establish tolerance (see TEPC-15 below). (ii) Although myeloma proteins are secreted by neoplastic plasma cells, they also occur on the surface of these cells (24). (iii) Each plasmacytoma produces a distinctive myeloma protein; in the uniqueness of association with particular clones of myeloma cells, myeloma proteins bear a rough resemblance to the individually unique TAAs of chemically induced tumors (see above).

We should like therefore to summarize our recent experience with the immune response of BALB/c mice to myeloma proteins from 8 myeloma tumors of BALB/c origin. The proteins are listed in Table 1. All proved to be capable of eliciting the formation of antibodies to their respective idiotypic (unique V domain) determinants (25,26). The specificity for an idiotype is illustrated in Fig. 1 with antibodies to M167 (anti-M167).

TABLE 1. *Immunogenicity of BALB/c myeloma proteins in BALB/c mice*

Myeloma tumor	Myeloma protein	Chains	Ligand	Immunization program	Formation of anti-idiotype	Ref.
MOPC-315	M315	α,λ_2	DNP	Freunds adjuvant	+	25
MOPC-460	M460	α,κ	DNP	Freunds adjuvant	+	25
MOPC-167	M167	α,κ	Phosphorylcholine	Freunds adjuvant	+	26
MOPC-603	M603	α,κ	Phosphorylcholine	Freunds adjuvant	+	26
MOPC-511	M511	α,κ	Phosphorylcholine	Freunds adjuvant	+	26
TEPC-15	T15	α,κ	Phosphorylcholine	Freunds adjuvant	+[a]	26
LPC-1	LPC-1	$\gamma 2a,\kappa$	?	Glutaraldehyde aggregated	+	26, 30
MOPC-21B	M21	$\gamma 1,\kappa$	?	Glutaraldehyde aggregated	+	26

[a] Only in neonatally suppressed or germ-free BALB/c.

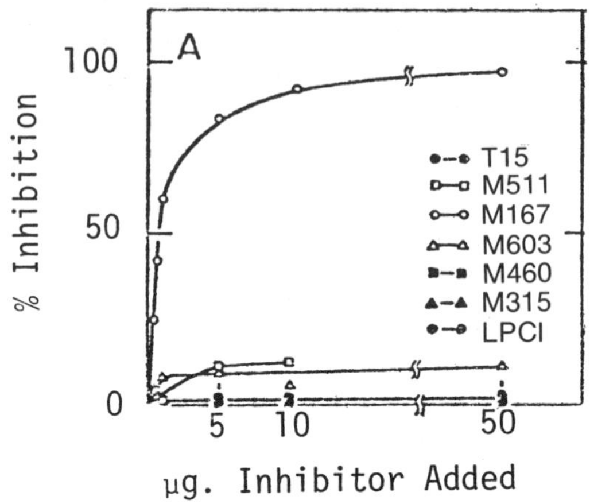

FIG. 1. Specificity of BALB/c antibodies for the unique determinants (idiotype) of M167, a BALB/c myeloma protein. BALB/c mice were immunized with M167; the resulting antiserum was reacted with [^{125}I]M167 in a radioimmunoassay. Of the 6 BALB/c myeloma proteins tested, significant inhibition was shown only with M167. (From ref. 26.)

With a conventional immunization program, employing Freund's complete adjuvant and rather high doses of myeloma proteins (25–27), sera of BALB/c mice immunized with M167 contained from 15 to 75 μg antibody (antiidiotype)/ml. The affinity of these antibodies for M167 (or its Fab fragment) fell in the range of 0.3 to 1 $\times$ 10^9 liters mole^{-1} and they displayed remarkably little heterogeneity with respect to affinity. By isoelectric focusing, the anti-167 molecules were also rather homogeneous (27), and by class analysis as well these antibodies were restricted: virtually all belonged to the IgG1 class.

All of the proteins listed in Table 1 were administered for immunization as purified proteins. For the IgAs, purification entailed mild reduction and alkylation to convert polydisperse serum IgA molecules into 7S protein (28). Injected in 7S form in complete Freund's adjuvant, all 6 IgAs of Table 1 elicited the formation of antiidiotypic antibodies (for T15, however, exceptional circumstances are required; see below). The same immunization procedure failed, however, with the LPC-1 (IgG2a) and M21 (IgG1) (26,29). These myeloma proteins are 7S molecules to begin with, and their isolation (by electrophoresis) did not entail reduction-alkylation. By injecting glutaraldehyde-aggregated LPC-1 and M21 it was, nevertheless, possible to produce antibodies to their respective idiotypes; coaggregation with carrier protein (such as ovalbumin) is especially effective (29).

Janeway showed some years ago (30) that isologous antiidiotypes to LPC-1 could be obtained by dinitrophenylating the protein and injecting it into Dnp-immunized BALB/c mice, and we have confirmed this finding (26). As he demonstrated subsequently, the key proved to be anti-DNP antibodies, rather than helper T cells; the helper antibodies in Janeway's experiments probably served simply to cross-link LPC-1 molecules. Why aggregation is such an enormous aid in the production of these immune responses is not entirely clear. Enhancement of binding of antigen to B and T cells and to macrophages, through multivalent interactions, may well be part of the explanation.

Of the 8 BALB/c myeloma proteins tested in BALB/c mice, 7 readily elicited the formation of antibodies to their respective idiotypes (Table 1). T15 was the exception: antibodies formed in normal adult BALB/c's against this protein appeared only after unusually prolonged immunization, and few if any were antiidiotypes (26). Instead, these antibodies were equally reactive with all α,κ myeloma proteins: they probably represent an autoimmune response to a class of immunoglobulins that is relatively infrequent in BALB/c mice. By injecting T15 aggregated (via glutaraldehyde) with itself—and especially with carrier proteins (ovalbumin, hemocyanin)—the antibody response could be accelerated, but its specificity remained directed to α,κ determinants, not to the T15 idiotype (26).

Why are normal, adult BALB/c mice unable to produce antibodies to the T15 idiotype? Antibodies to this idiotype can be raised with T15 in A/J mice, but not in A/J $\times$ BALB/c F_1 hybrids. Thus the ability to make the antiidiotype was inherited as a recessive trait, and it appeared that tolerance, rather than a deficient Ir gene, was responsible for the BALB/c unresponsiveness. This inference was subsequently supported by Lieberman and Potter's observation that serum from conventional BALB/c mice contains an extraordinarily high level of immunoglobulins with the T15 idiotype (perhaps as much as 1 in 500 molecules) (31). The high level was probably due to cryptic immune responses to inapparent infections, since sera from germ-free BALB/c mice did not contain immunoglobulins with this idiotype until the animals were allowed to become infected with commonplace bacteria (31).

To determine if BALB/c mice have the potential to make antiidiotypes to T15 before producing and becoming tolerant of immunoglobulins with this idiotype, the mice were raised under conditions that limited their synthesis of molecules with the T15 V region determinants. Two approaches were used (32). Newborn BALB/c's were injected with anti-T15 antisera (raised in A/J or CE/J mice and directed to the idiotype of T15). After about 3 weeks of injection some of these animals were found not to produce immunoglobulins with the T15 idiotype; and when immunized with T15 these neonatally suppressed BALB/c synthesized antibodies to the T15 idiotype. Thus the reaction between [^{125}I]T15 and sera from the immunized, suppressed mice was blocked only by T15 and not by other α,κ or other myeloma proteins. Furthermore, the reaction was about 95% inhibited by 10 mM phosphorylcholine, the ligand for T15 (Table 1).

Similar observations were made with germfree BALB/c mice: animals whose serum lacked immunoglobulins with the T15 idiotype responded to immunization with T15 by producing antibodies to the T15 idiotype. The specificity was the same as with the anti-T15 molecules made in the neonatally suppressed mice (32).

We conclude from these experiments that BALB/c mice contain B-cells that can recognize and respond to the idiotypes of any of their own myeloma proteins, and probably to any of their other immunoglobulins. In addition, these mice can also form "helper" T-cells that can recognize the idiotypes on at least many of these proteins (33). If myeloma proteins are reasonable models for some TAAs, there indeed ought to be a reservoir of B and T cells that are potentially able to respond to even those TAAs that are produced by some of the host's normal cells, providing the level of normal production is insufficient to establish solid or persistent tolerance. This fits with the presence of natural antibodies to glycoproteins of murine leukemia virus in most normal mice (6). Evidently the low level at which endogenous MLV is normally expressed suffices to stimulate an antibody response, rather than tolerance. Indeed, these viral antigens are able to elicit responses that can destroy large masses of MLV-producing tumor cells (34).

The foregoing comments focus almost entirely on specificity and therefore on V domains of immunoglobulins. C domains also deserve attention. Thus antiidiotype antibodies raised in BALB/c mice according to standard protocols (Table 1) are almost all IgG1 molecules. These antibodies do not fix complement, and it is therefore not surprising that they are not cytotoxic (in a C′ dependent reaction) for myeloma cells that produce and carry on their surface the corresponding myeloma proteins. [The surface localization of myeloma proteins has been demonstrated in direct rosette assay by Hannestad et al. (24). The surface localization explains the suppressive effects of immunization with some purified myeloma proteins on growth of the subsequent transplants of the corresponding myeloma tumor; in mice thus immunized a high proportion of the reduced number of tumors that finally appear are variants, producing only the myeloma protein's light or heavy chain, not both, and therefore lacking the idiotype of the myeloma protein used as immunogen (35)].

In the mouse, IgG1 molecules also do not bind to macrophages (36) and are unlikely to promote macrophage destruction of tumor cells. Mouse IgG2a antibodies, on the other hand, fix complement and have high affinity for an Fc receptor on macrophages (36). Hence to obtain antibodies of maximal tumoricidal activity the immunization program probably ought to elicit IgG2a antibodies of appropriate specificity. However, the current state of the art permits us to select for V but not for C domains: we can produce antibodies of almost any specificity but we are almost totally lacking in ability to generate at will antibodies of particular classes.

Incidental observations show the range of possibilities. As noted above, the antiidiotypes raised in BALB/c are almost all IgG1 molecules (27), but the high levels of antistreptococcal carbohydrate raised in the same strain of mice by Cramer and Braun (37) were almost all IgG2a molecules. Harris and Harris reported that with prolonged immunization they observed a shift in antibodies from IgG1 to IgG2a class (38). When it becomes possible to produce deliberately antibodies of particular classes, as well as of particular specificities the impact of immunology on the cancer problem (and other fields) may well be greatly extended.

What is the current status of the principal propositions listed at the outset? TAAs capable of eliciting autochthonous or syngeneic host responses are clearly demonstrable on many experimental animal tumors; but many other tumors seem to lack these antigens (15–17). For instance, if one ignores the special case of the myeloma proteins, we have so far been unable to demonstrate on BALB/c myeloma tumors TAAs that elicit cytotoxic serum or cellular responses, but it is not clear whether our negative results derive from resistance of the cells to immune destruction, deficiencies in detection assays, or a real lack of TAAs. (See note added in proof.)

With respect to the suppressive effects of autochthonous (or syngeneic) responses, it has long been clear that these responses do not invariably hinder tumor cell growth (39). Recently, moreover, Shearer and Parker (40) have shown that under certain circumstances trace levels of antibodies (plus complement) have a mitogenic effect on tumor cells, stimulating, rather than suppressing, their growth. But the most serious reason for questioning the universal validity of these propositions is the finding that closer scrutiny of immunodeficient humans and mice has revealed them not to be unusually susceptible to development of diverse cancers. For instance: some long-lived nude mice (genetically deficient in functioning T-cells) have developed reticulum cell sarcomas; but they do not develop these or other tumors in a high proportion of animals. Moreover, tumors transplanted into nude mice grow into large masses that remain encapsulated and do not become invasive or metastatic (41). In immunosuppressed humans the observed increased incidence of tumors seems also to be remarkably restricted to lymphoreticular cells; normal cells of this type are probably the targets for immunosuppressant agents, which may well be mutagenic and carcinogenic.

Since its rise to prominence about 15 years ago the idea of immune surveillance against cancer cells has been linked to cell-mediated immunity (42). With the current skepticism about the general importance of surveillance the significance of cell-mediated immunity to tumor cells may also be downgraded, and this would seem to be unwarranted. Nevertheless, these changing attitudes probably contribute to the current increase in interest in the role of antibodies in controlling tumor-cell survival. For more effective application of antibodies to the cancer problem, advances of two types are needed. (i) TAAs have to be treated as conventional antigens, instead of being approached as somewhat mystical trace substances in hopelessly complex mixtures. They will have to be isolated, purified, manipulated in diverse ways that permit augmentation of an antigen's immunogenicity. (ii) Immunization procedures will have to be developed so as to control the class of antibodies as well as their specificity. The realization of these aims would go a long way toward fulfilling the promises of immunological approaches to the cancer problem.

REFERENCES

1. Foley, E. (1953): *J. Cancer Res.*, 13:835.
2. Prehn, R. T., and Main, J. M. (1967): *J. Natl. Cancer Inst.*, 18:769.
3. Klein, G., Sjögren, H. O., Klein, E., and Hellström, K. E. (1960): *Cancer Res.*, 20:1561.
4. Baldwin, R. W. (1973): *Adv. Cancer Res.*, 18:1.
5. Embleton, M. J., and Heidelberger, C. (1972): *Int. J. Cancer*, 9:8.

6a. Oldstone, M. B. A., Aoki, T., and Dixon, F. J. (1972): *Proc. Natl. Acad. Sci. USA*, 69:134.

6b. Nowinski, R. C., Kaehler, S. L., and Burgess, R. R. (1974): *Cold Spring Harbor Symp. Quant. Biol.*, 39:1123.

7. Rohrschneider, L. R., Kurth, R., and Bauer, H. (1975): *Virology*, 66:481.

8a. Habel, K. (1962): *Cold Spring Harbor Symp. Quant. Biol.*, 17:433.

8b. Hellström, I., and Sjögren, H. O. (1962): *Int. J. Cancer*, 1:481.

9. Stephenson, J. R., and Aaronson, S. A. (1972): *J. Exp. Med.*, 135:503.
10. Kobayashi, H., Sendo, F., Kaji, H., Shirai, T., Saito, H., Takachi, N., Hosakawa, M., and Kodama, T. (1970): *J. Natl. Cancer Inst.*, 44:11.
11. Unkeless, J., Kirsch, L., Terres, G., and Eisen, H. N. (*work in progress*).
12. Coggin, J. H., and Anderson, N. G. (1974): *Adv. Cancer Res.*, 19:105.
13. Gold, P., and Freedman, S. O. (1965): *J. Exp. Med.*, 121:439.
14. Abelev, G. I. (1971): *Adv. Cancer Res.*, 14:295.
15. Baldwin, R. W., Glaves, D., and Vose, B. H. (1972): *Int. J. Cancer*, 10:233.
16. Prehn, R. T. (1969): *Ann. NY Acad. Sci.*, 164:449.
17. Klein, G. (1975): *Harvey Lectures Ser.*, 69:71.
18. Sato, H., Boyse, E. A., Aoki, T., Iritani, C., and Old, L. J. (1973): *J. Exp. Med.*, 138:593.
19. Potter, M. (1972): *Physiol. Rev.*, 52:631.
20. Warner, N. L. (1971): *J. Immunol.*, 107:937.
21. Potter, M., Sklar, M. D., and Rowe, W. P. (1973): *Science*, 182:592.
22. Rollinghoff, M., Rouse, B. T., and Warner, N. L. (1973): *J. Natl. Cancer Inst.*, 50:159.

23a. Hanafusa, H. W., Hayward, S., Chen, J. H., and Hanafusa, T., (1974): *Cold Spring Harbor Symp. Quant. Biol.*, 39:1139.

23b. Aaronson, S. A., and Stephenson, J. R. (1974): *Cold Spring Harbor Symp. Quant. Biol.*, 39:1129.

24. Hannestad, K., Kao, M. S., and Eisen, H. N. (1972): *Proc. Natl. Acad. Sci. USA*, 69:2295.
25. Sirisinha, S., and Eisen, H. N. (1971): *Proc. Natl. Acad. Sci. USA*, 68:3130.
26. Sakato, N., and Eisen, H. N. (1975): *J. Exp. Med.*, 141:1411.
27. Sakato, N., Hall, S. J., and Eisen, H. N. (*in preparation*).
28. Eisen, H. N., Simms, E. S., and Potter, M. (1968): *Biochemistry*, 7:4126.
29. Sakato, N., Fong, C., and Eisen, H. N. (*in preparation*).
30. Janeway, C. A., Jr.; and Paul, W. E. (1973): *Europ. J. Immunol.*, 3:340.
31. Lieberman, R., Potter, M., Mushiuski, E. B., Humphrey, W., Jr., and Rudikoff, S. (1974): *J. Exp. Med.*, 139:983.
32. Sakato, N., Janeway, C. A., Jr., and Eisen, H. N. (1976): *Cold Spring Harbor Symp. Quant. Biol. (in press)*.
33. Janeway, C. A., Jr., Sakato, N., and Eisen, H. N. (1975): *Proc. Natl. Acad. Sci. USA*, 72:2357.
34. Unkeless, J., Kirsch, I., Terres, G., and Eisen, H. N. (*in progress*).
35. Lynch, R. G., Graff, R. J., Sirisinha, S., Simms, E. S., and Eisen, H. N. (1975): *Proc. Natl. Acad. Sci. USA*, 69:1540.
36. Unkeless, J., and Eisen, H. N. (1975): *J. Exp. Med.*, 142:1520.

37. Cramer, M., and Braun, D. G. (1975): *Scand. J. Immunol.*, 4:63.
38. Harris, T. N., and Harris, S. (1975): *Transplantation*, 19:318.
39. Hellstrom, K. E., and Hellstrom, I. (1970): *Annu. Rev. Microbiol.*, 24:373.
40. Shearer, W. T., Atkinson, J. P., Frank, M. M., and Parker, C. W. (1975): *J. Exp. Med.*, 141:736.
41. Freedman, V. H., and Shin, S. (1974): *Cell*, 3:355, and *personal communication*.
42. Thomas, L. (1959): In: *Cellular and Humoral Aspects of the Hypersensitivity States*, p. 529, edited by H. S. Lawrence. Hoeber-Harper, New York.

NOTE ADDED IN PROOF

Since this paper was prepared, experiments in our laboratory have begun to yield considerable evidence for immunity to syngeneic myeloma tumors (J. H. Russell, G. Terres, and H. N. Eisen, *in preparation*). For instance, BALB/c mice immunized by several injections of irradiated cells from the BALB/c myeloma tumor MOPC-167 eliminate viable, labeled (^{125}I-iododeoxyuridine) MOPC-167 cells 4 to 6 times more rapidly than immunized controls or mice immunized with various other tumors, including some other BALB/c myeloma tumors. The immunized animals are also far more resistant than are the controls to the development of fatal tumors from MOPC-167 implants. In addition, peritoneal exudate cells from the immune mice are active in culture. They specifically lyse myeloma target cells (from the particular tumor used as immunogen) in an assay based on release of ^{125}I from cells labeled by incorporating ^{125}I-iododeoxyuridine into DNA. The effector cells are themselves destroyed by antisera (plus complement) to Thy-1 and to Ly2, but not by anti-Ly1; hence the killers are T cells (T_k) of the Ly2$^+$ subclass.

Why are T_k cells rather than antibodies elicited by immunization with myelomas and many other tumors? An understanding of the general mechanisms responsible for channeling immune responses to either synthesis of antibody molecules or production of killer or other effector T cells remains one of immunology's principal aims.

Antibodies in Human Diagnosis and Therapy,
edited by Edgar Haber and Richard M. Krause.
Raven Press, New York © 1977.

Induction of Immunity by Injection of Anti-Idiotypic Antibody

Klaus Eichmann

One can induce immunity in an animal by injection of an anti-idiotypic antibody, and this immunity is directed specifically to the antigen the original idiotype was raised against. Moreover, this immunity occurs in both the B- and the T-cell compartments.

Our experiments are done in mice, strain A/J. The immunogen is group A streptococci, one of the very useful antigens introduced by Krause into immunology. It is a rather complex antigen but we are only looking at the group-specific carbohydrate and its immunodominant group, N-acetyl-glucosamine. Our idiotype is an antibody called A5A which can be obtained by *in vivo* cloning of spleen cells in A/J mice. The anti-idiotype is a zenogeneic anti-idiotype raised in guinea pigs, which can be fractionated into the IgG-2 class and the IgG-1 class of guinea pig IgG. Anti-idiotypic antibody of the IgG-2 class, when injected into A/J mice, leads to suppression-type phenomena such as those first described by Nisonoff. In contrast, the IgG-1 class, when injected into A/J mice, leads to stimulation of immunity or immunization of A/J mice against streptococcal carbohydrate, both at the B-cell and the T-cell levels.

In order to show this type of immunization by anti-idiotype, adoptive transfer experiments were conducted. One mouse is primed with a hapten (NIP) and the other with a carrier. The carrier-primed mouse is given either Group A streptococci or anti-idiotypic antibody of the IgG-1 class. Spleen cells from both mice are mixed and injected into an irradiated host, which is then boosted with the hapten coupled to the streptococci.

The response of this host to the hapten is dependent on hapten-primed cells coming from one mouse and on carrier-primed cells coming from the other. Thus, if the host responds to NIP, this is an indication that the carrier-primed mouse has produced strep-specific helper cells. Therefore, when this mouse had been injected with anti-idiotype, it has produced these strep-specific helper cells in response to the injection of anti-idiotype.

The helper cells are further characterized by treating them with anti-Thy 1.2 to show that they are T-cells, or by running the cells through an anti-IgG column, a Wigzell-type column, which removes the B-cells to show that the helper effect is not diminished.

The results of such an experiment are shown in Table 1. Here two types of helper cells are used, strep-primed helper cells and anti-idiotype-primed helper cells. Two types of responses are determined: the anti-NIP response which is, as we said, indicative for helper-cell priming, and the response to streptococcal carbohydrate, which is indicative of B-cell priming in the donor.

It is clear from the anti-NIP titers in hosts of strep-primed cells that we have strep-specific helper cells. Anti-Thy 1.2 treatment removes this titer whereas treatment with an anti-IgG column does not alter this effect. So the strep-specific helper effect is mediated by T-cells. We obtain exactly the same result by anti-idiotype priming: positive anti-NIP titer. The helper effect is abrogated by anti-Thy 1.2

TABLE 1. *Anti-idiotypic antibody-induced helper function*

Donor 2[a]	Host	Antibody response	
Helper 1° Antigen	2° Antigen	ā NIP (×10⁻⁸M)	ā A-CHO (μg/ml)
—	NIP-CG	74 (1.4)	<2
—	NIP-Strep. A	1 (1.5)	4 (1.6)
Strep. A	NIP-Strep. A	65 (1.3)	172 (1.1)
" (ā Thy 1,2 + C′)	"	3 (1.5)	85 (1.3)
" (ā IgG Col.)	"	95 (1.2)	11 (2.3)
ā Id 1 (0.1 μg)	NIP-Strep. A	81 (1.5)	89 (1.2)
" (ā Thy 1,2 + C′)	"	2 (1.2)	38 (1.2)
" (ā IgG Col.)	"	69 (1.4)	6 (1.6)

[a] All groups received 1° NIP-CG cells from donor 1.
From: Eichmann, K., and Rajewsky, K. (1975): *Eur. J. Immunol.*, 5: 661.

treatment; it is not diminished by anti-IgG column treatment. On examination of antistreptococcal carbohydrate titers, it is apparent that recipients of both strep-primed and anti-idiotype primed cells respond. The response is abolished by running the cells through an anti-IgG column which removes the B-cells. This is evidence for B-cell priming by anti-idiotype antibody as well.

It is rather surprising that the same effects can be obtained with anti-idiotype as with antigen. A number of controls are needed. The most important is to show that the cells that are immunized or primed by the anti-idiotype are essentially idiotypically homogeneous as compared to the cells primed by the streptococci that ought to be somewhat heterogeneous. In Table 2, a comparison is made between anti-strep titers of recipients of cells primed with streptococci and those of recipients of cells primed with anti-idiotype. It is apparent that with strep-priming the proportion of idiotype-positive antibodies is about 25%, whereas idiotype-primed cells produce over 90% idiotype-positive antibodies.

TABLE 2. *Anti-idiotypic antibody-induced B cell priming and specific helper function*

Donor 2[a]	Host	Antibody response		
Helper 1° Antigen	2° Antigen	ā NIP (×10⁻⁸M)	ā A-CHO (μg/ml)	A5A (%)
—	NIP-CG	157 (1.2)	<2	
—	NIP-Strep. A	3 (1.2)	5 (1.5)	
Strep. A	NIP-Strep. A	36 (1.1)	413 (1.1)	26
ā Id 1 (0.1 μg)	NIP-Strep. A	152 (1.1)	180 (1.1)	91

[a] All groups received 1° NIP-CG cells from donor 1.
From: Eichmann, K., and Rajewsky, K. (1975): *Eur. J. Immunol.*, 5: 661.

TABLE 3. *Direct TNP-PFC/microculture*

	1° ā Id 1	1° Strep. A	N.C.
No antigen	2	4	2
TNP-Strep. A			
Total spleen	212	214	5
B cells[a] + NC	4	12	6
T cells[b] + NC	232		
Inhibitors			
A-CHO, 1 μg	17	44	
0.1 μg	51	173	
C-CHO, 1 μg	199	432	
0.1 μg	225	404	
ā Id, 0.1 μg	18	133	
0.01 μg	79	244	
NGPS	225	202	

Mean PFC/6 cultures, standard errors between 1.1 and 1.4.
[a] Treated with anti-Thy 1.2 and complement.
[b] Nylon-wool column passaged.
[c] Non fractionated anti-idiotype in IBC.
From: Black, S. J., et al. (1975): In: *Membrane Receptors of Lymphocytes,* edited by M. Seligmann et al., p. 117. North-Holland Publishing Co., Amsterdam.

This suggests that the anti-idiotype on the B-cell level picks out the idiotype-positive clones and expands them, which is indicative of the fact that anti-idiotype is indeed responsible for the priming.

The same is true for the helper cells, as shown in Table 3. This is a slightly different system which involves an *in vitro* boost with TNP-coupled streptococci; anti-TNP-PFC are measured. In this *in vitro* system we can test for the heterogeneity of the helper T-cells by including various inhibitors into the culture medium.

Upon strep-priming, more than 200 plaques against TNP are seen and the group C streptococcal carbohydrate does not inhibit these plaques. Consequently, T-cells can distinguish between C and A carbohydrate. The A carbohydrate also does not dramatically inhibit these plaques, which is not surprising because the strep is a very complex organism that has many different antigens in addition to the cell-wall carbohydrate. This expresses itself in only a low inhibition using either a carbohydrate or an anti-idiotype as inhibitor.

The situation is very different when one uses idiotype-primed helper cells. Both anti-idiotype and A carbohydrate inhibit the TNP plaques dramatically, indicating that on the helper-cell level, idiotype-positive helper cells have been selected by the anti-idiotypic antibody.

Another type of control (Fig. 1) shows that the degree of priming is dependent on the dose of anti-idiotypic antibody injected into the mouse. A comparison is shown of the degree of helper-cell priming as measured by the anti-NIP response, and the degree of B-cell priming as measured by the anti-carbohydrate response. Peak priming is effected with about 100 ng of anti-idiotype; the dose response curves are very similar. With these controls, it becomes very unlikely that an artifact is being observed.

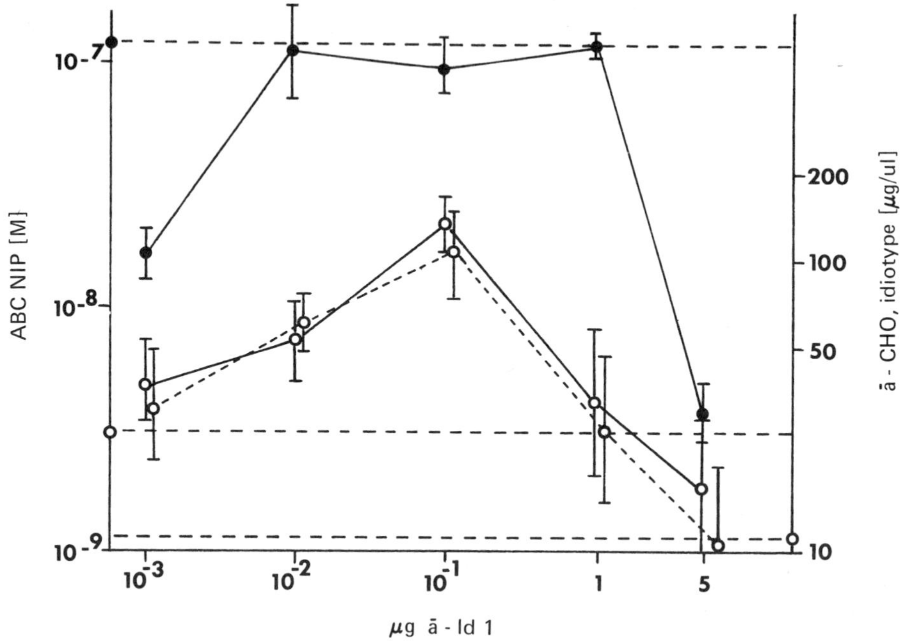

FIG. 1. T and B cell sensitization by anti-idiotypic antibody: dose response analysis. A/J mice were primed with various doses of ā-Id1 with specificity for the A5A idiotype (*abscissa*:doses refer to the IBC of ā-Id1). After 6 wk, spleen cells (25 $\times 10^6$ cells/host) from these animals were transferred into irradiated syngeneic hosts together with spleen cells (22 $\times$ 10^6 cells/host) sensitized to NIP-CG. Hosts were injected with 1 $\times$ 10^9 NIP-Strep. A particles and bled 10 days later. Sera were titrated for antibodies to NIP (*left ordinate*: molar NIP binding capacity) and A-CHO (*right ordinate*), and for the A5A idiotype (r.o.). ●——●, anti-NIP response; ○——○, response to A-CHO; ○---○, A5A idiotypic response. Also shown are (i) the level of a presumably maximal anti-NIP response in the system (secondary stimulation with 0.2 μg NIP-CG; broken horizontal line on top) and (ii) the response to NIP and A-CHO in the absence of cells primed with ā-Id1 (the two lower broken horizontal lines).

(From: Rajewsky, K., Hämmerling, G. J., Black, S. J., Berek, C., and Eichmann, K. (1975): T Lymphocyte receptor analysis by anti-idiotypic stimulation (*in press*).

Antibodies in Human Diagnosis and Therapy,
edited by Edgar Haber and Richard M. Krause.
Raven Press, New York © 1977.

Antilymphocyte Sera for Immunosuppression: A Powerful Class of Agents Awaiting Full Application to Patient Care

Paul S. Russell

Transplantation Unit, the Surgical Services, and the Department of Surgery of the Harvard Medical School at the Massachusetts General Hospital, Boston, Massachusetts 02114

It would appear to be a relatively simple matter to evaluate the efficacy of antilymphocyte serum for immunosuppression. Nevertheless, the seemingly straightforward transaction of preparing an antiserum in one species against leukocytes derived from another species calls into play variables which are difficult to measure and even more difficult, in an individual case, to anticipate. When one adds to this a third party, namely, the individual, animal, or patient, who is to be treated by the serum, the stage is set for biological variability of major and frustrating proportions. Nevertheless, interest in this subject has persisted. This has been so for two principal reasons. First, certain heterologous antilymphocyte sera prepared for use in animals have been remarkably successful in prolonging the survival of tissue transplants without apparently impairing the well-being of their recipients. Second, the need for better control of unwanted immune responses in patients, including rejection of life-sustaining tissue transplants, continues to be very great.

This presentation will be designed to summarize where we now stand with heterologous antilymphocyte cell sera (ALS), particularly as the subject relates to immunosuppression for tissue transplantation in clinical medicine. It will become clear that the present situation in regard to the application of existing fundamental knowledge to patient care is rudimentary, and much uncertainty concerning its applicability remains. A large proportion of the reliable information available is that coming from animal experiments so that no discussion of the clinical aspect of the subject could be countenanced without careful attention to the experiments upon which it must rest.

A short historical review will be presented first, to be followed by more detailed descriptions of recently employed methods of preparing ALS, and of their properties, both *in vitro* and *in vivo*. Factors known to potentiate or retard the *in vivo* effects of ALS will then be considered before going on to evaluate the possible mechanisms of action of these sera. Next, attention will be turned to the special problems of preparing active heterologous serum fractions for use in patients, their administration, side effects, and efficacy.

Finally, some of the possible lines for future progress will be set out and an effort will be made to support the view that there is considerable potential for a broad field of activity which has as its object the inactivation or destruction of various leukocytes, or even of cells of other morphological types, by heterologous antibody reagents directed against their characteristic antigens.

The subject of ALS is already large and has been dealt with in a number of general reviews and at several international meetings. Special attention is invited to some of these (3,6,34,61,69,74,77,86,102,133,186,189) in which a wealth of information can be found. Reviews emphasizing clinical features of the subject will be cited in the section "Some Clinical Applications of Antilymphocyte Serum."

HISTORY

The interest currently placed on the mechanisms of graft rejection and its control, which, in turn, was largely responsible for initiating the development of modern "cellular immunology," is scarcely 30 years old. It is not surprising, therefore, that sera made in one species against the leukocytes of another were not tested for possible effects on cell-mediated immune processes, including rejection, until about 20 years ago.

It is generally acknowledged that Elie Metchnikoff, in 1899, was the first to record observations of the immune destruction of leukocytes by sera prepared against them in another species; for example, he produced antisera in guinea pigs against rabbit or rat lymph node or spleen cells (104). These observations were extended by his student A. Besredka at the Pasteur Institute (14), who also demonstrated the hemolytic and highly toxic potential of many sera prepared by repeated injections of leukocytes between species. Of the many studies which followed, several were noteworthy for our purposes. Pappenheimer found that the cytotoxic property of an ALS *in vitro* was inactivated by heating to 58°C for 30 min, whereas its agglutinating capacity was not (121). Simon Flexner examined the pathological changes in the lymphoid structures of animals which had received multiple injections of an ALS (43). In general, he found hyperplasia of lymphoid centers, while in other experiments Bunting (23) found that a lymphopenia was maintained for 1 to 2 days in rabbits following an injection of a serum made in geese.

A well-planned series of experiments by Chew and Lawrence in 1937 (26) with rabbit anti-guinea pig lymph node cell sera, absorbed to eliminate Forssman and antierythrocyte antibodies, showed that lymphopenia could be prolonged for almost 2 weeks by repeated injections of serum. Steinberg and Martin's experiments were also important in that they showed that such sera could have not only a certain specificity for lymphoid cells as against granulocytes, but also that these sera could even detect differences between different morphological types of mononuclear cells (161).

Woodruff was the first to establish that this class of agents might influence the progress of allograft rejection. His first attempts were apparently unsuccessful (192), although he was later able to show, with Anderson, that the immunosuppressive effects of thoracic duct drainage in rats were significantly potentiated by an ALS made in rabbits against rat thoracic duct cells (193). Meanwhile, T. Inderbitzen, working in Humphrey's laboratory, had shown that delayed hypersensitivity reactions in guinea pigs could be clearly suppressed by ALS (64), and these findings had been extended by Waksman et al. (178) in the same species to include evidence of extension of skin allograft survival. From this time forward, quite a number of workers entered the field independently and a great deal of information was collected in a relatively short time, including some initial trials of antiserum fractions against human lymphoid cells in patients (see ref. 186). It would not serve our present purposes to detail the sequence of events hereafter. Rather we will select various operational headings, as mentioned above, under which the current state of knowledge, relevant mainly to the use of ALS to combat graft rejection, can be briefly considered. It is hoped that readers will be willing to consult the extensive reviews cited above for more comprehensive information, and that the author will be excused for having selected evidence to illustrate certain points with more attention to its adaptability to this account than with regard to the precise priority due to individual contributions.

PREPARATION AND GENERAL PROPERTIES OF HETEROLOGOUS ANTILYMPHOCYTE SERA AND THEIR FRACTIONS

Preparation

The sera we are referring to represent a class of reagents, and they may be prepared according to a number of different protocols. All are the result of the repeated injection of suspensions of lymphoid cells from one species into another. These cells may be derived from lymph nodes, spleen, thymus, or from peripheral blood or lymph. Some other cell types have also been tested as antigens.

The most familiar experimental system by which very active immunosuppressive sera can be raised is that in which lymph nodes, spleen, or thymus cells of rats or mice are injected into rabbits. In experiments using the rabbit antimouse combination, effectively immunosuppressive sera have been prepared either with or without the inclusion of Freund's adjuvant in the initial inoculum. In the former scheme, as used in our laboratory (52), about 100 ×

10^6 dissociated lymph node cells are emulsified with an equal volume of Freund's complete adjuvant and divided for injection into the four foot pads, delivering about 0.2 ml of the mixture in each injection. Booster injections of about 100×10^6 cells each are given intravenously on 3 successive days 4 weeks thereafter and the rabbits' sera are collected a week later. Sera are then decomplemented by heating to 56°C for 30 min and are stored in the frozen state.

The latter scheme, that of Medawar and his colleagues (86,91), involves simply two intravenous injections of murine lymphoid cells in suspension 2 weeks apart with 10^9 cells being delivered on each occasion. Serum is collected 7 days after the second injection and is decomplemented and stored.

Methods involving adjuvants generally yield sera with much higher titers. They include antibodies, not only against leukocytes, but often against erythrocytes and platelets. Furthermore, they may react with a variety of fixed tissue components, such as basement membrane, as well (see below). It is usually possible to administer sera prepared by the "two-pulse" intravenous method directly to mice, employing doses of 0.5 ml or more at a time, with surprisingly little evidence of toxicity, whereas some signs of toxicity with occasional deaths are seen in mice receiving like amounts of unabsorbed serum prepared by the adjuvant method. This toxicity can be virtually abolished by prior absorption with erythrocytes or suspensions of nonlymphoid tissues, for example, liver. Extending the period of immunization with repeated booster injections tends to result only in a gradual decline of *in vivo* immunosuppressive potency of sera collected as time progresses and some tendency toward increasing toxicity. In one study in which the two immunization methods mentioned above were carefully compared in several rabbits by Skamene and Russell (153), the titers of complement-dependent cytotoxins, leukoagglutinins, and hemagglutinins against mouse cells were maintained with continuing immunization. Meanwhile, the percentage of antibody in succeeding serum samples which bound to mouse lymphocytes gradually rose. (The latter determination was made using a ^{125}I-labeled IgG serum fraction and its significance will be discussed further below.) In spite of this, however, the immunosuppressive potency of sera, always considerably higher in sera prepared by the "adjuvant method," gradually declined in all cases (see Table 1).

Selection of Species

Essentially similar programs of immunization have been employed with other species combinations. In all cases considerable variability in immune responsiveness to stimulation has been encountered in individual "target" animals. Although sera with fairly high levels of cytotoxins and agglutinins to lymphoid cells can be raised in virtually all species tried, no clear princi-

TABLE 1. *Properties of rabbit antimouse ALS prepared by two immunization protocols*

	"IV" group (6 rabbits)				"Freund's adjuvant" group (12 rabbits)			
Time of serum collection (weeks after beginning immunization)	2	5	8	19	4	8	14	17
Number of booster injections	1	4	5	8	1	2	3	4
Cytotoxins	50	800	800	800	1,600	1,600	1,600	1,600
Leukoagglutinins	200	800	400	400	800	1,600	1,600	800
Hemagglutinins	80	160	80	80	80	80	80	160
Antilymphocyte antibody in IgG (%)	2.3	4.3	n.d.	5.2	5.3	7.6	8.8	9.6
M.S.T. skin allografts (control 10 ± 1.1 d C57 BL/6 × A)F_1 to CBA	16 ± 1.2	23 ± 1.3	22 ± 1.5	14 ± 1.3	35 ± 1.2	31 ± 1.2	n.d.	26 ± 1.1

Rabbits in "IV" group were immunized by 10^8 lymphoid cells i.v.; rabbits in the "Freund's adjuvant" group by 10^8 lymphoid cells in Freund's complete adjuvant s.c. Booster injections of 10^8 lymphoid cells i.v. were subsequently given to both groups about every 1 to 3 weeks and their cumulative number is indicated in the table. The figures displayed were arrived at by pooling the sera at each stage. For details, see Skamene and Russell (153).

ple regarding species selection has yet emerged. For example, one might suspect that a phylogenetically more distant relationship would favor strong immune responses and result in more active antisera. The fact that allospecific sera have feeble *nonspecific* immunosuppressive effects at best (162) (that is, effects which are to be sharply distinguished from those due to immunological enhancement) would appear to support this notion. It would also suggest that the important antigens concerned in the production of immunosuppressive antisera are in the ''species-specific'' category and thus only appreciable by individuals of a different species. Nevertheless, no clear stepwise relationship exists between immunogenetic disparity and species suitability for ALS production. This is illustrated in the results of Winn and Daddi (184) who found the guinea pig to be a much better source of antiserum to mouse cells than the dog. It is further substantiated by the fact that avian species appear to be relatively poor sources of immunosuppressive sera for mice (78). Pigs, goats, sheep, and calves have been tested as sources of antibodies for several other species (15,151). Actively immunosuppressive sera, as judged by prolongation of skin allograft survival, can be produced by animals in each combination selected, although more difficulty was apparently encountered with high levels of hemagglutinating activity in sheep and goats than in pigs following immunization with washed mouse thymocyte suspensions. The horse has also been employed frequently as a source of antilymphocyte and antithymocyte sera and seems to be at least a satisfactory, if not an optimal, source for antibody against dog (108), as well as mouse and human cells (see, for example, ref. 186).

In producing immunosuppressive sera, what is required is that the immunized individual be capable of detecting and responding to a certain antigen or antigens presented along with many other antigens by the immunizing cells. Much of the immune response to the cell injections—certainly that directed, for example, against erythrocytes—will be entirely undesirable, and by no means will all of the antibodies directed toward leukocytes be useful. This was emphasized particularly well in the experiments of Jeejeebhoy (76), who found high levels of serological reactivity against rat lymph node cells (agglutinins and cytotoxins) in dog antirat sera, but only weak immunosuppressive capacity.

Thus, a number of points must be taken into account in species selection for ALS production. Beyond the obvious, but important, ones of size, availability, and cost, a major quality must be the recipient's ability to respond strongly, and as selectively as possible, to the operative heterologous antigens concerned. As these antigens have not yet been precisely identified, and the magnitude and character of responsiveness to them is not predictable on a phylogenetic basis, no systematic procedure for species selection beyond pure empiricism is available. In later sections, we will examine the evidence regarding the antigens involved and how antibodies to them may have immunosuppressive effects.

Source of Antigen

Inhibition of the transplant rejection reaction can be accomplished to an impressive degree with heterologous antisera raised against the mixtures of lymphoid cells which are normally found in lymph nodes, spleen, or in the thymus. Although it has been suggested that the thymus provides cells which are superior to other lymphoid organs (115), in that the resultant antisera are regularly more potent as immunosuppressives, this has not been substantiated, and antilymph node or antithymus cell sera are roughly equivalent in potency with the former, even being superior under some conditions (188). The point seems to be that among the complex array of antigens presented to the target animals by such cell mixtures, a sufficient complement of those antigens most important to immunosuppression is contained in cell populations from any of these sources.

Of considerable theoretical import was the report by Levey and Medawar (92) that mouse epidermal cell suspensions, L cells from standard cultures, or mouse embryo fibroblasts could provoke antibodies in rabbits with immunosuppressive activity. This would lead one to believe that at least some of the requisite cellular antigens are widely disseminated throughout the body appearing on a considerable variety of anatomically different cell types. Although we concur that sera of feeble immunosuppressive potency might be produced in this way, and that, accordingly, it may be that some of the relevant antigens may be displayed on other than lymphoid cells, the main line of our argument will run strongly counter to this, as we believe that the important antigens are not only situated mainly, but probably exclusively, on lymphoid cells and, in particular, a certain subpopulation of lymphoid cells. This was first suggested by absorption experiments. For example, a highly immunosuppressive mouse antilymph node cell serum could be fully absorbed by mouse liver cell suspensions without the loss of *in vivo* potency, whereas lymph node cells themselves readily removed all activity (153). This result makes it likely that the most important antigens concerned are situated on lymphoid cells and are of the "species-specific" type. It also reinforces the point made above that allospecific antigens are of little importance in evoking immunosuppressive antibodies in interspecies combinations as these antibodies would be removed by such absorptions. There is no doubt, however, that antibodies to allospecific antigens are often generated when cells are transferred to another species (138). It is also noteworthy that the weight of opinion is against assigning any important degree of strain specificity of ALS when they are tested for this property in donor species individuals of differing genotypes (see, for example, ref. 78). More consideration regarding the "active antigens" in respect of ALS production will be given in later sections of this chapter.

Immunosuppressive sera can be induced not only by living cells but by freeze-thawed cells or by cells killed in certain other ways. It has been reported that active sera can also be raised by lymphoblasts maintained in large-scale cultures (149), although the potency of sera produced with certain cultured cells has been questioned (see ref. 81 and discussion in the section "Summary and Discussion of Future Possibilities for Antilymphocyte Cell Sera in Clinical Medicine"). Neoplastic lymphoid cells have also been used with success (185).

Some attention has been paid to the possibility of producing antisera by subcellular fractions [see ref. 194 and review by Lance et al. (86) for more complete description]. Cells disrupted by nitrogen decompression have been separated into rather crude mitochondrial, nuclear, plasma membrane, and supernatant fractions. Antisera to mitochondria contained high levels of hemagglutinins; the nuclear fraction produced sera of relatively low potency compared to the membrane fraction which was the best. The supernatant was virtually ineffective. At present, the subcellular fractions which have been tested offer little advantage over whole cells as a means of immunization for antilymphocyte sera. There is no reason, however, why this should continue to be true. Indeed, as we will point out below, this aspect of the problem should offer a distinct opportunity for further practical advances along lines which can now be foreseen.

In Vivo Properties

The injection of an active antilymphocyte serum into any member of the species donating the immunizing cells generally results in a prompt lymphopenia. We have observed this in particularly striking form in mice injected with rabbit antimouse lymph node (52) or thymus serum and in dogs with serum made in horses (108). Virtually all of the early observations with antisera to leukocytes suggested that certain of the effects observed following their use *in vivo*, such as lymphopenia, were short-lived (see, for example, ref. 178). With sera made more recently, this was found not to be the case. Indeed it was the apparent transience of these effects which made the potential utility of ALS appear to be so restricted at first. And, of course, this restriction was thought to be due to the rapid appearance of antibodies to the foreign protein which were thought to inactivate its effects by combining with it as it was administered in succeeding injections. It is now known that quite low peripheral lymphocyte counts can be maintained for some time in mice with continuing injections of rabbit antimouse serum and for many months in dogs with, for example, serum made in horses (108). A reduction in lymphocyte content in the thoracic duct lymph has also been recorded (42,49,176). Concomitant with this lymphopenia, the gradual disappearance of cells from paracortical areas of lymph nodes commences (see, for example, ref. 174). In the spleen, depletion of small lymphocytes, especially from the periarteriolar sheaths in the Malpighian bodies, has been noted (165,174), whereas there is fairly general agreement that cellular depletion from the thymus, even following persistent administration of sera raised with thymic cells, is minimal (135). Changes in the bone marrow are not striking and the numbers of polymorphonuclear cells in the peripheral blood are maintained or even somewhat elevated. In earlier experiments, the variation of effects on lymphoid centers from treatment by different types of ALS had been a source of confusion. Sera raised with adjuvants, particularly if given for several weeks in mice, could result in almost complete ablation of lymph nodes with large patches of necrosis (137), whereas those raised with the "two-pulse" intravenous technique usually produced mild lymph node hyperplasia (92). Furthermore, very distinct evidence of immunosuppression could be detected, especially in animals treated by sera in the latter category, long beyond the time when peripheral lymphocyte counts had returned to normal, and it has been known since the early experiments of Humphrey that suppression of delayed hypersensitivity reactions can occur without much apparent lymphocyte ablation (63). Careful examination of these events by a number of workers followed, in which depletion of cells from special regions in lymph nodes and spleen with properly absorbed sera were seen. Further studies were then carried out along the original direction of inquiry set by Everett et al. (42) and others. This was based upon the hypothesis that in the peripheral lymphocyte population at least two major classes of cells could be distinguished, a long-lived and slowly self-renewing type and a rapidly renewable, short-lived type. Many ALS apparently produce a transient panlymphocytopenia immediately after injection, but there is considerable evidence that the cells which return within a few days in the peripheral blood are largely of the short-lived type (176). Whether this is due merely to the fact that short-lived cells are restored from their precursors more rapidly following treatment, or whether selective destruction of long-lived cells is involved, will be considered in our later discussion of the mechanisms of action of these sera.

Effects on Graft Rejection and on Humoral Antibody Production

The impact of ALS treatment on the survival of skin allografts was studied first, and it is important to give special emphasis to the fact that serum treatment resulted in truly spectacular effects. Healthy appearing mice, bearing perfectly healed skin allografts several weeks beyond their normal survival time of about 10 days, could be regularly procured, following a brief course of serum treatment. It was also impressive that this hospitality to grafts could be extended to xenografts (52), so that rat skin grafts could be made to survive for over a month as a consequence of a course of treatment with rabbit antimouse lymph node cell serum which commenced a week before grafting and was extended for a week thereafter. Graft rejection, when it occurred, was usually fairly rapid in onset and progressed through its familiar stages to complete destruction of donor tissue within 24–48 hr in the presence of a fairly typical cellular inflammatory reaction, as determined by histological studies. The commencement of serum treatment before presentation of the antigen, the graft, adds to its effects. The doses used in experiments with mice by most workers range between 0.25 and 0.5 ml per injection. In Medawar's most commonly used treatment schedule, 0.5 ml of whole serum is given intraperitoneally on days +2 and +5; in our own, 0.25 ml of adjuvant-raised serum was given daily for 1 week with varying relationship to the time of grafting. In rats, Woodruff often gave 2 ml of serum per injection for 3 days prior to test grafting. Extending the period of treatment would generally result in extended graft survival. In our early experiments, continuing treatment (5 times per week) after grafting resulted in a gradual loss of weight of treated animals with obvious wasting and finally death at about 5 to 7 weeks with most skin grafts still in excellent condition (110). Very prolonged skin graft survival in mice can result, however, from continuing treatment with ALS preparations which are either rid of toxic antibodies by absorption (see below) or are prepared by the "two-pulse" method (78). A dose–response curve can be constructed for each preparation at each treatment frequency mainly by revising doses downward from those quoted above, but raising the individual dose above these amounts has only a slight further effect on graft survival.

Of special note is that the survival time of grafts placed upon specifically immunized recipients (that is, animals which had previously rejected a graft from a donor of the same genotype as the second donor) could be prolonged by ALS treatment for periods longer than expected even for normal, unsensitized recipients. The reversal of an obvious rejection reaction in progress is also frequently possible. Whether or not ALS treatment can result in complete elimination of a state of preexisting cellular immunity is somewhat unclear, although the studies of Levey and Medawar (91), carefully repeated by Lance (82), would strongly suggest that it can.

The situation with humoral antibody responses is apparently somewhat different in that ALS can inhibit the response to a considerable variety of soluble and particulate antigens on first exposure, especially if it is administered just prior to antigenic exposure. Once the machinery for humoral antibody production has been primed, however, it is more difficult to influence antibody production by ALS treatment (see especially 4,10,36,52,67). Suppression of responses to so-called thymus-dependent antigens has been thought by some to be relatively more difficult. Of special interest was the observation by Lance and Batchelor (84) that humoral antibodies directed against donor cells could be found in at least some mice bearing normal appearing allografts as a consequence of ALS treatment. This apparent disparity between the susceptibility of humoral responses as against primary or even secondary cell-mediated ones was looked to with hope in anticipating the use of ALS in patients where control of graft rejection, it was reasoned, might be achieved with less attendent risk of bacterial infection

(136). The degree of disparity between the suppression of cell-mediated versus humoral responses is quite unpredictable, however, and less comfort should probably be taken in this regard than originally seemed supportable.

Although it will be discussed further below, it should be mentioned here that a very high proportion, if not all, of the immunosuppressive activity of a given antiserum, no matter in what species it is prepared, or against what type of lymphoid cell, resides in its 7S or IgG fraction. Thus many animal experiments, as well as virtually all clinical trials in patients, have made use of globulin fractions of some sort.

Organ transplant survival in animals has been studied fairly extensively in the presence of ALS or ALG (antilymphocyte globulin) prepared in various ways. As mentioned above, uninterrupted ALS treatment, even for quite long periods in small animals (118), may be accompanied by few signs of toxicity if the serum is made by the "two-pulse" method or is fully absorbed with red cells. Adjuvant prepared sera tend to be somewhat more toxic and accordingly require more extensive absorption (52). In dogs, receiving horse antidog lymph node cell serum made by the adjuvant technique, acute toxic symptoms were marked, mainly from hemolysis, and required the careful absorption of crude serum with erythrocytes to prevent serious toxic symptoms. The subcutaneous route was found to be less dangerous than the intraperitoneal or intravenous, probably because it permits some degree of absorption of toxic antibodies locally and a slower release of the inoculum (108). Absorbed preparations and purified IgG fractions are, in general, well tolerated (1,29,66,89,123,158). In our experiments with kidney transplantation in dogs, some 10–20 ml of crude horse antidog lymph node serum was given daily to 7 animals for periods of from 1 to 3 weeks before transplantation and continuously thereafter. The survival times of these animals are displayed in Table 2. After 4 months, serum treatment was given only weekly to the two recipients surviving at that time and their peripheral blood lymphocyte counts returned gradually to a low normal range from the previous level of only 0–300 cells/mm^3. Finally, serum therapy was stopped

TABLE 2. *Survival of renal allografts in dogs given horse antidog lymphocyte serum s.c. daily*

Dog no.	Pretreatment days	Survival
1	38	13 days
2	7	20 days
3	14	26 days
4a	5	19 days
4b	32	52 days
5	7	76 days
6	7	1 year
7	21	10 years

completely over a year after transplantation. One dog was sacrificed shortly thereafter and the other not until over 10 years had elapsed. Both had normal kidney function and virtually normal appearing kidneys on histological examination. The majority of recipients, those succumbing earlier, showed definite evidence of rejection in the kidney transplants. Thickening of glomerular basement membrane and evidence of horse gamma globulin deposits have been found by some in the kidneys of dogs treated with ALS (123). The incidence of serious infections was not deemed to be increased in these animals, although no definite study of this important question has been accomplished.

Starzl found that the survival time of canine liver transplants could be extended from a control figure of about 7 days to a mean of 27 days with some survivors to several months (158), and there are a number of confirmatory reports. The use of purified IgG fractions, usually supplemented by suitable absorption with erythrocytes and platelets, greatly reduces the anemia and thrombocytopenia which can otherwise be seen. Acute serum sickness, anaphylaxis, and complex disease in the kidneys have been observed, although not commonly. Some of these points will be raised again below as they are closely pertinent to clinical application.

Effects on Other Immune Responses

Fully in concert with these effects on graft rejection are those on other types of delayed hypersensitivity reactions as found originally by Inderbitzin (64). Thus skin tests employing standard antigens can be converted to negative by ALS treatment, and the primary induction of contact sensitivity can also be prevented as well (86). ALS can likewise obviate the induction of sensitization by cellular antigens especially if given just prior to a "peripheral" injection of cells. In other words, as is emphasized by Levey and Medawar (92), ALS treatment can prevent the sensitization which will follow a skin graft or a subcutaneous injection of lymphocytes from donors of a given genotype, but is less capable of doing so if the cells are delivered intravenously. This, and the results of studies of the influence of ALS on the skin reactions produced by the intradermal injection of allogeneic lymphocytes described below, was to suggest to Medawar his concept of the predominantly "peripheral" action of ALS, that is, an action which favored the inactivation of lymphocytes, while in a peripheral site, such as the blood, rather than a central one, such as a lymph node or the spleen.

Manifestations of the graft-vs-host reaction, in its several forms, can also be suppressed by ALS. This includes the "normal lymphocyte transfer" reaction in guinea pigs, where courses of ALS treatment to recipients of intradermal allogeneic cell injections will abolish the local inflammatory reactions which they normally produce. Considerably longer courses of treatment of cell donors are required in order to render suspensions of cells which can be harvested from their lymph nodes incapable of producing a local graft-vs-host reaction in the skin on transfer to an allogeneic recipient (90). The capacity of lymph node cells from adult mice to produce "runt disease" on intravenous injection into newborn mice can be abolished in the same manner (177). Suppression of graft-vs-host disease in larger animals, as for example that which occurs in primates after radiation and allogeneic bone marrow replacement, has been less successful, at least in subhuman primates (see comments of Balner and van Bekkum in ref. 186), although the results in patients may be somewhat better (see "Transplantation of the Heart" under "Some Clinical Applications of ALS"). Rather remarkable suppression of a variety of experimental autoimmune processes has been reported by a number of investigators and this field has been ably reviewed (34).

The nonspecific suppression of certain types of inflammatory processes, along with transient depression in complement levels, which have been studied by Turk and his colleagues following treatment with serum directed against various types of lymphocyte preparations, does not appear to extend broadly to include such processes as passive cutaneous anaphylaxis (178) or immune complex disease involving antibodies other than the ALS itself (36). The effect of ALS fractions against purified cell subpopulations has not been carefully studied in such systems, but it appears unlikely that the primary or essential activity of ALS in respect of suppression of specific immune processes is mediated by directly influencing effector mechanisms such as the complement system. Turk's work, however, emphasizes the variety of physiological effects which can result from treatment with antisera to cellular as well as serum constituents (173).

Purification of Active Fractions from Whole Serum

As mentioned above, a broad range of antibodies may result from the repeated injection of crude inocula containing connective tissues and erythrocytes, such as those from some spleen cell preparations. Antierythrocyte, antiplatelet, and antibasement membrane activities serve as salient examples, and it is known that antibody to connective tissues of one portion of the body will react with similar structures elsewhere (80). The particularly wide cross-reactivity of antibody against basement membrane constituents has also been reaffirmed in the findings of Skamene et al. (152), who showed that an antiserum prepared in horses to whole human thymus homogenates contained antibody to kidney basement membranes of rats and rabbits. Considerable reduction in unwanted antibodies in a given ALS preparation can, accordingly, be accomplished by delivering as selected an antigenic inoculum as possible.

In this section, however, we will concentrate upon the potential of purifying or extracting from a given whole serum a more highly active fraction containing less extraneous material. Waksman (178) first showed that the immunosuppressive activity of his rabbit anti-guinea pig serum resided in the gamma globulin fraction by an ammonium sulfate precipitation technique. This finding was confirmed and extended by many others so that it is now quite clear that highly purified 7S antibody fractions carry virtually the full immunosuppressive potency of the whole serum. The 19S fraction from such sera, separated by column chromatography and confirmed as to purity by immunoelectrophoresis, has shown only the slightest ability to prolong the survival of skin grafts (86, 95). IgA in rabbit sera and the IgT fraction, found in horse sera, are both virtually inactive *in vivo* as well.

Thus, it has become standard practice, especially in clinical trials, to employ either a salt- or ethanol-precipitated gamma globulin preparation, or a more highly refined 7S fraction prepared by one or another chromatographic technique (ALSIgG). The range of methods used has been reviewed by James (68). But this step of isolation of the IgG fraction, while useful, obviously falls far short of the ideal mark of providing a preparation containing *only* active antibody molecules, since but a small proportion, even of a highly purified IgG fraction, is actually directed toward the desired cellular antigens. In a study employing highly purified ^{125}I-labeled IgG prepared from several rabbit antimouse lymph node and thymus cell sera by Skamene and Russell (153), it was found that no more than 9.6% of the total IgG in an adjuvant-stimulated serum would bind to lymph node cells *in vitro*. In the serum, 8.5% of the IgG would bind to liver cells. It was further determined that full absorption with liver would leave an IgG which retained its full immunosuppressive potency. Thus, only a little over 1% of the IgG in this highly active serum fraction appeared to carry its full immunosuppressive capacity. Woodruff (190) and Lance (83), in particular, have pursued this question by preparing antibody eluates from lymphocyte membrane fractions. In Lance's study, about 2.0% of the original IgG was eluted and about three-quarters of this would bind to lymphocytes *in vitro*. The eluted fraction was considered to be 50 times more potent in prolonging skin graft survival, milligram for milligram, than the parent IgG from which it had been derived.

On injection into members of the donor species, labeled IgG fractions of ALS have been found to localize in lymphoid structures with the significant exception of the thymus (37). The clearance from the blood of such material, radioactively labeled (^{131}I), has been studied by a number of workers (see James, ref. 69) and the rapid development of "immune" clearance of the heterologous protein can be demonstrated a few days after treatment is begun. The active fraction, however, seems to be cleared much more promptly, since a fraction of antithymocyte serum IgG, eluted from a mouse thymus cell membrane preparation by Lance, was largely cleared in 24 hr and had disappeared almost completely in 48 hr (83). This process of clearance is probably almost entirely due to the active binding of injected antibody to cells by its receptors as it occurs even in animals made tolerant of the foreign protein (86). One might expect that when a high degree of immunity to the foreign protein is generated, it would not only set the stage for dangerous reactions if serum administration is continued, including anaphylactic reactions as well as serum sickness and a "complex nephritis" of the Masugi type (66), but that since such subjects would eliminate the foreign protein molecules much more rapidly, their immunosuppressive effects would be reduced. Although reactions dangerous to the subject can certainly occur under these conditions, the immunosuppressive potency of ALS against cell-mediated reactions is reduced by only a modest degree in animals immunized to the appropriate heterologous protein (91,103,187) (see "Immunity or Tolerance to Foreign Protein" under "Effects of Other Treatments on the Impact of ALS").

Further subfractionation of the IgG portion of active rabbit sera has been performed by James et al. (70). The "fast" gamma 1 fraction was compared to the "slow" gamma 2, and the latter, also rich in cytotoxic activity, appeared to be more immunosuppressive as determined by responsiveness of rats to sheep red cells and BSA. The effect on immunosuppressive potency of actual dismemberment of the IgG molecule by enzymatic digestion has been studied with some thoroughness. Neither F(ab′)2 nor equivalent doses of Fab′ fragments were found to retain immunosuppressive activity *in vivo* by Riethmuller et al. (130). The F(ab′)2 fragments were found to have slight but definite activity, however, by Daddi et al. (32) in experiments involving several biological test systems, including graft–host reactions. After incubation of donor adult lymphocytes *in vitro* with F(ab′)2 fragments from ALSIgG, their injection into newborn mice failed completely to cause "runt disease," which invariably followed treatment with fragments prepared from normal serum. The greater activity of the divalent fragments may be attributable in part to their higher affinity for antigen than univalent fragments. Their rapidity of elimination from the body may also be important.

In Vitro Properties

Antilymphocyte antibody combines with lymphocytes *in vitro* and, in the presence of complement, will show varying titers of cytotoxic activity against lymphocytes. Leukoagglutinating capacity is also readily demonstrable. Some activity against the serum proteins of the donor species can often be identified, but this can be absorbed without changing the biological potency of the serum against cells (52). Wide cross-reactivity with cells from any member of the immunizing species has been demonstrated for the mouse (52), rat, dog (69), and man (186).

ALS can induce blast transformation of lymphocytes *in vitro* (51), and may enhance or retard lymphocyte transformation in mixed culture (53). A rather strong opsonizing effect can be demonstrated entirely *in vitro* (54), or can be shown by incubating lymphocytes *in vitro* with ALS, and subsequently injecting them into a test animal, where they are promptly sequestrated, particularly in the liver (98). This is reminiscent of the sequestration of labeled cells in liver and spleen which takes place promptly after the *in vivo* injection of ALS (164) and is one reason for its candidacy as a possible *in vitro* test for predicting the *in vivo* potency of sera (98). A number of other ways of detecting the binding of antibody molecules in ALS preparations to lymphoid cell surfaces are available. One of these can be found in the capacity of sera to inhibit the formation of rosettes with sheep red blood cells in the presence of complement (5). This phenomenon, studied extensively by J. F. Bach, is believed to be related to the amount of antibody binding to T cells which is present in the ALS, and has been advanced as a particularly good prospective test for immunosuppressive potency *in vivo* (see below). Suppression of the plaque-forming capacity of thymus-dependent cells has been described (99). Lymphocyte-mediated cellular cytotoxicity can also be inhibited by ALS *in vitro* (61) as can the release of migration inhibition factor by sensitized lymphocytes (129).

Thus, it is apparent that ALS, and even their IgG fractions, can be expected to contain antibodies with a great variety of specificities. At least some of these antibodies are detectable by almost as many different *in vitro* techniques as there are. But even from among those molecules of IgG which actually bind to lymphocytes [estimated by Woodruff as being $\sim 10^6$ per mouse lymphocyte in one system he has studied (190)], careful consideration must be given as to just which are the active molecules. Before we turn attention to the question of what sort of antibody in this complex mixture can be considered to have desirable biological effects, and by what mechanisms, we will consider some of the adjuvant treatments or maneuvers which can influence the impact of ALS as an immunosuppressive agent.

EFFECTS OF OTHER TREATMENTS ON THE IMPACT OF ANTILYMPHOCYTE SERA

Experiments concerning the influence of additional treatments of subjects on the impact of concomitant ALS treatment are of interest because of the light they may shed upon its mechanisms of action. Furthermore, some adjunctive treatments can, of course, be of immediate clinical utility.

Thoracic Duct Drainage

Woodruff and Anderson (193) observed that thoracic duct drainage had a distinctly potentiating effect on skin graft survival in rats when combined with ALS treatment. An interesting difference between these treatments, applied separately, is that thoracic duct drainage alone cannot prolong the survival of transplants to previously sensitized recipients, whereas ALS does.

Thymectomy

The removal of the thymus from adult rats (75) or mice (111) will prolong the lymphopenia resulting from a later dose of ALS and will also further prolong the survival of skin allografts resulting from ALS treatment, especially in some strains of animals. It is not clear what role thymic hormonal factors, or the lack thereof, play in bringing about ALS effects. Since the slight effects of adult thymectomy alone are increased greatly by ALS, it is likely that they are not mediated through the thymus, but evidence has been presented that antibody to a thymic hormone preparation ("thymosin") can have some immunosuppressive effects on intact animals (59). The addition of thymectomy to ALS treatment in mice can alter the immune state of adult animals sufficiently to allow the induction of a state of specific immunological tolerance of allogeneic cells following infusion of donor lymphoid cells into immunosuppressed animals (113). In this situation, a state of stable chimerism and specific lack of responsiveness against donor cells results. Although these attributes are held in common with mice which have received a cell-free donor antigen infusion along with ALS and *B. pertussis* treatment [see experiments of Brent and Kilshaw, for example (22)], it is uncertain as to whether the state of unresponsiveness created in these two circumstances is based on the same mechanism. Similar states of specific unresponsiveness, utilizing ALS in the treatment schedule, have been induced in respect of potassium dichromate as a skin-sensitizing agent in conjunction with cyclophosphamide (172). Other experiments with cellular antigens have been performed using various dosage schedules of cyclophosphamide, methylhydrazines, or small doses of irradiation [(44), and see below].

Chemical Immunosuppressives and Irradiation

The administration of agents, known to have nonspecific immunosuppressive properties on their own, has also been tested with ALS preparations *without* the addition of the test antigen until the response to it is to be determined. Corticosteroids of the hydrocortisone type are quite effective in potentiating prolongation of skin graft survival in conjunction with ALS (see ref. 86). Species differences in regard to steroid effects of this sort must always be taken into account, however, so it should probably not be surprising that Weil and Simmons (180) found that prednisolone treatment did not increase further the survival time of dog renal allotransplants when used to supplement a brief course of horse antidog lymph node cell globulin, whereas the addition of azathioprine (Imuran) did result in some additional immunosuppression.

Experiments employing sublethal doses of irradiation in mice (450 R) were thought by Levey and Medawar (91) to be particularly revealing in that radiation just prior to test skin grafting and ALS treatment resulted in a reinforcement of the ALS effect, whereas radiation given after ALS treatment (on days +2, +5, or +15) resulted in a curtailment of graft survival. This was interpreted as showing that a treatment which hastened lymphoid cell turnover after ALS would tend to restore a population of normally responsive cells more rapidly, whereas earlier treatment would allow the ALS to influence even more an attenuated and regenerating cell population. The potentiation of ALS effectiveness with *B. pertussis* treatment (127) has been attributed to the "peripheralization" or mobilizing influence of this agent on lymphocytes, thus making them more available to be influenced by ALS.

Route of Administration

In general, the route of injection of ALS has been found to be of relatively little importance to the degree of immunosuppression produced by a given dose, especially in small animals, although the local inflammatory reaction and resultant disability resulting from subcutaneous or intramuscular injections has been cause for considerable concern in larger animals and in clinical trials (see below). Evidence supporting some local influence of ALS has been advanced, in that direct injection of ALS into an extremity bearing a skin allograft has been stated to be more effective in the dog (93), the monkey (56), or the ear of the rabbit (58) than a similar dose of serum administered systemically. One report has also appeared recently suggesting that ALS can be effective in the rat in prolonging skin graft survival if given by the oral route (140).

Immunity or Tolerance to Foreign Protein

As mentioned previously in the section "Purification . . ." under "Preparation and General Properties . . . ," the prompt appearance of antibodies to the xenogeneic (heterologous) protein administered as ALS, or any of its fractions, has been observed very commonly and can be detected in a number of ways including the rate of elimination of labeled globulin (28). Although it was hoped initially that the immunosuppressive capacity of the ALS itself might result in little or no antibody being produced to the foreign globulin, the disparity between the suppressive effects to humoral as against cell-mediated responses appears to allow this, especially at lower dosage levels or if treatment is intermittent or discontinued.

A logical approach to circumvent any deleterious influence produced by such antibodies is the production of "high zone" tolerance of the foreign IgG, either by suitable injections of IgG prepared from the normal serum of a member of the species donating the ALS, or, perhaps, by the ALSIgG itself. Tolerance can be achieved fairly easily in mice to the serum proteins of foreign species even though the foreign IgG is quite immunogenic (85), and experiments to determine the effect of tolerance of the foreign IgG on the later impact of ALSIgG have been done. This has been tested both by prior injection of adult mice with about 5 mg of rabbit IgG, from which aggregates had been removed by "biological filtration" (103), and by the intravenous injection of rabbit IgG at birth (187). In both situations, a given dose of ALSIgG resulted in significantly, although not greatly, increased immunosuppression as evidenced by the duration of skin graft survival. Fairly good tolerance of IgG has also been reported to have been achieved in dogs (87) and in subhuman primates (30). Although no firm evidence for improved immunosuppression of ALSIgG or of reduced toxicity following such treatment is yet available in these larger species, it would be logical to expect it. The situation in regard to the use of this approach in man will be discussed below.

MECHANISMS FOR IMMUNOSUPPRESSION BY ANTILYMPHOCYTE SERA (INCLUDING A DISCUSSION OF *IN VITRO* TESTS FOR SERUM POTENCY)

Much attention has been paid to the question of the mechanism by which heterologous sera, or their IgG fractions, exert the striking biological effects which have been described above. The search for a better understanding of this process is encouraged by its considerable intrinsic interest and also by the fact that far more effective clinical applications of manipulation of the immune system by antibodies may well emerge from a full understanding of their mode of action. One critically needed advance has been the establishment of a reliable *in vitro* test which will predict the *in vivo* immunosuppressive potency of a given serum. As the questions of mechanism of action and the selection of an optimal test for serum activity are closely interrelated, they will be discussed together in this section.

A substantial literature now exists regarding the mechanism of action of antilymphocyte sera in various reviews, symposia, and articles (a partial list would include refs. 3,34,37,74,83,91,133,153,174,186,191). As mentioned previously, a particularly obvious fact related to ALS administration is that a striking peripheral lymphopenia is regularly produced within hours of beginning serum administration. This may not persist, however, and it was soon found, especially by Levey and Medawar (92), that continuing immunosuppressive effects did not require the continuance of a lymphopenia. It was accordingly thought that mechanisms, in addition to nonspecific or general elimination of lymphoid cells *in vivo,* should be involved, even though most sera will kill a wide variety of lymphoid cells in the presence of complement *in vitro*. Alteration of responsive cells, it was reasoned (91), could take place without their deletion, either by antiserum coating (resulting in a "blindfolding" of recognition by these cells), or by a process of blast cell transformation to a state in which the cells would be committed nonspecifically ("sterile activation"). The former idea became less attractive to its original proponents as it became clear that the state of immunosuppression outlasted a number of cell generations under certain conditions (90). The latter seemed less tenable for a number of reasons, but especially in view of three facts. First, that divalent antibody fragments could cause cellular transformation *in vitro,* but were quite inactive *in vivo* (see the section "Purification . . ." under "Preparation and General Properties . . ."). Second, that although marked cell transformation can be caused by several different agents, none possesses immunosuppressive powers even remotely similar in magnitude to ALS [see Landy and Chessin (88)]. Third, the effect of ALS on lymphocyte activation, either by itself or by other agents *in vitro,* is too unpredictable as to dosage and timing to support the contention that cell transformation is responsible for a major consistent *in vivo* effect (34,53). These points and the fact that some sera will cause T cell transformation (see below and ref. 72 for significance of this), make the use of lymphocyte activation by ALS *in vitro* of doubtful value as a predictive test for serum potency (54).

The possibility that ALS acts entirely by *antigenic competition,* a nonspecific process, can be ruled out by its effectiveness even in the presence of immunity to itself, while the possibility of specific *immunological enhancement* being a major factor is made unlikely by its broad range of specificity without regard to the cellular antigen concerned. Nevertheless, it would be unwise to dismiss these several forms of action completely as there may well be situations in which they could contribute to a given *in vivo* effect, though probably to a minor degree.

On the other hand, much evidence has now been accumulated to support the conclusion that by far the most important mechanism for the immunosuppressive action of antilymphocyte sera is by the selective deletion of long-lived recirculating lymphocytes. Before examining *how* such cells may be diluted, we will consider briefly the rather sizeable body of interrelated evidence which supports the conclusion that such cells are indeed deleted by ALS treatment. This evidence can be summarized as follows (see especially refs. 34,86,191 for general discussion):

1. ALS treatment results in a loss of cells from just those portions of the lymph nodes and spleen in which long-lived, recirculating thymus-dependent cells (T cells) are known to reside (122). The thymus is relatively spared (165,174).
2. The lymphopenia which follows an injection of ALS is associated with a similar depletion of small lymphocytes from the thoracic duct lymph, but this is slightly delayed and of greater duration than the depression of lymphocyte counts in the peripheral blood (2,35).
3. The population of cells remaining in lymph nodes and spleen after a brief course of ALS treatment, is capable of mounting normal cell-mediated responses, e.g., graft-vs-host responses (177). These remaining cells appear

to have a shorter average life span than the native cell population, as determined by following the fate of cells labeled *in vivo* with tritiated thymidine (35). Furthermore, it includes a lesser proportion of cells which can migrate normally to lymph nodes when injected intravenously into normal individuals (165); fewer respond to PHA (175); fewer form rosettes with sheep red blood cells (5); and fewer, under appropriate experimental conditions, are susceptible to the cytotoxic action of an anti-θ serum in the presence of complement (128,139), or migrate electrophoretically as T cells (13).

The means whereby this relatively selective elimination of a subpopulation of cells occurs could be traceable to one or more of three of their special properties. It might be because they are long-lived and are *replenished at a slower rate*; it might be because they recirculate and are accordingly *more available for repeated attack* in a peripheral location during periods of repeated exposure; or it might be because these cells bear *distinctive antigens* and that the ALS, or at least some sera, are able to destroy them selectively.

Selective depletion on the basis of less rapid replenishment would seem to be entirely plausible and consistent with well-established facts regarding the behavior and life span of this class of cell (50). The importance of a special susceptibility of cells while in a "peripheral" location has been suggested by Levey and Medawar (92), and their evidence, based mainly upon *in vivo* experiments with lymphocyte transfer reactions (see the section "Effects on Other Immune Responses" under "Preparation and General Properties . . ."), has received some support from the studies of others. Denman and Frenkel (37) found that labeled ALSIgG was taken up very little in lymphoid organs even after fairly prolonged periods of administration. Thus, the depletion of cells from "thymus-dependent" areas of lymph nodes and spleen observed after ALS treatment might be mainly attributable not to the selective destruction of these cells *in situ,* but to their failure to return in the cycle of their travels. Some reservations must be maintained regarding the physiology of local penetration of ALS into various portions of lymph nodes and its effects there, however, as the data are not entirely consistent (83). Still, the fact that selective destruction of T cells actually occurs seems well established. As Mitchison first suggested (106), however, T cell deletion may make itself felt in a number of ways. In particular, loss of the helper activity of T cells may be an important feature in determining the result of treatment (4).

The question of whether this selective destruction or inactivation of T cells takes place mainly on "physiological" grounds, i.e., because of its more prolonged effects on the slowly renewing recirculating cell population, or to some extent on immunologically specific grounds is of special interest. There is no doubt that characteristic cell surface isoantigens (alloantigens) are present on lymphocytes, and that increasing numbers of subpopulations of lymphoid cells can be distinguished amongst members of the same species by them (18). It has been stated that the important antigens on mouse lymphocytes to the production of ALS do not include those which segregate amongst strains, such as TL, Ly, or θ, in strongly immunosuppressive sera (146). Antibodies to other antigens, common to thymus-dependent cells in all strains, were found, however. There is also reason to believe that heterologous ALS can differentiate between populations of recirculating and nonrecirculating cells in the rat (65). In recent experiments, Ochiai et al. (120) have been able to absorb the globulin fraction of a rabbit antimouse thymocyte serum with "B cells" sufficiently to remove activity by several assays to B cells, but without reducing the ability of the antiserum to prolong skin graft survival (120). Most ALS preparations, unabsorbed, have no apparent ability to make such a distinction *in vitro,* however, where they will destroy a high proportion of lymphoid cells in the presence of complement.

In vivo, where antibody concentrations and the local presence of cofactors are quite different, T cells could be preferentially influenced by unabsorbed ALS because, for example, antibodies to certain T cell-specific antigens may have a higher affinity than those against other classes of antigens on lymphocyte surfaces or these antigens may be higher in density or clustered in a particularly propitious way to bind multiple antibody molecules. Accordingly, the presence of T cell-specific antigens could increase the chance of an "additional hit" by a complement-binding antibody molecule on the surface of a cell. This could make it more likely that T cells would be eliminated in the presence of a mixture of antibodies including some directed against specificities shared widely among lymphoid cells and others confined to T cells. All this would depend, of course, on the pattern of the immune response of the target species against the antigens presented to it. Some species apparently do not recognize and react strongly to these particular antigens in certain other species (see, for example, ref. 76), whereas others do. The relative contribution to immunosuppressive potency of ALS from antibodies directed to antigens shared widely among lymphoid cells as against those reacting only with T cells has not been studied extensively. There is some evidence, however, that absorption of sera with B cells may still leave a good deal of immunosuppressive activity behind (154), and we have found recently that ALS, prepared in rabbits against lymph node cells from nude (thymic-deficient) mice, is distinctly less potent than that against cells from normal individuals (7).

Finally, we come to a consideration of the mechanism of elimination of lymphocytes affected by ALS. Lysis of cells, through action of the full complement sequence, certainly may take place *in vivo* as it does *in vitro,* and it is known that complement is consumed *in vivo* after ALS infusion (173). The pattern of sequestration of labeled, infused cells mainly in the liver, but also in the spleen of ALS-treated subjects (100,164) suggests, however, that opsonization and phagocytosis may be a much more important mechanism for cell removal. This is supported further by the interesting observation that mice, congenitally deficient in C5, can support the immunosuppressive activity of ALS to about the same degree as congenic individuals possessing the full complement sequence (see refs. 11,27,183). Opsonization would require only the 1,4,2,3 components, at least in the classical scheme of complement action (114).

One might expect, accordingly, that tests of immunosuppressive potency which evaluate the opsonizing activity of a given serum on lymphocytes, either by treating them *in vitro* and injecting them into a test animal with suitable labeling techniques (97,98), or by an entirely *in vitro* technique (see, for example, ref. 54), might be useful. The correlation of results from the opsonization titers of several rabbit antimouse sera with mitogenic, agglutinating, and cytotoxic titers favored the opsonization test in one study (54). Nevertheless, this correlation has apparently not always been good (6). This could be because opsonization, although the principal route for cell destruction, might include many of the wrong cells, that is, cells less important to immunosuppression. This could also be true of lymphocyte agglutination as a test, which has been notoriously misleading in predicting immunosuppressive potency (76) as has simple complement-dependent cytotoxicity. The difficulty with these tests, then, is probably due to the great heterogeneity both of the cells and of the antibodies in the reaction mixtures used. This might be improved by using a purified preparation of T cells as the target for the IgG antibody fraction of unknown potency to be tested.

Given the present status of the antibody preparations under study for potency, some kind of assessment of their activity against T cells, even in a mixed population, would appear to be the most logical alternative. At present, the most appropriate test in use appears to be the SRBC rosette inhibition assay of J. F. Bach (5). Even though the exact mechanism by which rosetting occurs is as yet unknown, this test appears to yield a reasonable estimate of T cell numbers, a measure which is already being put to use in certain clinical trials as we shall see below. With careful attention to technical details, such as the elimination of lymphocyte clumping *in vitro* by the use of a Ca- and Mg-free medium, and by DNase treatment of cell suspensions, and the reduction of extraneous impurities which can cause inhibition of rosettes on their own, variables can be reduced (119). Nevertheless, as will be seen below, the rosette inhibition assay may still leave something to be desired in predicting immunosuppressive potency of sera for clinical use.

SOME CLINICAL APPLICATIONS OF ANTILYMPHOCYTE SERUM

History and General Remarks

Shortly after the striking efficacy of ALS had been demonstrated in animals in the middle 1960's, there was naturally a strong interest on the part of clinicians, especially those dealing with transplant rejection and autoimmune diseases, to employ antisera against lymphoid cells in clinical trials. Starzl and his colleagues were the first to do this with a therapeutic intent (158). A globulin fraction (prepared by ammonium sulfate precipitation) from the serum of horses which had been immunized with human thymus, lymph node, and spleen cells was used in the early trials (66). We also tested a column purified IgG fraction of rabbit antihuman lymph node cell serum in a small number of volunteer individuals for the evaluation of immunosuppressive effects (112).

In the former study, ALG was employed in the treatment of 20 renal transplant recipients, along with azathioprine and prednisone as regularly used, in doses ranging from 14 to 50 mg protein/kg body wt per week in the early weeks, later descending to 2–6 mg/kg per week over a total treatment period of 3 to 4 months. During the few days of treatment with ALG alone just prior to transplantation, several delayed hypersensitivity skin tests were evaluated, and preexisting sensitivities were found to be "invariably abolished" (159). Nineteen of the 20 patients received transplants from living related donors. The results were analyzed on the basis of survival of transplants, appearance of renal biopsies, and the dosage requirements for other immunosuppressive agents to maintain patients free of active rejection. The appearance of antibody to the intramuscularly injected horse protein was also examined by an electroimmunodiffusion method, and the effects of serum treatment on the formed elements of the blood were evaluated. The protein content of the different batches of ALS used ranged from 4.6 to 9.3 mg/100 ml and the leukoagglutinating titers against human leukocytes from 1:4,096 to 1:32,768. In assessing the effect of ALG treatment on rejection, three groups of prior recipients of consanguineous donor kidneys were selected for comparison. These groups differed from one another mainly in the protocol for administration of steroids in each group. In the treated group, only one patient died (after receiving two kidney transplants). This compared somewhat favorably with the control groups. Less prednisone was required in their treatment. Lymphopenia was variable and not striking before transplantation and could not be interpreted thereafter because of the complexity of events. Kidney function was variable but, in general, reportedly satisfactory. All patients developed local pain at injection sites, often severe, and usually accompanied by fever. Four patients developed systemic reactions of some kind including alterations in blood pressure and dyspnea. Eight patients developed thrombocytopenia, one requiring platelet infusions. All patients showed rises in precipitin levels to horse protein. Treatment was continued in most instances and, in some, the titers to horse protein fell in the face of continuing treatment. Histological studies of renal biopsies revealed a number of changes in the glomeruli, including pathological localization of human serum fractions, but no localization of horse globulin. Thus, in this early study, it was concluded that some evidence for a favorable influence on the management of transplant recipients had been gained and that dangerous side effects had not been found to be prohibitive of further trials.

In the other study, in which rabbit ALSIgG had been used in five volunteer patients, tests for immunosuppressive effects consisting of delayed hypersensitivity skin tests and the survival of small test skin allografts were carried out. A single batch of IgG (protein concentration 10 mg/ml) was used in a course of 3 injections over a period of 5 days (delivering 9, 5, and 5 ml, respectively, by the subcutaneous route). The preparation had a leukoagglutinin titer of 1:2,048 and a saline hemagglutinin titer of 1:32. Local pain, fever, and one instance of joint pain consistent with serum sickness were encountered. Striking reduction in cutaneous responses to trichophyton, monilia, and mumps antigens (see Fig. 1) occurred with a tendency for return to normal responsiveness by 6 weeks after treatment. A lymphopenia in the range of 10–20% of preinjection values was detectable within 12 hr of commencing treatment and this remained for from 2 to 6 days after injections ceased. Skin allograft survival was assessed by placing from 1 to 3 grafts on each of 5 recipients. A modest but definite prolongation of survival time was observed in the group of recipients by comparison with 39 normal controls.

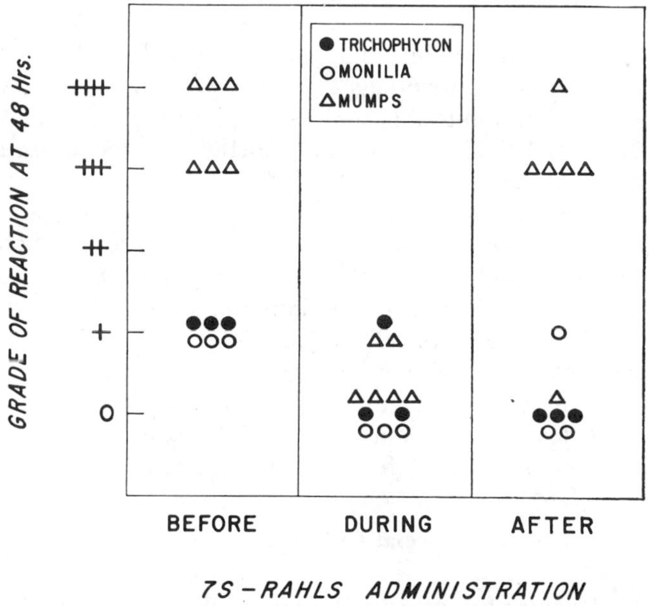

FIG. 1. Delayed skin reactions in 10 human subjects before, during, and 6 weeks after a course of 3 injections (90, 50, and 50 mg on alternating days of a 7S fraction of rabbit anti-human lymphocyte serum). Note that ALS treatment greatly reduced these preexisting immune responses, and that evidence of return of reactivity was only partial at 6 weeks. (For details see ref. 112.)

These early efforts are recounted in some detail as they illustrate the problems which have been encountered in clinical trials of ALS therapy. The immediate obstacles in the way of trials which can give truly reliable results in the complex set of circumstances presented by patients requiring organ transplants include the following:

1. The availability of a *single, uniform preparation* in sufficient quantity to allow treatment of a group of patients large enough to permit sound statistical analysis of results. This preparation must be available in sufficient quantity to allow delivery of an adequate dose (see below).
2. A serum IgG fraction which has been tested for toxic and extraneous antibodies (with maximal removal of such antibodies by absorption) and which shows favorable signs of having immunosuppressive potency for human beings.
3. A thoroughly *standardized study protocol* into which patients are entered systematically with consideration of the histocompatibility differences obtained between donors and recipients and with careful attention to uniformity of patient management and analysis of results. One of the most difficult, but most needed, provisions in regard to a study protocol has been the necessity for *prospective randomized* trials with as similar patients as possible being admitted to each study.

These requirements have been met to only a limited degree in the past decade and progress in the clinical application of ALS has been disappointing. Several international meetings and reviews of various sorts have devoted major consideration to this subject. These include the following references: 3,6,21,31,34,116,134,136,141,143,156,186, and much useful information is to be found in them.

It can readily be understood from the foregoing account of animal experiments with ALS how perplexing inconsistencies could be encountered in clinical trials using multiple batches of sera prepared and administered in quite different ways. This can only compound the familiar challenges in clinical research of evaluating objectively the course of a very complex disease process in concert with other means of treatment. Before attempting a brief summary of the current status of clinical trials with ALS, we will take up some of the leading problem areas which have introduced important variables in order to examine their current status individually.

Special Problem Areas

Preparation of Anti-Human Lymphoid Cell Sera

Most efforts have been devoted to sera prepared in the horse (30,66,116,141,170) and the rabbit (33,112), with A. G. R. Sheil's group in Australia being particularly interested in the testing of sera made in the goat (see, for example, ref. 101). Some others have tested certain other species, although no sera from other species have been employed in major clinical trials. In more recent years, several industrial concerns have engaged in the preparation and testing of large batches of horse antihuman lymphoid cell IgG. These have included the Hoechst-Behringwerke and the Upjohn pharmaceutical companies. Lymph nodes, thymus, thoracic duct cells, normal peripheral blood lymphocytes (101), chronic lymphatic leukemia cells (41), and large-scale cultures of lymphoblasts (149) have been used as antigens. Although, as mentioned in the section "Source of Antigen" under "Preparation and General Properties . . .", it has been suggested by some that sera of less than optimal potency are produced by the lymphoblasts from cultures which have been employed, since they are composed of B cells, they have been shown to have definite potency in human beings (125). We would expect, however, that B cell cultures would be undesirable antigen sources.

Large quantities of IgG are generally produced by salt or ethanol precipitation, followed by chromatographic separation of the IgG fraction. Since hemagglutinins reside mainly in the 19S fraction, these undesirable antibodies are excluded to a considerable degree during fractionation, but absorption with erythrocytes and/or platelets may be required. Absorption with nonlymphoid cells to eliminate extraneous antibodies, which will bind with other cell types, has not been performed with sera for clinical trials, and it is easy to understand why this is true in view of the large quantities of cells which would be required for the production of a sizeable batch of material. The same is the case for subcellular fractions as potential antigens, and no full clinical trial has made use of a serum prepared against anything but intact cells.

Testing of Serum Fractions for Potency and Toxicity

Since the *in vivo* potency of ALS preparations cannot be routinely evaluated by testing in human volunteers, it has been difficult to establish a reliable standard against which the results of various *in vitro* tests might be compared. Accordingly, an *in vivo* assay system for antihuman sera has been devised by Balner in which skin *allograft* survival on subhuman primates, which have received antihuman cell serum products, is used as the measure of serum potency (9). With this system, ALS preparations may be tested in normal subjects and in the absence of other immunosuppressive measures. In general, somewhat longer extension of graft survival time is achieved in chimpanzees than in lower primates, but it has been felt that the comparative results of different sera with chimpanzees and several species of monkey, e.g., *Rhesus*, are similar to one another especially if higher doses are used. The rosette inhibition titers for antihuman IgG preparations have correlated fairly well with their ability to prolong the survival of primate skin allografts (5), although this correlation may be less close if whole serum, rather than its IgG fractions, is tested (125). Systematic proof of a reliable correlation between the primate test results and actual potency in patients is still lacking (8). Such a connection may often be obtained, but if one accepts the contention that an appreciable proportion of the selective effect of ALS may depend upon recognition of particular cell surface antigens on human T cells by the antiserum, it must be recognized that subhuman primates could be imperfect models of human beings.

Another reason for utilizing the primate test has been its value as a screening test for toxicity. This was probably more important formerly when sera were fractionated less thoroughly and absorbed less completely with erythrocytes. But the large proportion of extraneous protein contained in all preparations now in clinical use can have unpredictable toxic effects.

Most sera now under clinical trial have been prepared by the adjuvant technique in horses with 2 to 3 subsequent intravenous booster injections of human thymus cells in suspension. The thymus is used as it provides a source of cells which can be rid of connective tissue contaminants and red cells which contain sufficient numbers of "thymus dependent" cells. A study of this approach to ALS production has been performed in our Unit by Cosimi et al. (30). It fully confirms earlier work with rabbit antimouse sera in that continuing booster injections result in modest increases in serological (lymphocytotoxin) titers, but lessening immunosuppressive potency, at least as determined in the primate assay using *Rhesus* monkeys.

Typical sera raised by this method, as now carried out by the Upjohn Company (179), will have *in vitro* properties as displayed in Table 3. Careful tests for sterility and for endotoxin activity are also performed.

Side Effects of ALSIgG Administration and Their Management

In early trials, ALS was administered entirely by the deep subcutaneous or intramuscular routes in relatively small dosage, (i.e., about 20 mg protein/week (158). Quite severe local pain, swelling, and erythema fol-

TABLE 3. *Some properties of two lots of Upjohn antihuman ATG tested*

	Lot number	
Assay	16,138-16	16,138-18
Hemagglutination	1:80	1:20
Cytotoxicity	1:2,790	1:2,790
Ab (platelets)	<1:33	1:46
Rosette inhibition	1:11,600	1:20,000
Total protein	49.3 mg/ml	49.9 mg/ml
% 9 and IOS	4.8%	4.6%
In vivo (primate) (control = 8 days)	16 days	17 days

lowed such injections. These were often associated with fever spikes (to 103°F, or so) within a few hours and, not infrequently, by signs of serum sickness, especially joint pain and erythematous skin rash. All these symptoms were consistent with previous experiences in which horse or rabbit sera were used for other purposes, usually for the passive transfer of immunity in the treatment of active infections (see, for example, ref. 62). Severe anaphylactic reactions have been very uncommon, although occasional hypotensive episodes have occurred and caution is required, especially shortly after the administration of the first doses in a course of treatment, even though careful preliminary skin testing with the intradermal injection of a diluted serum sample has been carried out. No marked difference in toxicity between horse and rabbit serum fractions has been apparent. Three deaths have been reported. One followed the intramuscular injection of a crude ammonium sulfate precipitated horse globulin preparation containing about 5 mg protein. This occurred 12 days after kidney transplantation when the patient had received 11 previous doses during an intended 3-week course of treatment (182). The two other deaths occurred after anaphylactic reactions in similar circumstances (57). In our experience, these reactions, even the minor ones, are reduced by preliminary treatment with antihistaminics at ordinary doses, by the accompanying steroid treatment which is normally given, and by slow infusion of the protein intravenously.

The appearance of local reactions, while rarely dangerous, obviously limit greatly the dose of foreign protein which can be administered subcutaneously or intramuscularly. It has been estimated that the amount of serum which should be required for a patient on the basis of body weight could be about 3,000 times greater than that for a mouse (86). In the mouse a single dose in a treatment schedule which will maintain a skin allograft is about 0.25 ml whole serum. This would mean in a patient that the IgG from about 750 ml serum, or about 7.5 g, would be delivered per dose. The point here is not, of course, the exact quantities, but that an order of magnitude separates the amount of serum used in animals from that employed in patients so far.

With these considerations in mind, a number of groups began trials using the intravenous route for higher dose serum administration (for example, see refs. 20,30,71,116). This requires the placement of a catheter into a central vein and its maintenance with special care as to sterile precautions. Using *Rhesus* monkeys as subjects for testing an antihuman thymus cell IgG preparation from horses, Cosimi et al. (30) found that stepwise increases of the intravenous dose to 100 mg/kg, delivered on three occasions over the first week after skin grafting, progressively increase survival time. Higher doses in this protocol failed to extend survival further and were met with increasing toxic reactions. In tests using patients with multiple sclerosis, Najarian found that increasing doses resulted in steadily increasing survival times of test skin allografts to 20 mg of antilymphoblast IgG/kg (116). No higher doses were tested in this trial, although these investigators have employed doses as high as 30 mg/kg-day in some kidney transplant recipients (150).

Considerable discussion has centered about the desirability of intentionally inducing high zone tolerance in patients of the foreign gamma globulin in order to reduce undesirable immune reactions to it as well as, perhaps, to increase its immunosuppressive effects (see the section "Immunity or Tolerance to Foreign Protein" under "Effects of Other Treatments . . ."). Taub et al. (163) reported the induction of unresponsiveness to horse gamma globulin in a patient by administering a single dose of 70 mg/kg of aggregate free material intravenously. This obviated the immune clearance of labeled gamma globulin subsequently. Brendel and his colleagues (132) have advocated this approach, and have suggested that a dose of 10 mg/kg of ultracentrifuged normal IgG be given intravenously 6 days and a second dose of 30 mg/kg 3 days before beginning ALSIgG treatment. This treatment, it is stated, reduced allergic side effects of various types. This approach has not been widely adopted as yet, however, as most workers employing relatively high-dosage intravenous treatment with purified IgG find remarkably low incidences of severe side reactions (30) and rare evidence of antibody formation. The University of Minnesota group now feels that reactions to their IgG protein preparation are rare (148), although in their earlier experience, perhaps with somewhat lower doses of material, cessation of treatment because of side effects was forced in 24% of kidney transplant patients and as many as 62% of nonkidney transplant patients who were receiving ALSIgG treatment for various autoimmune conditions (48). Although tolerance induction was tried by this group, they discontinued its use because it proved, in their hands, to have uncertain and not definitely beneficial results (107). There seems to be little doubt, however, that the incidence of serious side effects including antigen–antibody complex disease affecting the kidney, has been less in recent years than formerly (79). This may be due to the use of higher doses of protein from which more extraneous antibodies have been removed. The combined use of ALGIgG with other immunosuppressives, especially steroids, appears to reduce the severity of reactions. The presence of some antibody to the foreign IgG, however, could probably be detected by sufficiently sophisticated techniques in many patients (71). It is also a matter of some interest that the level of antibodies to the foreign protein, should they form, may actually decline with continuing intravenous treatment (71), as mentioned above.

The question of increased incidence of infections with various pathogens in association with ALG treatment merits attention. As yet, no definite trend has been reported, and it will be a difficult point to settle since infection rates are quite high with other methods of immunosuppression. The same uncertainty remains in regard to the incidence of malignant tumors in patients receiving ALG.

Present Status of Some Clinical Applications of ALS as Therapy

Kidney transplantation

The principal use for which ALS preparations have been under trial has been in the management of kidney transplant rejection. Rejection reactions are difficult to quantify and they are determined by a number of important factors in addition to the immunosuppressive treatment given the recipient. These include the immunogenetic difference between donor and recipient, the presence of preimmunization of the recipient against donor antigens, the immune responsiveness of the recipient (whether genetically determined or otherwise) to the antigens concerned, and the "suppressability" of the immune reaction to be generated by the recipient.

The results so far are consistent with the expectation one must have from the entire foregoing account. Most ALG preparations have been "homemade" and have been tested thereafter with different tests of potency. Because of this, and because of the considerable expense and effort involved in producing batches of ALSIgG of sufficient size to permit testing, most reports have been based upon observations using small and multiple lots of material. In the United States, regulations of the Food and Drug Administration prevent the exchange of such materials across state borders unless the drug or agent has met certain preliminary tests and is certified as being safe to undergo more advanced clinical trials (has been the subject of a "New Drug Application," N.D.A.). Under these circumstances, it has become almost impossible for material to be exchanged for anything but *in vitro* testing or testing in animals. Fairly recently, several larger programs of study have been embarked upon with the indispensable cooperation and interest of a central agency providing certain necessary resources, including the ability to prepare large amounts of purified serum fractions. These have included the Medical Research Council of Canada (Dr. H. E. Taylor, Director, Antilymphocyte Serum Program), the Hoechst-Behringwerke concern (Dr. R. H. Rousell, Associate Medical Director, Biological in the United Kingdom), and the Upjohn Company, Kalamazoo, Michigan (Drs. N. B. Marshall and W. Wechter). One additional study which is now actively proceeding is that sponsored by the National Institute of Allergy and Infectious diseases and conducted by Drs. K. Sell and F. Belzer. This has made use of a well-characterized rabbit antihuman thymus cell serum (IgG fraction) which has been employed in "double-blind" fashion in kidney transplant recipients (142). Other groups, of whom we are not aware, may also be embarking upon programs in which standardized protocols for patient management are agreed to by all the clinical groups participating and a single large batch of uniformly prepared and tested ALGIgG is employed throughout. On the date of this writing, none have reached the stage of a final, definitive report.

It would be beyond the scope of this article to attempt to reproduce a detailed account and evaluation of all published reports concerning the use of ALG preparations in clinical transplantation, even in transplantation of the kidney. Nevertheless, by examining Table 4, one can gain some impression of the range of experience revealed so far.

An examination of Table 4 will serve to show that even where immunosuppression for kidney transplantation is concerned, a wide variety of different methods of preparation, plans of dosage, and means of evaluation have been employed. In only a few instances were patients assigned to be treated or not with the ALS preparation concerned on a randomized basis, and rarely was a single batch of uniformly prepared material tested throughout a given trial. Given the very large number of variables involved, known and unknown, an ALS would have to produce a major effect for it to be discernible in a small series of patients. In examining the reports summarized in Table 4 and others, the following general conclusions are suggested about ALS preparations for kidney transplantation:

1. ALS is remarkably well tolerated in general and can be delivered over a short course, more conveniently and in higher dosage by the intravenous route.
2. Long-term side effects, as, for example, renal damage from immune complex disease, are uncommon.
3. During ALS administration, and shortly thereafter, rejection reactions are less common and are less severe. As a corollary of this less steroid medication can suffice in the early postoperative period.
4. Improvement in long-term survival of transplanted kidneys has been reported only from a minority of the trials.
5. No definite increase in incidence of infection or of spontaneous neoplasms has emerged.

The experience in our unit may be useful as an example. Initially, we produced our own antihuman sera both in rabbits (112) and horses (30). As it became apparent that the availability of much larger quantities of purified IgG was going to be essential to a clinical trial using the doses of material indicated, we sought the aid of the Upjohn Company and have since participated in a trial including a number of institutions in which this material is being tested under a fairly standardized plan in a randomized prospective fashion. Our own initial results in this program, which will be published in full elsewhere, emphasize the difficulties which arise in evaluating the data of such studies, but we believe that they also define some approaches to their solution. In this study, all adult recipients of cadaver allografts were randomized to one of two treatment groups. Both groups of patients received azathioprine and prednisone according to a standard protocol. In addition, after entry into the study, randomly selected patients received horse antihuman thymocyte globulin (supplied by the Upjohn Company) which was administered intravenously in doses of 30 mg/kg-day for the first 2 weeks after transplantation. This serum was prepared and fractionated according to techniques very similar to those described by Cosimi et al. (30), and its production and use will be published in full elsewhere.

Kidney survival in the first 20 patients can now be evaluated for a full 24 months and is depicted in Fig. 2. This figure reveals an improvement in cadaver-donor kidney survival in those patients receiving ALS, although the size of the patient groups makes the difference inconclusive. However, supplementing our data with that being observed in over 100 patients, similarly randomized at other centers where the same batch of ALG has been employed, appears to reveal almost the same survival difference which we have observed. Thus a statistically significant beneficial effect from this lot of ALG has been established (179). The *in vitro* and subhuman primate skin grafting data of this batch of ALG (No. 138-16) are summarized in Table 3.

A subsequent group of cadaver kidney recipients has since been randomized similarly and treated with a different lot of ALG (No. 138-18), the

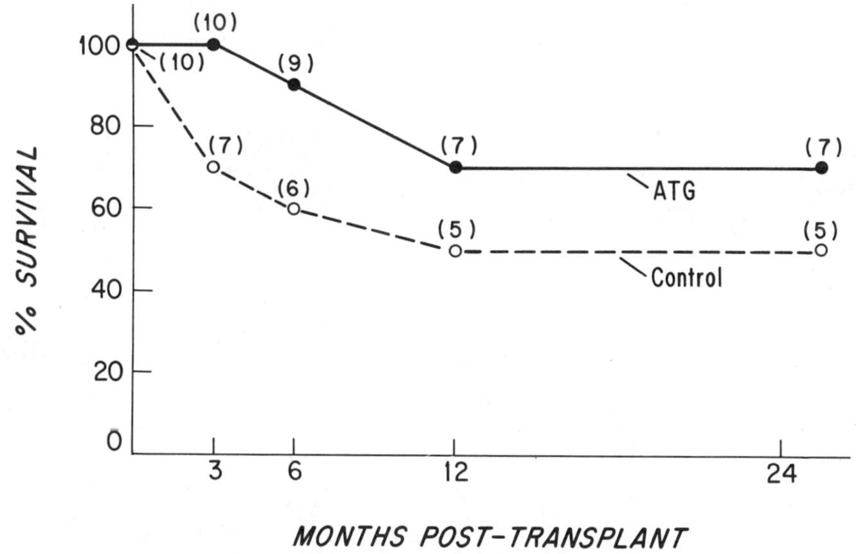

FIG. 2. Survival of transplanted kidneys in MGH-controlled series to test ALG. These curves compare the "actuarial" survival of kidneys transplanted to 20 randomly selected recipients who had either received standard treatment with azathioprine and prednisone alone, or had received this treatment *plus* ALSIgG (Upjohn ALG, Lot 138-16) at a dosage of 30 mg/kg-day intravenously for 14 days. A distinct improvement in survival is suggested but the number of patients in this portion of the study is insufficient to reach a reliable conclusion. A statistically significant difference between groups of patients treated in just this way in the larger study conducted by the Upjohn Company has confirmed the statistical significance of these differences in which the same lot of ALSIgG was used in a larger group of patients from multiple centers.

TABLE 4. *Partial summary of results using ALG in human kidney transplantation*

Investigator, institution, and ref.	Number of patients (treated/ controls)	Immuno-genetic class of donor	Plan of trial	Types of ALG preparation	*In vitro* properties of ALG
Starzl et al., Univ. of Colorado, 1967 (159)	20/-	19 related (5 parents, 13 siblings, 1 uncle); 1 unrelated	"Historic controls"	Horse antithymus; lymph node and spleen; ammonium sulfate ppt used; several batches	Leukagglutinins 1:4,096 to 1:32,708; protein content 4.6–9.3%
Doak et al., Aukland Hospital, New Zealand, 1969 (40)	10/20	All unrelated, nonliving	"Historic controls"	Horse antilymphocyte; fresh frozen spleen cells; no adjuvant; IgG fraction; several batches	Leukagglutinins 1:1024 to 1:32,000; hemagglutinins ≤ 1:128; platelets 1:32 to 1:128; protein content 5 g%
Fries, Traeger, et al., Lyon, 1969 (46, 169)	71/-	41 related; 30 unrelated, nonliving	"Historic controls"	Horse antilymphocyte; thoracic duct lymphocytes; adjuvant used; ammonium sulfate ppt used; several batches	Lymphotoxicity 1:1024 to 1:2048; hemagglutinins 1:16; protein content 50 mg/ml; prolongation of monkey skin grafts (80 mg/kg^3 × p. week) and chimpanzee skin grafts (15 mg/kg^3 × p. week)
Brendel, Univ. of Munich, 1969 (19)	40/-	All unrelated, nonliving	No controls	Horse antilymphocyte; (? of some rabbits used); thoracic duct, thymus; no adjuvant; ammonium sulfate ppt used; multiple batches	Lymphotoxicity 1:1000 to 1:8000; confirmed "high immunosuppressive effect" by testing on chimpanzees
Shorter et al., Mayo Clinic, 1970 (147)	36/-	25 related (10 parents; 15 siblings); 11 unrelated, nonliving	No controls	Horse antithymus; no adjuvant; ammonium sulfate ppt used; multiple batches	Lymphotoxicity averaged 1:256; protein content 2–3 g%
Deodhar et al., Cleveland Clinic, 1971 (38)	53/-	25 related; 28 unrelated, nonliving	"Historic controls"	Horse antispleen; fresh frozen spleen cells; no adjuvant; IgG fraction; several batches	Leukagglutinins 1:2000 to 1:4000; lymphotoxicity 1:4000 to 1:8000; protein content 4 g%; 2–3-day enhancement of rhesus monkey skin allograft survival
Starzl et al., Univ. of Colorado, 1972 (160)	77/183	131 related; 52 unrelated	"Historic controls"	Horse antilymphocyte; spleen used; adjuvant used; ammonium sulfate ppt used; multiple batches	Lymphotoxicity and leukoagglutinin titers ≥1:4000; anti-GBM not demonstrable in Dixon rat model
Monoco et al., Boston City Hospital, 1972 (109)	60/-	15 related; 45 unrelated, nonliving	"Good" ALS vs "bad" ALS	Horse antilymphocyte; lymph node; adjuvant used; IgG fraction; 8 batches from 2 horses	Passive hemagglutination 1:1000 to 1:128,000; "good" ALS with hemagglutinins >1:16,000
Wilson et al., Peter Bent Brigham Hospital, 1972 (182)	58/37	⅓ related; ⅔ unrelated, nonliving	First 47 treated if not allergic; next 48 randomized	Horse antilymphocyte; thoracic duct; ammonium sulfate ppt used; several batches	Rosette inhibition 1:2000 to 1:16,000; protein content 40 mg/ml

TABLE 4. (Part II) (*Continued*)

Dosage	Route of administration and duration of treatment	Special observations or problems	Results
–50 mg/kg-wk; in final months 2–6 mg/kg	IM; total of 4 months treatment; 5–6 days pre-op for living donor recipients	Precipitins to horse protein seen; no "foreign protein nephritis" encountered in renal biopsies; some thrombocytopenia; four patients had systemic reactions to treatment	All but one patient survived observation period (≥3 months); requirement for steroid treatment considered to be less; rejection reaction less and more easily reversed
–100 mg/dose	IM; total of 4 months treatment	Lymphopenia in all ALG cases; thrombocytopenia in 4 cases; less local reaction with ALG absorbed with human liver and muscle powders than with red cells and serum; ALG discontinued in 8 cases due to side effects	Fewer rejection episodes in ALG group without evidence of reduced renal function although given less prednisone by protocol
-5 mg/kg-day	IM; total of 3 months up to 2 years; 15 days pre-op for living donor recipients	Precipitins to horse protein seen; both groups histologically similar; *in vivo* fixation of C′ on antigen antibody complexes suggested; lymphopenia not significant after 2 weeks	Decrease in number and intensity of rejection episodes—only 19% had significant rejection in first 3 months; reduction in dose of steroids; low incidence of infection (17%) in first 3 months
0 ml/kg-day	13 patients received IM therapy; 27 patients received IV therapy; total of 4 weeks	Anaphylaxis in 2 patients (1 IM, 1 IV); no serum sickness	No serious rejection activity during ALG treatment; IV route recommended
0–300 mg/day	IM or SC; up to 21 days; 2 days pre-op for living donor recipients	Precipitins to horse protein seen; urticaria in 3 patients necessitating discontinuance, but no other reactions; no evidence of Musugi or serum sickness nephritis; followup, 6–18 months	12 patients in related group had no rejection with 1 immunological failure; 2 patients in unrelated group had no rejection with 2 immunological failures and 3 technical or nonimmunological failures
0–200 mg/dose	IM; 35 doses given up to 5½ months	ALG patients given ½ prednisone by protocol; 6 complications of anaphylactoid nature; development of reticulum cell sarcoma at site of injection in 1 patient; majority of ALG patients had splenectomies; lymphopenia in first 10 days	No difference in survival for unrelated group relative to ALG at 1 or 2 years; significantly better survival at 2 years for related group getting ALG, but not specifically related to ALG; no increase in infections
-8 mg/kg-dose	IM; 4 months total	Minimum followup, 3 years	Increased survival with ALG; ALG deemed a useful adjuvant to immunosuppressive therapy
mg/kg-day	IM; 3 weeks or 6 months	Only uniformity in production of ALS was that same animals were used over 2½ years; no significant difference noted with treatment duration; minimum followup, 1 year	1-year kidney survival 80% in unrelated group and 86% in related group; results less favorable in sensitized patients than nonsensitized; incidence of rejection with "good" ALS significantly lower
0 mg/dose	IM; total of 21 days, except first 5 related recipients who received 5 doses pre-op	9 exclusions (4 hyperacute rejections and 5 "A" matched siblings with no rejection); ALG discontinued because of anaphylactic reaction or intolerable local reactions in 14 patients; 1 death from anaphylaxis; ALS protocol discontinued; 2 year followup	No significant difference between ALG- and non-ALG-treated patients at 1 year; significantly greater number of ALG patients had no rejection through 1st year with less severe episodes; longer treatment slightly better; no increase in infections with ALG

TABLE 4. (PART I) (*Continued*)

Investigator, institution, and ref.	Number of patients (treated/ controls)	Immuno-genetic class of donor	Plan of trial	Types of ALG preparation	*In vitro* properties of ALG
Simmons et al., Univ. of Minnesota, 1972 (117, 149)	169/-	100 related; 69 unrelated, nonliving	High dose vs low dose within high- or low-risk categories	Horse antilymphoblast; (some goat antilymphoblast); adjuvant in some batches; IgG fraction; several batches	Leukagglutinins 1:512 or >1:640; no hemagglutinins detectable; protein content 2–12 mg/ml
Turcotte et al., Univ. of Michigan, 1973 (171)	36/35	36 related; 35 unrelated, nonliving	Randomized	Horse antithymocyte; IgG fraction (Upjohn ATG lot #16, 138-3) to all but 2 patients; adjuvant	Lymphotoxicity 1:65,536; hemagglutinins 1:5,120; rosette inhibition 1:153,600; protein concentration 51.5 mg/ml; primate skin allograft survival 19–26 days
Sheil et al., Univ. of Sydney, 1973 (144, 145)	59/48	All unrelated, nonliving	Randomly controlled in first 3 months; then selected to prevent imbalance in typing, etc.	Horse and goat antilymphocyte; thymocytes; no adjuvant; IgG fraction; 1 batch horse; 3 batches goat	Lymphotoxicity: horse 1:512, goat 1:1024; rosette inhibition: horse 1:4000, goat 1:32,000; hemagglutinins negative; average monkey skin graft survival was 18.5 for horse ALG and 23.3 for goat ALG vs 12.8 for control
Davis et al., Boston Univ., 1973 (33, 96)	80/-	All unrelated, nonliving	"Historic controls"	Rabbit antithymus; adjuvant used; ammonium sulfate ppt used; several batches	Hemagglutinins ≤1:128; thymocyte agglutinins ≥1:512; rosette inhibition >1:64,000
Jacobsen, Oslo, 1975 (134)	15/15	All unrelated, nonliving	Controlled, randomized	Horse antilymphocyte	Behringwerke AHLG
Sanson, Birmingham, UK, 1975 (134)	100 (total)	All unrelated, nonliving	Controlled, randomized	Horse antilymphocyte;	Private preparation
Taylor et al., Medical Research Council of Canada, 1975 (166)	87/92	All unrelated, nonliving	Controlled, randomized	Horse antithymocyte; membrane- and microsome- rich fraction; one batch	Monkey skin graft survival of 32.7 ± 6.4 days; chimpanzee skin graft survival of 49 and 66+ days
Cosimi et al., Massachusetts General Hospital, 1975 (179)	22/19	All unrelated, nonliving	Controlled, randomized	Horse antithymocyte; adjuvant used; IgG fraction; 2 batches	Hemagglutinins 1:80; cytotoxicity 1:2,790; rosette inhibition 1:11,600 and 1:20,000

TABLE 4. (PART II) (*Continued*)

Dosage	Route of administration and duration of treatment	Special observations or problems	Results
–40 mg/kg-day; "high dose" was 20 mg/kg; "low dose" was 10 mg/kg	IV; 2 weeks with some pre-op doses for related recipients	2 cases of thrombosed renal artery with ALG prepared by multiple adjuvant technique with high titers of cytotoxic antibody; 1 case of severe temporary arthritis with joint effusion (? serum sickness); minimum followup 10 months	2-year survival in related group $>80\%$ regardless of risk or dose; survival in low-risk unrelated group up to 80% with high dose and 50% with low dose; better survival in high-risk group with high dose vs low dose
.5 mg/kg-day prior to and for 7 days post-op; 1.8 mg/kg-day on days 8–21; 0.9 mg/kg-day on days 22–35	IM; 35 days for unrelated; 38 days for related	All ATG patients spiked high fevers intermittently; one episode of significant serum sickness; ATG stopped on days 32 and 33 in 2 patients due to complications; non-ATG patients received twice as much prednisone; followup of at least 1 year	Significantly better survival with ATG for 2 years in "incompatible" related grafts only; no significant difference in survival for unrelated or "compatible" related grafts; fewer rejection episodes with unrelated grafts; no significant difference in infections
.75 mg/kg of horse ALG; 3 mg/kg of goat ALG	IM for horse (39 patients); IV for goat (20 patients); total course of 2–3 months; IM after 3 weeks post-op	Horse ALG resulting in severe local and general reactions and thrombocytopenia in ½ patients; infrequent febrile reaction with goat ALG; followup of at least 2⅔ years	1-year graft survival for ATG group 75% with 62% for non-ATG group; improvement by ATG least evident in badly mismatched grafts; effect greatest in first 6 months; frequency of acute rejection 7% in ATG group and 21% in control ($p < 0.05$)
0–100 mg/dose	Subcutaneous; 5 doses on alternate days	All patients received azathioprine, steroids, and total of 450 rads of local irradiation; minimum followup, 1 year	1 year functional survival was 70% with 12% mortality; no significant difference in survival of presensitized vs nonsensitized, although strongly sensitized ($>30\%$) lost more kidneys due to rejection
0 mg/kg-day	Daily for 3 weeks; alternate days 7 weeks		1-year graft survival for ALG group marginally better; fairly identical in onset of function, number and pattern of rejection episodes
g/day changed to 500 mg/day	10 days, initially 1 g IV/day then 500 mg sc/day	1 major anaphylactic reaction with first dosage of 1 g/day IV	Only marginal better success rate with ALG group; oliguric phase shorter in ALG group; % of rejection crises successfully treated in ALG group higher
0 mg/kg-day	IV; 10 days	12 independent centers participating; thrombocytopenia seen in 30% of patients during ALS treatment with ALS discontinued in some cases	At 3 months, graft survival significantly better in ALS group; ½ as many rejection episodes in ALS group and less prednisone given; no differential in survival after 3 months
0 mg/kg-day	IV; 14 days	Well tolerated; no patients had course discontinued; only one rejection episode seen during therapy but frequent rejection seen 1–2 weeks after discontinuing therapy; no apparent increased incidence of infection	2-year survival 70% for treated group, 50% for controls from 1 batch; no apparent difference in survival curves between treated and control groups in second batch (followup not yet 1 year for some patients receiving second batch)

characteristics of which are also set out in Table 3. Although many of these patients have not yet completed a 12-month follow-up period, the kidney survival curves thus far do not appear to show significant improvement for the ALG-treated group. Moreover, the initial results obtained from the other participating centers with this lot of ALG seem to corroborate this finding. Thus, the second lot of material, although showing some favorable *in vitro* properties, may not prove to be as satisfactory as the first.

In observing the patients, it was quite clear that early rejection reactions, that is during the first 4 weeks, were much less marked in the ALG-treated groups, especially with the first batch of material. Rejection reactions often occurred shortly after the discontinuance of ALG treatment (see Fig. 3). Side effects of heterologous serum administration were minimal. Occasional febrile reactions and skin reactions were seen, but in no instance was it necessary to interrupt serum therapy because of reactions. Antibody to horse protein was sought repeatedly in the first few patients by a passive hemagglutination method employing erythrocytes coated with a sub-agglutinating dose of ALG. As antibody was not found, this test has not been continued except in patients who developed possible serum sickness reactions. The incidence of serious infections with a variety of organisms was not higher in the ALG-treated group.

This experience suggests to us that ALG treatment, even at this dosage and for this brief a period of time, can have a beneficial effect on the survival of transplanted kidneys in patients. Our ability to anticipate the immunosuppressive potency of a given serum is still unsatisfactory and the optimal duration of treatment has not been established. Individualized dosage on the basis of the requirements of the patient is desirable and we have been using this approach in skin transplantation for burns (see below) by monitoring the depression of numbers of rosetting cells in the peripheral blood of treated patients and adjusting the dose of ALSIgG appropriately. A similar plan should be applicable to kidney transplant recipients.

Transplantation of the liver

Liver transplantation has been performed much less commonly than transplantation of the kidney and the complexity of the individual circumstances surrounding each patient are greater than with kidney transplantation. Accordingly, results with this organ are still quite fragmentary. Starzl has reported his impression that the likelihood of success following liver transplantation with ALG treatment, much as it is administered by him after kidney transplantation (see above), is clearly increased (157). Fortner agrees (45). Calne, of Cambridge, is apparently less sure of its importance to a successful outcome (see ref. 31).

Transplantation of the heart

Various ALG preparations have been used sporadically by heart transplant teams in different parts of the world. Brendel (20) has reported favora-

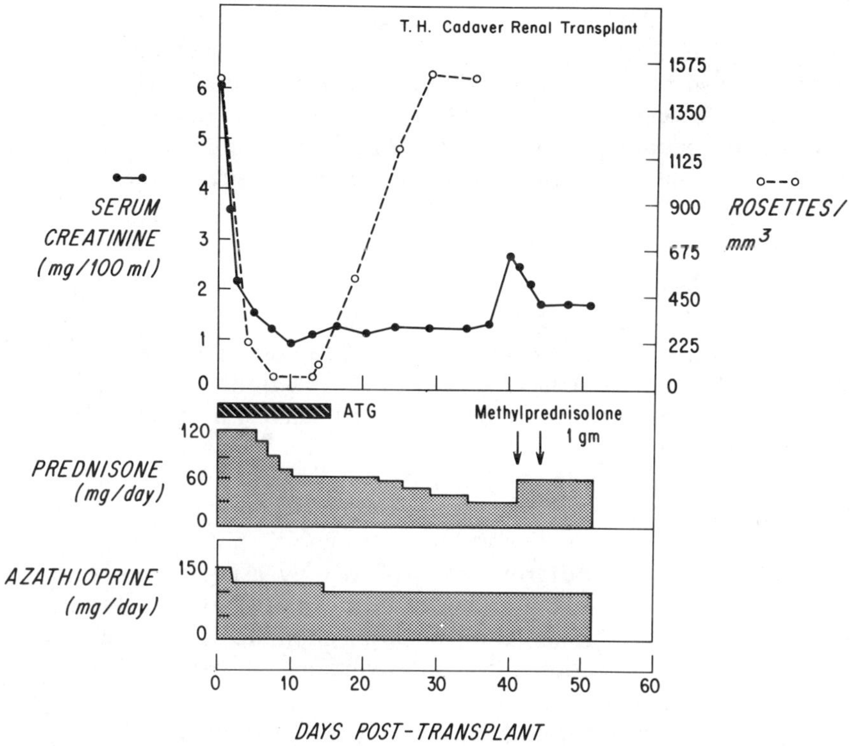

FIG. 3. Rosetting cells related to ALG (ATG) treatment and to a delayed early rejection reaction. Early postoperative course of a 63-year-old man who received a kidney allograft from a nonliving, unrelated donor (donor antigens HL-A 2, 12, W29, W16; recipient antigens HL-A 1, 8, W29). This patient was randomized into the ALG group in our study (see text) and received a dose of 30 mg/kg-day of IgG of Lot 138-16 on each of the first 14 postoperative days. Renal function was excellent during the period of ALG treatment and absolute counts of sheep red blood cell rosetting lymphocytes were depressed to very low levels during treatment. After completing treatment, rosetting cell numbers returned promptly to normal and a distinct rejection episode, responsive to increased steroid dosages, occurred. Renal function remains very satisfactory (creatinine 1.5 mg/100 ml) in this patient at the present time 32 months later.

ble results in reversing ongoing rejection of heart transplants in conjunction with intravenous infusion of ALG for varying periods. In the United States, the Stanford group have now used ALG for some time. In one report, they have described a comparison between two groups of 13 patients (55). One group received a 3-month course of equine ALG (ammonium sulfate precipitate; lymphocytotoxicity titer ranging from 1:2,000 to 1:16,000 and protein content from 50 to 60 mg/ml; patients were treated daily with 2–5 ml intramuscularly for 3 weeks, every other day for 3 weeks, and every third day for an additional 3 weeks) along with their other standard immunosuppressive measures. The second group of 13 patients received an equine antithymus cell globulin (ethanol precipitate; lymphocytotoxicity of

1:65,000, rosette inhibition titer of 1:50,000 in one batch and 1:150,000 in the second; protein content of 50 mg/ml; patients were treated with 2–5 ml intravenously each day for 2 weeks, every other day for 2 weeks, and every third day for 2 weeks). The time of onset of the first rejection episode was delayed in the second group (from day 7 to 12; $p < 0.05$) and the amount of additional immunosuppression needed to control rejection was less (especially actinomycin D).

Bone marrow transplantation

ALG might be expected to be beneficial, not only in controlling the rejection of transplanted marrow cells, as it can do nicely in animals (155), but also in the management of graft-vs-host reactions which have been so trying a problem even in HL-A identical bone marrow transfers. Mathé (101) has reported the most encouraging results with the former aspect of the transaction in patients with aplastic states, and recent reports from Hobbs of London and Storb of Seattle (134) have been encouraging in regard to the beneficial effect which ALG can have on ongoing graft-vs-host reactions. For example, the Seattle group reports that by using a course of ALG, commencing as soon as symptoms of graft vs host disease are manifest, consisting of 7 mg IgG/kg subcutaneously on alternate days, 12 of 19 patients showed complete resolution and 5 improvement. Six patients became long-term survivors. This result was considered distinctly superior to previous experience and was substantiated by animal experiments.

Skin transplantation for burns

As burned patients with extensive open wounds are extremely prone to life-threatening infection, the addition of immunosuppression to their management has been attempted only in cases in which experience has shown that there is virtually no chance of recovery by the use of currently available methods of treatment (that is, in excess of 70% of the body surface involved in full thickness burn). In our experience, azathioprine (Imuran) was first used (24) in 5 patients to permit survival of skin, taken both from living family donors and from cadavers, for periods long enough to permit gradual replacement of grafted areas with autologous skin before the grafts underwent active rejection. Subsequently, 6 additional children with burns of this magnitude have been treated using ALG (25) in conjunction with relatively low doses of cortisone treatment (0.25 mg/kg-day). Of the total group of 11 patients, 7 have survived, including 4 of the 6 who received ALG immunosuppression.

Once again the number of variables involved in clinical situations of this kind is so large that it is difficult to determine what role should be ascribed to the use of immunosuppression. In our hands, skin allografts placed even to extensively burned patients without immunosuppression are all rejected by 2 to 3 weeks. Thus, the survival of life-sustaining skin grafts in these patients until their intentional removal as long as 6 weeks after injury is deemed to have been due to immunosuppression. Diethelm has treated two children who suffered 75% surface burns by a somewhat different strategy with one survivor (39).

A special aspect of our approach has been the repeated enumeration of the number of cells in the peripheral blood which will form rosettes with sheep red blood cells (16). This "T cell" count has been used to regulate the daily dose of intravenous ALSIgG with an effort being made to maintain the number of T cells at about 10% of the pretreatment level. This percentage has been selected on the basis of a small empirical experience, which suggests that skin graft rejection is much less likely while T cell numbers are depressed to this degree and with at least some justification from studies of dose–response relationships in mice (94). Rejection has been associated with those instances in which T cell numbers have been allowed to return to normal (see Fig. 4). Another point of importance is the fact that the dose of ALSIgG required to produce a given reduction in T cell numbers in one patient differs considerably from that required for another (see Fig. 5).

Autoimmune diseases

The background and rationale for using ALG in the treatment of autoimmune conditions has been well set out in Denman's review (34). In many conditions in which an immune component is believed to exist because of the histological character of some of the lesions seen or because of actual detection of abnormal immune responses in certain patients, the immunological pathogenesis of the primary disease state remains unproved. Accordingly, immunosuppressive treatment is still on uncertain ground in many putatively autoimmune conditions. Multiple sclerosis is a particularly interesting example. Brendel and his colleagues (131) have reported the combined use of ALG, thoracic duct drainage, azathioprine, and steroid treatment in 20 patients with multiple sclerosis in various stages. "Significant" clinical improvement was observed in 11 patients as evaluated by an objective scale of factors with followup for 1 to 5 years at the time of this report. Pirofsky et al. (124) have tested the effect of goat antihuman lymphocyte globulin on a number of conditions which may have an autoimmune component, including lupus nephritis, Goodpasture's syndrome, Sjogren's syndrome, and autoimmune hemolytic anemia. Some improvement was reported to have occurred in 65% of patients. Even if these disorders are mediated by ongoing immune reactions, their responsiveness to ALG treatment would, of course, be expected to vary considerably on the basis of the type of immune process which is primarily involved (i.e., cellular or humoral) as well as many other factors. Thus, it is difficult to judge the possible long-term future of more refined immunosuppressive sera in many of these important disease states as yet.

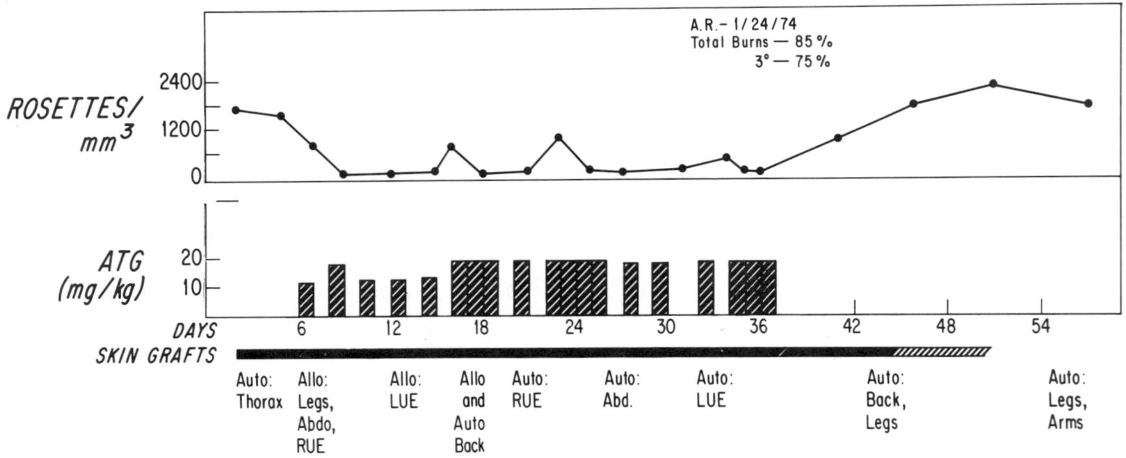

FIG. 4. Rosetting cells related to dose of ATG. This figure relates the number of rosetting cells in the peripheral blood of an 11-year-old child who sustained lethal thermal burns and was subsequently treated with ALSIgG for immunosuppression in order to allow the temporary survival of skin allografts from a related living donor to replace burned areas in gradual progression over different parts of the body. Varying daily doses of ATG were required to maintain the number of sheep red blood cell rosetting lymphocytes at approximately 10% of control values. On ceasing ALG treatment, the number of rosetting cells returned steadily to the normal range (about 70% by the method used) and the skin allografts were gradually rejected.

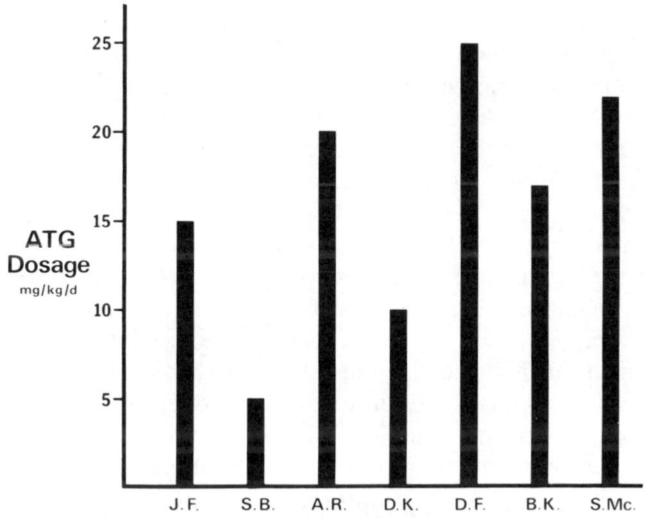

FIG. 5. This figure shows the average dose of ATG per day in 7 burned patients in order to maintain rosetting cells in the peripheral blood at about 10% of pretreatment levels over periods of ~1 month. The variation in dosage required to achieve the same degree of immunosuppression, at least according to this estimate, is apparent.

SUMMARY AND DISCUSSION OF FUTURE POSSIBILITIES FOR ANTILYMPHOCYTE CELL SERA IN CLINICAL MEDICINE

From a practical standpoint, ALS must be compared with other immunosuppressive agents in terms of *potency, safety*, and *cost*. At present, those serum fractions being tested seem to be only marginally preferable to other agents in their potency, reasonably acceptable in terms of safety, and distinctly less acceptable economically. Nevertheless, we believe that the evidence presented above supports the potential value of further efforts in this field, and that there is reason to expect that antisera directed against immunologically competent cells can become therapeutic tools of major value.

It has become gradually more certain that the main immunosuppressive property of ALS in respect of tissue transplants depends upon their capacity to inactivate T cells selectively. The peripheral availability and slow rate of renewal of these cells probably contribute considerably to their vulnerability. Nevertheless, there is increasing evidence favoring the involvement of serological specificity in the process of cell deletion. The fact that IgG fractions can retain high immunosuppressive potency after absorption, not only with nonlymphoid cells but apparently even with non-T cells, suggests that there are "species-specific" antigens on T cells which are sufficient to evoke antibodies with strong immunosuppressive activity (120). The alloantigens on T cells, for example, θ antigen in the mouse, are not necessarily involved in this xenogeneic recognition process, although anti-θ antibodies may have a mild immunosuppressive effect by quite a different mechanism (47). Other antibodies, even those shared with B cells, may supplement this "T cell immunosuppressive effect" in certain situations, but it seems likely that for the suppression of the primary rejection reaction to transplanted cells, antibodies against T cell-specific cell membrane antigens can suffice. This suggests a careful reexamination of the use of purified T cells as the antigen, or, even better, of membrane fractions prepared from such T cell preparations, because it may be possible to eliminate undesirable antigens even more completely by selecting cell fractions. IgG fractions of sera prepared against purified T cells, and fully absorbed against other cell types, could provide the necessary guide in cell fractionation processing by identifying the most important cellular antigens as purification proceeds.

Testing of serum potency should be much simpler in such a system as antibody binding to a standard antigen preparation should correlate well with prolongation of survival of grafts to normal recipients. Antibody affinity may be an important consideration. At present, this variable defies quantification.

In order to eliminate the extraneous IgG molecules, which would still flood preparations made, even against the most highly selected antigens, absorption of the desired antibody molecules with subsequent elution would be desirable. Drawbacks include the tendency for the highest affinity antibody to be lost during elution and for the addition of toxic materials to the serum fraction during processing. Still the techniques for these operations on a large scale appear to be emerging steadily (181). Large-scale cultures of cells bearing appropriate T cell-specific antigens will be of immense help in the various steps outlined, and should make more uniform and more convenient cell sources for immunization than human thymuses in which only a portion of cells probably carry the appropriate antigens. Such cultures should be quite possible to produce and some success in this regard has been reported with mouse cells (12), although efforts with human cells have been frustrating (154).

A similar approach should, in principal, be workable for other types of leukocytes bearing characteristic cell surface markers. The discovery of such antigens is proceeding rapidly. For example, new alloantigenic systems are being found on human (17) and "species-specific" antigens on mouse B cells (167), characteristic of different cell types, may well be present. Plasma cells are also known to possess characteristic antigens whose relationship with antigens on other cell types is being analyzed (60). The biological effects of treatments with sera containing antibodies reactive against these newly discovered antigens await elucidation. New applications in various disease states in patients would emerge rapidly if these experiments succeed.

A similar line of progress can be foreseen in the production of sera against "activated" lymphoid cells. W. Baldwin and I have confirmed the observations of others (105) that rabbit antisera to lymph node cells from mice which have just rejected a skin graft, are more effective in prolonging skin graft survival on mice which have been preimmunized to their donors than sera made against "resting" cells.

The *dosage* of ALSIgG should be regulated on an individualized basis. Although this is presently being done by the enumeration of T cells (24), the methods for such tests are subject to considerable improvement. If sera can be prepared as outlined above, the amount of foreign protein required to produce the desired effect should be very much less than currently used. This should, in turn, reduce the danger of toxic side effects. So far, it seems quite likely that a major cause of the apparent low level of immunosuppressive activity of many of the serum fractions which have been tested in patients is that simply not enough actively immunosuppressive molecules have been delivered to the patient.

The use of antilymphocyte antibodies in the induction of immunological tolerance would still seem to us to be as valuable a goal as it did some years ago. Indeed, it was with this objective in mind that our very first experiments in the field were conducted. The likelihood that synthetic chemical agents will be derived which can selectively inactivate various subclasses of lymphoid cells seems remote, at present, although perhaps not impossible eventually.

The tremendous growth potential of the field to which this article has been directed offers a major challenge to clinicians and to industrial laboratories as well. The demands for relatively large-scale production of antibody fractions for clinical trial will exceed the capacity of most hospital and university-based laboratory facilities. Thus it is hoped that a flexible and continuing alliance may emerge between industry and the academic sectors which will bring these possibilities to fruition.

CONCLUSIONS

1. Xenospecific antisera against lymphoid cells with powerful immunosuppressive properties in animals can be made.
2. They depend, for their activity, on the deletion or inactivation of certain lymphoid cells (for graft retention these appear to be T cells).
3. Reliable, quantitative, *in vitro* tests of immunosuppressive potency applicable to a wide variety of sera are not available.
4. This is probably due to the complexity of the reaction mixtures used in which a range of cell types and a variety of antibodies reactive with different cells are included.
5. It is certain that immunosuppressive antibodies have also been pro-

duced for use in human beings as well, and it seems very likely that much more potent and more specific reagents can be produced in the future.

6. It is known that patients can tolerate cautious treatment with even quite large doses of foreign globulins remarkably well.

7. It is suggested that xenospecific antibodies prepared against various types of leukocytes, and even against normal or neoplastic cells of other morphological classes, may prove to be very useful in clinical medicine, either for the deletion of these cells *in vivo* or to aid in their localization as part of the clinical management of various conditions.

ACKNOWLEDGMENTS

This work was supported in part by grants from the National Institutes of Health, USPHS, and from the Upjohn Company. Special gratitude is due my colleague, Dr. A. B. Cosimi, for useful discussions and for the use of data from studies for which he is primarily responsible. Thanks are also due to Dr. William Wechter of the Upjohn Company whose constant support has been invaluable. Finally, it is a pleasure to acknowledge the expert assistance of Miss Mary King and Miss Cathy Padyk in the preparation of the manuscript.

REFERENCES

1. Abaza, H. M., Nolan, B., Watt, J. G., and Woodruff, M. F. A. (1966): Effect of antilymphocytic serum on the survival of renal homotransplants in dogs. *Transplantation*, 4:618–632.
2. Agnew, H. D. (1968): The effect of heterologous anti-lymphocytic serum on the small lymphocyte population of rats. *J. Exp. Med.*, 128:111–119.
3. Amos, D. B., Billingham, R. E., Lawrence, H. S., and Russell, P. S. (editors) (1970): Proceedings of the conference on antilymphocyte serum. *Fed. Proc.*, 29:95–230.
4. Araneo, B. A., Marrack (Hunter), P. C., and Kappler, J. W. (1975): Functional heterogeneity among the T-derived lymphocytes of the mouse. II. Sensitivity of subpopulations to anti-thymocyte serum. *J. Immunol.*, 114:747–751.
5. Bach, J. F. (1970): Mechanism and significance of rosette inhibition by antilymphocyte serum. In: *International Symposium on Antilymphocyte Serum; Symposia Series in Immunobiological Standardization*, Vol. 16, pp. 189–198, edited by J. F. Bach, J. Dormont, A. Eyquem, M. Raynaud, and R. H. Regamey. Karger, Basel.
6. Bach, J. F., Dormant, J., Eyquem, A., Raynaud, M., and Regamey, R. H. (editors) (1970): *International Symposium on Antilymphocyte Serum; Symposia Series in Immunobiological Standardization*, Vol. 16. Karger, Basel.
7. Baldwin, W. M., and Russell, P. S. (1975): *Unpublished observations.*
8. Balner, H. (August 1975): *Personal communications.*
9. Balner, H., Dersjant, H., Betel, I., and van Bekkum, D. W. (1970): Current state of evaluating anti-human lymphocyte sera by *in vivo* testing. In: *International Symposium on Antilymphocyte Serum: Symposia Series on Immunobiological Standardization*, Vol. 16, pp. 179–188, edited by J. F. Bach, J. Dormont, A. Eyquem, M. Raynaud, and R. H. Regamey. Karger, Basel.
10. Barth, R. F. (1969): Studies on heterologous antilymphocyte and antithymocyte sera. IV. Further evidence for a preferential effect on antigen-sensitive cells. *J. Immunol.*, 130:648–654.
11. Barth, R. F., and Carroll, G. F. (1970): Immunosuppressive effects of antilymphocyte serum in complement-deficient mice: evidence that immune cytolysis is not essential for ALS activity. *J. Immunol.*, 104:522–524.

12. Ben-Sasson, S. Z., Paul, W. E., Shevach, E. M., and Green, I. (1975): *In vitro* selection and extended culture of antigen-specific T lymphocytes. I. Description of selection culture procedure and initial characterization of selected cells. *J. Exp. Med.*, 142:90–105.
13. Bert, G., Forrester, J. A., and Davies, A. J. S. (1971): Differential response of B and T cells to ALS revealed by electrophoretic mobility determinations. *Nature*, 234:86–87.
14. Besredka, A. (1900): La leucotoxine et son action sur le systeme leucocytaire. *Ann. Inst. Pasteur (Paris)*, 14:390–401.
15. Binns, R. M., Simpson, E., Nehlsen, S. L., and Ruszkiewiez, M. (1971): A method for producing immunosuppressive antilymphocyte serum in pigs and ruminants. *Transplant. Proc.*, 3:784–787.
16. Bishop, G., Cosimi, A. B., Voynow, N. K., Whelchel, J. D., and Wortis, H. H. (1975): Effect of immunosuppressive therapy for renal allografts on the number of circulating sheep red blood cell rosetting cells. *Transplantation*, 20:123–129.
17. Bodmer, W. F. (1975): Histocompatibility testing international. *Nature*, 256:696–698.
18. Boyse, E. A., Miyazama, M., Aoki, T., and Old, L. J. (1968): Ly-A and Ly-B: two systems of lymphocyte isoantigens in the mouse. *Proc. R. Soc. Lond.* [*Biol.*], 170B: 175–193.
19. Brendel, W. (1969): Clinical experiences with antilymphocyte globulin in kidney transplants: essential advantages of the intravenous route. In: *Proceedings of the 4th International Congress of Nephrology, Stockholm, 1969*, Vol. 3, pp. 245–255. Karger, Basel.
20. Brendel, W., and Land, W. (1968): Unexpected results of intravenous therapy with antilymphocyte serum in organ transplantation. *Dtsch. Med. Wochenschr.*, 93:2309–2313.
21. Brendel, W., Ring, J., and Seifert, J. (1974): Experimental and clinical aspects of ALG. In: *Progress in Immunology II; Clinical Aspects II,* pp. 245–252, edited by L. Brent and J. Holborow. North-Holland and American Elsevier, Amsterdam and New York.
22. Brent, L., and Kilshaw, P. J. (1970): Production of skin allograft survival with spleen extracts and antilymphocytic serum. *Nature*, 227:898–900.
23. Bunting, C. H. (1903): The effects of lymphotoxins and myelotoxins on the leucocytes of the blood and on the blood-forming organs. *Univ. Penn. Bull.*, 16:200–216.
24. Burke, J. F., May, J. W., Jr., Albright, N., Quinby, W. C., and Russell, P. S. (1974): Temporary skin transplantation and immunosuppression for extensive burns. *N. Engl. J. Med.*, 290:269–271.
25. Burke, J. F., Quinby, W. C., Bondoc, C. C., Cosimi, A. B., Russell, P. S., and Szyfelbein, S. K. (1975): Immunosuppression and temporary skin transplantation in the treatment of massive third degree burns. *Ann. Surg.*, 182:183–195.
26. Chew, W. B., and Lawrence, J. S. (1937): Antilymphocytic serum. *J. Immunol.*, 33:271–278.
27. Cinader, B., Jeejeebhoy, H. F., Koh, S. W., and Rabbat, A. G. (1971): Immunosuppressive and graft-rejecting antibodies in heterologous anti-lymphocyte serum. *J. Exp. Med.*, 133:81–99.
28. Clark, J. G., James, K., and Woodruff, M. F. A. (1967): Elimination of normal horse IgG labelled with iodine-131 in rats receiving horse anti-rat lymphocyte IgG. *Nature,* 215:869–870.
29. Clunie, G. J. A., Nolan, B., James, K., Watt, J. G., and Woodruff, M. F. A. (1968): Prolongation of canine renal allografts with antilymphocyte serum. *Transplantation*, 6:459–467.
30. Cosimi, A. B., Skamene, E., Bonney, W. W., and Russell, P. S. (1970): Experience with large-dose intravenous antithymocyte globulin in primates and man. *Surgery*, 68:54–61.
31. Editorial (1975): Current status of antilymphocyte globulin, *Br. Med. J.*, 1:644–645.
32. Daddi, G., Winn, H. J., and Russell, P. S. (1969): Studies on RAMLS: Biological activity of Fab and $F(ab')_2$ fragments. In: *Pharmacological Treatment in Organ and Tissue Transplantation; Proceedings of an International Symposium Milan, 1969*, pp. 194–199, edited by A. Bertelli and A. P. Monaco. Williams and Wilkins, Baltimore, Md., and Excerpta Medica, Amsterdam.
33. Davis, R. C., Nabseth, D. C., Olsson, C. A., Idelson, B. A., Schmitt, G. W., Cho, S. I., and Mannick, J. A. (1974): Use of rabbit ATG in cadaver kidney transplantation: an updated report with consideration of presensitized recipients. *Transplant. Proc.*, (4 Suppl 1), 6:67–70.
34. Denman, A. M. (1969): Anti-lymphocytic antibody and autoimmune disease: A review. *Clin. Exp. Immunol.*, 5:217–249.

35. Denman, A. M., Denman, E. J., and Embling, P. H. (1968): Changes in the life-span of circulating small lymphocytes in mice after treatment with anti-lymphocyte globulin. *Lancet*, 1:321–325.
36. Denman, A. M., Denman, E. J., and Holborow, E. J. (1966): Effect of anti-lymphocyte globulin on kidney disease in (NZB × NZW)F1 mice. *Lancet*, 2:841–843.
37. Denman, A. M., and Frenkel, E. P. (1968): Mode of action of anti-lymphocyte globulin. I. The distribution of rabbit anti-lymphocyte globulin injected into rats and mice. *Immunology*, 14:107–113.
38. Deodhar, S. D., Konomi, K., Nakamoto, S., and Kuruvila, K. C. (1971): Clinical experience with antilymphocyte globulin (ALG) in renal transplantation. *Transplant. Proc.*, 3:758–761.
39. Diethelm, A. G., Dimick, A., Shaw, J. F., Baker, H., and Wilson, S. P. (1972): Management of the severely burned child utilizing skin allografts modified by equine antithymocyte globulin. In: *ALG Therapy and Standardization Workshop; Behring Institute Research Communications No. 51*, pp. 144–148, edited by F. R. Seiler and H. G. Schwick. Behringwerke AG, Marburg.
40. Doak, P. B., Dalton, N. T., Meredith, J., Montgomerie, J. Z., and North, J. D. K. (1969): Use of antilymphocyte globulin after cadaveric renal transplantation. *Br. Med. J.*, 4:522–525.
41. Dormont, J., Hors, J., Lamy, R., Eyquem, A., Lavergne, M., Wallon, C., Raynaud, M. (1970): Preparation of horse antilymphocyte sera using blood lymphocytes, thymocytes, and cells from chronic lymphatic leukemia as antigens. In: *International Symposium on Antilymphocyte Serum; Symposia Series in Immunobiological Standardization*, Vol. 16, pp. 1–8, edited by J. F. Bach, J. Dormont, A. Eyquem, M. Raynaud, and R. H. Regamey. Karger, Basel.
42. Everett, N. B., Schwartz, M. R., Tyler, R. W., and Perkins, W. D. (1970): Observations relative to the mechanism of action of antilymphocyte serum. *Fed. Proc.*, 29:212–219.
43. Flexner, S. (1902): The pathology of lymphotoxic and myelotoxic intoxication. *Univ. Penn. Med. Bull.*, 15:287–295.
44. Floersheim, G. L., and Brune, K. (1971): Tolerance to skin allografts by immunosuppressive "pulse" of methylhydrazine derivatives and antilymphocyte serum. *Transplant. Proc.*, 3:688–692.
45. Fortner, J. G., Shiu, M. H., Balner, H., Wilson, C. B., Sichuk, G., Kawano, N., Holmes, J. T., and Beattie, E. J. (1971): Observations on prolonged immune suppression for human liver homografts. *Transplant. Proc.*, 3:383–386.
46. Fries, D., Traeger, J., Banssillon, V., Dubernard, J. M., Touraine, J. L., Betuel, H., and Mamelle, C. (1970): Criteres cliniques, biologiques et histologiques de l'activite des globulines antilymphocytaires en transplantation renale. In: *International Symposium on Antilymphocyte Serum; Symposia Series on Immunobiological Standardization*, Vol. 16, pp. 363–370, edited by J. F. Bach, J. Dormont, A. Eyquem, M. Raynaud, and R. H. Regamey. Karger, Basel.
47. Gelfand, M. C., and Paul, W. E. (1975): Prolongation of allograft survival in mice by administration of anti-thy 1 serum. I. Mediation by *in vivo* activation of regulatory T cells. *J. Immunol.*, 115:1–4.
48. Gewurz, H., Pickering, R. J., Moberg, A., Simmons, R. L., Good, R. A., and Najarian, J. S. (1970): Reactivities to horse antilymphocyte globulin. II. Serum sickness nephritis with complement alterations in man. *Int. Arch. Allergy Appl. Immunol.*, 39:210–220.
49. Girardet, R., Friedman, E. A., and Karlson, K. E. (1974): Rabbit anti-rat lymphocyte serum-induced lymphopenia in thoracic duct lymph and peripheral blood. *Surgery*, 76:267–271.
50. Gowans, J. L., and Knight, E. J. (1964): The route of re-circulation of lymphocytes in the rat. *Proc. R. Soc. Lond.* [*Biol.*], 159B:257–282.
51. Grasbeck, R., Nordman, C. T., and de la Chapelle, A. (1964): The leucocyte-mitogenic effect of serum from rabbits immunized with human leukocytes. *Acta Med. Scand.*, 412:39–47.
52. Gray, J. G., Monaco, A. P., Wood, M. L., and Russell, P. S. (1966): Studies on heterologous anti-lymphocyte serum in mice. I. *In vitro* and *in vivo* properties. *J. Immunol.*, 96:217–228.

53. Greaves, M. F., Roitt, I. M., Zamir, R., Carnaghan R. B. A. (1967): Effect of antilymphocyte serum on responses of human peripheral-blood lymphocytes to specific and non-specific stimulants *in vitro*. *Lancet*, 2:1317–1319.
54. Greaves, M. F., Tursi, R., Playfair, J. H. L., Torrigiani, G., Zamir, R., and Roitt, I. M. (1969): Immunosuppressive potency and *in vitro* activity of anti-lymphocyte globulin. *Lancet*, 1:68–72.
55. Griepp, R. B., Stinson, E. B., Dong, E., Phillips, R. C., Morrell, R. M., and Shumway, N. E. (1972): Use of antithymocyte globulin in human heart transplantation. *Circulation*, Suppl. 1, 45:147–153.
56. Grobbelaar, B. G., Brain, P., and Rogers, A. J. (1970): Regional administration of antilymphocyte serum in monkeys. *Lancet*, 1:1120.
57. Groth, C. G., Kashiwagi, N., Moore, G. E., Husberg, B., Bjorken, C., Putnam, C. W., and Starzl, T. E. (1972): Production of a standardized anti-lymphocyte globulin. In: *ALG Therapy and Standardization Workshop; Behring Institute Research Communications No. 51,* pp. 86–89, edited by F. R. Seiler and H. G. Schwick. Behringwerke AG, Marburg.
58. Halasz, N. A., Amonic, R., Devin, J., and Miller, S. (1969): Local effects of antilymphocyte serum. *Transplant. Proc.*, 1:442–444.
59. Hardy, M. A., Quint, J., Goldstein, A. L., State, D., and White, A. (1968): Effect of thymosin and an antithymosin serum on allograft survival in mice. *Proc. Natl. Acad. Sci. USA*, 61:875–882.
60. Harris, N. S. (1974): Plasma cell surface antigen on human blood lymphocytes. *Nature* 250:507–509.
61. Holm, G., and Perlmann, P. (1970): Inhibition of lymphocyte mediated cytotoxicity by antilymphocyte serum and antisera to other lymphocyte associated antigens. In: *International Symposium on Antilymphocyte Serum; Symposia Series in Immunobiological Standardization*, Vol. 16, pp. 267–274, edited by J. F. Bach, J. Dormont, A. Eyquem, A. Raynaud, and R. H. Regamey. Karger, Basel.
62. Horsfall, F. L., Goodner, K., MacLeod, C. M., and Harris, A. H. (1937): Antipneumococcus rabbit serum as a therapeutic agent in lobar pneumonia. *JAMA*, 108:1483–1490.
63. Humphrey, J. H. (1967): Chairman's opening remarks. In: *Antilymphocytic Serum, Ciba Foundation Study Group No. 29*, pp. 1–3, edited by G. E. W. Wolstenholme and Maeve O'Connor. Little, Brown, Boston.
64. Inderbitzin, T. (1956): The relationship of lymphocytes, delayed cutaneous allergic reactions and histamine. *Int. Arch. Allergy Appl. Immunol.*, 8:150–159.
65. Iversen, J. G. (1970): Specific antisera against recirculating and non-recirculating lymphocytes in the rat. *Clin. Exp. Immunol.*, 6:101–108.
66. Iwasaki, Y., Porter, K. A., Amend, J. R., Marchioro, T. L., Zuhlke, V., and Starzl, T. E. (1967): The preparation and testing of horse antidog and antihuman anti-lymphoid plasma or serum and its protein fractions. *Surg. Gynecol. Obstet.*, 124:1–24.
67. James, K. (1970): Effect of antilymphocyte antibody on humoral antibody formation. *Fed. Proc.*, 29:160–166.
68. James, K. (1973): Preparation of antilymphocytic antibodies. In: *Handbook of Experimental Immunology*, 2nd ed., Vol. 2, Chap. 31, edited by D. M. Weir. Blackwell Scientific Publications, Oxford.
69. James, K. (1969): The preparation and properties of anti-lymphocytic sera. *Progr. Surg.*, 7:140–216.
70. James, K., Pullar, D. M., and Brydon, J. (1971): The immunosuppressive properties of sub-fractions of antilymphocytic IgG of equine origin. *Immunology*, 21:247–257.
71. James, K., Pullar, D. M., Morton, J. B., Dalton, R. G., Nolan, B., and Woodruff, M. F. A. (1971): The intravenous administration of equine antilymphocytic globulin in renal transplant recipients and the detection of circulating antibodies to equine globulin. *Clin. Exp. Immunol.*, 8:529–542.
72. Janossy, G., and Greaves, M. F. (1972): Lymphocyte activation. II. Discriminating stimulation of lymphocyte subpopulations by phytomitogens and heterologous antilymphocyte sera. *Clin. Exp. Immunol.*, 10:525–536.
73. Jeejeebhoy, H. F. (1965): *Lancet*, 2:106.
74. Jeejeebhoy, H. F. (1971): Heterologous anti-lymphocyte serum: a review. *Can. J. Surg.*, 14:5–18.

75. Jeejeebhoy, H. F. (1965): Immunological studies on the rat thymectomized in adult life. *Immunology*, 9:417–425.
76. Jeejeebhoy, H. F. (1967): Studies on the mode of action of heterologous anti-lymphocyte plasma. I. A comparison of the immunosuppressive properties of dog and rabbit anti-rat lymphocyte plasma. *Transplantation*, 5:273–282.
77. Jooste, S. V. (1970): Immunological effects of heterologous antilymphocyte sera. *Lymphology*, 3:79–103.
78. Jooste, S. V., Lance, E. M., Levey, R. H., Medawar, P. B., Ruszkiewiez, M., Sharman, R., and Taub, R. N. (1968): Notes on the preparation and assay of anti-lymphocytic serum for use in mice. *Immunology*, 15:697–705.
79. Kashiwagi, N., Brantigan, C. O., Brettschneider, L., Groth, C. G., and Starzl, T. E. (1968): Clinical reactions and serologic changes after the administration of heterologous antilymphocyte globulin to human recipients of renal homografts. *Ann. Intern. Med.*, 68:275–286.
80. Katz, D. H., Unanue, E. R., and Dixon, F. J. (1967): Nephritogenic properties of cross-reacting kidney-fixing antibodies to heart, spleen, and muscle. *J. Immunol.*, 98:260–268.
81. Kayhoe, D. E., Darrow, C. C., and Scott, R. C. (1972): The *in vivo* activity of antilymphocyte serum. In: *ALG Therapy and Standardization Workshop; Behring Institute Research Communications No. 51*, pp. 12–14, edited by F. R. Seiler and H. G. Schwick. Behringwerke AG, Marburg.
82. Lance, E. M. (1968): Erasure of immunological memory with anti-lymphocyte serum. *Nature*, 217:557–558.
83. Lance, E. M. (1969): The mechanism of action of anti-lymphocyte serum. Studies of antibody eluate. *J. Exp. Med.*, 130:49–76.
84. Lance, E. M., and Batchelor, J. R. (1968): Selective suppression of cellular immunity by antilymphocyte serum. *Transplantation*, 6:490–491.
85. Lance, E. M., and Dresser, D. W. (1967): Antigenicity in mice of antilymphocyte gamma globulin. *Nature*, 215:488–490.
86. Lance, E. M., Medawar, P. B., and Taub, R. N. (1973): Antilymphocyte serum. *Adv. Immunol.*, 17:1–92.
87. Land, W., Seifert, J., Fatch-Moghadam, A., Hopf, U., and Brendel, W. (1969): Immunological tolerance induced in adult dogs by small amounts of horse IgG. *Transplantation*, 8:748–750.
88. Landy, M., and Chessin, L. N. (1969): The effect of plant mitogens on humoral and cellular immune responses. *Antibiot. Chemother.*, 15:199–212.
89. Lawson, R. K., Ellis, L. R., Kirchheim, D., and Hodges, C. V. (1967): The prolongation of canine renal homograft function using antilymphocyte serum as an immunosuppressive agent. *Transplantation*, 5:169–183.
90. Levey, R. H., and Medawar, P. B. (1967): Further experiments on the action of antilymphocytic antiserum. *Proc. Natl. Acad. Sci. USA*, 58:470–477.
91. Levey, R. H., and Medawar, P. B. (1966): Nature and mode of action of antilymphocytic serum. *Proc. Natl. Acad. Sci. USA*, 56:1130–1137.
92. Levey, R. H., and Medawar, P. B. (1967): The mode of action of antilymphocytic serum. In: *Antilymphocytic Serum, Ciba Foundation Study Group No. 29*, pp. 72–80, edited by G. E. W. Wolstenholme and Maeve O'Connor. Little, Brown, Boston.
93. Malek, P., Kolc, J., Haskova, V., and Zastava, V. (1969): Antilymphocyte serum and lymphatic system. I. The effect of regional administration on the survival of skin allografts. *Transplantation*, 8:333–337.
94. Mandel, M. A., and Asofsky, R. (1968): The effects of heterologous anti-thymocyte sera in mice. I. The use of a graft-versus-host assay as a measure of homograft reactivity. *J. Immunol.*, 100:1319–1325.
95. Mandel, M. A., and Asofsky, R. (1968): The effects of heterologous anti-thymocyte sera in mice. II. Quantitative evaluation of IgM, IgG, and their subunits in suppressing homograft reactions. *J. Immunol.*, 100:1259–1267.
96. Mannick, J. A., Davis, R. C., Cooperband, S. R., Glasgow, A. H., Williams, L. F., Harrington, J. T., Cavallo, T., Schmitt, G. W., Idelson, B. A., Olsson, C. A., and

Nabseth, D. C. (1971): Clinical use of rabbit antihuman lymphocyte globulin in cadaver-kidney transplantation. *N. Engl. J. Med.*, 284:1109–1115.
97. Martin, W. J. (1969): Assay for the immunosuppressive capacity of antilymphocyte serum. III. Opsonizing activity of anti-human lymphocyte serum. *J. Immunol.*, 103:1000–1005.
98. Martin, W. J., and Miller, J. F. A. P. (1969): Assay for the immunosuppressive capacity of antilymphocyte serum based on its action on thymus-derived cells. *Int. Arch. Allergy Appl. Immunol.*, 35:163–178.
99. Martin, W. J., and Miller, J. F. A. P. (1968): Cell to cell interaction in the immune response. IV. Site of action of antilymphocyte globulin. *J. Exp. Med.*, 128:855–874.
100. Martin, W. J., and Miller, J. F. A. P. (1967): Site of action of antilymphocyte globulin. *Lancet*, 2:1285–1287.
101. Mathé, G., Amiel, J. L., Schwarzenberg, L., Choay, J., Trolard, P., Schneider, M., Hayat, M., Schlumberger, J. R., and Jasmin, C. (1970): Bone marrow graft in man after conditioning by antilymphocytic serum. *Br. Med. J.*, 2:131–136.
102. Medawar, P. B. (1968): Biological effects of heterologous antilymphocyte sera. In: *Human Transplantation*, pp. 501–509, edited by F. T. Rapaport and J. Dausset. Grune and Stratton, New York.
103. Mergenhagen, S. E., and Howard, R. J. (1970): Induction of tolerance to antilymphocytic globulin: a method for increasing the effectiveness of antilymphocytic serum. In: *International Symposium on Antilymphocyte Serum; Symposia Series in Immunobiological Standardization*, Vol. 16, pp. 241–246, edited by J. F. Bach, J. Dormont, A. Eyquem, M. Raynaud, and R. H. Regamey. Karger, Basel.
104. Metchnikoff, E. (1899): Etudes sur la reesorption des cellules. *Ann. Inst. Pasteur (Paris)*, 13:737–769.
105. Miller, J., Tanner, D., and Cohn, R. (1970): Variations in rabbit antimouse lymphocyte sera caused by lymphocytes of differing origins in the immunizing inocula. *Transplantation*, 9:352–365.
106. Mitchison, N. A. (1970): Mechanism of action of antilymphocyte serum. *Fed. Proc.*, 29:222–223.
107. Moberg, A., Gewurg, H., Simmons, R. L., and Najarian, J. S. (1970): Pretreatment with deaggregated horse gamma-globulin; failure to enhance tolerance to horse antilymphoblast globulin in man. *Transplantation*, 10:344–346.
108. Monaco, A. P., Abbott, W. M., Othersen, H. B., Simmons, R. L., Wood, M. L., Flax, M. H., and Russell, P. S. (1966): Antiserum to lymphocytes: prolonged survival of canine renal allografts. *Science*, 153:1264–1267.
109. Monaco, A. P., Schlesinger, R., and Sahyoun, A. (1972): Experience with equine anti-human lymph node lymphocyte serum in renal transplantation. In: *ALG Therapy and Standardization Workshop; Behring Institute Research Communications, No. 51*, pp. 86–89, edited by F. R. Seiler and H. G. Schwick. Behringwerke AG, Marburg.
110. Monaco, A. P., Wood, M. L., Gray, J. G., and Russell, P. S. (1966): Studies on heterologous anti-lymphocyte serum in mice. II. Effect on the immune response. *J. Immunol.*, 96:229–238.
111. Monaco, A. P., Wood, M. L., and Russell, P. S. (1965): Adult thymectomy: Effect on recovery from immunologic depression in mice. *Science*, 149:432–435.
112. Monaco, A. P., Wood, M. L., and Russell, P. S. (1967): Some effects of purified heterologous antihuman lymphocyte serum in man. *Transplantation*, 5:1106–1114.
113. Monaco, A. P., Wood, M. L., and Russell, P. S. (1966): Studies of heterologous anti-lymphocyte serum in mice. III. Immunologic tolerance and chimerism produced across the H-2 locus with adult thymectomy and anti-lymphocyte serum. *Ann. NY Acad. Sci.*, 129:190–209.
114. Muller-Eberhard, H. J. (1968): Chemistry and reaction mechanisms of complement. *Adv. Immunol.*, 8:1–80.
115. Nagaya, H., and Sieker, H. O. (1965): Allograft survival: effect of antiserums to thymus glands and lymphocytes. *Science*, 150:1181–1182.
116. Najarian, J. S., and Simmons, R. L. (1971): The clinical use of antilymphocyte globulin. *N. Engl. J. Med.*, 285:158–166.
117. Najarian, J. S., Simmons, R. L., Gewurz, H., Moberg, A., Merkel, F., and Moore, G. E.

(1969): Anti-serum to cultured human lymphoblasts: preparation, purification and immunosuppressive properties in man. *Ann. Surg.*, 170:617–629.
118. Nehlsen, S. L. (1971): Prolonged administration of antithymocyte serum in mice. I. Observations on cellular and humoral immunity. II. Histopathological investigation. *Clin. Exp. Immunol.*, 9:61–98.
119. Nelson, J. W., Parcells, A. J., Cosimi, A. B., and Esser, R. E. (1975): Studies on the rosette inhibition assay. Improvement of the method and correlation with monkey skin graft prolongation. *Transplantation*, 20:3–12.
120. Ochiai, T., Ahmed, A., Strong, D. M., Scher, I., and Sell, K. W. (1975): Specificity and immunosuppressive potency of rabbit antimouse T cell-specific antiserum. *Transplantation*, 20:198–210.
121. Pappenheimer, A. M. (1917): Experimental studies upon lymphocytes. II. The action of immune sera upon lymphocytes and small thymus cells. *J. Exp. Med.*, 26:163–180.
122. Parrott, D. M. V., de Sousa, M. A. B., and East, J. (1966): Thymus-dependent areas in the lymphoid organs of neonatally thymectomized mice. *J. Exp. Med.*, 123:191–203.
123. Pichlmayr, R., Brendel, W., and Zenker, R. (1967): Production and effect of heterologous anti-canine lymphocyte serum. *Surgery*, 61:774–783.
124. Pirofsky, B., Reid, R. H., Bardana, E. J., and Bayracki, C. (1971): Antithymocyte antisera therapy in nonsurgical immunologic disease. *Transplant. Proc.*, 3:769–772.
125. Prince, G. H. (1970): Rabbit anti-human lymphocyte serum. In: *International Symposium on Antilymphocyte Serum: Symposia Series on Immunobiological Standardization*, Vol. 16, pp. 93–96, edited by J. F. Bach, J. Dormont, A. Eyquem, M. Raynaud, and R. H. Regamey. Karger, Basel.
126. Ptak, W., and Asherson, G. L. (1969): Contact and delayed hypersensitivity in the mouse. II. The role of different cell populations. *Immunology*, 17:769–775.
127. Ptak, W., Festenstein, H., Asherson, G. L., and Denman, A. M. (1969): Improved graft survival after treatment with Bordetella and anti-lymphocyte serum. *Nature*, 222:1083–1085.
128. Raff, M. C. (1969): Theta isoantigen as a marker of thymus-derived lymphocytes in mice. *Nature*, 224:378–379.
129. Ranlov, P., and Hardt, F. (1970): The migration-inhibitory factor (MIF). *Transplantation*, 10:438–442.
130. Riethmuller, G., Riethmuller, D., Stein, M., and Hausen, P. (1968): *In vivo* and *in vitro* properties of intact and pepsin-digested heterologous anti-mouse thymus antibodies. *J. Immunol.*, 100:969–973.
131. Ring, J., Lob, G., Angstwurm, H., Brass, B., Backmund, H., Seifert, J., Coulin, K., Frick, E., Mertin, J., and Brendel, W. (1974): Intensive immunosuppression in the treatment of multiple sclerosis. *Lancet*, 2:1093–1096.
132. Ring, J., Seifert, J., Lob, G., Hopf, U., Land, N., and Brendel, W. (1974): Allergic reactions to a horse globulin therapy and their prevention by induction of immunological tolerance. *Allergol. Immunopathol. (Madr.)*, 2:93–98.
133. Rolland, J. M., and Nairn, R. C. (1972): Anti-lymphocyte serum: a review of its immunological effects and therapeutic value. *Pathology*, 4:85–122.
134. Rousell, R. H. (editor) (1975): Symposium on antilymphocyte globulin in clinical practice. *Postgrad. Med.*, (*in press*).
135. Russe, H. P., and Crowle, A. J. (1965): A comparison of thymectomized and antithymocyte serum-treated mice in their development of hypersensitivity to protein antigens. *J. Immunol.*, 94:74–83.
136. Russell, P. S. (1968): Antilymphocyte serum as an immunosuppressive agent. *Ann. Intern. Med.*, 68:483–486.
137. Russell, P. S., and Monaco, A. P. (1967): Heterologous antilymphocyte sera and some of their effects. *Transplantation*, 5:1086–1099.
138. Sachs, D. H., Winn, H. J., and Russell, P. S. (1971): The immunologic response to xenografts. Recognition of mouse H-2 histocompatibility antigens by the rat. *J. Immunol.*, 107:481–492.
139. Schlesinger, M., and Yron, I. (1970): Serologic demonstration of a thymus-dependent population of lymph-node cells. *J. Immunol.*, 104:798–804.

140. Seifert, J., Ring, J., and Brendel, W. (1974): Prolongation of skin allografts after oral application of ALS in rats. *Nature*, 249:776.
141. Seiler, F. R., and Schwick, H. G. (editors) (1972): *ALG Therapy and Standardization Workshop; Behring Institute Research Communications No. 51.* Behringwerke AG, Marburg.
142. Sell, K. (1975): *Personal communication.*
143. Sell, S. (1969): Antilymphocytic antibody: effects in experimental animals and problems in human use. *Ann. Intern. Med.*, 71:177–196.
144. Sheil, A. G. R., Kelly, G. E., Storey, B. G., May, J., Kalowski, S., Mears, D., Rogers, J. H., Johnson, J. R., Charlesworth, J., and Stewart, J. H. (1971): Controlled clinical trial of antilymphocyte globulin in patients with renal allografts from cadaver donors. *Lancet*, 1:359–363.
145. Sheil, A. G. R., Mears, D., Johnson, J. R., Kelly, G. E., May, J., Ibels, L. S., and Stewart, J. H. (1973): Antilymphocyte globulin in patients with renal allografts from cadaveric donors: Late results of a controlled trial. *Lancet*, 2:227–228.
146. Shigeno, N., Hammerling, Y., Arpels, C., Boyse, E. A., and Old, L. J. (1968): Preparation of lymphocyte-specific antibody from anti-lymphocyte serum. *Lancet*, 2:320–323.
147. Shorter, R. G., Hallenbeck, G. A., Johnson, W. J., and Danielson, G. K. (1970): Further studies of antilymphoid sera in clinical renal allotransplantation. *Arch. Surg.*, 100:132–135.
148. Simmons, R. L. (1975): *Personal communications.*
149. Simmons, R. L., Condie, R., and Najarian, J. S. (1972): Antilymphoblast globulin for renal allograft prolongation. *Transplant. Proc.*, 4:487–490.
150. Simmons, R. L., Moberg, A. W., Gewurz, H., Sole, R., and Najarian, J. S. (1971): Immunosuppression by antihuman lymphocyte globulin: correlation of human and animal assay systems with clinical results. *Transplant. Proc.*, 3:745–748.
151. Simpson, E., Binns, R. M., Nehlsen, S. L., and Ruszkiewicz, M. (1972): Sources of ALS. In *ALG Therapy and Standardization Workshop; Behring Institute Research Communications No. 51*, pp. 54–60, edited by F. R. Seiler and H. G. Schwick. Behringwerke AG, Marburg.
152. Skamene, E., Hawkins, D., Gold, P., Shuster, J., Freedman, S. O., and Taylor, H. E. (1972): Studies on nephrotoxic antibody in anti-lymphocyte globulin. *Transplantation*, 13:9–14.
153. Skamene, E., and Russell, P. S. (1971): A quantitative study of the binding of ALS to various cell types. *Clin. Exp. Immunol.*, 8:195–204.
154. Smith, R. W., and Woody, James N. (1974): The immunosuppressive potency of antilymphocyte serum is related to activity against human thymic lymphocyte-specific antigens. *Transplantation*, 17:503–507.
155. Speck, B., and Kissling, M. (1971): Successful bone marrow grafts in experimental aplastic anemia using antilymphocytic serum for conditioning. *Rev. Eur. Etud. Clin. Biol.*, 16:1047–1051.
156. Starzl, T. E. (1968): Heterologous antilymphocyte globulin. *N. Engl. J. Med.*, 279:700–703.
157. Starzl, T. E., Brettschneider, L., Penn, I., Giles, G., Picache, R., and Putnam, C. W. (1969): Clinical liver transplantation. *Transplant. Rev.*, 2:3–68.
158. Starzl, T. E., Marchioro, T. L., Porter, K. A., Iwasaki, Y., and Cerilli, G. J. (1967): The use of heterologous antilymphoid agents in canine renal and liver homotransplantation and in human renal transplantation. *Sur. Gynecol. Obstet.*, 124:301–318.
159. Starzl, T. E., Porter, K. A., Iwasaki, Y., Marchioro, T. L., and Kashiwagi, N. (1967): The use of heterologous antilymphocyte globulin in human renal homotransplantation. In: *Antilymphocytic Serum; Ciba Foundation Study Group No. 29*, pp. 4–34, edited by G. E. W. Wolstenholme and Maeve O'Connor. Little, Brown, Boston.
160. Starzl, T. E., and Putnam, C. W. (1972): Clinical experience with ALG. In *ALG Therapy and Standardization Workshop; Behring Institute Research Communications No. 51*, pp. 124–125, edited by F. R. Seiler and H. G. Schwick. Behringwerke AG, Marburg.
161. Steinberg, B., and Martin, R. A. (1946): Leukoagglutination differentiation of normal leukocytic and leukemic cell-types. *J. Immunol.*, 52:71–75.

162. Taub, R. N. (1968): Prolongation of skin homografts in mice using homologous antilymphocytic antisera. *Transplant. Proc.*, 1:445–447.
163. Taub, R. N., Kochwa, S., Brown, S. M., Rubin, A. D., and Dameshek, W. (1969): Antigen-induced immunological tolerance in man to equine anti-human-thymocyte gamma-globulin. *Lancet*, 2:521–523.
164. Taub, R. N., and Lance, E. M. (1968): Effects of heterologous anti-lymphocyte serum on the distribution of ^{51}Cr-labelled lymph node cells in mice. *Immunology*, 15:633–642.
165. Taub, R. N., and Lance, E. M. (1968): Histopathological effects in mice of heterologous antilymphocytic serum. *J. Exp. Med.*, 128:1281–1308.
166. Taylor, H. E. (1975): *Personal communication.*
167. Tonkonogy, S., and Winn, H. J. (1975): A serologically detectable Mls locus antigen (*to be published*).
168. Toussaint, C., van Geertruyden, J., Govaerts, A., Plan, R., Latour, M., Kinnaert, P., Vereerstraeten, P., Depelchin, A., Bonneau, M., Buchin, R., and Wybran, J. (1970): Administration intraveineuse d'une globuline antilymphocytaire chez cinq patients poeturs d'une greffe renale. In: *International Symposium on Antilymphocyte Serum; Symposia Series on Immunobiological Standardization*, Vol. 16, pp. 371–378, edited by J. F. Bach, J. Dormont, A. Eyquem, M. Raynaud, and R. H. Regamey. Karger, Basel.
169. Traeger, J., Carraz, M., Fries, D., Perrin, J., Saubier, E., Bernhardt, J. P., Revillard, J. P., Bonnet, P., Archimbaud, J. P., and Brochier, J. (1969): Studies of anti-lymphocyte globulins made from thoracic duct lymphocytes. *Transplant. Proc.*, 1:455–459.
170. Traeger, J., Perrin, J., Fries, D., Saubier, E., Carraz, M., Bonnet, P., Archimbaud, J.-P., Bernhardt, J.-P., Brochier, J., Beteul, H., Veysseyre, C., Bryon, P.-A., Prevot, J., Jouvenceau, A., Bansillon, V., Zech, P., and Rollet, A. (1968): Utilisation chez l'homme d'une globuline antilymphocyte: resultats cliniques en transplantation renale. *Lyon Med.*, 219:307–369.
171. Turcotte, J. G., Feduska, N. J., Haines, R. F., Freier, D. T., Gikas, P. W., McDonald, F. D., Johnson, A. G., Morrell, R. M., and Thompson, N. W. (1973): Antithymocyte globulin in renal transplant recipients. *Arch. Surg.*, 106:484–488.
172. Turk, J. L. (1970): Pathological effects and mode of action of antilymphocyte serum treatment. *Fed. Proc.*, 29:136–141.
173. Turk, J. L., and Polak, L. (1969): Action of antilymphocytic sera on reactions caused by immune-complex formation. *Lancet*, 1:130–131.
174. Turk, J. L., and Willoughy, D. A. (1967): Central and peripheral effects of anti-lymphocyte sera. *Lancet*, 1:249–251.
175. Tursi, A., Greaves, M. F., Torrigiani, G., Playfair, J. H. L., and Riott, I. M. (1969): Response to phytohaemagglutinin of lymphocytes from mice treated with anti-lymphocyte globulin. *Immunology*, 17:801–811.
176. Tyler, R. W., Everett, N. B., Schwartz, M. R. (1969): Effect of antilymphocytic serum on rat lymphocytes. *J. Immunol.*, 102:179–186.
177. van der Werf, B. A., Monaco, A. P., Wood, M. L., and Russell, P. S. (1967): Immune competence of mouse lymph node cells after *in vivo* and *in vitro* contact with rabbit anti-mouse lymphocyte serum (RAMLS). In: *Advances in Transplantation*, pp. 133–140, edited by J. Dausset, J. Hamburger, and G. Mathe. Munksgaard, Copenhagen.
178. Waksman, B. H., Arbouys, S., and Arnason, B. (1961): The use of specific "lymphocyte" antisera to inhibit hypersensitive reactions of the "delayed" type. *J. Exp. Med.*, 114:997–1022.
179. Wechter, William (Upjohn Co., Kalamazoo, Michigan) (1975): *Personal communication.*
180. Weil, R., and Simmons, R. L. (1968): Combined immunosuppression for canine renal allograft prolongation: antilymphocyte serum plus prednisone or azathioprine. *Ann. Surg.*, 167:239–245.
181. Wilson, B. J., Malley, A., Cook, M. W., and Mackler, B. F. (1971): Isolation of rabbit anti-rhesus lymphocyte antibody using a specific immunoadsorbent. *J. Immunol.*, 106:402–406.
182. Wilson, R. E., Carpenter, C. B., Birtch, A. G., Murray, J. E., and Merrill, J. P. (1972): Peter Bent Brigham Hospital experience with controlled clinical trial of antilymphocyte globulin for human renal allografts. In: *ALG Therapy and Standardization Workshop;*

Behring Institute Research Communications No. 51, pp. 86–89, edited by F. R. Seiler and H. G. Schwick. Behringwerke AG, Marburg.

183. Winn, H. J., Baldamus, C. A., Jooste, S. V., and Russell, P. S. (1973): Acute destruction by humoral antibody of rat skin grafted to mice: The role of complement and polymorphonuclear leukocytes. *J. Exp. Med.*, 137:893–910.
184. Winn, H. J., and Daddi, G. (1970): Properties of antimouse lymphocyte sera produced in several species. *Fed. Proc.*, 29:104–105.
185. Witz, I., Yagi, Y., and Pressman, D. (1968): Immunosuppressive activity of rabbit antisera directed against mouse lymphocytic leukemia L 1210. *Proc. Soc. Exp. Biol. Med.*, 127:562–565.
186. Wolstenholme, G. E. W., and O'Connor, Maeve (editors) (1967): *Antilymphocytic Serum; Ciba Foundation Study Group No. 29*. Little, Brown, Boston.
187. Wood, M. L. (1970): Effect of pretreatment of mice with normal rabbit immunoglobulin G on the action of anti-mouse lymphocyte gammaglobulin. In: *International Symposium on Antilymphocyte Serum; Symposia Series in Immunobiological Standardization*, Vol. 16, pp. 247–252, edited by J. F. Bach, J. Dormont, A. Eyquem, M. Raynaud, and R. H. Regamey. Karger, Basel.
188. Wood, M. L., and Vriesendorp, H. M. (1969): A comparative study of heterologous antimouse lymphocyte and antimouse thymocyte sera prepared by two different immunization methods. *Transplantation*, 7:522–533.
189. Woodruff, M. F. A. (1967): Immunological properties of anti-lymphocytic serum. *J. Clin. Pathol.*, 20:466–470.
190. Woodruff, M. F. A. (1968): Purification of antilymphocytic antibody. *Nature*, 217:821–824.
191. Woodruff, M. F. A. (1970): The mode of action of antilymphocytic antibody. In *International Symposium on Antilymphocyte Serum: Symposia Series on Immunobiological Standardization*, Vol. 16, pp. 379–388, edited by J. F. Bach, J. Dormont, A. Eyquem, M. Raynaud, and R. H. Regamey. Karger, Basel.
192. Woodruff, M. F. A. (1960): *The Transplantation of Tissue and Organs*, p. 101, Charles C Thomas, Springfield, Ill.
193. Woodruff, M. F. A., and Anderson, N. F. (1963): Effect of lymphocyte depletion by thoracic duct fistula and administration of antilymphocytic serum on the survival of skin homografts in rats. *Nature*, 200:702.
194. Zola, H., Mosedale, B., and Thomas, D. (1970): The preparation and properties of antisera to subcellular fractions from lymphocytes. *Transplantation*, 9:259–272.

Antibodies in Human Diagnosis and Therapy,
edited by Edgar Haber and Richard M. Krause.
Raven Press, New York © 1977.

On the Prevention of Rh Hemolytic Disease of the Newborn by Hyperimmune Anti-Rh Globulin with Comments on the IgG Effector Site for Opsonic Activity

Fred S. Rosen

*Children's Hospital and Harvard Medical School,
Boston, Massachusetts 02115*

During the past decade there has been a remarkable medical achievement in the prevention of Rh hemolytic disease of the newborn by preventing maternal isoimmunization with hyperimmune Rh globulin (1). The successful demonstration of the efficacy of this procedure was arrived at independently by investigators at Liverpool, England and at Columbia University in New York City. Between 3,000 to 4,000 perinatal deaths per year have been prevented in the United States.

HISTORICAL ASPECTS

The Rh antigen of red cells was first described by Landsteiner and Wiener (2) and was demonstrated shortly thereafter by Levine et al. (3) to be the cause of erythroblastosis fetalis. By the mid-1950s it was appreciated from experimental studies in man that red cells could be quickly cleared from the circulation by antibody with resultant splenic or hepatic sequestration and destruction of incompatible erythrocytes (4,5). Attempts were then made to prevent Rh isoimmunization in male volunteers who were Rh-negative with plasma from Rh-sensitized mothers (6) or with gamma globulin (Cohn Fraction II) separated from such plasma (7). Successful clinical trials followed soon thereafter and prevention of Rh isosensitization of postpartum mothers was achieved (8).

APPLICATION OF ANTI-Rh IMMUNE GLOBULIN

Anti-Rh immune globulin is presently administered in a dose of 1 ml, containing at least 300 μg of anti-Rh_0 or anti-D antibody as IgG. The criteria for its postpartum administration are:

1. The mother must be Rh negative and D^u-negative.
2. The mother must have no preexisting anti-Rh antibodies.
3. The infant must be Rh- or D^u-positive.
4. The infant must have a negative direct Coombs test, or if positive, it must result from an antibody other than anti-Rh.

PATHOGENESIS OF Rh ISOSENSITIZATION

A critical step in the evaluation of fetal hemorrhage into the maternal circulation was the development of the Kleihauer–Betke acid elution technique for the detection of fetal hemoglobin (9). Approximately 50% of postpartum women delivered of an ABO-compatible infant have demonstrable fetal cells in their circulation and in 0.2% the fetal hemorrhage into the mother is sizeable (25–30 ml) (10).

Although 1–2% of Rh-mothers have Rh antibody during their first pregnancy, probably due to prior sensitization from an abortus, the incidence of Rh antibody rises to 8% during a second pregnancy. Certainly ABO compatibility between mother and fetus increases the risk of sensitization, up to fivefold. The earlier suggestion of Bramwell and Mitchison that Rh− mothers might be tolerized by Rh+ grandmothers and thus have less risk of sensitization has not been statistically validated (11).

EFFICACY OF Rh PROPHYLAXIS

Studies have now been conducted in several centers in Australia, the United States, and western Europe and they reveal a 90% efficacy of administration postpartum of Rh-immune globulin. The overall incidence of Rh sensitization is now under 1%. An irreducible minimum of Rh sensitization may well have been achieved.

The rapid ingestion by macrophages of antibody-coated red cells and the subsequent intracellular digestion of antigen probably explains the extraordinary efficacy of passive antibody administration. Since the Rh antibody is not complement-fixing, intravascular lysis of sensitized erythrocytes is probably irrelevant to this prophylaxis. Mollison estimates that 10 molecules of anti-Rh per erythrocyte or 25 μg antibody/ml packed red cells will result in the rapid internalization of sensitized erythrocytes into macrophages (12).

CYTOPHILIC ANTIBODY

The ability of antibody bound to red cells to form rosettes with macrophages was attributed by Berken and Benacerraf (13) to guinea pig gamma G_2. Subsequently Lay and Nussenzweig (14) demonstrated that mouse 7S gamma G_2 is also cytophilic for mononuclear cells. LoBuglio, Cotran, and Jandl (15) used anti-Rh antibody to form rosettes with Rh-positive erythrocytes and human blood monocytes. The usual method of quantitating rosette formation is as follows: Human peripheral blood drawn into heparin (100 units/4 ml) is sedimented for 1 hr at room temperature by adding 1 vol of 3% dextran in saline to 2 vol blood. Leukocytes centrifuged from plasma at 300 × *g* for 10 min and then washed twice in Hanks' balanced salt solution are poured into 16 mm Linbro disposable tissue culture trays, and nonadherent cells subsequently removed by washing four times with Hanks' solution. Human erythrocytes, type O Rh$^+$, are labeled with ^{51}Cr. Then, 1 vol of diluted antiserum is incubated with 3 vol of a 16% suspension of ^{51}Cr-labeled erythrocytes (E) at 37°C for 1 hr. Sensitized erythrocytes (EA) are washed three times in Hanks' solution and finally suspended at a concentration of 4% (v/v). Then, 0.4 ml of EA added to each monolayer is incubated for 2 hr at room temperature. Unbound EA are removed from the monolayers by aspiration of the supernatant medium, and the monolayers are assayed for radioactivity after being washed four times with Hanks' solution.

In man, only IgG_1 and IgG_3 exhibit this cytophilic property (16,17). Minute activity can be detected in IgG_4, but none with IgG_2, IgM, IgA, or IgD.

LOCALIZATION OF THE IgG EFFECTOR SITE

LoBuglio et al. showed that the binding site of IgG to monocytes is in the Fc fragment which effectively and completely inhibits rosette formation. $F(ab)_2$ gives about 15% inhibition maximally (18). Peptic and tryptic digests of the Fc region furthermore revealed the inhibitory activity to be in the CH_3 region (19,20). In further attempts to delineate the effector site it was found that cyanogen bromide treatment of papain derived Fc fragments yielded dialyzable peptides with inhibitory activity (21). Fc fragments were allowed to react with 10% cyanogen bromide in 70% formic acid prior to dialysis of the reaction mixture. Peptides in the dialysate were chromatographed into seven fractions on paper with a butanol:pyridine:acetic acid:H_2O mixture in a ratio of 15:10:3:12. Peptide fractions, which were functionally active, were labeled with ^{125}I iodinated 3-(4-hydroxy)phenylpropionic acid *N*-hydroxysuccinimide ester. Radiolabeled peptides were further fractionated by chromatography on paper and thin-layer silicic acid in the same solvent system. Seven peptide fractions were separated from the peptide mixture by chromatography on paper. Each fraction could be resolved into four or five component peptides by high-voltage electrophoresis. However, only fractions with an R_f of 0.35 to 0.70 resulted in inhibition of rosette formation, whereas the other peptide fragments enhanced the formation of rosettes. Synthetic peptides containing glycine, such as glycyl-alanine, glycyl-lysine, or glycyl-glutamic acid, also produced comparable enhancement in the formation of rosettes on a standard monolayer. The peptides eluted were labeled with ^{125}I-3-(4-hydroxy)phenylpropionic acid

N-hydroxysuccinimide ester and applied to monocyte monolayers. The monolayers were incubated for 20 min at 37°C and then washed with Hanks' balanced salt solution three times. IgG (15 mg/ml) was then added to the monolayers to elute adherent, radiolabeled peptides. The eluate was dialyzed, and the dialysate was rechromatographed on paper. The radioactive peptide in the R_f ranges of 0.65–0.8 was further chromatographed on silicic acid in order to separate the radiolabeled IgG peptide from nonlabeled peptides from the eluates of the monolayers and the chromatography paper. While contaminating peptides remained at the origin in the silicic acid chromatograms, a single radioactive spot was eluted that had an R_f of 0.8. The radioactive peptide eluted from silicic acid was tested for its capacity to inhibit rosette formation in four separate experiments. Inhibition was 88–90% at a concentration of 0.6 nmol of peptide per monolayer of 8×10^5 monocytes.

The active peptide consisted of 10 amino acids residues: Tyr, Ser_2, Lys_2, Leu, Thr, Val, Asp, and Arg. This is the composition of residues 407–416 of the Eu myeloma protein.

This localization was rendered feasible by the anomalous papain cleavage, presumably in a small portion of IgG molecules, of a Leu-Tyr bond at position 406–407 and acid degradation of tryptophan at position 417, yielding a dialyzable peptide which inhibits rosette formation by cytophilic antibody. Attempts are now under way to synthesize the active peptide in order to be certain of this localization of the IgG effector site.

REFERENCES

1. Woodrow, J. C. (1970): Rh immunization and its prevention. *Ser. Haematol.*, 3:3.
2. Landsteiner, K., and Wiener, A. S. (1940): An agglutinable factor in human blood recognized by immune sera for rhesus blood. *Proc. Soc. Exp. Biol.*, 43:223.
3. Levine, P., Katzin, E. M., and Burnham, L. (1941): Isoimmunization in pregnancy, its possible bearing on the etiology of erythroblastosis fetalis. *JAMA*, 116:825.
4. Jandl, J. H., Jones, A. R., and Castle, W. B. (1957): The destruction of red cells by antibodies in man. *J. Clin. Invest.*, 36:1428.
5. Hughes-Jones, N. C., Mollison, P. L., and Veall, N. (1957): Removal of incompatible red cells by the spleen. *Br. J. Haematol.,* 3:125.
6. Finn, R., et al. (1961): Experimental studies on the prevention of Rh hemolytic disease. *Br. Med. J.*, 1:1486.
7. Freda, V. J., Gorman, J. G., and Pollack, W. (1964): Successful prevention of experimental Rh sensitization in man with an anti-Rh gamma globulin: a preliminary report. *Transfusion*, 4:26.
8. Freda, V. J. et al. (1967): Prevention of Rh isoimmunization. Progress report of the clinical trial in mothers. *JAMA*, 199:390.
9. Kleihauer, E., Braum, H., and Betke, K. (1957): Demonstration von fetalem Hamoglobin in den Erythrozyten eines Blutausstriches. *Klin. Wochenschr.*, 35:637.
10. Woodrow, J. C. (1974): Rh immunization and its prevention. *Pathobiology Annual*, pp. 65–86. Appleton-Century-Crofts, New York.
11. Race, R. R., and Sarger, R. (1968): *Blood Groups in Man*. Blackwell, Oxford.
12. Mollison, P. L., et al. (1965): Rate of removal from the circulation of red cells sensitized with different amounts of antibody. *Br. J. Haematol.*, 11:461.

13. Berken, A., and Benacerraf, B. (1966): Properties of antibodies cytophilic for macrophages. *J. Exp. Med.*, 123:119.
14. Lay, W. H., and Nussenzweig, V. (1968): Receptors for complement on leukocytes. *J. Exp. Med.*, 128:991.
15. LoBuglio, A. F., Cotran, R. S., and Jandl, J. H. (1967): Red cells coated with immunoglobulin G: Binding and sphering by mononuclear cells in man. *Science*, 158:1582.
16. Abramson, N., et al. (1970): The interaction between human monocytes and red cells. Specificity for IgG subclasses and IgG fragments. *J. Exp. Med.*, 132:1207.
17. Huber, H., et al. (1971): IgG subclass specificity of human monocyte receptor sites. *Nature*, 229:419.
18. Abramson, N., et al. (1970): The interaction between human monocytes and red cells. Binding characteristics. *J. Exp. Med.* 132:1191.
19. Yasmeen, D. et al. (1973): Evidence for the domain hypothesis: Location of the site of cytophilic activity toward guinea pig macrophages in the C_H3 homology region of human immunoglobulin G. *J. Immunol.*, 110:1706.
20. Okafor, G. O., Turner, M. W., and Hay, F. C. (1974): Localization of monocyte binding site of human immunoglobulin G. *Nature*, 248:228.
21. Ciccimarra, F., Rosen, F. S., and Merler, E. (1975): Localization of the effector site of IgG for binding to monocytes. *Proc. Natl. Acad. Sci. USA*, 72:2081.

Antibodies in Human Diagnosis and Therapy,
edited by Edgar Haber and Richard M. Krause.
Raven Press, New York © 1977.

Cardiac Glycoside-Specific Antibodies in the Treatment of Digitalis Intoxication

Thomas W. Smith, Vincent P. Butler, Jr., and Edgar Haber

Cardiac Unit, Massachusetts General Hospital, Cardiovascular Division, Peter Bent Brigham Hospital, Department of Medicine, Harvard Medical School, Boston, Massachusetts, 02114; and Columbia University College of Physicians and Surgeons, New York, New York 10032

Although exogenous antibodies have been used for therapeutic purposes for many decades, only relatively recently have efforts been directed toward the use of hapten-specific antibodies to reverse the effects of a drug or nonmacromolecular toxin. The purpose of this discussion is to describe ongoing studies of the use of specific antibodies and their Fab fragments to reverse digitalis toxicity.

Despite widespread clinical use of digitalis glycosides for nearly 200 years, toxicity observed in a number of studies ranges up to 20% or more of patients chronically treated with these drugs (3), due in large part to the unusually narrow margin between therapeutic and toxic doses (57). Massive, life-threatening or fatal toxicity is by no means uncommon (46,60), and currently available therapeutic approaches are nonspecific and sometimes ineffective.

A new approach to this problem was suggested by the work of Butler and Chen (7), who demonstrated that digoxin-specific antibodies could be elicited in rabbits challenged with digoxin coupled as a hapten to appropriate carrier proteins. Recognition of the high affinity and specificity of selected antibody populations for cardiac glycoside haptens has led to widespread application of radioimmunoassay techniques for measurement of subnanogram amounts of these substances in tissues and biological fluids (6,55,58), and evidence has been presented indicating that use of serum digoxin concentration measurements can significantly reduce the incidence of clinically overt digitalis intoxication (17). Recent progress suggests that direct therapeutic use of digoxin-specific antibodies or their Fab fragments in cases of advanced digitalis toxicity may also assume an important clinical role. The evidence supporting this statement will be reviewed in the discussion that follows, and the principal problems remaining to be overcome will be considered.

SYNTHESIS OF IMMUNOGENIC DIGOXIN-PROTEIN CONJUGATES

The digitalis glycoside digoxin consists of an aglycone, digoxigenin, and three glycosidic digitoxose residues (Fig. 1). The pharmacological activity of the glycoside resides in the aglycone, which consists of a steroidal nucleus to which is attached an unsaturated lactone ring at C17; the digitoxose residues are linked to the aglycone at C3 and are believed to modify the water solubility, distribution properties, and potency of the resulting glycoside (40).

The molecular weight of digoxin is only 781, and it is not immunogenic by itself. Digoxin was therefore coupled as a hapten to immunogenic protein carriers by the periodate oxidation method of Erlanger and Beiser (18), as shown schematically in Fig. 1. When sodium metaperiodate is added to digoxin, the adjacent hydroxyl groups of the terminal digitoxose residue are oxidized to form a dialdehyde, as illustrated in the upper portion of Fig. 1. Ethylene glycol is added to inactivate any excess periodate and the dialdehyde derivative is then added to a solution of protein (bovine serum albumin, in this instance) in a mildly alkaline (pH = 9.5) aqueous solution. At this pH, the dialdehyde reacts with free amino groups in the protein (principally ϵ-amino groups of lysine residues) with the formation of an

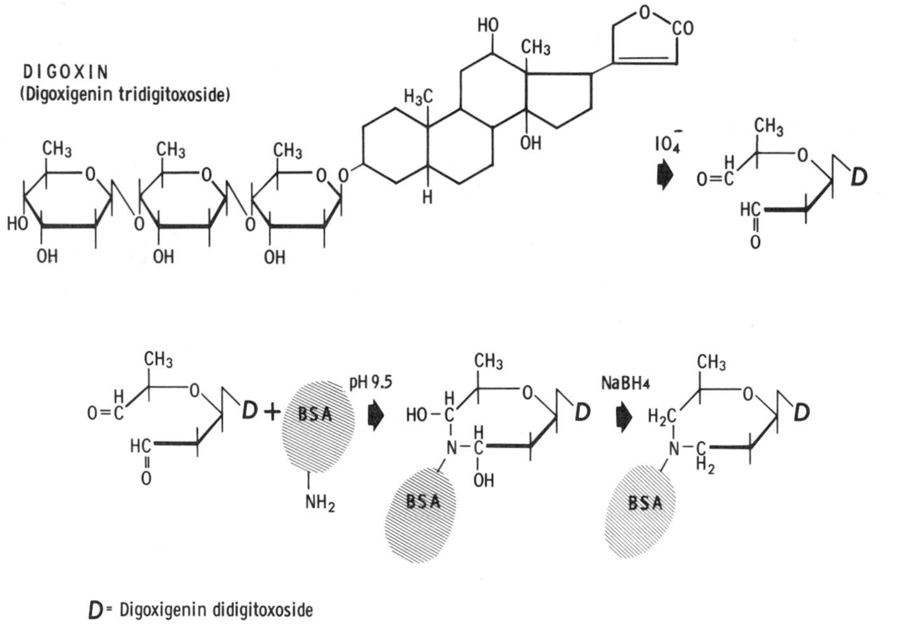

FIG. 1. Suggested scheme of conjugation of digoxin to bovine serum albumin (BSA). See text for details. [Reprinted by permission from the *Annals of the New York Academy of Sciences* (9).]

unstable Schiff base-type product. After reduction by sodium borohydride, the final, more stable, structure shown at the lower right of Fig. 1 is formed. Following removal of unconjugated digoxin and of other low molecular weight reagents and reaction products by a series of precipitation and dialysis steps, the digoxin-protein conjugate is lyophilized and stored until employed as an immunogen. When bovine or human serum albumin is used as the protein carrier, the final product usually contains from 3 to 7 molecules of digoxin per molecule of albumin carrier (7,56).

IMMUNIZATION OF ANIMALS AND HARVESTING OF ANTIBODIES

Rabbits, dogs, and sheep have been immunized by the repeated injection of digoxin-protein conjugates, 1 mg/ml in complete Freund's adjuvant mixture. Various immunization schedules have been successfully employed, but the most commonly used schedules have involved the weekly intramuscular injection of antigen (0.4 ml in rabbits, 1.0 ml in sheep) for 6 to 8 weeks, followed by injections at 2- to 4-weekly intervals for 6 to 36 months thereafter (the initial 3 injections in rabbits are often given by the toe-pad route) (17,48,49,56,68).

Most rabbits and sheep immunized in this manner have formed satisfactory antidigoxin antibodies (dogs, in general, have not produced large quantities of antidigoxin antibodies); however, during the first 3 months of immunization, the affinity and specificity of these antibodies for digoxin is often less satisfactory than the affinity and specificity of antibodies harvested later (56). Hence, bleeding of animals (40–50 ml from rabbits, 400–500 ml from sheep) is usually begun after 3 months of immunization and continued at 2- to 4-week intervals thereafter.

Antibodies of high affinity and specificity for the cardiac glycoside ouabain have also been raised in rabbits (53). The antigen used for this purpose was a conjugate of ouabain linked through its rhamnose moiety to terminal α-amino groups of poly D,L-alanyl-human serum albumin, a maneuver intended to place antigenic determinants of the ouabain molecule at a greater distance from the carrier protein backbone than would be the case if ouabain were coupled directly to the carrier protein through its single sugar moiety. Immunologic techniques were otherwise analogous to those used to obtain digoxin-specific antibodies.

IMMUNOCHEMICAL CHARACTERIZATION OF CARDIAC GLYCOSIDE-SPECIFIC ANTIBODIES

Detailed characterization of rabbit digoxin-specific antisera showed remarkably high average intrinsic affinity constants for selected bleedings, with K_0 values of the order of 10^{10} M^{-1} found after prolonged immunization (56). Also of interest was the high degree of specificity of selected antibody populations, which showed up to 40- to 50-fold higher affinity for digoxin than digitoxin, despite a difference between the homologous and cross-reacting hapten of only a single hydroxyl group. Endogenous steroids from mammalian species such as cholesterol, cortisol, dehydroepiandrosterone, 17β-estradiol, progesterone, and testosterone showed minimal or no tendency to cross-react in hapten displacement experiments, even when present in concentrations 1,000 to 10,000-fold greater than the homologous hapten.

Ouabain-specific antibody populations raised in rabbits have similarly shown high affinity and specificity (53). All cardiac glycoside-specific antibody populations from rabbits or sheep undergoing immunization and repeated boosting studied in detail to date have shown relatively restricted heterogeneity of binding site affinities, with a values for Sips plots typically being in the 0.9–1.0 range (53,56).

The kinetic basis for the specificity of these antibody–hapten interactions has recently been studied in detail, using newly devised techniques for the determination of both association and dissociation kinetics of antibody–hapten interactions (52). Use of tritiated haptens of high specific activity, together with the high affinity of the antibody populations under study, allowed forward (association) reactions to be run at initial hapten and antibody concentrations of the order of 10^{-9} to 10^{-10} M, well below the level at which mixing becomes the rate-limiting step. Separation of antibody-bound from free hapten was achieved within 2 sec by brief exposure to dextran-coated charcoal, thus avoiding appreciable disturbance of the systems under study. Interactions of 12 different combinations of four antibody populations (rabbit digoxin-specific antibody, sheep digoxin-specific antibody, rabbit ouabain-specific antibody, and rabbit digitoxin-specific antibody) with three tritiated haptens ([^{3}H]digoxin, [^{3}H]ouabain, and [^{3}H]digitoxin) were studied, both in terms of association and dissociation kinetics (59). Despite marked (100-fold range) differences in equilibrium constants, association rate constants determined under both second-order and pseudo-first-order conditions were quite similar for all 12 antibody–hapten pairs studied, ranging from 0.87×10^7 to 1.7×10^7 M^{-1} sec^{-1}, and were comparable to values previously estimated by stopped-flow techniques for dye haptens of markedly lower affinity.

In contrast to the similarity of association rates, dissociation rate constants varied markedly from 1.9×10^{-4} to 1.7×10^{-2} sec^{-1}, and fully accounted for observed differences in equilibrium binding of the various antibody–hapten combinations. Figure 2 shows dissociation kinetics for combinations of rabbit digoxin-specific antibody with [^{3}H]digoxin, [^{3}H]digitoxin, and [^{3}H]ouabain. As expected, first-order kinetics are evident, with some heterogeneity of dissociation rates for [^{3}H]digitoxin and [^{3}H]ouabain at later points on the curve. [^{3}H]Digoxin dissociation data demonstrate good agreement between two different methods, one using prolonged exposure to

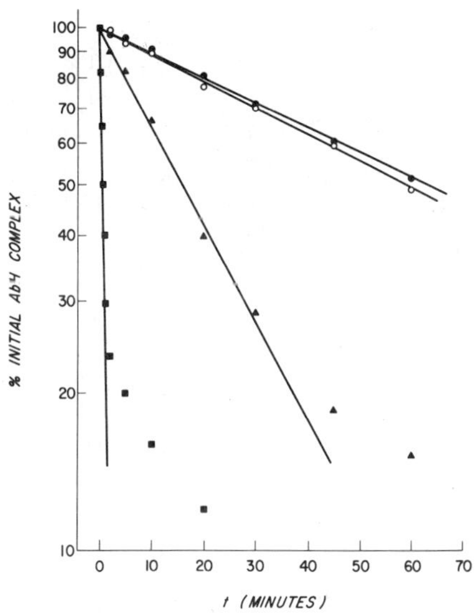

FIG. 2. Dissociation kinetics for complexes of rabbit digoxin-specific antibody with [^{3}H]digoxin (circles), [^{3}H]digitoxin (triangles), and [^{3}H]ouabain (squares). Data are shown for [^{3}H]digoxin dissociation both as determined by prolonged charcoal exposure (closed circles) and by addition of a 10^3-fold excess of unlabeled digoxin (open circles). [Reprinted by permission from *Biochemistry* (59).]

dextran-coated charcoal and the other using a "cold chase" with a large excess of unlabeled digoxin added to an equilibrated mixture of [^{3}H]digoxin and antibody.

Ratios of association to dissociation rate constants were in satisfactory agreement with K_0 values measured at equilibrium. Thus, the major kinetic variable governing these antibody–hapten interactions is the rate of dissociation of the complex. It follows that the strength of antibody–hapten association is determined principally by the activation energy for dissociation. Current studies (Skubitz and Smith, *unpublished observations*) indicate that kinetics of interactions of Fab fragments with cardiac glycoside haptens are very similar to those of the parent immunoglobulin molecule. The rapidity of the antibody–hapten association reaction, which is within about one order of magnitude of theoretical diffusion limitation, is of obvious potential importance in terms of the kinetics of competition for cardiac glycoside molecules between antibody and cellular receptors.

EFFECTS OF ACTIVE IMMUNIZATION ON DIGOXIN TOLERANCE AND PHARMACOKINETICS

Rabbits which have been actively immunized with digoxin-protein conjugates and whose sera contain antidigoxin antibodies are protected from the toxic cardiac effects of a lethal dose of digoxin. As shown in Fig. 3, 15 nonimmunized rabbits given a single intravenous dose of digoxin (0.5 mg/kg body wt) exhibited cardiac rhythm disturbances, accompanied usually by bradycardia, and associated with death within 1 to 2 hr. In contrast, 10

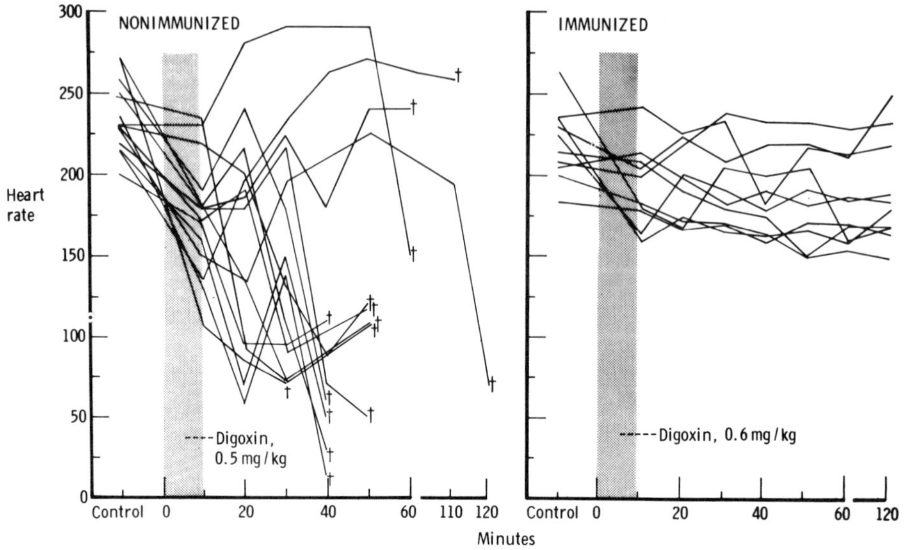

FIG. 3. Effect of digoxin on heart rates of 15 nonimmunized rabbits given 0.5 mg digoxin/kg body wt and on heart rates of 10 rabbits immunized with digoxin-albumin conjugates and given 0.6 mg digoxin/kg body wt. Heart rates were determined during a 30-min control period prior to digoxin administration, during the 10-min intravenous infusion of digoxin (shaded areas), and for up to 2 hr thereafter. Crosses indicate times of deaths. [Reprinted by permission from the *Journal of Clinical Investigation* (48).]

digoxin-immunized rabbits showed no significant electrocardiographic abnormalities during 2 hr of monitoring following a single intravenous dose of 0.6 mg digoxin/kg body wt (48).

When digoxin is administered to actively immunized rabbits, the presence of antidigoxin antibodies results in a profound alteration in the pharmacokinetics of the cardiac glycoside (Fig. 4). Thus, although the antidigoxin antibodies protect the myocardium from the toxic effects of digoxin these antibodies prevent renal excretion and initially produce 90-fold higher serum digoxin concentrations, principally in the form of protein-bound, inactive glycoside. For example, 12 hr after a single intravenous dose of [^{3}H]digoxin (0.4 mg/kg body wt), the mean serum digoxin concentration in nine nonimmunized rabbits was 92 ng/ml, while the mean serum digoxin level in five digoxin-immunized animals was 8,300 ng/ml. The mean serum half-life of digoxin was prolonged 21-fold from 3.4 days in nonimmunized animals to 72 days in the digoxin-immunized group; significant concentrations of circulating protein-bound digoxin were present in the sera of actively immunized rabbits 14 months after a single injection of the drug (50).

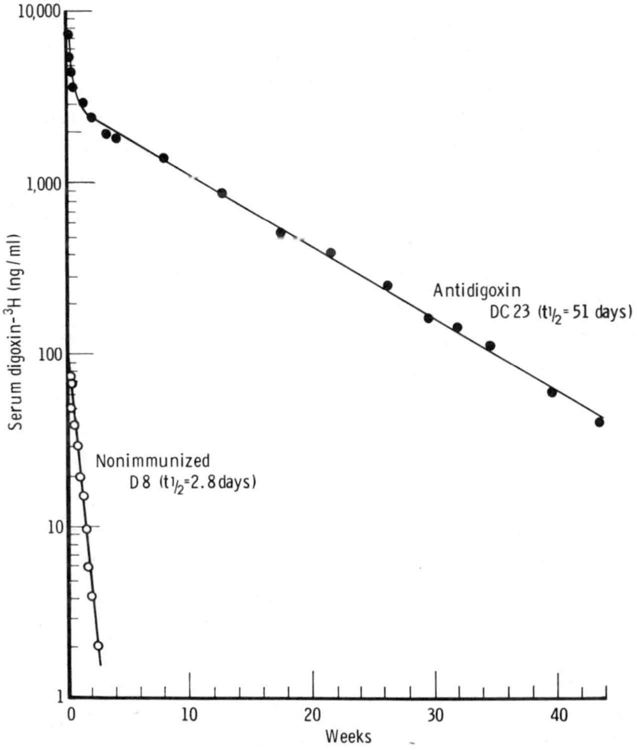

FIG. 4. Comparison of representative computer-fitted serum [^{3}H]digoxin disappearance curves in rabbit D8, a nonimmunized animal, and in a digoxin-immunized rabbit, DC23. [Reprinted by permission from the *Journal of Experimental Medicine* (50).]

ANTIBODY EFFECTS ON CELLULAR DIGOXIN CONCENTRATION *IN VITRO*

Renal cortical slices of rats accumulate digoxin against a concentration gradient. At a concentration of 10^{-5} M, digoxin is rapidly taken up and achieves a steady-state distribution ratio (of tissue-bound to extracellular glycoside concentration) in the 1.8 range within 1 hr. This capacity of rat renal cortical slices to take up digoxin against a concentration gradient provided a useful system in which the ability of digoxin-specific antibodies to remove tissue-bound glycoside could be evaluated. Accordingly, after 1 hr of incubation of slices with [^{3}H]digoxin, 0.5 ml of various sera were added to individual slices. When control serum from a nonimmunized rabbit was added a slight fall in distribution ratio occurred due to dilution. In contrast, when antidigoxin serum was added, rapid removal of [^{3}H]digoxin occurred; 50% of the glycoside was removed within 3 min (68).

Antidigoxin antibodies have also been shown to be capable of removing digoxin rapidly from human erythrocytes. It was originally thought that the rapid removal of digoxin was almost complete (68), but subsequent studies have revealed that this rapid removal is restricted to that fraction of the red cell digoxin which has been taken up across the cell membrane into the erythrocyte (26). For example, in the experiment shown in Fig. 5, the concentration of intracellular digoxin (defined as that fraction of total cellular digoxin which is not membrane-bound after cell lysis) was decreased from 23.2 pmol/ml cells to 0.2 pmol/ml cells within 6 min after the addition of antidigoxin serum, whereas antiserum to an antigen unrelated to digoxin had no significant effect. Analysis of these data suggests that the antibodies remove the intracellular digoxin by lowering the effective extracellular concentration of free glycoside, since repetitive washing of the cells produces a similar effect on intracellular (but not on membrane-bound) digoxin. Membrane-bound digoxin is also removed from red cells by antidigoxin antibodies, but at a much slower rate; removal is only partial after 5 hr, the longest period studied (Fig. 5). Since in the presence of excess ouabain (10^{-3} M), membrane-bound [^{3}H]digoxin decreases at a rate very similar to its rate of decrease in the presence of antibody, it has been postulated that digoxin-specific antibodies bind this fraction of erythrocyte digoxin in the extracellular space after dissociation from membrane-binding sites, and that the mechanism by which antibodies decrease membrane digoxin concentrations is by preventing reassociation of digoxin with these binding sites (26).

ANTIBODY REVERSAL OF CARDIAC GLYCOSIDE EFFECTS *IN VITRO*

Studies in a number of experimental systems have demonstrated the ability of cardiac glycoside-specific antibodies to block the actions of these drugs and, more importantly, to reverse established effects. For example, in removing digoxin from human erythrocytes, antidigoxin antibodies reverse the inhibitory effect of the glycoside on the active influx of monovalent cations into these cells (13,26,68). In one experiment, $^{42}K^+$-influx had been decreased from 2.41 mmol/liter red cells-hr in control erythrocytes to 0.93 mmol/liter red cells-hr in erythrocytes which had been allowed to incubate for 3 hr in the presence of 1.53×10^{-7} M digoxin. After addition of antidigoxin antibodies, progressive restoration of $^{42}K^+$-influx occurred but it was incomplete (1.54 mmol/liter red cells-hr) at the termination of the experiment 5 hr later. This reversal of a demonstrable effect of the glycoside on cells appeared to be correlated temporally with the removal of the membrane-bound fraction of erythrocyte digoxin (Fig. 5). These immunologic studies lend strong support to other experiments which suggest that there are at least two fractions of red cell digoxin but that only one of these fractions, namely the membrane-bound, contains the digoxin molecules which are

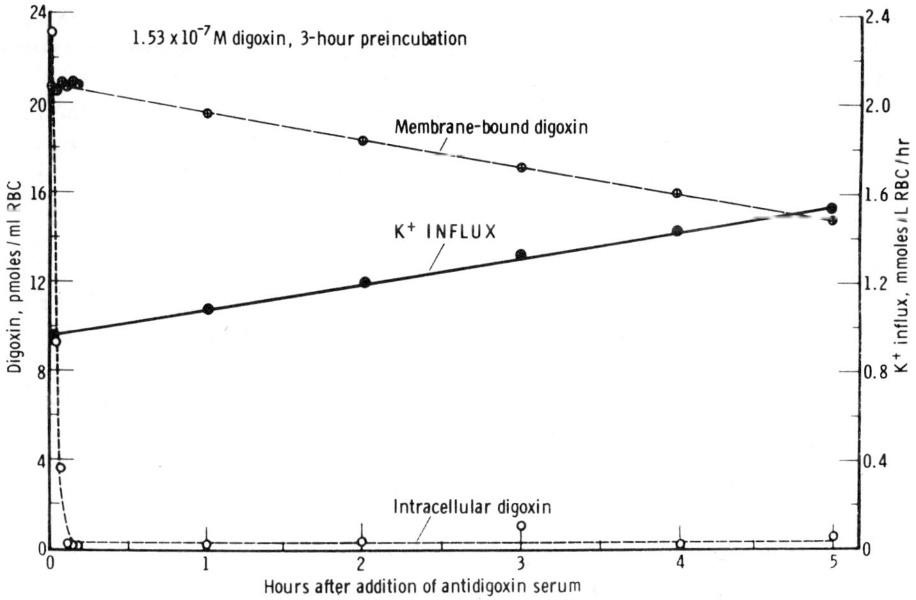

FIG. 5. Effect of antidigoxin serum on loss of erythrocyte digoxin and restoration of potassium influx. Human erythrocytes were incubated for three hours at 37°C in the presence of 1.53×10^{-7} M [^{3}H]digoxin. Aliquots were removed for determination of total cellular digoxin, the amount of membrane-bound digoxin, and potassium influx just before and at various times after the addition, at zero time, of sheep antidigoxin serum. Membrane-bound digoxin is depicted as scored circles (⊕). Intracellular digoxin (O) was calculated as total cellular digoxin minus membrane-bound digoxin. Potassium influx (●) was determined over a 20-min period with ^{42}K and a final K^+ concentration of 15.4 mM; potassium influx in control red cells not treated with digoxin was 2.41 mmol K^+/liter cells-hour. The values noted at zero time did not change significantly over the 5-hr study period after the addition of control antiserum containing antibodies to an antigen structurally unrelated to digoxin. [Reprinted by permission from the *Annals of the New York Academy of Sciences* (9).]

responsible for the inhibitory effect of the glycoside on monovalent cation transport (26).

Antibodies to digoxin and to other cardiac glycosides are also capable of reversing certain pharmacologic actions and certain toxic effects of these drugs on isolated cardiac preparations *in vitro*. Digoxin-specific antibodies have been shown to be capable of reversing toxic electrophysiological effects on isolated canine Purkinje fibers and on rabbit atrioventricular node preparations. In these studies, antibodies reversed digoxin-induced electrical inexcitability, marked phase 4 depolarization in Purkinje fibers, and prolongation of atrioventricular nodal refractoriness (39).

Studies in isolated cardiac muscle preparations have provided further evidence of the ability of specific antibodies to block and reverse cardiac glycoside effects. Following the demonstration by Curd et al. (13) that digoxin-induced increases in active tension development by guinea pig atrial strips could be reversed by digoxin-specific antibody, further studies were undertaken to delineate the time course and certain aspects of the mechanism of reversal. Reversal of ouabain and acetylstrophanthidin effects was investigated in normal and failing cat papillary muscles after exposure to ouabain-specific antibody (27). Antibodies with K_0 values of the order of 10^9 M^{-1} for ouabain and acetylstrophanthidin, when present in 1.1- to 1.5-fold molar excess over cardioactive steroid concentrations, blocked positive inotropic effects of these agents and gradually reversed established contractile effects. Papillary muscles from cats with right ventricular failure induced by chronic pulmonary artery constriction responded similarly, with levels of contractile function returning to, but not below, baseline values. Presence of cardioactive steroid–antibody complex did not alter the myocardial contractile response to subsequently added drug, suggesting that clinical reversal of inotropy could potentially be overcome by the judicious administration of additional digitalis (10). Because of this lack of alteration of response to subsequently added drug, we consider it unlikely that a stable antibody–hapten complex forms at the cardiac glycoside receptor site of the muscle, and this inference is supported by evidence that the dominant chemical determinants of both cardiac activity and antibody binding of cardioactive steroids reside in the C–D rings and unsaturated lactone portions of the molecule, making simultaneous receptor and antibody binding unlikely.

Also of note, as shown in Fig. 6, was the fact that near-stoichiometric amounts of specific antibody reversed inotropic effects of ouabain and acetylstrophanthidin more rapidly than removal of these agents from the muscle bath by washout. The mean times for half-reversal of cardioactive steroid-induced inotropy by antibodies were 124 min and 37 min, respectively, while removal of these two agents by washout of the baths required 181 min and 63 min, respectively, for half-reversal of inotropy. We interpret these studies as indicating that specific high-affinity antibody acts to lower the free cardioactive steroid concentration to near-zero levels, with progressive removal of drug from receptor sites as the drug–receptor equilibrium is displaced in the direction of dissociation, a mechanism supported by the studies of [^{3}H]digoxin removal from erythrocytes by digoxin-specific antibody alluded to previously (26). The rate of reversal of effect by antibody would depend on the rate constant for drug–receptor dissociation and the relative effective concentrations and association rate constants of cardiac receptors and antibodies. Antibody reversal of effect at rates more rapid than drug washout alone could thus result from diffusion of antibody throughout the interstitial space, shortening diffusion distances for effective removal of drug and preventing rebinding to another receptor site. Use of Fab fragments would be expected to have the advantage of more rapid diffusion into the interstitial space compared with intact antibody, and the

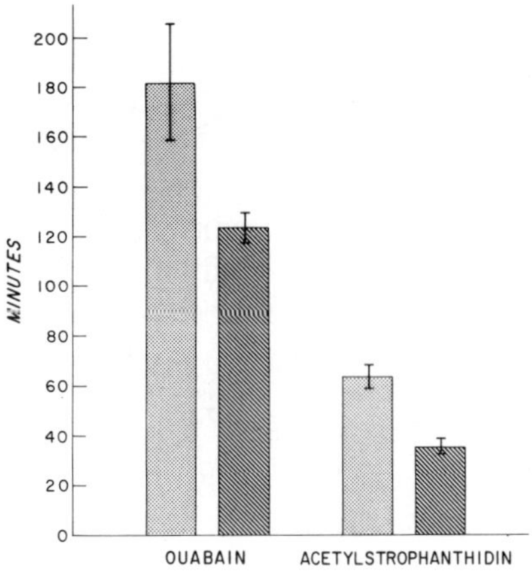

FIG. 6. Comparison of mean times for half-reversal of ouabain and acetyl strophanthidin-induced inotropy by bath washout (strippled bars) and by addition of antibody (cross-hatched bars). Differences in mean time to half-reversal of inotropy are significant for both ouabain ($p < 0.05$) and acetylstrophanthidin ($p < 0.001$). Brackets depict 1 SEM above and below the mean. [Reprinted by permission from the *Journal of Clinical Investigation* (27).]

rapid kinetics of antibody–hapten complex formation have been commented upon previously.

Experiments with isolated cat papillary muscle demonstrated another interesting phenomenon, that of antibody reversal of spontaneous automaticity occurring as a toxic response to ouabain or acetylstrophanthidin (27). As shown in Fig. 7, spontaneous automaticity was rapidly reversed by the addition of a small excess of specific antibody, and this effect occurred in each of 8 experiments at a time when positive inotropic effects were still fully manifest. Early mechanical toxicity resulting in progressive contracture could also be reversed by addition of specific antibody to the bath.

A possible mechanism for these cellular effects has been suggested by studies documenting both blockade and reversal of cardiac glycoside inhibition of ($Na^+ + K^+$)-ATPase, a putative receptor for the pharmacologic and toxic effects of cardiac glycosides (51). Established ouabain inhibition of partially purified canine myocardial ($Na^+ + K^+$)-ATPase could be reversed

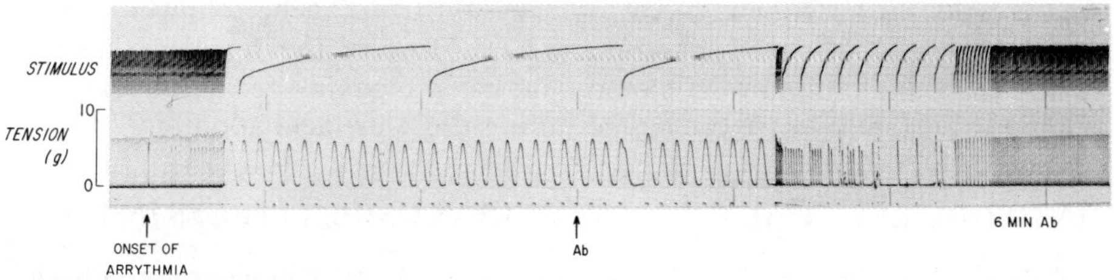

FIG. 7. Reversal of ouabain-induced automaticity in an isolated cat papillary muscle. In this continuous recording, paper speed is varied to show the relationship between stimulus artifact (upper tracing) and muscle contraction. Exposure to 4×10^{-7} M ouabain resulted in development of spontaneous automaticity as peak inotropic effect was reached. Addition of specific antibody is followed, after 4 min, by a return to contractile activity only when a stimulus is applied. Positive inotropic effect is still fully manifest at the time when spontaneous automaticity ceases. [Reprinted by permission from the *Journal of Clinical Investigation* (27).]

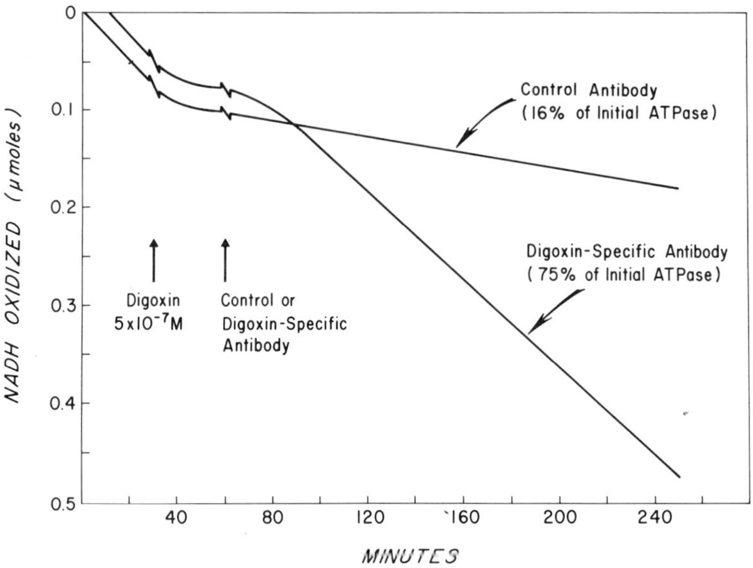

FIG. 8. Reversal of established digoxin inhibition of (Na^+ + K^+)-ATPase by digoxin-specific antibody. Rate of ATP cleavage is proportional to the slope of the line relating NADH-NAD^+ conversion to time. Addition of 5×10^{-7} M digoxin reduces enzymatic activity in both cuvettes to 16% of initial activity. Subsequent addition of digoxin-specific (but not control) antibody produces gradual reversal of inhibition. [Reprinted by permission from the *Annals of the New York Academy of Sciences* (54).]

by addition of ouabain-specific antibody, but not by control antibody (53). Figure 8 illustrates a similar experiment, in which (Na^+ + K^+)-ATPase inhibition by 5×10^{-7} M digoxin is reversed by digoxin-specific antibody (but not control antibody). Continuous monitoring of ATPase activity was accomplished by the use of a linked enzyme assay in which cleavage of ATP was coupled to an NADH-NAD^+ conversion that can be followed at 340 nm in a recording spectrophotometer (1). Most of the residual (Na^+ + K^+)-ATPase inhibition in the experiment using digoxin-specific antibody is related to instability of the enzyme during the 4-hr incubation at 37°C.

ANTIBODY REVERSAL OF CARDIAC GLYCOSIDE EFFECTS *IN VIVO*

Passively administered antibodies are also capable of reversing the toxic effects of the cardiac glycoside in digoxin-intoxicated dogs (49). This property of digoxin-specific antibodies was first demonstrated in a study in which 17 dogs were given large doses of digoxin (0.09 mg/kg body wt, intramuscularly, once daily for 3 days). All 17 dogs developed vomiting, weakness, lethargy and, within 1 to 3 hr after the final dose, a toxic arrhythmia. Figure 9 depicts the clinical course of these arrhythmias in the 17 dogs. In dogs 1

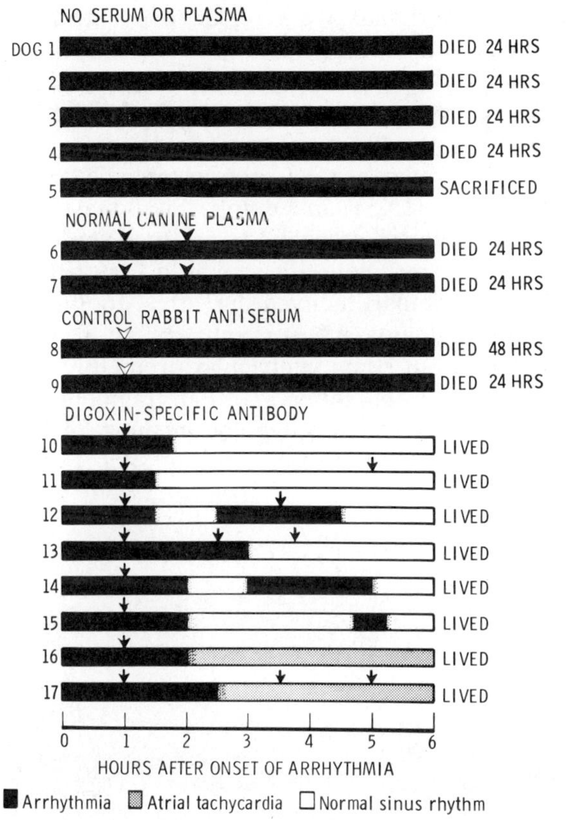

FIG. 9. Course of digoxin-induced arrhythmias in 17 nonimmunized dogs. Black bars indicate the presence of a digoxin-toxic arrhythmia (atrioventricular dissociation with ectopic ventricular beats; atrioventricular dissociation with a junctional rhythm; various forms of atrioventricular block; slow idioventricular rhythm with no evidence of atrial activity; and ventricular tachycardia). Shaded bars denote atrial tachycardia with variable atrioventricular block and a ventricular rate of 80 to 100. White bars indicate normal sinus rhythm. Normal canine plasma, control rabbit antisera (containing antibodies unrelated to digoxin), and antidigoxin plasma and sera were infused at the times indicated by the symbols. The horizontal axis represents the duration of the study period in hours after the onset of the arrhythmia. [Reprinted by permission from the *Journal of Clinical Investigation* (49).]

through 9, none of which received antidigoxin antibodies, the arrhythmias persisted throughout a 6-hr period of continuous electrocardiographic monitoring; normal sinus rhythm was not observed at any time. Seven of the nine control dogs were dead within 24 hr and all had died by 48 hr (one moribund animal was sacrificed at 24 hr). In contrast, in six of eight dogs given antidigoxin plasma or serum, antidigoxin antibodies reversed the toxic rhythm disturbances and restored normal sinus rhythm during the 6-hr study period. In the two remaining animals given antidigoxin antibodies, serious ventricular arrhythmias were replaced by atrial tachycardia with atrioventricular block and a relatively slow ventricular rate; both dogs later reverted to normal sinus rhythm (Fig. 9). All eight antibody-treated dogs were also promptly relieved of their gastrointestinal and other overt manifestations of digitalis toxicity (49).

Digoxin-specific Fab fragment reversal of established digoxin toxicity has also been demonstrated (13,54) in pentobarbital-anesthetized dogs given digoxin intravenously in doses sufficient to produce ventricular tachycardia. The results of such an experiment are shown in Figure 10. A toxic dose of digoxin produced ventricular tachycardia after 16 minutes. Infusion of digoxin-specific Fab fragments at this point resulted in return of conducted sinus beats after 2 min, but some ventricular premature beats persisted for 15 min, when a small additional infusion of Fab fragments produced a stable sinus rhythm, which persisted through the ensuing 24 hr of observation. Data shown in Fig. 11 summarize a number of such experiments. Control

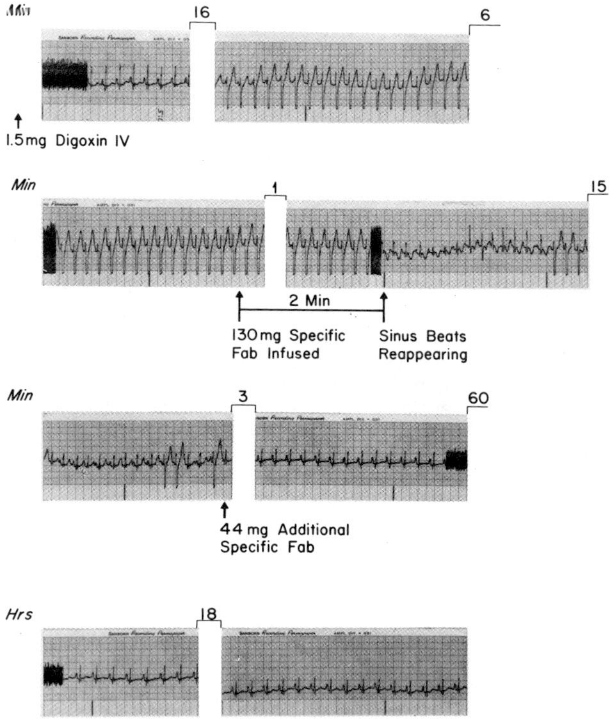

FIG. 10. Continuous ECG recording from a dog in which overt digoxin toxicity (ventricular tachycardia) was induced by intravenous administration of 0.15 mg digoxin/kg body wt. See text for details. [Reprinted by permission from the *Annals of the New York Academy of Sciences* (54).]

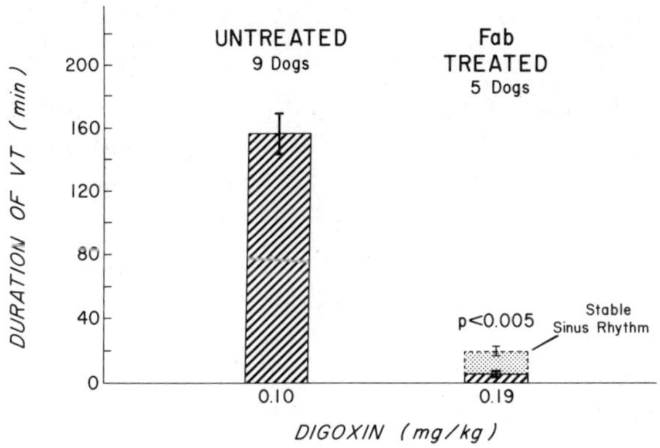

FIG. 11. Duration of ventricular tachycardia in dogs given 0.10 mg/kg digoxin, compared with dogs given an average of 0.19 mg/kg and subsequently treated with digoxin-specific Fab fragments. [Reprinted by permission from the *Annals of the New York Academy of Sciences* (54).]

animals given digoxin, 0.1 mg/kg body wt, developed ventricular tachycardia that lasted for a mean of 157 min. Dogs receiving nearly twice the mean digoxin dose (0.19 mg/kg) were treated with near-stoichiometric amounts of specific Fab fragments following onset of ventricular tachycardia. This resulted in reversal of sustained ventricular tachycardia in an average of 6 min, with return of stable sinus rhythm with no evidence of residual toxicity after 19 min.

Taken together, these experiments provide strong evidence that passively administered digoxin-specific antibodies or Fab fragments are capable of reversing the toxic cellular effects of excessive digoxin in canine experimental models.

EFFECTS OF PASSIVELY ADMINISTERED ANTIBODY ON DIGOXIN PHARMACOKINETICS

Recently, studies have been completed of the effect of passively administered sheep antidigoxin antibodies on the pharmacokinetics of nontoxic doses of [^{3}H]digoxin (0.02 mg/kg body wt) in dogs. Antibodies, given in the form of whole antidigoxin serum, caused a 33- to 57-fold increase in serum digoxin concentrations, presumably due to the removal of digoxin from tissues by antibody. More than 90% of the serum digoxin is antibody-bound in these passively immunized dogs. Thus, as was the case in actively immunized rabbits, the urinary excretion of digoxin in passively immunized dogs becomes negligible and the biologic half-life of the drug is significantly prolonged (8).

APPROACHES TO POTENTIAL PROBLEMS IN CLINICAL USE OF ANTIDIGOXIN ANTIBODIES

Despite the foregoing evidence that digoxin-specific antibodies are able to reverse established toxic effects of the drug, several potential problems must be overcome or minimized before the use of heterologous antibodies can be considered in human therapy. Such problems include the possibility of hypersensitivity reactions to foreign protein, manifest acutely as anaphylaxis, or as a delayed serum sickness response. Furthermore, rapid reversal of therapeutic as well as toxic effects might result in exacerbation of rhythm disturbances or congestive heart failure. Finally, late release of digoxin from the inactive, antibody-bound state might occur during immune degradation of the foreign protein, and might result in recrudescence of toxicity, particularly if the total body burden of digoxin were large and degradation of infused antibody rapid.

Two approaches to these problems have been pursued in our laboratory. The first has been purification of digoxin-specific antibody to remove all extraneous protein. This has been achieved by an affinity chromatographic approach in which the cross-reacting ligand ouabain is coupled to bovine pancreatic ribonuclease and thence to the solid support bromoacetylcellulose (13). More recently, we have extended this approach to the purification of digoxin-specific Fab fragments. Fab fragments provide several advantages over intact antibody, including absence of antigenic and complement-binding determinants of the Fc fragment. Their smaller size should permit more rapid diffusion into the interstitial space, and also results in relatively rapid excretion via glomerular filtration, a process known to occur with a half-life of about 4–5 hr (61,67).

The ouabain-ribonuclease-bromoacetylcellulose immunoadsorbent can be used to purify digoxin-specific Fab fragments as shown schematically in Fig. 12. After a preliminary ammonium sulfate fractionation of crude antiserum from sheep immunized with a digoxin-serum albumin conjugate, cleavage with papain into Fab and Fc fragments is carried out and the digest allowed to interact with the immunoadsorbent, resulting in binding of Fab fragments with cardiac glycoside specificity. Other proteins are eluted with buffer, followed by recovery of digoxin-specific Fab fragments with 25 mM free ouabain. The majority of free ouabain is then removed by dialysis or gel filtration. Residual ouabain bound to Fab fragments is released by guanidine denaturation and removed by dialysis or gel filtration in the presence of 6 M guanidine. Renaturation of the Fab is accomplished by dialyzing away the guanidine, resulting in a preparation that sediments as a single, symmetrical 3.5S peak in the analytical ultracentrifuge and as a single band with Fab mobility on cellulose acetate electrophoresis. [^{3}H]Digoxin binding studies demonstrated binding of 0.8 to 1.0 moles of digoxin per mole of Fab fragments after this sequence of steps, and the affinity of the purified Fab

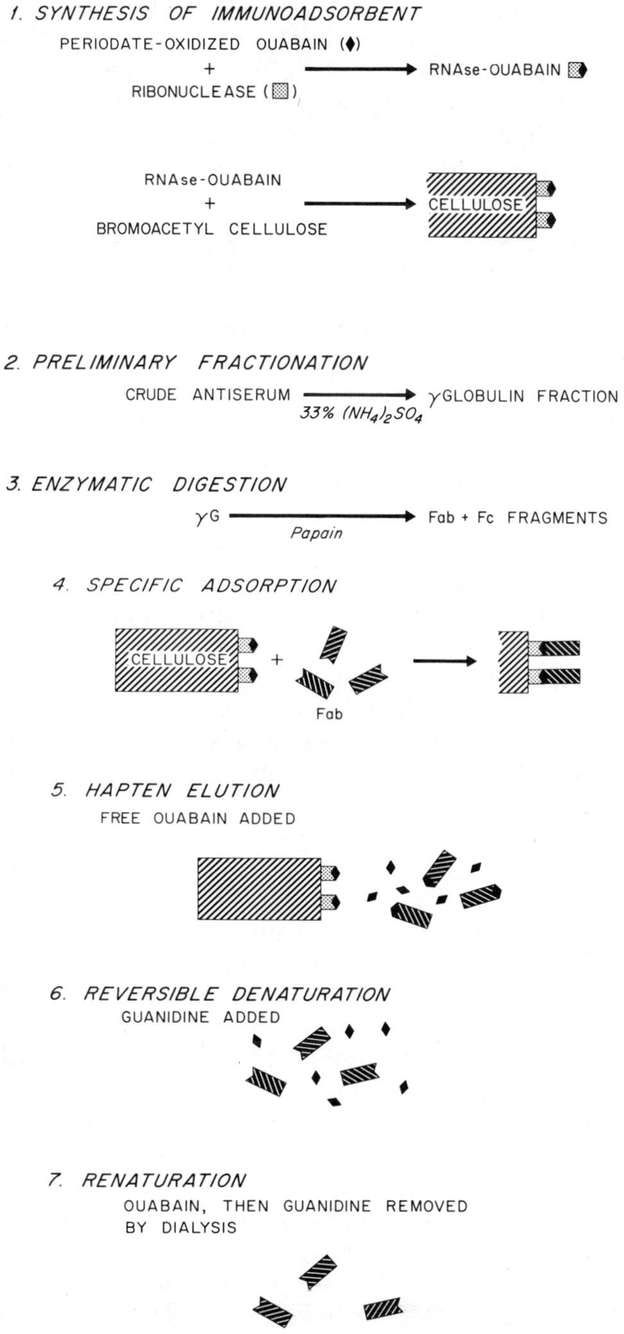

FIG. 12. Schematic outline of purification of digoxin-specific Fab fragments from sheep antibody. See text for discussion. [Reprinted by permission from the *Annals of the New York Academy of Sciences* (54).]

fragments for digoxin is similar to that of the intact antibody population prior to purification.

Preliminary results of immunogenicity studies (Smith, Spicer, and Haber, *unpublished observations*) indicate that purified Fab fragments of ovine digoxin-specific antibodies are very weakly immunogenic compared with intact γ-globulin in both rabbit and baboon experimental models.

EFFECTS OF ANTIDIGOXIN FAB FRAGMENTS ON DIGOXIN PHARMACOKINETICS

When administered to digoxin-treated dogs, antidigoxin Fab fragments produce a significant increase in serum concentrations analogous to the rises observed when intact antidigoxin antibodies are administered. In contrast to dogs treated with intact antibodies, urinary excretion of digoxin in Fab-treated animals is comparable to that observed in control dogs, and hence serum digoxin concentrations fall more rapidly in Fab-treated dogs than in animals receiving intact antibodies; much of the urinary digoxin is Fab-bound (8). These findings indicate that antidigoxin Fab fragments share with intact antidigoxin antibodies the capacity to remove glycoside from tissue and that, in addition, the Fab fragments are capable of promoting the relatively prompt urinary excretion of bound digoxin, a capacity not present in the large intact antibody molecule. Further experimental studies will be required, however, to determine the best approach to the reversal of toxic effects of digoxin with antidigoxin Fab fragments in patients who may require the maintenance of at least some digoxin effect for therapeutic purposes.

INITIAL CLINICAL USE OF DIGOXIN-SPECIFIC FAB FRAGMENTS

We have recently treated the first patient with purified digoxin-specific Fab fragments with gratifying results. A 39-year-old man maintained on digoxin for management of atrial fibrillation 5 years after mitral valve replacement ingested 90 0.25 mg digoxin tablets (22.5 mg) with suicidal intent. The initial serum digoxin concentration at the time of hospital admission was greater than 25 ng/ml. A pervenous endocardial pacemaker catheter was placed because of bradycardia and rising serum potassium concentration. During several ensuing hours of close observation, the serum potassium concentration rose steadily despite aggressive treatment with glucose and insulin, bicarbonate, and polystyrene sulfonate cation exchange resin. Over this period the heart rate continued to slow to a rate of 14/min when the pacemaker was briefly turned off. The QRS complex progressively widened, and the threshold for pacing steadily rose. When the serum potassium concentration reached 8.7 meq/liter, in view of the grave prognosis for massive digoxin intoxication with these manifestations (46,60), treatment with purified digoxin-specific Fab fragments was undertaken. A total of 1,100 mg of Fab fragments were infused over a period of about 2 hr. Within 10 min of completion of the infusion, the patient was in a stable sinus rhythm at a rate of 75/min and the serum potassium concentration fell rapidly back to the normal range. The patient showed no signs of hemodynamic instability, and temperature remained normal. Followup over ensuing weeks has shown no demonstrable immunologic or other ill effects.

FUTURE APPLICATIONS OF DIGOXIN-SPECIFIC ANTIBODIES

The availability of a preparation of digoxin-specific antibody characterized by minimal or absent immunogenicity may be far more widely applied than in the small number of patients who are subject to overwhelming digitalis intoxication. The physician often encounters intoxication in the course of ordinary therapy (3,5). There is no specific antidote and treatment is essentially directed at the electrophysiological manifestations of toxicity. While most instances of intoxication are successfully dealt with simply by withholding the drug, the presence of life-threatening ventricular arrhythmias requires more prompt action. Antiarrhythmic drugs are not always successful and electric countershock in the presence of digitalis intoxication is fraught with danger (57). The availability of digoxin-specific antibody for routine use in the reversal of digitalis effects could add considerably to the physician's therapeutic armamentarium.

In the patient with cardiovascular disease, the relationship of serious arrhythmia to digitalis overdose may be uncertain. All arrhythmias caused by digitalis intoxication may also arise spontaneously in the absence of the drug; there are no absolute criteria for differential diagnosis (57). While the present availability of methods for determining serum or plasma levels of digitalis glycosides (55) has helped in making appropriate therapeutic judgments, assay results are not always available in sufficient time. It would be most convenient to be able to reverse digitalis effects rapidly as a diagnostic maneuver in order to determine whether an arrhythmia is of toxic origin. This would have evident advantages over the acetylstrophanthidin tolerance test, which in essence superimposes an additional dose of digitalis in a clinical situation where an excess may already be present. A safe, well-tolerated antibody or antibody fragment preparation would be expected to meet this need. The body's total content of digoxin can be estimated, either from the history regarding prior dosage or from the serum digoxin concentration. An appropriate quantity of antibody could then be administered therapeutically to "back-titrate" digitalis effects. If this resulted in worsening of the clinical situation because of the need for the beneficial electrophysiological or inotropic actions of the drug, an additional increment of digoxin could be administered so as to restore the patient to the prior level of digitalization. Such applications clearly require a form of antibody or its fragments which can be used repetitively and with safety.

OTHER POTENTIAL THERAPEUTIC APPLICATIONS OF SPECIFIC ANTIBODIES

The digitalis glycosides are but one example of a number of therapeutic opportunities wherein the specific inactivation by antibody of an exogenously administered drug or toxin, or an endogenous hormone may be undertaken. Such approaches have not yet found wide application in human therapy because of the absence of a safe, nonimmunogenic antibody preparation. Their efficacy in experimental circumstances, however, indicates their ultimate potential. Antibodies have been utilized to inactivate renin and thereby correct renin-dependent hypertension in the experimental animal (47). One could envision the use of such antibodies in man to identify renin-dependent hypertensives. This could circumvent some of the problems associated with the use of competitive inhibitors of angiotensin or of converting enzyme which have broader and more complex pharmacologic actions. The inhibitory capacity of antibodies to carcinogens (23,44), gastrin (38), histamine (15), morphine (4,16,65), steroid hormones 19–22,28,29,37,41,45,63), and oxytocin and vasopressin (64) has been demonstrated. Attractive applications which have not been explored are the use of antibodies to insecticides such as DDT (11,31) and paralytic shellfish poison associated with the red tide (35). A condition characterized by hormone excess which often is refractory to definitive surgical cure, hyperparathyroidism, may well be an example of a situation in which antibodies could be effectively applied.

Organ transplantation and cancer chemotherapy provide obvious opportunities. Drugs may be directed at cells bearing unique antigenic determinants on their surfaces (24,25) or unblocking antibodies may be utilized which allow the normal host cellular immune mechanism to attack tumor cells (32). A more potent or more specific antilymphocyte antibody might be developed for use in organ transplantation.

SOURCES FOR NONIMMUNOGENIC ANTIBODIES FOR USE IN HUMAN THERAPY

Two general approaches may be envisioned for the production of antibodies which are likely to be nonimmunogenic when used in substantial doses or repeatedly in human therapy: chemical synthesis and synthesis by cells growing in tissue culture.

Chemical Synthesis

Recently available X-ray crystallographic models of ligand-binding myeloma proteins indicate that there is remarkable preservation of the general structure of variable domains among immunoglobulins of different species of origin (14). Padlan and Davies (43) have shown that the framework regions of REI, a human protein, and MCPC, a mouse protein, are largely superimposable in three dimensions, even though their amino acid sequences are quite different. The only region of marked difference in three-dimensional structure is the hypervariable portion of the sequence, which in an antibody comprises the antigen-binding cavity. Thus, it appears that different complementarity regions corresponding to different antigen-binding specificities are engrafted upon the same basic framework structure. Antibodies to complementarity regions (idiotypic antibodies) are, at best, difficult to raise. Structural variations in this part of the molecule do not seem to lead to immunogenicity. Species-specific substitutions in the framework portion of the variable region, on the other hand, are likely to be significant immunogenic determinants which lead to recognition of variable regions between species.

The commonality of the three-dimensional structures of the framework suggests that a given framework may functionally support any number of hypervariable regions. It may thus be possible to engraft upon a single nonimmunogenic synthetic framework sequence, based on the variable sequence of a human myeloma protein, a variety of different hypervariable region sequences, each specific for a different antigen. Such a direct approach to the synthesis of nonimmunogenic antibodies does not seem remote, since advances in the technology of peptide chemistry now make the chemical synthesis of relatively long polypeptide chains feasible. Examples of biologically active proteins of substantial size which now have been synthesized successfully by the solid-phase technique include acyl-carrier protein (33), ribonuclease A (30), trypsin inhibitor (42), lysosyme (2), and ribonuclease T_1 (34,66).

Synthesis in Cell Culture

While both human and murine plasmacytomas previously have been shown to produce immunoglobulin continuously *in vitro* in culture, it has not been possible to commit such cultures to the production of antibodies of desired specificity. More recently, however, it has been feasible to initiate a continuous culture secreting type III pneumococcal polysaccharide-specific antibody by transforming normal spleen cells from a hyperimmune rabbit with SV40 virus (12,62). This approach is discussed extensively in the chapter by Black and colleagues in this volume. An alternative method, somatic cell fusion (36), has also resulted in the continuous propagation of clones of cells which secrete specific antibody. This approach is discussed in detail by Milstein and colleagues in this volume.

One of these approaches may ultimately provide us with a nonimmunogenic antibody or antibody fragment suitable for regular use in human therapy. The potential applications, as suggested here and by many other chapters in this volume, are very broad indeed.

ACKNOWLEDGMENTS

The studies described herein were supported in part by grants from DHEW, USPHS (HL18003 and HL10608) and from the New York and American Heart Associations.

Dr. Smith is an Established Investigator of the American Heart Association. Dr. Butler is the recipient of an Irma T. Hirschl Career Scientist Award.

The assitance of Mrs. Janifer Johnson in preparation of this manuscript is gratefully acknowledged.

REFERENCES

1. Albers, R. W., Koval, G. J., and Siegel, G. J. (1968): Studies on interaction of ouabain and other cardioactive steroids with sodium—potassium-activated adenosine triphosphatase. *Mol. Pharmacol.*, 4:324.
2. Barstowle, Hruby, V. J., Robinson, A. B., Rupley, J. A., Sharp, J. J., and Shimota, T. (1971): Preparation of protein with lysosyme activity. *Fed. Proc.*, 30:1292.
3. Beller, G., Smith, T. W., Abelmann, W. H., Haber, E., and Hood, W. B., Jr. (1971): Digitalis intoxication. A prospective clinical study with serum level correlations. *N. Engl. J. Med.*, 284:989.
4. Berkowitz, B., and Spector, S. (1972): Evidence for active immunity to morphine in mice. *Science*, 178:1290.
5. Boston Collaborative Drug Surveillance Program (1972): Relation between digoxin arrhythmias and ABO blood groups. *Circulation,* 45:352.
6. Butler, V. P., Jr. (1972): Assays of digitalis in the blood. *Prog. Cardiovasc. Dis.* 14:571.
7. Butler, V. P., Jr., and Chen, J. P. (1967): Digoxin-specific antibodies. *Proc. Natl. Acad. Sci. USA*, 57:71.
8. Butler, V. P., Jr., Schmidt, D. H., Smith, T. W., Haber, E., Raynor, B. D., and Demartini, P. (1973): Effect of Fab fragments of sheep antidigoxin antibody (Dig Ab) on serum concentrations and urinary excretion of digoxin (Dig). *Circulation*, 48:IV-130 (abstract).
9. Butler, V. P., Jr., Schmidt, D. H., Watson, J. F., and Gardner, J. D. (1974): Production and properties of digoxin-specific antibodies. *Ann. N.Y. Acad. Sci.*, 242:717.
10. Butler, V. P., Jr., Watson, J. F., Schmidt, D. H., Gardner, J. D., Mandel, W. J., and Skelton, C. L. (1973): Reversal of the pharmacological and toxic effects of cardiac glycosides by specific antibodies. *Pharmacol. Rev.*, 25:239.
11. Centeno, E. R., Johnson, W. J., and Sehon, A. H. (1970): Antibodies to two common pesticides, DDT and malathion. *Int. Arch. Allergy Appl. Immunol.*, 37:1.
12. Collins, J. J., Black, P. H., Strosberg, A. B., Haber, E., and Bloch, K. J. (1974): Transformation by simian virus 40 of spleen cells from a hyperimmune rabbit: Evidence for synthesis of immunoglobulin by the transformed cells. *Proc. Natl. Acad. Sci. USA*, 71:260.
13. Curd, J., Smith, T. W., Jaton, J. C., and Haber, E. (1971): The isolation of digoxin-specific antibody and its use in reversing the effects of digoxin. *Proc. Natl. Acad. Sci. USA*, 68:2401.

14. Davies, D., Padlan, E., and Segal, D.: Immunoglobulin structures at high resolution. In: *Contemporary Topics in Molecular Immunology,* Vol. 4. Plenum Press, New York.
15. Davis, T. R. A., and Meade, K. M. (1970): Biologically active antibodies to histamine. *Nature*, 226:360.
16. DeCato, L, Jr., and Adler, F. L. (1973): Neutralization of morphine activity by antibody. *Res. Commun. Chem. Pathol. Pharmacol.*, 5:775.
17. Duhme, D. W., Greenblatt, D. J., and Koch-Weser, J. (1974): Reduction of digoxin toxicity associated with measurement of serum levels. *Ann. Intern. Med.*, 80:516.
18. Erlanger, B. F., and Beiser, S. M. (1964): Antibodies specific for ribonucleosides and ribonucleotides and their reaction with DNA. *Proc. Natl. Acad. Sci. USA*, 52:68.
19. Ferin, M., Dyrenfurth, I., Cowchock, S., Warren, M., and Vande Wiele, R. L. (1974): Active immunization to 17β-estradiol and its effects upon the reproductive cycle of the rhesus monkey. *Endocrinology*, 94:765.
20. Ferlin, M., Tempone, A., Zimmering, P. E., and Vande Wiele, R. L. (1969): Effect of antibodies to 17β-estradiol and progesterone on the estrous cycle of the rat. *Endocrinology*, 85:1070.
21. Ferin, M., Zimmering, P. E., and Vande Wiele, R. L. (1969): Effects of antibodies to estradiol-17β on PMS-induced ovulation in immature rats. *Endocrinology*, 84:893.
22. Ferin, M., Zimmering, P. E., Lieberman, S., and Vande Wiele, R. L. (1968): Inactivation of the biological effects of exogenous and endogenous estrogens by antibodies to 17β-estradiol. *Endocrinology*, 83:565.
23. Franks, W. R., and Creech, H. J. (1939): Chemoantigens and carcinogenesis. *I. Am. J. Cancer*, 35:203.
24. Gahose, T., Norvell, S. T., Guclu, A., Cameron, D., Bodurtha, A., and MacDonald, A. S. (1972): Immunochemotherapy of cancer with chlorambucil-carrying antibody. *Br. Med. J.*, 3:495.
25. Gahose, T., Path, M. R. C., and Nigam, S. P. (1972): Antibody as a carrier of chlorambucil. *Cancer*, 29:1398.
26. Gardner, J. D., Kiino, D. R., Swartz, T. J., and Butler, V. P., Jr. (1973): Effects of digoxin-specific antibodies on accumulation and binding of digoxin by human erythrocytes. *J. Clin. Invest.*, 52:1820.
27. Gold, H. K., and Smith, T. W. (1974): Reversal of ouabain and acetyl strophanthidin effects in normal and failing cardiac muscle by specific antibody. *J. Clin. Invest.*, 53:1655.
28. Goldman, A. S., Shapiro, B. H., and Root, A. W. (1973): Inhibition of fetal masculine differentiation in the rat by maternal administration of antibodies to bovine LH, cyanoketone, or antibodies to testosterone-3-bovine serum albumin. *Proc. Soc. Exp. Biol. Med.*, 143:422.
29. Goodfriend, L., and Sehon, A. H. (1961): Antibodies to estrone-protein conjugates. II. Endocrinological studies. *Can. J. Biochem. Physiol.*, 39:961.
30. Gutte, E., and Merrifield, R. V. (1971): The synthesis of ribonuclease. *J. Biochem.*, 246:1922.
31. Haas, G. J., and Guardia, E. J. (1968): Production of antibodies against insecticide-protein conjugates. *Proc. Soc. Exp. Biol. Med.*, 129:546.
32. Hakala, T. R., and Lange, P. H. (1974): Serum induced lymphoid cell mediated cytotoxicity to human transitional cell carcinoma of the genito-urinary tract. *Science*, 184:795.
33. Hancock, W. S., Marshall, G. R., and Vagelos, P. R. (1973): Acyl carrier protein. Chemical synthesis and characterization of analogs of acyl-carrier protein. *J. Biochem.*, 248:2424.
34. Izumiya, N., Waki, M., Kato, T., Ohno, M., Aoyagi, H., and Mitsuasu, N. (1972): Solid phase synthesis of ribonuclease T_1. In: *Chemistry and Biology of Peptides*, p. 269, edited by J. Meienhofer. Ann Arbor, Mich.
35. Johnson, H. M., Frey, P. A., Angelotti, R., Compbell, J. E., and Lewis, K. H. (1964): Haptenic properties of paralytic shellfish poison conjugated to proteins by formaldehyde treatment. *Proc. Soc. Exp. Biol. Med.*, 117:425.
36. Kohler, G., and Milstein, C. (1975): Continuous cultures of cells secreting antibody of predetermined specificity. *Nature*, 256:495.
37. Lieberman, S., Erlanger, B. F., Beiser, S. M., and Agate, F. J., Jr. (1959): Steroid-protein conjugates: Their chemical, immunochemical, and endocrinological properties. *Recent Prog. Horm. Res.*, 15:165.

38. Lipshutz, W., Hughes, W., and Cohen, S. (1972): The genesis of lower esophageal sphincter pressure: Its identification through the use of gastrin antiserum. *J. Clin. Invest.*, 51:522.
39. Mandel, W. J., Bigger, J. T., Jr., and Butler, V. P., Jr. (1972): The electrophysiologic effects of low and high digoxin concentrations on isolated mammalian cardiac tissue: Reversal by digoxin-specific antibody. *J. Clin. Invest.*, 51:1378.
40. Moe, G. K., and Farah, A. E. (1975): Digitalis and allied cardiac glycosides. In: *The Pharmacological Basis of Therapeutics*, 5th ed., pp. 653–682, edited by L. S. Goodman, and A. Gilman. Macmillan, New York.
41. Neri, R. O., Tolksdorf, S., Beiser, S. M., Erlanger, B. F., Agate, F. J., Jr., and Lieberman, S. (1964): Further studies on the biological effects of passive immunization with antibodies to steroid-protein conjugates. *Endocrinology*, 74:593.
42. Noda, K., Terada, S., Mitsuasu, N., Waki, N., Kato, T., and Izeumiya, N. (1971): Synthesis of a protein with basis pancreatic trypsin inhibitory activity. *Naturwissenschaften*, 58:147.
43. Padlan, E., and Davies, D. R. (1975): Variability of three-dimensional structure in immunoglobulins. *Proc. Natl. Acad. Sci. USA.*, 72:819.
44. Peck, R. M., and Peck, E. B. (1971): Inhibition of chemically induced neoplasia by immunization with an antigenic carcinogen-protein conjugate. *Cancer Res.*, 31:1550.
45. Raziano, J., Ferin, M., and Vande Wiele, R. L. (1972): Effects of antibodies to estradiol-17β and to progesterone on nidation and pregnancy in rats. *Endocrinology*, 90:1133.
46. Reza, M. J., Kovick, R. B., Shine, K. I., and Pearce, M. L. (1974): Massive intravenous digoxin overdosage. *N. Engl. J. Med.*, 291:777.
47. Romero, J. C., Hubler, S. W., Cosak, T. J., and Warzynski, R. J. (1973): Effect of anti-renin on blood pressure of rabbits with experimental renin hypertension. *Am. J. Physiol.*, 225:810.
48. Schmidt, D. H., and Butler, V. P., Jr. (1971): Immunological protection against digoxin toxicity. *J. Clin. Invest.*, 50:866.
49. Schmidt, D. H., and Butler, V. P., Jr. (1971): Reversal of digoxin toxicity with specific antibodies. *J. Clin. Invest.*, 50:1738.
50. Schmidt, D. H., Kaufman, B. M., and Butler, V. P., Jr. (1974): Persistence of hapten-antibody complexes in the circulation of immunized animals after a single intravenous injection of hapten. *J. Exp. Med.*, 139:278.
51. Schwartz, A., Lindenmayer, G. E., and Allen, J. C. (1975): The sodium-potassium adenosine triphosphatase: Pharmacological, physiological and biochemical aspects. *Pharmacol. Rev.*, 27:3.
52. Skubitz, K. M., and Smith, T. W. (1975): Determination of antibody-hapten association kinetics: A simplified experimental approach. *J. Immunol.*, 114:1369.
53. Smith, T. W. (1972): Ouabain-specific antibodies: Immunochemical properties and reversal of Na^+, K^+-activated adenosine triphosphatase inhibition. *J. Clin. Invest.*, 51:1583.
54. Smith, T. W. (1974): Use of antibodies in the study of the mechanism of action of digitalis. *Ann. NY Acad. Sci.*, 242:731.
55. Smith, T. W., Butler, V. P., Jr. and Haber, E. (1969): Determination of therapeutic and toxic serum digoxin concentrations by radioimmunossay. *N. Engl. J. Med.*, 281:1212.
56. Smith, T. W., Butler, V. P., Jr., and Haber, E. (1970). Characterization of antibodies of high affinity and specificity for the digitalis glycoside digoxin. *Biochemistry*, 9:331.
57. Smith, T. W., and Haber, E. (1973): Medical Progress. Digitalis. *N. Engl. J. Med.*, 289:945; 1010; 1063; 1125.
58. Smith, T. W., and Haber, E. (1973): The current status of cardiac glycoside assay techniques. *Prog. Cardiol.*, 2:49.
59. Smith, T. W., and Skubitz, K. M. (1975): Kinetics of interactions between antibodies and haptens. *Biochemistry*, 14:1496.
60. Smith, T. W., and Willerson, J. T. (1971): Suicidal and accidental digoxin ingestion: Report of five cases with serum digoxin level correlations. *Circulation*, 44:29.
61. Spiegelberg, H. L., and Weigle, W. O. (1965): The catabolism of homologous and heterologous 7S gamma globulin fragments. *J. Exp. Med.*, 121:323.
62. Strosberg, A. D., Collins, A. J., Black, P. H., Malamud, D., Wilburn, S., Bloch, K. H., and Haber, E. (1974): Transformation by simian virus 40 of spleen cells from a hyperimmune rabbit: Demonstration of production of specific antibody to the immunizing antigen. *Proc. Natl. Acad. Sci. USA*, 71:263.

63. Sundaram, K., Tsong, Y. Y., Hood, W., and Brinson, A. (1973): Effect of immunization with estrone-protein conjugate in rhesus monkeys. *Endocrinology*, 93:843.
64. Vorherr, H., and Munsick, R. A. (1970): Identification of neurohypophyseal hormones with their antisera. *J. Clin. Invest.*, 49:828.
65. Wainer, B. H., Fitch, F. W., Rothbert, R. M., and Schuster, C. R. (1973): *In vitro* morphine antagonism by antibodies. *Nature*, 241:537.
66. Waki, M., Mitsuyasu, N., Terada, S., Matsuura, S., Kato, T., and Izumuji, N. (1974): Synthesis of polypeptides corresponding to ribonuclease T_1 and its analog. *Biochem. Biophys. Res. Commun.*, 61:576.
67. Waldmann, T. A., and Strober, W. (1969): Metabolism of immunoglobulins. *Prog. Allergy*, 13:1.
68. Watson, J. F., and Butler, V. P., Jr. (1972): Biological activity of digoxin-specific antisera. *J. Clin. Invest.*, 51:638.

Antibodies in Human Diagnosis and Therapy,
edited by Edgar Haber and Richard M. Krause.
Raven Press, New York © 1977.

The Immune Responses to Bacterial Polysaccharides in Man

Emil C. Gotschlich, Irving Goldschneider,
Martha L. Lepow, and Ronald Gold

*The Rockefeller University, New York, New York, 10021;
and the Department of Pathology and Pediatrics,
University of Connecticut Health Center, Farmington, Connecticut 06032*

The conference that produced this volume represented a remarkable synthesis of the state of the art of structural aspects of the antibody molecule. In addition, the possibility of producing antibody, either in tissue culture or by actual synthesis, was reviewed, and several of the more recent applications of antibody to human therapy were discussed. However, little was discussed concerning one of the major functions of humoral antibody, namely, the prevention of microbial disease. Except under unusual circumstances, these antibodies are actively produced by the host as a result of either natural exposure to antigen or a purposeful vaccination.

I would like to describe the recent development of vaccines to bacterial meningitis and use this story to illustrate the present trends in the preparation, standardization, and administration of vaccines.

Bacterial meningitis can be subdivided into that which occurs during the first few weeks of life and the disease that occurs in infants, children, and adults. Neonatal meningitis is most often caused by *Escherichia coli* K1, group B streptococci, and listeria (1). After the first 2 months of life, other etiologic agents become predominant, namely, *Hemophilus influenzae* type b, the meningococcus, and, in certain parts of the world, the pneumococcus. Meningococcal meningitis at this time is relatively uncommon in the United States, with approximately 2,500 to 3,000 cases per year. The mortality is approximately 10%, and the rate of neurological sequelae as a result of the disease is low. However, the meningococcus is the organism capable of causing epidemics, and a recent instance is a grim reminder of the potential of this organism. For the last 4 years in the greater city of Sao Paolo, Brazil, with a population of 10,000,000 inhabitants, there has been an epidemic of meningococcal meningitis, first with group C and then, last year, with the group A meningococcus. There were over 40,000 cases of meningitis last year, an incidence of more than 1 per 300 inhabitants. Furthermore, keep in mind that the majority of cases under these circumstances occur in young school-age children of the lower socioeconomic class. *Hemophilus* b is the other major cause of meningitis, and there are approximately 30,000 cases per year in the United States. This disease occurs almost exclusively in children between the ages of 2 months and 4 years, and the mortality is approximately 10%. However, about one-third of the children who do survive have severe neurological deficits, such that they are either blind, deaf, or require institutionalization for the rest of their lives. Approximately 1 child in 300 live births is ruined by this disease. The pathogenicity of these organisms has been shown to depend in great part on the presence of capsular polysaccharide, and the chemistry of these is outlined in Table 1.

Listed here are the majority of the meningococcal serogroups of which the first three, A, B, and C, are clinically important, causing nearly 90–95% of the disease. The others are very commonly found as normal inhabitants of the nasopharynx but only rarely prove invasive. The polysaccharides can be obtained in a high molecular weight form and in high purity so that they essentially contain no other demonstrable components except those listed in Table 1 (8). Vaccines have been developed for group A and group C and have now been used in well over 100,000,000 people. The products are licensed in several countries, and it is noteworthy that these are the first vaccines where the standards for licensure and release depend on physicochemical criteria. Group C polysaccharide has been shown to be an effective vaccine in Army recruits in the United States (9), and there is some evidence from Brazil that the vaccine is effective in children above the age of 2, but is not effective in younger children (10). The group A vaccine has been shown to be effective in four independently conducted field trials (11–14), and a fifth one is presently underway to test its effectiveness in very young infants (14). In regard to the *Hemophilus* b vaccine, an extensive amount of work has been carried out by two groups: David Smith, Porter Anderson, and their associates in Boston (15,16); and John Robbins, Rachel Schneerson, and their associates at the NIH (17,18). This polysaccharide is a polymer of polyribose and polyribitol phosphate (PRP) (19).

TABLE 1. *Chemistry of the meningococcal group specific polysaccharides*

Serogroup	Polysaccharide	Ref.
A	*N*-acetyl-3-*O*-acetyl, mannosamine phosphate (α 1-6)	2, 3
B	*N*-acetylneuraminic acid (α 2-8)	4, 5
C	*N*-acetyl, *O*-acetylneuraminic acid (α 2-9)	4, 5
D	No chemical information	
X	*N*-acetylglucosamine phosphate (α 1-4)	3
Y	*N*-acetylneuraminic acid: glucose	6
Z	Chemistry unknown	
Z	(29-E) KDO	7

Immunity to these diseases has been shown to be due to the presence of circulating antibodies (20,21). Thus, to study the applicability of these vaccines to young infants, the population which in terms of incidence is most in need of these vaccines, serological studies have been carried out to determine their response to primary and secondary vaccination with polysaccharides. These studies have been conducted in collaboration with Drs. Goldschneider, Lepow, and Gold (22–25). In essence, there is a maturation of the immune response which can best be described as a linear increase in antibody response versus the logarithm of the age up to young adulthood. The geometric mean of the adult human response to the meningococcal polysaccharide is approximately 20 μg antibody protein/ml, with a number of individuals producing responses in excess of 100 μg/ml (22). Generally, the adult human response to polysaccharides is primarily a 7S response (26), and it remains at or near peak levels for a period generally lasting for several years (27). The immune response to group C meningococcal polysaccharide is presented in Fig. 1. Infants at the age of 3 months, while maternal antibodies are still demonstrable in the circulation, are able to produce an immune response to this polysaccharide, but the level of the response is quite meager. At 7 months of age, the response is considerably higher, and

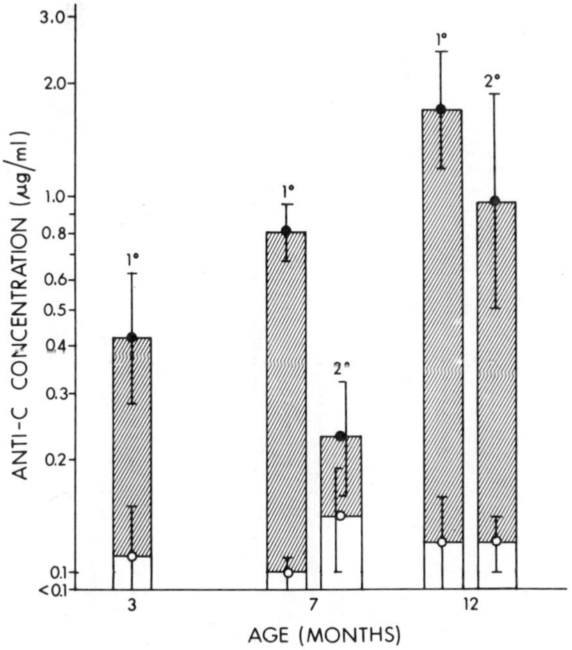

FIG. 1. Geometric mean and 95% confidence limits of the anti-C antibody concentration after primary (1°) and booster (2°) immunization of infants with 25 μg of group C meningococcal polysaccharide. The blank area of the histogram indicates the preimmune response; the hatched area, the immune response.

even higher at the age of 13 months. By the time the children are 7 years old, they produce approximately 10 μg/ml in response to vaccination. If a child who received the vaccine at 3 months is revaccinated at 7 months, the secondary response is lower than the primary response at that age. This hyporesponsiveness is also seen at 13 months, but by the age of 2 years the difference between primary or secondary immunization disappears.

Figure 2 illustrates the response to group A meningococcal polysaccharide. The response of 3-month-old children is essentially nil, but thereafter the same maturation of the immune response can be seen with low titers at 7 months and considerably greater ones at 13 months. However, in direct contrast to the results obtained with the group C polysaccharide, secondary immunization with the A polysaccharide produces a booster rather than a decreased response. This anamnestic response is seen at 7 months and 13 months but is not significant later. In the case of the *Hemophilus* b, the responses seen before the age of 1 year are very feeble and inconstant (16,18).

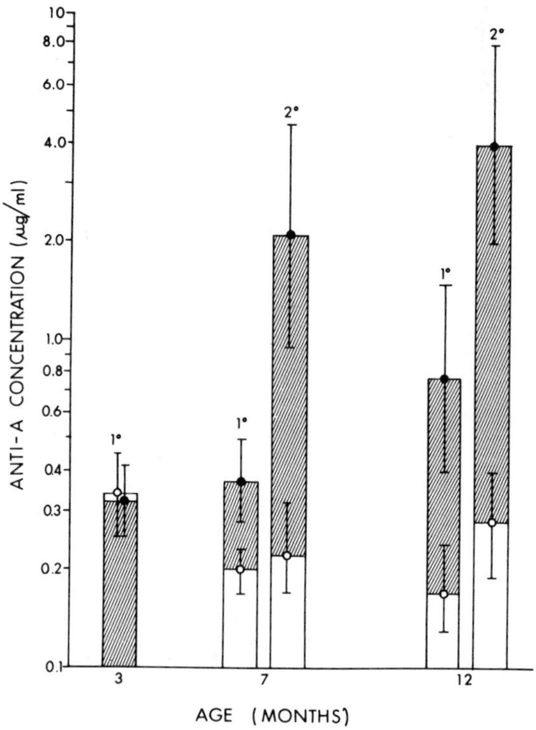

FIG. 2. Geometric mean and 95% confidence limits of the anti-A antibody concentration after primary (1°) and booster (2°) immunization of infants with 25 μg of group A meningococcal polysaccharide. The blank area of the histogram indicates the preimmune response; the hatched area, the immune response.

To study the immune response of children to these polysaccharides in an intelligent manner, it is necessary to try to relate these data to the wealth of information that is available on the response of animals to various antigens. However, it must be kept in mind that the validity of such extrapolations is always a moot point. For instance, two situations to which I have already alluded have no known counterpart in animal experiments; these are the duration and ease of induction of the adult immune response to polysaccharides and the booster effect seen with the group A meningococcal polysaccharide.

Let me first deal with the aspect of the maturation of the immune response to these polysaccharides. This is a process that takes at least 7 years (23,24) and, if viewed from the point of view of populations, occurs in a gradual fashion. This is not the case for the maturation of the immune response to other antigens that have been studied carefully in children. The response to diphtheria toxoid is deficient within the first 2–4 months of life but thereafter is essentially indistinguishable from the primary response of adults (28). The maturation of the immune response to tetanus toxoid seems to be even a trifle earlier (29). Maturational events (Table 2) can be loosely grouped in a number of headings. Some aspects of the maturation appear to be intrinsically programmed and are antigen-specific. These programmed changes are exemplified particularly by the work of Silverstein et al., which indicates that the capacity to produce antibodies to different antigens in fetal lambs occurs at very definite points in time (30). For instance, the ability to produce antibodies to $\phi \times 174$ occurs at 35 days, to ferritin at 66 days, and to ovalbumin at 125 days. Is this form of maturation the principal event occurring in the human response to polysaccharides? I think not, although it cannot, strictly speaking, be excluded. First, the maturation seems too slow and steady. Second, the infant is capable of producing much higher levels of antibodies in response to the disease (31,32); in short, the immune response is accessible to manipulation, whereas the responses in fetal lambs are not.

There appears to exist a second category of maturational events which would fall under the general heading of programmed but unspecific, and by this I would understand the decay of humoral immunosuppressor factors in the infant. Pertinent to this point is the recent demonstration by Murgita and Tomasi of the immunosuppressive activity of α-fetoprotein (33,34). Again,

TABLE 2. *Maturation of the immune system*

- I. Programmed maturation
  - A. Antigen-specific
  - B. Unspecific
- II. Unprogrammed maturation
  - A. Loss of inhibitory maternal antibodies
  - B. Natural "priming"

this would not seem to be the major cause of the hyporesponsiveness to the polysaccharide antigens, because one would then expect the immune response to different antigens to be similar rather than markedly divergent. Programmed maturation, both antigen-specific and -unspecific, may also be due to the loss of excessive T-suppressor cell function.

The next two changes listed in Table 2 are unprogrammed ones, and the first one would refer to the loss of inhibition with the decay of maternal antibodies. This may well explain the maturation between 3 and 7 months of age, but would not be pertinent to the period thereafter. The last one listed as unprogrammed "priming" would refer to the contact that the child unavoidably has with other microorganisms containing antigens cross-reactive with the meningococcus or the *Hemophilus*, and the putative expansion of the B-cell memory population to these antigens. This is probably the most satisfactory explanation for the maturation of the polysaccharide immune response. For instance, it can be shown that there is an age-related increase of natural antibodies to the group A meningococcal polysaccharide, even though this serogroup of meningococcus has been, for all practical purposes, absent from the United States for 25 years. From the work of Dr. John Robbins' group, cross-reactive organisms which potentially could fulfill this role have been identified. For instance, *E. coli* with a capsular antigen consisting of polyribitol phosphate and *Staphylococcus aureus* with a similar antigen have been shown to cross-react strongly with *Hemophilus* b PRP (35). These organisms are very common in normal stool. There are also *E. coli* which cross-react with the group B meningococcal polysaccharide (36–38) and others that cross-react with the group C meningococcal polysaccharide (39). In addition, a gram-positive organism has been recognized, *Bacillus pumilis,* which produces a teichoic acid strongly cross-reactive with the group A meningococcal polysaccharide (39).

Turning for a moment to the question of T- and B-cell interactions in the response to polysaccharide antigens, one major point must be emphasized. Polysaccharides are usually considered to be thymus-independent antigens; i.e., they do not require the presence of helper T cells to induce an antibody response. However, the antibody response to polysaccharide antigens is affected by modulator T cells, which are distinct from helper T cells. Modulator T cells have been shown to amplify or suppress the B cell response to pneumococcal type III capsular polysaccharide (S III) and other antigens. This is best illustrated by the following experiment. If nude mice and their normal litter mates are immunized with S III and their immune response is assessed by the number of plaque-forming cells 5 days after an optimal dose of antigen, the response of the athymic mice is slightly but significantly higher than that of the normal mice. However, if antilymphocyte serum (ALS) is given at the same time as the immunization, the response of the normal mice greatly exceeds that of the athymic mice. Inasmuch as the antibody response of nude mice is unaffected by ALS treatment, the increased response of the normal mice is thought to be due to selective depletion of a suppressor T-cell population by ALS and the unopposed expression of an amplifier T-cell population. It should be noted that helper T cells are also depleted by ALS treatment. A related and very pertinent point, particularly to the infant's response to group C antigen, is that low zone tolerance induced by polysaccharide antigens in mice is due to the creation of a T cell-suppressor population and can be abolished by appropriate use of ALS. Furthermore, this low zone tolerance phenomenon is not seen in nude mice (41).

I would now like to summarize the existing information in the form of an action plan for future practical work to be done in regard to the problem of obtaining sufficient immune responses in infants. One of the simplest explanations for the difference in reactivity to the group A and the group C meningococcal polysaccharide to a second injection might be that the two antigens are cleared differently by infants. The group A polysaccharide is chemically quite unstable and probably rapidly degraded *in vivo,* whereas the group C antigen, like many polysaccharides, is a rather stable molecule. I should hasten to add that experiments to prove this point are technically feasible but have not yet been carried out. In the case of PRP naturally administered to infants in the course of *Hemophilus* septicemia, it has been shown that there are two patterns of clearance: one rapid, followed by antibody production, and the other very prolonged, followed either by a rather poor antibody response or none at all (31). It is, however, not known whether inability to clear or inability to make antibodies is the primary event. An approach to dealing with this problem comes from the finding of Dr. Apicella that there exists a variant of the group C meningococcal polysaccharide which is serologically slightly different from the classic group C polysaccharide and which is readily degraded by neuraminidase, whereas normally group C meningococcal polysaccharide is resistant to the action of neuraminidase (43). We have confirmed these findings and, in collaboration with Dr. Jennings' group in Ottawa, have demonstrated that this polysaccharide differs from the normal one by not being *O*-acetylated. Thus studies are underway to determine the response of adults to the neuraminidase-sensitive group C polysaccharide, which we would expect to be degradable *in vivo* and thereafter the response of infants, to evaluate whether this polysaccharide would be a more appropriate primary immunogen.

Other approaches to modify the infant's immune response direct themselves at attempting to expand the B memory-cell population and perhaps also T helper-cell function. We are talking here about the purposeful use of nonpathogenic organisms that bear identical or cross-reactive antigens on their surface, and which may prove more immunogenic because of the presentation of the antigen in a complex form. Springer and Horton (44) showed that feeding of *E. coli O*86 will raise antibodies to human group B red blood cells. The increase in isoagglutinins occurs both in adults and also in very young infants. It was particularly effective if the feeding of the organisms was carried out at a time when the infants had some unrelated mild diarrheal disorder. These studies were done by feeding dead organisms. The work of John Robbins and his associates has indicated that one can raise antibodies in young rabbit, subhuman, and human primates by feeding of live *E. coli* serotype K-100, which is the one that contains PRP as a capsule (45). In experiments to date, priming of the animal to subsequent parenteral vaccination by feeding live organisms has been only partially successful. The animals do respond somewhat better to injection with live *Hemophilus* b organisms but still do not produce an immune response to the purified polysaccharide itself (45). Even though these first results are not wildly encouraging, this approach deserves a high priority because it is very likely that this is the main event in the natural maturation of the immune function. This will be a very difficult area, requiring judicious choice of cross-reactive organisms, very convincing demonstration of their nonpathogenicity, and a great deal of experience in establishing the best dose and time to administer this type of immunization.

The next tactic that must be considered is the administration of polysaccharide coupled with an immunogenic substrate and administered possibly with adjuvants to ensure a good immune response. There is a great deal of precedent for this approach. Nearly 50 years ago, Landsteiner and Scheer demonstrated that tartaric acid linked to protein was an immunogenic material (46). This work was continued particularly by Goebel and Avery with various monosaccharides, disaccharides, or polysaccharides conjugated to protein (47,48), and culminated in a demonstration that it was possible to produce immunity to type III pneumococcal infection with an artificial antigen consisting of cellobiuronic acid coupled to protein (49). Braun and Plescia (50) have demonstrated that various polyanionic substances, such as nucleic acid, did not need to be covalently linked to protein, but that an electrostatic complex with methylated bovine serum albumin together with complete Freund's adjuvant proved immunogenic. In fact, this approach has been used by Jennings and his associates in Ottawa to raise antisera in rabbits to the purified group A and group C meningococcal polysaccharide. In children, therefore, one would be first faced with the choice of the protein or polyamino acid to which this material should be coupled. A cardinal requirement is that this material be an excellent immunogen. Diphtheria or tetanus toxoid would be good candidates because they are given to children anyway. Technically, this would probably be difficult because the modification of these proteins would have to be small enough so that the antibodies raised to them would remain protective for those diseases. Additional problems would include that these proteins, in their native state, are exceedingly toxic, but that after formalin detoxification, the amino groups are no longer readily available to coupling reactions. Thus one of the factors in choosing a material is the ease of standardizing the coupling; and in this regard the simplest would appear to be some kind of electrostatic complex, such as with methylated bovine serum albumin or some other methylated protein. The possibility of using viruses or bacteriophages as a carrier should also be kept as an option because these have been shown to be remarkably immunogenic even in premature infants (51).

In animal experiments, almost invariably, complete Freund's adjuvant has proved necessary. This material is obviously inappropriate for use in infants because the mineral oil persists indefinitely in tissues and because the inflammatory reaction to the mycobacteria is rather severe. However, biodegradable oil adjuvants have been developed, notably Adjuvant 65 studied by Dr. Hilleman's group at Merck (52). This adjuvant consists of biodegradable peanut oil and, to obtain stable water-in-oil emulsions, mannoside monooleate and aluminum monostearate are added. The immune responses obtained with influenza vaccines suspended in this adjuvant are nearly equal to those seen with incomplete Freund's adjuvant. A considerable literature on the safety and lack of long-term effects of this adjuvant has and is accumulating (53).

Frequently in animal experiments, successful immunization is not obtained with incomplete Freund's adjuvant, and the inclusion of mycobacteria greatly enhances the immune response. In recent years, a great deal of work has been carried out on the adjuvant effect of mycobacteria, and one of the exciting developments in this area is the demonstration that intact tubercle bacilli are unnecessary. It has been shown that the murein or peptidoglycan of mycobacteria has this adjuvant property. In addition, the murein of gram-negative and of certain gram-positive organisms, even after solubilization by lysozyme digestion, retains adjuvant properties (54,55). In fact, synthetic *N*-acetylmuramic acid linked with the dipeptide L-alanyl-D-isoglutamine exhibits strong adjuvant properties and appears to be the minimal structure necessary for adjuvancy (54,55). Thus, it would appear possible to reproduce adjuvants equally as effective as Freund's complete adjuvant, but lacking its objectionable qualities.

In closing this chapter, I would like to remind the reader that the quantitative and qualitative difference of information concerning the cellular immunology of the mouse and of the human, particularly of the infant, is immense. Nevertheless, the lessons learned with mice are becoming applicable to man as the approach of using purified defined reagents as vaccinating agents allows more imaginative intervention in the human immune system at an acceptable risk (56).

ACKNOWLEDGMENTS

This work was supported in part by Contracts (No1-AI-25502 and N01-AI-12502) from the National Institute of Allergy and Infectious Diseases and by a Contract (DAPH-17-70-C-0027) from the U.S. Army Research and Development Command.

REFERENCES

1. Wehrle, P. F., Mathies, A. W., Jr., and Leedom, J. M. (1969): The critically ill child: Management of acute bacterial meningitis. *Pediatrics*, 44:991.
2. Liu, T. Y., Gotschlich, E. C., Jonssen, E. K., and Wysocki, J. R. (1971): Studies on the meningococcal polysaccharides. I. Composition and chemical properties of the Group A polysaccharide. *J. Biol. Chem.*, 246:2849.
3. Bundle, D. R., Smith, I. C. P., and Jennings, H. J. (1974): Determination of the structure and conformation of bacterial polysaccharides by ^{13}C nuclear magnetic resonance. Studies on the group-specific antigens of *Neisseria meningitidis* serogroups A and X. *J. Biol. Chem.*, 249:2275.
4. Liu, T. Y., Gotschlich, E. C., Dun, F. T., and Jonssen, E. K. (1971): Studies on the meningococcal polysaccharide. II. Composition and chemical properties of the Group B and Group C polysaccharide. *J. Biol. Chem.*, 246:4703.
5. Bhattacharjee, A. K., Jennings, H. J., Kenny, C. P., Martin, A., and Smith, I. C. P. (1975): Structural determination of the sialic acid polysaccharide antigens of the *Neisseria meningitidis* serogroups B and C with ^{13}C nuclear magnetic resonance. *J. Biol. Chem.*, 250:1926.
6. Branche, W. C., Jr. (1970): Immunochemical analysis of a meningococcal antigen from strain Boshard. Doctoral dissertation, Catholic University, Washington, D.C.
7. Bhattacharjee, A. K., Jennings, H. J., and Kenny, C. P. (1974): Characterization of 3-deoxy-D-manno octulosonic acid as a component of the capsular polysaccharide antigen from *Neisseria meningitidis* serogroup 29-E. *Biochem. Biophys. Res. Commun.*, 61:489.
8. Gotschlich, E. C. (1975): Development of polysaccharide vaccines for the prevention of meningococcal diseases. *Monogr. Allergy,* 9:245.
9. Gold, R., and Artenstein, M. S. (1971): Meningococcal infections. 2. Field trial of Group C meningococcal polysaccharide vaccine in 1969–1970. *Bull. WHO,* 45:279.
10. Taunay, A. de E., Galvao, P. A., de Morais, J. S., Gotschlich, E. C., and Feldman, R. A. (1974): Disease prevention by meningococcal serogroup C polysaccharide in pre-school children: Results after 11 months in Sao Paulo, Brazil. *Pediatr. Res.*, 8:429.
11. Wahdan, M. H., Sanborn, W. R., Sippel, J. E., Gotschlich, E. C., El Ghoroury, A. A., Hablas, R., Jirgis, N. I., Amer, A., Boctar, W., and Cvjetanovic, B. (1973): A controlled field trial in Egypt of a serogroup A meningococcal polysaccharide vaccine. *Bull. WHO*, 48:667.
12. Erwa, H. H., Haseeb, M. A., Edris, A. A., Lapeyssonnie, L., Sanborn, W. R., and Sippel, J. (1973): A serogroup A meningococcal polysaccharide vaccine. Studies in the Sudan to combat cerebrospinal meningitis caused by *Neisseria meningitidis* Group A. *Bull. WHO* 49:301.
13. Austrian, R. (1975): *personal communication.*
14. Mäkelä, H. (1975): *personal communication.*
15. Anderson, P., Peter, G., Johnston, R. B., Jr., Wetterlow, L. H., and Smith, B. H. (1972): Immunization of humans with polyribophosphate, the capsular antigen of *Hemophilus influenzae* type b. *J. Clin. Invest.*, 51:39.
16. Smith, D. H., Peter, G., Ungram, D. L., Harding, A. L., and Anderson, P. (1973): Responses of children immunized with the capsular polysaccharide of *Hemophilus influenzae* type b. *Pediatrics*, 52:637.
17. Schneerson, R., Rodrigues, L. P., Parke, J. C., Jr., and Robbins, J. B. (1971): Immunity to disease caused by *Hemophilus influenzae* type b II. Specificity and some biological characteristics of "natural" infection-acquired and immunization-induced antibodies to the capsular polysaccharide of *Hemophilus influenzae* type b. *J. Immunol.*, 107:1081.
18. Robbins, J. B., Parke, J. C., Jr., Schneerson, R., and Whisnant, J. K. (1973): Quantitative measurement of "natural" and immunization-induced *Hemophilus influenzae* type 6 capsular polysaccharide antibodies. In: *Hemophilus influenzae*, p. 293, edited by S. H. W. Sell and D. T. Karzon. Vanderbilt University Press, Nashville, Tenn.

19. Robbins, J. B., *personal communication.*
20. Fothergill, L. D., and Wright, J. (1933): Influenzal meningitis: The relation of age incidence to the bactericidal power of blood against the causal organism. *J. Immunol.*, 24:273.
21. Goldschneider, I., Gotschlich, E. C., and Artenstein, M. S. (1969): Human immunity to the meningococcus. I. The role of humoral antibodies. *J. Exp. Med.*, 129:1307.
22. Gotschlich, E. C., Rey, M., Triau, R., and Sparks, K. J. (1972): Quantitative determination of the human immune response to immunization with meningococcal vaccines. *J. Clin. Invest.*, 51:89.
23. Goldschneider, I., Lepow, M. L., Gotschlich, E. C. (1972): Immunogenicity of the Group A and Group C meningococcal polysaccharides in children. *J. Infect. Dis.*, 125:509.
24. Goldschneider, I., Lepow, M. L., Gotschlich, E. C., Mauck, F. T., Bachl, F., and Randolph, M. (1973): Immunogenicity of the Group A and Group C meningococcal polysaccharides in human infants. *J. Infect. Dis.*, 128:769.
25. Gold, R., Lepow, M. L., Goldschneider, I., Draper, T. L., and Gotschlich, E. C. (1975): Clinical evaluation of Group A and Group C meningococcal polysaccharide vaccines in infants. *J. Clin. Invest.*, 56:1536.
26. Yount, W. J., Dorner, M. M., Kunkel, H. G., and Kabat, E. A. (1968): Studies on human antibodies. VI. Selective variation in sub-group composition and genetic markers. *J. Exp. Med.*, 127:633.
27. Allen, P. Z., and Kabat, E. A. (1958): Persistence of circulating antibodies in human subjects immunized with dextran, levan, and blood group substances. *J. Immunol.*, 80:495.
28. Osborn, J. J., Dancis, J., and Julia, J. F. (1952): Studies on the immunology of the newborn infant. 1. Age and antibody production. *Pediatrics*, 9:736.
29. Evans, D. G. (1963): Response of the young infant to active immunization. *Br. Med. J.*, 19:225.
30. Sterzl, J., and Silverstein, A. M. (1967): Developmental aspects of immunity. *Adv. Immunol.*, 6:337.
31. O'Reilly, R. J., Anderson, P., Ingram, D. L., Peter, G., and Smith, D. H. (1975): Circulating polyribophosphate in *Hemophilus influenzae* type b meningitis. *J. Clin. Invest.*, 56:1012.
32. Gotschlich, E. C. (1974): *unpublished observations.*
33. Murgita, R. A., and Tomasi, T. B., Jr. (1975): Suppression of the immune response by α-fetoprotein. I. The effect of mouse α-fetoprotein on the primary and secondary antibody response. *J. Exp. Med.*, 141:269.
34. Murgita, R. A., and Tomasi, T. B., Jr. (1975): Suppression of the immune response by α-fetoprotein. II. The effect of mouse α-fetoprotein on mixed lymphocyte reactivity and mitogen-induced lymphocyte transformation. *J. Exp. Med.*, 141:440.
35. Bradshaw, M., Schneerson, R., Parke, J. C., and Robbins, J. B. (1971): Bacterial antigen cross-reactive with the capsular polysaccharide of *Hemophilus influenzae* type b. *Lancet*, 1:1095.
36. Grados, O., and Ewing, W. H. (1970): Antigenic relationship between *Escherichia coli* and *Neisseria meningitidis*. *J. Infect. Dis.*, 122:100.
37. Kasper, D. L., Winkelhake, J. L., Zollinger, W. D., Brandt, B. L., and Artenstein, M. S. (1973): Immunochemical similarity between the polysaccharide antigens of *Escherichia coli* 0-7: K1(L): NM and *Neisseria meningitidis*. *J. Immunol.*, 110:262.
38. Robbins, J. B., McCracken, G. H., Gotschlich, E. C., Ørskov, F., Ørskov, I., and Hanson, L. A. (1974): *Escherichia coli* K1 capsular polysaccharide associated with neonatal meningitis. *N. Engl. J. Med.*, 290:1216.
39. Robbins, J. B., Myerowitz, R. L., Whisnant, J. K., Argaman, M., Schneerson, R., Handzel, Z. T., and Gotschlich, E. C. (1972): Enteric bacteria cross-reactive with *Neisseria meningitidis* Groups A and C and *Diplococcus pneumoniae* types I and III. *Infect. Immun.*, 6:651.
40. Humphrey, J. H., Parrott, D. M. V., and East, J. (1964): Studies on globulins and antibody production in mice thymectomized at birth. *Immunology*, 7:419.
41. Baker, P. J., Prescott, B., Stashak, P. W., and Amsbaugh, D. F. (1974): Regulation of the antibody response to the type III pneumococcal polysaccharide by thymic-derived cells. In: *The Immune System: Genes, Receptors, Signals*, edited by E. E. Sercarz, A. R. Williamson, and C. F. Fox. Academic Press, New York and London.

42. Gershon, R. K. (1974): T-cell, control of antibody production. *Contemp. Top. Immunobiol.*, 3:1.
43. Apicella, M. A. (1974): Identification of a sub-group antigen on the *Neisseria meningitidis* Group C capsular polysaccharide. *J. Infect. Dis.*, 129:147.
44. Springer, G. F., and Horton, R. E. (1969): Blood group iso-antibody stimulation in man by feeding blood group-active bacteria. *J. Clin. Invest.*, 48:1280.
45. Handzel, Z. T., Argaman, M., Parke, J. C., Jr., Schneerson, R., and Robbins, J. B. (1975): Heteroimmunization to the capsular polysaccharide of *Hemophilus influenzae* type b induced by enteric cross-reacting bacteria. *Infect. Immun.*, 11:1045.
46. Landsteiner, K., and Scheer, J. (1929): Serological differentiation of steric isomers (antigens containing tartaric acids). *J. Exp. Med.*, 50:407.
47. Avery, O. T., and Goebel, W. F. (1931): Chemo-immunological studies on conjugated carbohydrate-proteins. V. The immunological specificity of an antigen prepared by combining the capsular polysaccharide of type III pneumococcus with foreign protein. *J. Exp. Med.*, 54:437.
48. Avery, O. T., and Goebel, W. F. (1929): Chemo-immunological studies on conjugated carbohydrate-proteins. II. Immunological specificity of synthetic sugar-protein antigens. *J. Exp. Med.*, 50:533.
49. Goebel, W. F. (1939): Studies on antibacterial immunity induced by artificial antigens. I. Immunity to experimental pneumococcal infection with an antigen containing cellobiuronic acid. *J. Exp. Med.*, 69:353.
50. Plescia, O. J., and Braun, W. (1967): Nucleic acids as antigens. *Adv. Immunol.*, 6:253.
51. Uhr, J. W., Dancis, J., Franklin, E. C., Finklestein, M. S., and Lewis, E. W. (1962): The antibody response to bacteriophages $\phi \times 174$ in newborn premature infants. *J. Clin. Invest.*, 41:1509.
52. Woodhour, A. F., Metzgar, D. P., Stim, T. B., Tytell, A. A., and Hilleman, M. R. (1964): New metabolizable immunologic adjuvant for human use. I. Development and animal immune response. *Proc. Soc. Exp. Biol. NY*, 116:516.
53. Weibel, R. E., McLean, A., Woodhour, A. F., Friedman, A., and Hilleman, M. R. (1973): Ten-year follow-up study for safety of Adjuvant 65 influenza vaccine in man. *Proc. Soc. Exp. Biol. NY*, 143:1053.
54. Adam, A., Ellouz, F., Ciorbaru, R., Petit, J. F., and Lederer, E. (1975): Peptidoglycan adjuvants: Minimal structure required for activity. *Z. Immunitaetsforsch.*, 149:341.
55. Kotani, S., Watanabe, J., Shimono, T., Narita, T., Kato, K., Stewart-Tull, D. E. S., Kinoshita, F., Yokogawa, K., Kawata, S., Shiba, T., Kusumoto, S., and Tarumi, Y. (1975): Immunoadjuvant activities of cell walls, their water-soluble fractions and peptidoglycan subunits, prepared from various gram-positive bacteria, and of synthetic *N*-acetyl muramyl peptides. *Z. Immunitaetsforsch.*, 149:302.
56. The Oath of Hippocrates.

Antibodies in Human Diagnosis and Therapy,
edited by Edgar Haber and Richard M. Krause.
Raven Press, New York © 1977.

Problems in Priming

Joseph Davie

Joseph Davie: We have two pieces of information that might be pertinent to this discussion. The first pertains to the study of anti-dextran antibodies. By serendipity, Dr. Hansburg found that a combination protocol involving priming of animals with soluble B1355 dextran and secondary immunization with *E. coli* resulted in large amounts of homogeneous anti-dextran antibodies. The mean concentration of anti-dextran antibodies in these mice was 15 mg/ml with 10% of the animals producing more than 40 mg/ml, all of it specific for α (1–3) linkages. We have a trivial explanation for this in that we have shown now that the *E. coli,* in fact, contains in its wall an α (1–3) linkage. This demonstrates what Krause and Haber have been telling us for years: carbohydrates, when incorporated into the surface of bacterial vaccines, are very immunogenic.

We know that these antibodies are identical to the antibodies stimulated with dextran alone in their isoelectric focusing profile, idiotypic makeup, and class distribution. The requirements for priming are relatively strict: B1355 dextran primes, whereas *E. coli* does not.

The other point I would like to make is quite different and possibly more bothersome in terms of the use of soluble carbohydrates for priming. We have been studying the immune response to Ficoll with DNP coupled to it. We have found, as McMaster and Paul demonstrated, that DNP-Ficoll is a very potent immunogen which appears to be thymus-independent. The problem with DNP-Ficoll as an immunogen is that it leaves the animal tolerant following the massive primary response. We think that it may do this by inducing terminal differentiation of precursor cells without proliferation of those precursor cells.

This may cause difficulties in the use of at least some carbohydrates in terms of the generation of memory, and it is conceivable that these kinds of things are behind the variation seen between group C and group A meningococcal priming.

Subject Index

B

C

D

E

X